Benchmark Papers in Soil Science Series

Editor: Charles W. Finkl, Jnr., Florida Atlantic University

SOIL CLASSIFICATION / *Charles W. Finkl, Jnr.*
CHEMISTRY OF IRRIGATED SOILS / *Rachel Levy*
PODZOLS / *Peter Buurman*
ANDOSOLS / *Kim H. Tan*

Related titles
THE ENCYCLOPEDIA OF SOIL SCIENCE, PART I / *Rhodes W. Fairbridge and Charles W. Finkl, Jnr.*
EROSION AND SEDIMENT YIELD / *Jonathan B. Laronne and M. Paul Mosley*

ANDOSOLS

Edited by
KIM H. TAN
University of Georgia

A Hutchinson Ross Benchmark® Book

 **VAN NOSTRAND REINHOLD COMPANY**

Dedicated to the late Dr. Norman H. Taylor, Soil Bureau director, Taita Experimental Station, Lower Hutt, New Zealand, and Professor of Soils, Victoria University of Wellington

Copyright © 1984 by **Van Nostrand Reinhold Company Inc.**
Benchmark Papers in Soil Sciences, Volume 4
Library of Congress Catalog Card Number: 84-10370
ISBN: 0-442-28282-6

All rights reserved. No part of this work covered by the copyrights hereon may be reproduced or used in any form or by any means—graphic, electronic, or mechanical, including photocopying, recording, taping, or information storage and retrieval systems—without permission of the publisher.

Manufactured in the United States of America.

Published by Van Nostrand Reinhold Company Inc.
135 West 50th Street
New York, New York 10020

Van Nostrand Reinhold Company Limited
Molly Millars Lane
Wokingham, Berkshire RG11 2PY, England

Van Nostrand Reinhold
480 Latrobe Street
Melbourne, Victoria 3000, Australia

Macmillan of Canada
Division of Gage Publishing Limited
164 Commander Boulevard
Agincourt, Ontario MIS 3C7, Canada

15 14 13 12 11 10 9 8 7 6 5 4 3 2 1

Library of Congress Cataloging in Publication Data
Main entry under title:
Andosols.
  (Benchmark papers in soil science ; v. 4)
  "A Hutchinson Ross Benchmark book."
  Bibliography: p.
  Includes indexes.
  1. Volcanic ash soils—Addresses, essays, lectures.
I. Tan, Kim H. (Kim Howard), 1926-   II. Series.
S592.17.V65A53   1984   631.4   84-10370
ISBN 0-442-28282-6

CONTENTS

Series Editor's Foreword ... ix
Preface ... xi
Contents by Author ... xiii

Introduction ... 1

PART I: DEFINITION OF ANDOSOLS

Editor's Comments on Papers 1, 2, and 3 ... 10

1   SIMONSON, R. W.: Origin of the Name "Ando Soils" ... 15
     Geoderma **22**:333-335 (1979)

2   KANNO, I.: Genesis and Classification of Humic Allophane Soil in Japan ... 18
     International Soil Conference, Joint Meeting Commissions IV and V, Transactions, International Society of Soil Science, New Zealand, 1962, pp. 422-427

3   DUDAL, R.: Correlations of Soils Derived from Volcanic Ash ... 24
     Meeting on the Classification and Correlation of Soils from Volcanic Ash, Tokyo, Japan, June 11-27, 1964, World Soil Resources Report No. 14, Food and Agriculture Organization of the United Nations, 1964, pp. 134-138

PART II: GEOGRAPHIC DISTRIBUTION OF ANDOSOLS

Editor's Comments on Papers 4 Through 7 ... 30

4   LEAMY, M. L., G. D. SMITH, F. COLMET-DAAGE, and M. OTOWA: The Morphological Characteristics of Andisols ... 34
     Soils with Variable Charge, B. K. G. Theng, ed., New Zealand Society of Soil Science, Lower Hutt, N. Z., 1980, pp. 17-34

5   SIMONSON, R. W., and S. RIEGER: Soils of the Andept Suborder in Alaska ... 52
     Soil Sci. Soc. Am. Proc. **31**:692-699 (1967)

6   TAN, K. H.: The Andosols in Indonesia ... 60
     Soil Sci. **99**:375-378 (1965)

7   FREI, E.: Andepts in Some High Mountains of East Africa ... 64
     Geoderma **21**:119-131 (1978)

Contents

PART III: GENESIS, MORPHOLOGY, AND CLASSIFICATION

Editor's Comments on Papers 8, 9, and 10   78

8   FLACH, K. W., C. S. HOLZHEY, F. DE CONINCK, AND R. J. BARTLETT: Genesis and Classification of Andepts and Spodosols   84
Soils with Variable Charge, B. K. G. Theng, ed., New Zealand Society of Soil Science, Lower Hutt, N. Z., 1980, pp. 411-426

9   MOHR, E. C. J., F. A. VAN BAREN, and J. VAN SCHUYLENBORGH: Andosols   100
Tropical Soils: A Comprehensive Study of Their Genesis, 3rd ed., Mouton—Ichtiar Baru—Van Hoeve, The Hague, The Netherlands, 1972, pp. 397-418

10   SMITH, G. D.: A Preliminary Proposal for Reclassification of Andepts and Some Andic Subgroups   122
Unpublished paper, 1978, 20p.

PART IV: PHYSICAL CHARACTERISTICS

Editor's Comments on Papers 11 and 12   142

11   MAEDA, T., H. TAKENAKA, and B. P. WARKENTIN: Physical Properties of Allophane Soils   147
Advances Agronomy **29**:229-264 (1977)

12   SHOJI, S., and T. ONO: Physical and Chemical Properties and Clay Mineralogy of Andosols from Kitakami, Japan   183
Soil Sci. **126**:297-312 (1978)

PART V: CHEMICAL CHARACTERISTICS

Editor's Comments on Papers 13 Through 17   200

13   ESPINOZA, W., R. G. GAST, and R. D. ADAMS, JR.: Charge Characteristics and Nitrate Retention by Two Andepts from South-Central Chile   205
Soil Sci. Soc. Am. Proc. **39**(5):842-846 (1975)

14   BIRRELL, K. S., and M. GRADWELL: Ion-Exchange Phenomena in Some Soils Containing Amorphous Mineral Constituents   210
Jour. Soil Sci. **7**:130-147 (1956)

15A   GEBHARDT, H., and N. T. COLEMAN: Anion Adsorption by Allophanic Tropical Soils: I. Chloride Adsorption   228
Soil Sci. Soc. Am. Proc. **38**:255-259 (1974)

15B   GEBHARDT, H., and N. T. COLEMAN: Anion Adsorption by Allophanic Tropical Soils: II. Sulfate Adsorption   233
Soil Sci. Soc. Am. Proc. **38**:259-262 (1974)

15C   GEBHARDT, H., and N. T. COLEMAN: Anion Adsorption by Allophanic Tropical Soils: III. Phosphate Adsorption   237
Soil Sci. Soc. Am. Proc. **38**:263-266 (1974)

16   AMANO, Y.: Phosphorus Status of Some Andosols in Japan   241
Japan Agric. Research Quart. (JARQ) **15**(1):14-21 (1981)

Contents

17	**EGAWA, T.:** Properties of Soils Derived from Volcanic Ash *Soils Derived from Volcanic Ash in Japan*, Y. Ishizuka and C. A. Black, eds., Centro Internacional de Mejoramiento de Maiz y Trigo, Mexico City, 1977, pp. 10–63	**249**

PART VI: MINERALOGICAL CHARACTERISTICS

Editor's Comments on Papers 18 Through 22		**304**
18	**BIRRELL, K. S., and M. FIELDES:** Allophane in Volcanic Ash Soils *Jour. Soil Sci.* **3**(2):156-166 (1952)	**310**
19	**YOSHINAGA, N., and S. AOMINE:** Allophane in Some Ando Soils *Soil Sci. Plant Nutr. (Japan)* **8**(2):6-13 (1962)	**322**
20	**YOSHINAGA, N., and S. AOMINE:** Imogolite in Some Ando Soils *Soil Sci. Plant Nutr. (Japan)* **8**(3):22-29 (1962)	**330**
21	**BESOAIN, E.:** Imogolite in Volcanic Soils of Chile *Geoderma* **2**(2):151-169 (1968)	**338**
22	**ESWARAN, H.:** Morphology of Allophane, Imogolite and Halloysite *Clay Minerals* **9**:281-285 (1972)	**357**

PART VII: BIOLOGICAL CHARACTERISTICS

Editor's Comments on Papers 23 and 24		**368**
23	**MURAYAMA, S.:** The Monosaccharide Composition of Polysaccharides in Ando Soils *Jour. Soil Sci.* **31**:481-490 (1980)	**373**
24	**MARTINEZ, A. T., and C. RAMIREZ:** Study of the Microfungal Community of an Andosol *Jour. Ecology* **67**:305-319 (1979)	**383**

Bibliography	**399**
Author Citation Index	**407**
Subject Index	**413**
About the Editor	**418**

SERIES EDITOR'S FOREWORD

The Benchmark Papers in the Soil Science series attempt to provide cogent summaries of the field by reproducing classical and modern papers, ones that provide keys to understanding of critical turning points in the development of the discipline. Scientific literature today is so vast and widely dispersed, especially in a multifaceted discipline like soil science, that much valuable information becomes ignored by default. Many pioneering works are now coveted by libraries, and retrieval from the archives is not easy. In fact, many important papers published in the ephemeral literature are no longer available to serious or committed researchers through interlibrary loan. Other professionals devoted to teaching or burdened with administrative duties must be hard pressed to keep up with comprehensive arrays of technical literatures spread through scores of journals. Most of us can, at best, skim only a few select journals to make copies to tables of contents, abstracts and summaries, and reviews in order to remain abreast of specialized and often limited aspects of the robust field of soil science as a whole.

The Benchmark Papers in Soil Science series, developed as a practical solution to this problem, reprints key papers and investigative landmarks that relate to a common theme. The papers are reproduced in facsimile, either in their entirety or in significant part, so readers can follow major original events in the field, not peruse paraphrased or abbreviated versions of others. Some foreign works have been especially translated for use in the series. Occasionally short, foreign language articles are reproduced from French or German journals.

Essays by the volume editor provide running commentaries that introduce readers to highlights in the field, provide critical evaluation of the significance of the various papers, and discuss the development of selected topics or subject areas. It is hoped that the volume editor's comments will ease the transition for the seasoned investigator who wishes to step into a new field of research as well as provide students and professors with a compact working library of most important scientific advances in soil science.

Areas of specialization in soil science are divided by the International Society of Soil Science into seven divisions or "commissions." The first six commissions cover soil physics, chemistry, mineralogy, biology, fertility, and technology. Because the scope of the field is so great, we concentrate initially on topics traditionally devoted to the seventh commission: soil morphology, genesis, classification, and geography. The series thus begins with volumes dealing with the major soils of the world: their recognition,

Series Editor's Foreword

characteristics, formation, distribution, and classification. Other volumes concentrate on topics in agronomy, soil-plant relationships, soil engineering topics, or melds of pure science with soil systems. Benchmark Papers in Soil Science plow deeply through the field, picking significant but timely topics on an eclectic basis.

Each volume in the series is edited by a specialist or authority in the area covered by the book. The volume editor's efforts reflect a concerted worldwide search, review, selection, and distillation of the primary literature contained in journals and monographs and in industrial and governmental reports. Individual volumes thus represent an information-selection and repackaging program of value to libraries, students, and professionals.

Benchmark books contain a preface, introduction, and highlight commentaries by the volume editor. Many volumes contain rare papers that are hard to locate and obtain, as well as landmark papers published in English for the first time. All volumes contain author citation and subject indexes of the contained papers, usually twenty to fifty key papers in a given subject area.

This fourth volume in the series deals with embryonic soils with few diagnostic features that have developed from volcanic material. Colloquially referred to as volcanic forest or ash soils, these immature soils have been reported from mountainous regions ranging from the Kamchatka Peninsula (Siberia), Japan, and Alaska, to more southern realms in Latin America, Africa, Indonesia, and New Zealand. In tropical regions such as the Hawaiian archipelago these soils are most extensive on the younger volcanic islands where they are used for the production of sugar cane, truck crops, coffee, and macadamia nuts. In higher latitudes they are used for pasture and rangeland.

The name "Ando soils" was introduced in 1947 by American soil scientists conducting soil surveys in Japan after World War II. As then conceived, the name identified soils with thick, dark A1 horizons that were derived from volcanic ash and were acid in reaction. This name was largely replaced by the adoption of a new nomenclature in the American *Soil Taxonomy*. In the American soil classification system these soils are identified as Andepts in the order of Inceptisols. In the legend for the FAO-UNESCO soil map, the compound name "Ando soils" was converted to the single term *Andosols*.

The unique feature of Andosols is the amorphous nature of the mineral fraction. These soils inherit amorphous material from the pyroclastic parent material that readily weathers to allophane (an amorphous aluminum silicate clay mineral) under a humid climatic regime. Other characteristic properties include high organic matter content, high cation-exchange capacity, high phosphorus-fixing capacity (and low phosphorus availability), and low bulk densities (commonly 0.4 to 0.8 grams per cubic centimeter).

The reprinted papers in this volume cover the definition of Andosols, their geographic distribution, genesis, morphology and classification. Physical, chemical, mineralogical, and biological characteristics of Andosols are also reviewed. Editorial comments introduce the different subject areas and provide an historical perspective for changing concepts. Critical turning points in the advancement of understanding of Andosols are indicated in benchmark papers dealing with these unique soils.

CHARLES W. FINKL, JNR.

PREFACE

Over the past few years, the need for a comprehensive volume on Andosols has become eminent and urgent. Historically, these soils were considered of local importance only in Japan and New Zealand, where volcanic activity is prevalent. But since the international workshop on classification and correlation of soils from volcanic ash held in Tokyo, Japan, June 11-27, 1964, under the auspices of the Food and Agriculture Organization of the United Nations, these soils have gained prominence in world opinion. It became apparent that Andosols were also major soils in other parts of the world, and the resulting increase in research activity has led to the proliferation of articles.

The Tokyo meeting, perhaps the first of its kind on Andosols, was followed by other meetings. The latest to date were the Narino conference held in 1972 by the Instituto Interamericano de Ciencas Agricolas Turrialba, Costa Rica, and the conference on Soils of Variable Charge held in 1980 at Palmerston North, New Zealand, under the sponsorship of the New Zealand Soil Science Society. By then a large number of publications had accumulated, scattered through a variety of local and international journals. From time to time a bibliography on Andosols has been compiled for an overview of the progress made. These include J. Gautheyrou, M. Gautheyrou, and F. Colmet Daage (1976), P. Quantin and G. G. C. Claridge (1974), and New Zealand Commonwealth Bureau of Soils (1978). Some of the listed articles were difficult to find, however, and others were printed in local or national journals with limited accessibility. Moreover, a list of references does not supply the information usually provided by a book, and titles are often misleading.

The purpose of this Benchmark volume on Andosols is to bring together reprints of key papers representing landmarks in the progress of knowledge on Andosols. The organization of the book is such that readers can both follow original events in the field and also have a scientific text or comprehensive treatise on Andosols. The volume starts with a few papers on the definition or concept of Andosols. Part II discusses the geographic distribution of the soils in the world. Part III is composed of benchmark papers on past and present classification of Andosols. The remaining parts are about physical, chemical, and mineralogical properties, respectively. The book closes with a part on biological characteristics.

This volume will be useful for a wide reader distribution. Soil scientists,

Preface

crop and plant scientists, people working in irrigation, forestry, conservation and ecology, microbiologists, geologists, and others associated with the soil may be in need of this book.

In the development of a book of this nature, such factors as availability of reprints, copyrights, and economics have to be taken into consideration. These factors, economics especially, imposed limits on the selection and numbers of papers in this book. It is with regret that I was unable to include many other excellent papers.

I wish to acknowledge a number of persons for their generous contributions and excellent suggestions. Thanks are due to: Y. Amano (Japan), S. Aomine (Japan), K. S. Birrell (New Zealand), R. Dudal (Italy), T. Egawa (Japan), H. Eswaran (United States), K. W. Flach (United States), E. Frei (Switzerland), R. G. Gast (United States), I. Kanno (Japan), K. Kawai (Japan), K. Kyuma (Japan), M. L. Leamy (New Zealand), Walter Luzio L. (Chile), T. Maeda (Japan), A. T. Martinez (Spain), S. Murayama (Japan), P. F. Pratt (United States), R. W. Simonson (United States), S. Shoji (Japan), K. Wada (Japan), B. P. Warkentin (United States), and N. Yoshinaga (Japan). Special thanks are also due to Mrs. Yolanda Smith (Belgium) and Mrs. A. Van Schuylenborgh (the Netherlands) for allowing the use of their late husbands' articles. Sincere appreciation is extended to the various publishers, organizations, and scientific societies, among them the FAO, United Nations, International Society of Soil Science, American Society of Soil Science, New Zealand Soils Bureau, the Williams and Wilkins Co., Mouton Publishers, Royal Tropical Institute, *Japan Agricultural Research Quarterly (JARQ)*, Centro Internacional de Mejoramiento de Maiz y Trigo, Japanese Society of Soil Science and Plant Nutrition, and the British Mineralogical Society for their permission to reproduce articles, figures, photographs, diagrams, and tables. Finally, I wish to express my gratitude to Charles W. Finkl for his cooperation and effectiveness in editing the *Benchmark Papers in Soil Science* and to the many unnamed people who have assisted in the development of this volume.

KIM H. TAN

CONTENTS BY AUTHOR

Adams, Jr., R. S., 205
Amano, Y., 241
Aomine, S., 322, 330
Bartlett, R. J., 84
Besoain, E., 338
Birrell, K. S., 210, 310
Coleman, N. T., 228, 233, 237
Colmet-Daage, F., 34
De Coninck, F., 84
Dudal, R., 24
Egawa, T., 249
Espinoza, W., 205
Eswaran, H., 357
Fieldes, M., 310
Flach, K. W., 84
Frei, E., 64
Gast, R. G., 205
Gebhardt, H., 228, 233, 237
Gradwell, M., 210
Holzhey, C. S., 84

Kanno, I., 18
Leamy, M. L., 34
Maeda, T., 147
Martinez, A. T., 383
Mohr, E. C. J., 100
Murayama, S., 373
Ono, T., 183
Otowa, M., 34
Ramirez, C., 383
Rieger, S., 52
Simonson, R. W., 15, 52
Shoji, S., 183
Smith, G. D., 34, 122
Takenaka, H., 147
Tan, K. H., 60
Van Baren, F. A., 100
Van Schuylenborgh, J., 100
Warkentin, B. P., 147
Yoshinaga, N., 322, 330

INTRODUCTION

Andosols are soils derived from volcanic ash. The consensus is that the weathering products of the ash give to the soil the distinct features of Andosols. Compared to other soils, such as Oxisols, Spodosols, and Ultisols, they are young soils since most volcanic ash, which gives rise to the development of Andosols, originated from eruptions in the Quaternary period (Ohmasa, 1964; Tan and Van Schuylenborgh, 1961). Volcanic eruptions have intermittently or continuously occurred in regions occupied by Andosols through the Quaternary period, depositing either ash or other pyroclastic materials on the earth's surface (Wada and Aomine, 1973). The ash may have accumulated on the earth's surface following an eruption and may sooner or later form consolidated layers called tuffs. On very highly contoured terrain, rain and erosion may form mud flows from the ash deposits, filling valleys, canyons, and low-lying areas. In Indonesia these deposits, called *lahar*, contain not only fine-particle-size ash material but also stone, gravel, and boulders (Tan and Van Schuylenborgh, 1961). Most of the Andosols are derived from the finely comminuted ejecta from magma of a composition between acid and basic magma. Wright (1964) indicated that basaltic ash seldom developed into Andosols, because of its relatively high Fe and low Al contents. However, the U.S. *Soil Taxonomy* (1975) states that in climates with pronounced dry seasons, basic volcanic ash is also an important parent material for Andosols. At the other end of the scale, acidic ash contains large amounts of Si and on weathering produces sufficient amounts of allophane to give to the soil the unique properties of Andosols. Most of the ashes in Japan are andesitic of origin, whereas in Indonesia and New Zealand, they may vary in type from rhyolite, dacite to andesite, or their intergrades.

Large areas of volcanic ash soils also occur in arid (desert) regions, but under conditions of low soil humidity, the soils do not develop the special features of Andosols (Wright, 1964).

Introduction

DISTINGUISHING FEATURES

At first believed to occur mainly in Japan (Kanno, 1956; 1961) and New Zealand (Taylor and Cox, 1956), Andosols are currently found to be more widespread than expected. Known under different names in various countries, the soils are located throughout a wide range of climatic conditions, from subalpine regions to the humid tropics. Regardless of the differences in climate in which they have been formed, Andosols seem to constitute of a group of soils with similar morphological, physical, and mineralogical characteristics. The most conspicuous features that Andosols have in common are the thick black surface horizon and the soil mineralogy dominated by amorphous or noncrystalline clay. The latter is considered to be allophane, a product of the weathering of volcanic glass. The soil colloidal fraction may also include hydrated silica and alumina, which together with allophane possess highly reactive surfaces. The black color was the reason that U. S. soil scientists called the soils *Ando soils* (from the Japanese *An,* "dark" or "black," and *do,* "soil") (Thorp and Smith, 1949). Most Japanese scientists, however, know the soils under the name *Kuroboku (Kuro,* "black" or "dark," and *boku,* "soft, friable"), although some proposed the use of the term *humic allophane* soils (Wright, 1964). The first three papers, making up Part I, relate the origin of the name and discuss the definition of Andosols. The genesis, morphology, and classification of these soils are the topics of Part III, in which Papers 8 through 10 illustrate the general concepts followed in many countries around the world.

SOIL ORGANIC MATTER CONTENT

The black color of Andosols is attributed to accumulation of high amounts of organic matter in the surface horizon. Such an accumulation requires a soil environment with the proper moisture condition. The geographic distribution, as illustrated by Papers 4 through 7, shows Andosols to occur in a wide variety of climatic conditions, where the soil moisture regimes are expected to be quite different from one to the other region. In a number of instances, the differences in climate are so drastic that it is difficult to perceive how soils with similar properties can occur in, for example, the arctic and temperate regions, as well as in the warm humid tropics, especially when one of the properties pertains to accumulation of high amounts of soil organic matter. Even under tropical conditions, Andosols can occur in areas with large differences in climate—for example, in the cool regions of the mountains and in the hot, humid lowlands, where the climate factors favor a rapid decomposition of soil organic matter. Organic carbon contents of 6 to 7 percent are common in lowland Andosols in Indonesia (Tan and Van Schuylenborgh, 1961), while in Japan some Andosols may even contain 15 to 20 percent C_{org} (Wada and Aomine, 1973).

A number of hypotheses have been presented for the relatively large accumulation of organic matter. Although some differences exist among them, all agree that the presence of amorphous or noncrystalline clay, especially allophane, is the major reason for the high organic matter content (Fitzpatrick, 1980; Wada and Aomine, 1973; Jackson, 1964). The contention that poor drainage and anaerobic conditions are the reasons for the humus accumulation in Andosols has not been supported by the fact that most Andosols are biologically dry and aerobic in condition. Kanno (1961) and Wada and Aomine (1973) tend to endorse the idea that allophane and/or allophane-like material enters into a reaction with humic matter, forming clay-humic complexes and chelates. In this way, organic matter would be protected against microbial attack, making possible its accumulation in the soils. Other poorly crystalline materials, such as alumina, may also contribute to organic matter accumulation in Andosols. Because of the presence of Fe- and Al-bearing minerals, the soils are kept supplied with Fe and Al ions, which form stable bonds with humic substances. The subsequent formation of metal-humic acid complexes and chelates increases the resistance of the organic fraction to microbial attack (Tate and Theng, 1980; De Coninck, 1980). The resistance is perhaps due to the fact that many heavy metals—in the case of Andosols, aluminum—are toxic to microorganisms. Another reason is that in a complex or chelate molecule, the organic component is physically and sterically less accessible to microbial enzymes (Sen, 1961).

This hypothesis for accumulation of soil organic matter requires a condition in which the Al concentration in the soil must be sufficiently high to cause saturation and neutralization of the humic compounds to such an extent that the compounds become immobilized in the surface soil before downward migration can take place. In Andosols, weathering of the primary minerals, which abound in volcanic ash, ensures the release of sufficient amounts of Si and Al, which may give rise to the formation of metal-organo complexes with high Al/organic ratios. Evidence has been presented that the higher the values for the metal/organic ratios, the greater will be the chances for immobilization of the humic fraction (De Coninck, 1980; Kodama and Schnitzer, 1977). On the other hand, solubility and subsequent mobility of metal-humates increase with a decrease in metal/humic

Introduction

acid ratio. In conclusion, it can be stated that by virtue of having a high Al/organic ratio, the organic complexes in Andosols become insoluble and precipitate in the surface horizon.

SOIL MINERALOGY

The other property that makes Andosols so unique is the presence of allophane in their clay fractions. Most scientists agree that the predominating presence of allophane coincides with maximum development of Andosol's characteristics. The occurrence of allophane and allophane-like material is usually connected with areas of recent volcanic activity, but in a few instances allophane has also been detected in soils from basaltic rocks (Wada, 1977).

Although several other clay minerals have recently been found ind Andosols, (among them, imogolite, halloysite, and kaolinite), the majority opinion is that it is the allophane that gives to the soil the distinctive properties: high organic matter content, high waterholding capacity, high porosity, low bulk density, and high phosphate fixation capacity. The physical properties are the topics of Papers 11 and 12; the chemical properties are discussed in Papers 13 through 17. The Papers in Part VI outline the clay mineralogy of Andosols in several countries.

The term *allophane* has different meanings to many people. In Andosols, allophane refers to hydrous aluminosilicate clays, generally considered to be the weathering product of volcanic glass. It has a chemical composition characterized by a molar SiO_2/Al_2O_3 ratio between 1 and 2 and by the predominance of Si-O-Al bonds in its structure (Van Olphen, 1971). This mineral is usually amorphous to X-ray diffraction analysis, meaning that X-ray diffraction analysis yields featureless curves. It has a high specific surface area, which equals or frequently exceeds that of montmorillonite. The shape and size of allophane units indicate that they may exhibit high porosity. The mineral is known for its high variable charges and cation exchange capacity. It has also a considerable anion exchange capacity and is very effective in specific and nonspecific adsorption of organic anions (Wada, 1977).

The reactivity of allophane is considered to be attributed to its acidic character due to the presence of tetrahedrally coordinated aluminum (Swindale, 1964). However, the surface acidity of allophane can vary according to humidity. Wada (1977) reported weak acidities for allophane in moist conditions and higher acidities in dry environments. Wada also indicated that the acid sites developed on allophane by changes in coordination of surface Al atoms.

Included in soil mineralogy is the mineralogy of the sand fraction, an area that generally has received less attention than clay mineralogy. Volcanic ash is composed of an assortment of primary minerals, many of which are the usual primary minerals found in soils. In general, feldspars, pyroxenes, hornblende, quartz, and volcanic glass make up the sand fraction of volcanic ash. The percentage of these primary minerals varies, of course, according to their acidic or basic origin (Tan and Van Schuylenborgh, 1961). Volcanic ash from basic magma contains more ferromagnesians and basic volcanic glass but little quartz, whereas ash from acidic magma has more quartz and acid volcanic glass but less ferromagnesians. Birrell (1964) reported that volcanic ash in New Zealand also contained significant amounts of magnetite but was low in micas. Since primary minerals are the principal sources of clay formation and many of the plant nutrients, differences in primary mineral composition are expected to play an important role in the kinds of Andosols formed and in their potential fertility.

MAJOR SOIL CHARACTERISTICS

Many authors believe that the physical and chemical properties of Andosols are attributed to the presence of allophane. Not only do Andosols exhibit high organic matter content, low bulk density values, high water-holding capacity, and high total porosity, but the soils are also friable and have low plasticity and stickiness. When wet they are greasy and smeary. Generally they yield moisture when squeezed between the fingers. These physical characteristics of Andosols change with moisture conditions. In Indonesia, it was observed that on drying, pronounced changes occurred in the physical properties of Andosols, which manifested themselves in a phenomenon called "mountain granulation" (Tan, 1959; Tan and Van Schuylenborgh, 1961). On drying, the soil usually becomes very finely grained and acquires a dusty appearance. The soil is then very difficult to remoisten and on disturbance produces black clouds. This is the reason why in the past Dutch scientists called these soils black dust soils (Druif, 1939).

The phenomenon of irreversible drying presents problems in particle size analysis. The soils are usually difficult to disperse, perhaps, according to Birrell (1964), because of the presence of (1) amorphous silicate clays with ZPC (zero point of charge) values higher than the usual crystalline minerals and (2) hydrous oxides, which induce co-precipitation.

The chemistry of the soils reflects the influence of high amounts

Introduction

of variable charges and the lack of permanent charges. Andosols in New Zealand have been reported to possess base saturation values that are inconsistent with their pH values because of the presence of a strong buffer capacity dominated by allophane (Taylor, 1964). Generally these soils are considered to have high phosphate fixation capacities and tend to become rapidly deficient in potassium.

Not much is known about the biological properties of Andosols. In general, soils contain a large variety of microorganisms, which are responsible for the many biochemical reactions in soils. The numerical value of the population of many of the soil organisms shows figures exceeding our expectations and imagination. Bacterial counts in a well-drained soil can amount to 2 million per gram of soil.

Since each group of microorganisms is known to require specific environmental conditions and since the different kinds of soils on earth are characterized by different pedological, physical, and chemical properties, it can be expected that the different groups of soils may also contain different kinds of microorganisms (Ishizawa, 1964). Consequently Alfisols, Mollisols, Spodosols, Oxisols, Ultisols, and so on may well be characterized by different biological properties. The discovery of a characteristic microflora in Andosols, with a high content of aerobic organisms, was the reason for stating that Andosols are aerobic and biologically dry (Ishizawa, 1964). However, more detailed studies are required to support the hypothesis, and an urgent need is apparent to collect more data on biological properties of major soil groups, including Andosols. The Papers in Part VI reflect our present knowledge of some of the biological properties of Andosols.

REFERENCES

Birrell, K. S., 1964, Some Properties of Volcanic Ash, Meeting on the Classification and Correlation of Soils from Volcanic Ash, Tokyo, Japan, June 11-27, 1964, *Food and Agric. Org., United Nations, World Soil Resources Rept.* **14:**74-81.

De Coninck, F., 1980, Major Mechanisms in Formation of Spodic Horizons, *Geoderma* **24:**101-128.

Druif, J. H., 1939, De Bodem van Deli. III. Toelichting bij de Agrogeologische Kaarten en Beschrijving der Grondsoorten van Deli, *Mededelingen Deli Proefstation* (N. Sumatra) 3-4:1-74.

Fitzpatrick, E. A., 1980, *Soils: Their Formation, Classification and Distribution,* Longman, London, 353p.

Ishizawa, I., 1964, Microbiology of Volcanic Ash Soils in Japan, Meeting on the Classification and Correlation of Soils from Volcanic Ash, Tokyo, Japan, June 11-27, 1964, *Food and Agric. Org., United Nations, World Soil Resources Rept.* **14:**87-88.

Jackson, M. L., 1964, Chemical Composition of Soils, in *Chemistry of the Soils,* F. E. Bear, ed., Van Nostrand Reinhold, New York, pp. 71-141.

Kanno, I., 1956, A Pedological Investigation of Japanese Volcanic Ash Soils, *Kyushu Agric. Expt. Sta. Bull.* **4**:81-84.

Kanno, I., 1961, Genesis and Classification of Main Genetic Soil Types in Japan, *Kyushu Agric. Expt. Sta. Bull.* **7**:1-185.

Kodama, H., and M. Schnitzer, 1977, Effect of Fulvic Acid on the Crystallization of Fe (III) Oxides, *Geoderma* **19**:279-291.

Ohmasa, M., 1964, Genesis and Morphology of Volcanic Ash Soils, Meeting on the Classification and Correlation of Soils from Volcanic Ash, Tokyo, Japan, June 11-27, 1964, *Food and Agric. Org., United Nations, World Soil Resources Rept.* **14**:56-60.

Sen, B. C., 1961, Studies on the Bacterial Decomposition of Humic Acid in the Clay-Humus Mixture, *Jour. Indian Chem. Soc.* **38**:737-740.

Swindale, L. D., 1964, The Properties of Soils Derived from Volcanic Ash, Meeting on the Classification and Correlation of Soils from Volcanic Ash, Tokyo, Japan, June 11-27, 1964, *Food and Agric. Org., United Nations, World Soil Resources Rept.* **14**:82-86.

Tan, K. H., 1959, Classification of Black Colored Soils in the Humid Regions of Indonesia (in Indonesian), *Tehnik Pertanian* **8**(5):217-222.

Tan, K. H., and J. Van Schuylenborgh, 1961, On the Classification and Genesis of Soils Developed over Acid Volcanic Materials under Humid Tropical Conditions. II, *Netherlands Jour. Agric. Sci.* **9**:41-54.

Tate, K. R., and B. K. G. Theng, 1980, Organic Matter and Its Interactions with Inorganic Soil Constituents, in *Soils with Variable Charge*, B. K. G. Theng, ed., New Zealand Soil Science Society, Soils Bureau, Lower Hutt, New Zealand, pp. 225-249.

Taylor, N. H., 1964, The Classification of Volcanic Ash Soils in New Zealand, Meeting on the Classification and Correlation of Soils from Volcanic Ash, Tokyo, Japan, June 11-27, 1964, *Food and Agric. Org., United Nations, World Soil Resources Rept.* **14**:101-110.

Taylor, N. H., and J. E. Cox, 1956, *The Soil Pattern of New Zealand,* New Zealand Soil Bureau, Lower Hutt, New Zealand, Publ. 113.

Thorp, J., and G. D. Smith, 1949, Higher Categories of Soil Classification: Order, Suborder, and Great Soil Groups, *Soil Sci.* **67**:117-126.

U. S. Soil Survey Staff, 1975, *Soil Taxonomy. A Basic System of Soil Classification for Making and Interpreting Soil Surveys,* Soil Conservation Service, U. S. Dept. Agric., Agric. Handbook 436, 754p.

Van Olphen, H., 1971, Amorphous Clay Materials, *Science* **171**:90-91.

Wada, K., 1977, Allophane and Imogolite, in *Minerals in Soil Environments,* J. B. Dixon and S. B Weed, eds., Soil Sci. Soc. Am., Madison, Wisconsin, pp. 603-638.

Wada, K., and S. Aomine, 1973, Soil Development on Volcanic Materials during the Quaternary, *Soil Sci.,* **116**:170-177.

Wright, A. C. S., 1964, The "Andosols" or "Humic Allophane" Soils of South America, Meeting on the Classification and Correlation of Soils from Volcanic Ash, Tokyo, Japan, June 11-27, 1964, *Food and Agric. Org., United Nations, World Soil Resources Rept.* **14**:9-21.

Part I

DEFINITION OF ANDOSOLS

Editor's Comments
on Papers 1, 2, and 3

1   SIMONSON
        Origin of the name "Ando Soils"

2   KANNO
        Genesis and Classification of Humic Allophane Soil in Japan

3   DUDAL
        Correlation of Soils Derived from Volcanic Ash

Volcanic ash soils in the category of Andosols have been known in different countries under different names. The characteristics may also exhibit considerable variations to the soils originally described in Japan as Ando Soils. Simonson (Paper 1) reported that this name was introduced to the scientific community in 1945 as a result of a reconnaissance soil survey in Japan by U.S. soil scientists. The striking black feature of the soils, due to its high organic matter content, and the country of origin were the reasons for choosing the Japanese terms. Ando soils were delineated as soils with thick dark A1 horizons, had strongly acid reaction, low exchangeable bases, and silica-sesquioxide ratios of the colloidal fraction ranging from 1.3 to 2.0. This definition does not recognize the presence of allophane or other amophous materials as an important characteristic, although indications of this aspect are perhaps given by the silica-sesquoxide ratios; however, the values of these ratios—from 1.3 to 2.0—are large enough to include other soil colloids, such as 1:1 type of clays. Low base saturation and pH values from 4.5 to 5.5 apparently are common in soils in Japan, but the use of low levels of exchangeable bases together with pH values below 5.0 as criteria for distinguishing Ando Soils from other soils may exclude many of these kinds of soils in New Zealand, where base saturations are commonly high (Taylor, 1964). Swindale (1964) and Birrell (1964) noted that the pH of New Zealand Ando Soils was generally above 5.0. Due to the presence of allophane, it was unusual to find pH values much below 5.0. Even when base saturations were low, Swindale (1964) indicated that the pH remained comparatively high.

Editor's Comments on Papers 1, 2, and 3

The name *Ando Soils* apparently has gained acceptance in the scientific community throughout the world, but the definition has left open many questions and arguments. With the increased knowledge on these soils, the concept of Ando Soils has been subjected to some refinement. Kanno (Paper 2) started to include more detailed pedological criteria and humus characteristics in terms of carbon/nitrogen (C/N) ratio in the unwritten definition of these soils. He reported that the soils were characterized by three horizons: a thick A horizon, rich in humus, over a yellowish-brown B, underlaid by weathered volcanic ash corresponding to the C horizon. The humus content was not only exceptionally high, but the C/N ratio of this humus was also high. For the first time in the history of Ando Soils, emphasis was placed on the role of aluminum and allophane in the accumulation of humus. Kanno indicated that the character of the humus should therefore be clearly distinguished from other soils, such as Spodosols, Alfisols, Ultisols, and Mollisols. The dominant influence of humus and allophane on the development of the characteristic features of Ando Soils was the reason for Kanno to introduce the name *humic allophane soils*. Although several other scientists accepted the new nomenclature (Wright, 1964), the name *Ando Soils* was preferred and remained entrenched.

The concept as proposed by Kanno provides some improvement, but a definition of Ando Soils encompassing all soils in this category is still lacking. The criteria used appear to be valid in Japan, but may exclude many of these soils in other countries. In Indonesia, these kind of soils contain humus with a much narrower C/N ratio. Depending on climatic variations, due to differences in elevation above sea level, the C/N ratio may vary from 8.0 to 15.0. The composition of the soil pedon also can vary from a pedon with A, B, and C horizons, as Kanno proposed, to pedons with A and C horizons. In Indonesia Ando Soils with A-C profiles are more common, and are a reflection of the prevailing intermittent volcanic activity. Usually a sharp decrease in organic matter content can be noticed from surface to subsurface horizons. Figures 1 and 2 are illustrations of Ando Soils in Indonesia. Figure 1 represents an Ando Soil at relatively high elevation in west Sumatra near Lake Maninjau. The soil exhibits a thick black A horizon over a yellowish C horizon, underlaid by a buried A horizon. Figure 2 shows an Ando Soil representative of the mountain regions in west Java. The soil has a brown A over a thin B horizon, underlaid by a yellowish layer of weathered volcanic tuff. This pedon lies on top of a buried Ando Soil. The younger pedon on the surface fits the description of the yellow brown loams in New Zealand (Taylor and Cox, 1956).

An attempt to correlate Ando Soils around the world and amend existing differences so that a common definition can be drawn was made at a meeting on the classification and correlation of soils from volcanic ash, held in Tokyo, Japan, June 11–27, 1964, under the sponsorship of the Food and Agriculture Organization of the United Nations. The paper by Dudal (Paper 3) tries to coordinate the differences and similarities present among these kinds of soils. This meeting, attended by major countries where Ando Soils can be found, adopted the term *Andosols* as the official name. The definition formulated for the soils is as follows: "Andosols are mineral soils in which the active fraction is dominated by amorphous materials (minimum 50%). These soils have a high sorptive capacity, a relatively thick friable dark A horizon, are high in organic matter, have a low bulk density and a low stickiness. They may have a (B) horizon and [do] not show significant clay movement. These soils occur under humid and sub humid conditions" (Dudal, 1964).

The name *Andosols* has taken precedence over all the other names. Although since the introduction of the U. S. *Soil Taxonomy* in

Figure 1. Andosol near Lake Maninjau (West Sumatra) developed from volcanic tuff.

Editor's Comments on Papers 1, 2, and 3

1975 U.S. soil scientists have used the name *Andepts*, the tendency in many other countries is to maintain the name *Andosols*.

REFERENCES

Birrell, K. S., 1964, Some Properties of Volcanic Ash Soils, Meeting on the Classification and Correlation of Soils from Volcanic Ash, Tokyo, Japan, June 11-27, 1964, *Food and Agric. Org., United Nations, World Soil Resources Rept.* **14:**74-81.

Dudal, R., 1964, Summary of the Technical Discussion, Meeting on the Classification and Correlation of Soils from Volcanic Ash, Tokyo, Japan, June 11-27, 1964, *Food and Agri. Org., United Nations, World Soil Resources Rept.* **14:**139-141.

Swindale, L. D., 1964, The Properties of Soils Derived from Volcanic Ash, Meeting on the Classification and Correlation of Soils from Volcanic Ash, Tokyo, Japan, June 11-27, 1964, *Food and Agric. Org., United Nations, World Soil Resources Rept.* **14:**82-86.

Figure 2. Andosol in the mountain regions north of Bandung (west Java). The pedon on top of the buried Andosol, derived from younger volcanic ash, fits the description of yellow brown loams in New Zealand.

Editor's Comments on Papers 1, 2, and 3

Taylor, N., 1964, The Classification of Ash-derived Soils in New Zealand, Meeting on the Classification and Correlation of Soils from Volcanic Ash, Tokyo, Japan, June 11-27, 1964, *Food and Agric. Org., United Nations, World Soil Resources Rept.* **14:**101-110.

Taylor, N. H., and J. E. Cox, 1956, *The Soil Pattern of New Zealand,* New Zealand Soil Bureau, Lower Hutt, New Zealand, Publ. 113.

Wright, A. C. S., 1964, The "Andosols" or "Humic Allophane" Soils of South America, Meeting on the Classification and Correlation of Soils from Volcanic Ash, Tokyo, Japan, June 11-27, 1964, *Food and Agric. Org., United Nations, World Soil Resources Rept.* **14:**9-21.

ORIGIN OF THE NAME "ANDO SOILS"

ROY W. SIMONSON

College Park, Maryland 20740 (U.S.A.)

(Accepted for publication May 17, 1979)

ABSTRACT

Simonson, Roy W., 1979. Origin of the name "Ando soils". Geoderma, 22: 333—335.

The name "Ando soils" was introduced in 1947 during reconnaissance soil surveys in Japan by American soil scientists. The name identified a great soil group of the intrazonal order in the then current U.S. system. The group consists of soils that had rather thick, dark A1 horizons, were derived from volcanic ash and were acid in reaction. The great soil group was later recognized in places as far apart as Alaska, France and New Guinea. Although the original name is being phased out at the present time, it has been the source of two other names that are in current use, viz., Andepts and Andosols.

The purpose of this communication is to record the circumstances surrounding the origin of the name "Ando soils". Part of that history has been reported by Simonson and Rieger (1967) in a paper on soils of the Andept suborder in Alaska, but the record is incomplete. The full history may be of interest because the original name was the source of two others in current use.

The name "Ando soils" originated as a result of reconnaissance soil surveys made in Japan by American soil scientists between 1945 and 1949, inclusive. Mapping units in those surveys were associations of phases of great soil groups as the latter had been defined in *Soils and Men* (Baldwin et al., 1938). From the beginning of the program, however, it was recognized that soils derived from volcanic ash fit poorly into any of the great soil groups in the current American system. Such soils had been called Brown forest soils, Black forest soils, prairie-like Brown forest soils, Black soils, Onji soils, and volcanic ash soils in Japan. In their morphology, the soils were most like those of the Prairie (Hapludoll) and Brown Forest (Eutrochrept) groups recognized at the time in the United States. In their composition and chemistry, however, the soils were markedly different from the Prairie and Brown Forest groups and were closer kin to Podzols.

During the first year and a half of the survey program, the question of an appropriate name or names for the soils from volcanic ash was discussed frequently. Should a name such as Brown Forest soils be used with an explanation that the soils did not fit well into that group as currently defined in the

United States? Should a new great soil group be proposed? Eventually a decision was reached to propose a new great soil group in the American system. Consequently, W.S. Ligon, in Japan at the time as the principal soil scientist in the Natural Resources Section, GHQ, SCAP, wrote a memorandum on January 8, 1947 to J.K. Ableiter in Beltsville, Maryland, U.S.A., summarizing available information on the soils derived from volcanic ash and recommending that they be set apart as a new great soil group in the intrazonal order. Two possible names were suggested, viz., Ando Podzolic soils and Anshoku Podzolic soils. The first of the two names was recommended because Ando was not a commonly used term in Japan, although derived from words meaning "dark soil". Hence, it could be specific for the great soil group. The memorandum continues, "Anshokudo is a descriptive term in common use, meaning dark (an), colored (shoku), and soil (do)". The second name would therefore be less specific for the group but might be more readily understood in Japan.

Principal characteristics of these soils from volcanic ash, as summarized in the memorandum from Ligon to Ableiter, are given below.

(1) The A1 horizons are usually thick and dark, and the soils lack the A2 horizons expected in Podzols. Amounts of organic matter in A1 horizons commonly range up to 15% and reach 30% in extreme cases. The C/N ratio of the organic matter ranges from 13 to 25.

(2) Where not cultivated, the soils are moderately to strongly acid with pH ranges of 4.5 to 5.6 in A horizons and 5.0 to 5.7 in deeper profiles. Levels of exchangeable bases are low, ranging from 2 to 9 mequiv./100 gm of soil material, whereas exchangeable Al (soluble in M NH_4Cl) exceeds 3 mequiv./100 gm generally and ranges up to 8 mequiv./100 gm.

(3) The silica—sesquioxide ratios of colloid fractions range from 1.3 to 2.0 in A horizons and from 0.75 to 0.90 in deeper profiles of these soils on Honshu. In Kyushu, the corresponding ranges in ratios are 0.4 to 1.13 in A horizons and 0.67 to 1.03 in deeper profiles.

A copy of the memorandum from Ligon was sent by Ableiter to Roy W. Simonson in Knoxville, Tennessee, chairman of a committee on great soil groups in the Division of Soil Survey, BPISAE, USDA. After some exchange of correspondence, Ableiter and Simonson agreed that recognition of a new great soil group in the intrazonal order would be appropriate but that it would be best not to include the word *podzolic* in a name for soils without the general morphology of Podzols. It was further agreed that the more specific name would be the better choice of the two that were suggested. A cablegram was therefore sent by Ableiter to Ligon on February 3, 1947 approving recognition of a new great soil group in the intrazonal order to be identified as *Ando soils*.

First publication of the name Ando soils followed in about a year in the initial report of the reconnaissance soil survey of Japan (Austin, 1948). The name was also used in all nine later reports, published from 1948 to 1951, inclusive. During that same interval, the name appeared for the first time in a journal (Thorp and Smith, 1949). The Ando group, including most of the soils derived from volcanic ash, had a total extent of about 3,000,000 ha or 8.41% of Japan (Ritchie, 1951).

Following introduction of the name in 1947 and its publication in the next two years, the Ando group was identified in such widely separated places as Alaska (Rieger and Wunderlich, 1960), France (Duchaufour and Souchier, 1966) and New Guinea (Rutherford, 1962).

The name Ando soils was largely but not completely replaced between 1960 and 1970. A new system of soil classification with a new nomenclature was then being developed in the United States (Soil Survey Staff, 1975). During that same period, a general soil map of the world was being prepared by the Food and Agriculture Organization, and that effort also included the development of a number of new names (FAO, 1971). Part of the original name of *Ando soils* was, however, retained in those coined for the new classification system and for the map legend. In the American system, a suborder of *Andepts* was defined in the order of Inceptisols. In the legend for the FAO-Unesco soil map, the two words *Ando soils* were converted to a single term *Andosols*. Thus, although the name as originally proposed is now passing out of the picture, elements of that name and the concept it represented are still in use.

REFERENCES

Austin, M.E., 1948. Reconnaissance soil survey of Japan — Kanto plain area. Gen. Headquarters, SCAP, Natural Resources Section Report 110-A, pp. 19—22.

Baldwin, M., Kellogg, C.E. and Thorp, J., 1938. Soil classification. In: G. Hambidge (Editor), Soils and Men, Yearbook of Agriculture, 1938. U.S. Gov. Printing Office, Washington, D.C., pp. 979—1001.

Duchaufour, P. and Souchier, B., 1966. Sols andosoliques et roches volcaniques des Vosges. Sciences de la Terre, 11: 345—365.

Food and Agriculture Organization, 1971. Soil Map of the World, Vol. IV, South America, Unesco, Paris, pp. 119—120.

Rieger, S. and Wunderlich, R.E., 1960. Soil survey and vegetation, northeastern Kodiak Island Area, Alaska. U.S. Dept. Agric. Soil Surv. Ser. 1956, No. 17, pp. 27—29.

Ritchie, T.E., 1951. Reconnaissance soil survey of Japan — Summary. Gen. Headquarters, SCAP, Natural Resources Section Report 110-I, 64 pp., illus., maps.

Rutherford, G.K., 1962. The yellow-brown soils of the highlands of New Guinea. Int. Soil Sci., Commissions IV and V, Trans. Joint Mtg., New Zealand, Nov. 13—22, 1962, pp. 343—349.

Simonson, R.W. and Rieger, S., 1967. Soils of the Andept suborder in Alaska. Soil Sci. Soc. Am. Proc., 31: 692—699.

Soil Survey Staff, 1975. Soil Taxonomy: A Basic System for Making and Interpreting Soil Surveys. U.S. Dept. Agric. Handbook No. 436, pp. 230—236.

Thorp, J. and Smith, G.D., 1949. Higher categories of soil classification: Order, suborder, and great soil groups. Soil Sci., 67: 117.

GENESIS AND CLASSIFICATION OF HUMIC ALLOPHANE SOIL IN JAPAN

ICHIRO KANNO,

National Kyushu Agricultural Experiment Station,
Chikugo, Fukuoka Prefecture, Japan

Volcanic ash (andesitic, basaltic, and rhyolitic) is widely distributed on terraces, hills, and mountains in Japan, extending from the southern boreal zone to the humid subtropical zone. It is distributed also along the Circumpacific volcanic zone and on South Pacific islands. Classification problems of the soils derived from volcanic ashes elsewhere in the world as well as in Japan are still unsettled. Therefore, it is extremely important to settle the taxonomic position of the volcanic ash soils.

Seki (1934) and Kamoshita (1958) regarded the soils derived from volcanic ash in Japan as Brown Forest soils and Prairie-like Brown Forest soils respectively, whereas Kawamura (1950) is of the opinion that the volcanic ash soils do not belong to the Brown Forest soils. Thorp and Smith (1949) designated volcanic ash soils in Japan as Ando soils. Kanno (1956) reported that volcanic ash soils in Japan should be regarded as a new genetic soil type (Japanese Volcanic Ash soils). Gerasimov (1958, 1959) expressed his view that Japanese Volcanic Ash soils belong to the subtropical black soil class and closely resemble Humified Allaphanic soils derived from basalt in Hainan Island of South China. In his opinion, Humified Allaphanic soils resemble in many respects Margalite soils in Indonesia designated by Dames (1950). In the writer's opinion, the Yellow Brown Loam of New Zealand (Taylor and Cox, 1956) closely resembles the Japanese Volcanic Ash soil.

According to a new system of soil classification in the United States (Soil Survey Staff, 1960), Ando soils have been classified as a member of Inceptisols, which are a soil order, and the term "Andept" as a substitute for Ando soil has been adopted. Recently, Kanno (1961) proposed a new name—Humic Allophane soil—in place of Japanese Volcanic Ash soil as one of the intrazonal soil types. Although the name proposed has been greatly influenced by the name "Humified Allaphanic soil" designated by Gerasimov (1959), Humic Allophane soil should be distinguished from Humified Allaphanic soil at the level of genetic soil type.

The object of the present paper is to record fundamental data on the Humic Allophane soils of Japan and their related soil types.

Genesis

1. Characteristics of Volcanic Ash and its Deposits as Parent Rock and Parent Material Respectively

Primary ash derived directly from magma is regarded as composed of powdered fragments of andesite, basalt, or rhyolite which have not yet been subjected to any weathering. Immediately after deposition the ash begins to weather and gradually changes to parent material as weathering proceeds. Therefore, the ash deposits belong to an allophanic weathering crust (orthoeluvium according to Polynov's concept) in the upper part of which Humic Allophane soil has developed.

With respect to weathering, a remarkable feature of unconsolidated ash particles differentiated from consolidated crystalline rocks is the fact that ash particles weather more rapidly than crystalline rocks because of the former's enormous surface areas exposed to weathering. The ash deposits are characterised by high porosity and permeability, which are considered to play the most important role in the formation of allophane under humid conditions. Thus, severe leaching throughout the deposit caused as a result of the high porosity and permeability of the ash accelerates losses of bases and silica and hinders the formation of layered silicates.

Secondary ash, which has already been weathered in the craters and which contains varying amounts of clay minerals other than allophane, also occurs. Humic Allophane soil has not developed on such secondary ash deposits. Volcanic ash ejected from Shimmoé-dake of the Kirishima volcanic cluster on 17 February 1959 is an example of this kind of ash. It contains approximately 20 per cent of clay, which consists mainly of beidellite-like minerals (Kanno, 1961). It is important to distinguish secondary ash from primary ash and to know the degree of contamination by secondary ash in any ash deposit.

2. Make-up of Soil Profile

Typically Humic Allophane soils consist of three horizons, namely, humic, thick A horizons, and pale yellow, yellowish brown to brown B horizons, followed by weathered ash deposits corresponding to C horizons. Several types of A horizons and the intercalation of heterogeneous ash layers reflecting differences in mode of deposition can be recognised. Furthermore, slightly developed bleached horizons or slightly gleyed horizons have been formed in the profiles of Humic Allophane soils by the influence of other pedogenetic processes. Such soils with heterogeneous horizons could not be placed in the same category as Humic Allophane soil and should be classified as subtypes between Humic Allophane soil and other genetic soil types. Thus Red-Yellow soil–Humic Allophane soil, Weakly gleyed soil–Humic Allophane soil, and Weakly podzolised soil–Humic Allophane soil are found in Japan.

3. Genetic Characteristics of Humus

Humic Allophane soils are characterised by abundant accumulation of humus (8 to 30 per cent) with high C/N ratios (15 to 30 or more) in the surface. Grasses, especially *Miscanthus sinensis*, which is a solfatara plant of wild grassland, have a profound influence on the accumulation of humus, whereas forest litters seem not to be essential for the development of thick, humic A horizons. Although the composition of humus in Humic Allophane soils has not yet been elucidated thoroughly, some chemical data obtained by the methods of Tiurin (1951) and Ponomareva (1957) indicate that the ratios of humic acid to fulvic acid are not definite, but are usually unity or less, and that the soils contain highly polymerised and condensed black humic acids with high C/N ratios (35 or more) and high molecular weights of about 6,000 or more (Hosoda and Takata, 1957; Kosaka et al., 1961). The degrees of polymerisation and condensation of humic acids and the humus content increase with the pedogenetic process.

In spite of the fact that drying out of the soil body for some time is essential for the polymerisation and condensation of humic acids (Kononova, 1951), water-holding capacities and water contents of Humic Allophane soils are usually very high. This contradiction may be explained by the fact that there is a periodic spell of dry weather with high temperatures in the Japanese summer.

Aluminium and allophane are able to combine strongly with humic and fulvic acids and suppress activity of micro-organisms. Thus it may be considered that aluminium and allophane are important factors responsible for the accumulation of humus. These characteristics of humus in Humic Allophane soils are unique and should be distinguished from those of humus in Podzols, Brown Forest soils, Red-Yellow soils, Calcareous Dark-coloured soils in the Tropics, and Chernozems.

4. The Formation of Clay Minerals

Notwithstanding the high base-status of primary volcanic ash, Humic Allophane soils are characterised by low degrees of base-saturation (40 per cent or less) and acidic reactions. River waters flowing through volcanic ash regions contain much more bases and silica than do river waters flowing through other regions. These facts indicate that intensive leaching of bases and silica has taken place in the ash deposits, leaving the products with low silica/alumina ratios. As a result of clay mineralogical investigations, it has been shown that allophane is predominant in Humic Allophane soils, especially in the surface horizons, and that these soils may be divided into three groups: namely, (1) those with abundant allophane, (2) those where allophane, hydrated halloysite, and gibbsite predominate; and (3) those with abundant allophane and hydrated halloysite. The young member of the soils generally belongs to the first group, whereas the old member, including buried layers, belongs to the third group. The old member with strongly acidic reaction belongs to the second group. In addition to allophane and kaolin minerals, varying amounts of gibbsite, goethite, hematite, illite, 14 A minerals (mainly Al-vermiculite and Al chlorite series), and amorphous silica and sesquioxides could be detected in clay separates of Humic Allophane soils. Investigations on weathering of pumice (rhyolitic in composition) showed that the main allophane-supplying minerals are volcanic glasses and plagioclases in part (Kanno, 1961).

The formation of the main clay minerals in Humic Allophane soils is represented by the following schema in which the reactions are indicated by arrows and letters:

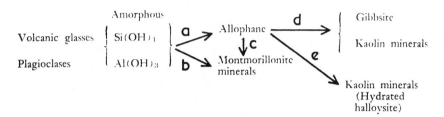

A Schema for the Formation of Main Clay Minerals from Volcanic Ash in Japan.

Remarks: Illite, 14 A minerals, kaolins, and Fe_2O_3 minerals may be formed from ferromagnesian minerals and biotite. In general, reactions (b) and (c) have not taken place in Humic Allophane soils.

Reaction (a) is due to the characteristics of volcanic ash as parent rock and to well drained conditions. Under such conditions the early weathering products of ash are unable to form kaolin minerals by crystallisation. Reaction (b) was observed in a closed crater lake under conditions inhibiting desilication and leaching of bases. The formation of montmorillonite minerals from allophane (reaction c) was observed in pumice beds in which drainage is intermittently imperfect and the addition of soluble silica and bases takes place.

Reaction (d) was observed in some old members of Humic Allophane soils with strongly acid reactions where gibbsite and hydrated halloysite increase with depth. The formation of gibbsite and kaolin minerals from allophane and their accumulation in the B horizon may be due to partial destruction of allophane by aggressive fulvic acids in the surface horizon. Reaction (e) was observed in the lower part of the buried type of ash deposits. The formation of kaolin minerals from allophane is due to the influence of percolating water containing soluble silica and to the ageing of allophane. The buried Musashino Loam layer (Pleistocene sediment) of the Kanto Loam Bed, which is a siallitic weathering crust derived from volcanic ash, is of this type. It is obvious that the Musashino Loam layer and Brown Granular Clays of New Zealand (Fieldes and Swindale, 1954) should be excluded from the category of Humic Allophane soil.

Classification

As stated before, Humic Allophane soils are subjected to intensive leaching of silica and bases by which allophane has been formed. Chemical weathering of the mineral mass in the soils is profoundly affected by the fragmentary nature of ash particles and the high permeability and porosity of ash deposits. The soils are characterised by abundant accumulation of peculiar organic matter in the surface horizon. The nature of the humus in the soil is neither podzolic nor chernozemic, but is unique. Therefore, Humic Allophane soils are an independent genetic soil type belonging to an intrazonal soil group.

Although it belongs to "Dark-coloured soil class in the subtropical zone", Humic Allophane soil should be distinguished from "Dark-coloured soil subclass in the semi-arid subtropical zone", which is characterised by alternate wet and dry climate, high content of montmorillonite clays firmly united to dark-coloured humus, abundance of exchangeable Ca and Mg, and by basic rocks, as has been stated by Gerasimov (1960). Consequently, the writer arrives at the following conclusion: Humic Allophane soil, which is a dark-coloured soil subclass in the humid subtropical zone, occupies a taxonomic position closely resembling Type 5 (automorphic, biolithogenic; Subclass 2) of Class X (Jeltozem-forest soils class) in the soil classification schema proposed by Ivanova and Rozov (1960). Provided related soils have developed in the southern boreal and the humid tropical zones, elucidation of genesis and classification of such soils requires further investigation.

Establishment of taxonomic units lower than genetic soil type for Humic Allophane soils is a matter of great urgency for detailed soil surveys in Japan. Although this problem is still under consideration, the following should be considered as criteria of lower taxonomic units: degrees of intervention of accessory pedogenetic processes coexistent with main pedogenetic process; thickness of humus horizons; composition of humus, especially degrees of polymerisation and condensation of humic acids; degrees of allitisation; clay mineralogical composition; lithological characteristics of parent rock (volcanic ash); ages of ash deposits; modes of deposition; textures; influences of human activity, and so on. The selection of the most appropriate criteria for a given categorical unit is of extreme importance for resolving the problem.

References

Dames, T. W. G. 1950: *Trans. 4th int. Congr. Soil Sci.* 2: 180–2.
Fieldes, M.; Swindale, L. D. 1954: *N.Z. J. Sci. Tech.* B36: 140–54.
Gerasimov, I. P. 1958: *Izv. ANSSSR, seriya geograficheskaya* (2): 54–63.
—— 1959: *Vestnik Moskovskogo Universiteta* (1): 183–6.
—— 1960: *Dokl. Sovetskikx Pochvovedov k VII Mezhdunarod. Kongr. v SShA*: 275–9.
Hosoda, K.; Takata, H. 1957: *J. Sci. Soil, Tokyo* 28: 64–68.

IVANOVA, YE. N.; ROZOV, N. N. 1960: *Dokl. Sovetskikx Pochvovedov k VII Mezhdunarod. Kongr. v SShA*: 280-93.

KAMOSHITA, Y. 1958: *Misc. Publ. Nat. Inst. agric. Sci. B(5)*. 58 pp.

KANNO, I. 1956: *Trans. 6th int. Congr. Soil Sci. E*: 105-9.

―― 1961: *Kyushu agric. Expt. Sta. Bull.* 7: 1-185.

KAWAMURA, K. 1950: *Agric. & Hort.* 25: 11-4.

KONONOVA, M. M. 1951: Problema Pochvennogo Gymusa i Sovremennye Zadachi ego Izucheniya. str. 390. ANSSSR, Moskva.

KOSAKA, J.; HONDA, C.; ISEAKI, A. 1961: *J. Sci. Soil, Tokyo* 32: 333-7.

PONOMAREVA, V. V. 1957: *Pochvovedenie* (8): 66-71.

SEKI, T. 1934: *J. Sci. Soil, Tokyo* 8: 3-18.

SOIL SURVEY STAFF 1960: Soil Classification, a Comprehensive System, 7th Approximation. U.S. Dept. Agric. Soil Cons. Serv. Washington D.C. 265 pp.

TAYLOR, N. H.; COX, J. E. 1956: *Proc. N.Z. Inst. Agric. Sci.*: 28-44.

THORP, J.; SMITH, G. D. 1949: *Soil Sci.* 67: 117-26.

TIURIN, I. V. 1951: *Trud. Pochv. in-ta Dokuchayeva*, 38: 5-21.

CORRELATION OF SOILS DERIVED FROM VOLCANIC ASH

by R. Dudal
Soil Correlator, World Soil Resources Office
FAO, Rome

The soils dealt with here, under the general heading of "Soils derived from Volcanic Ash", are limited to those which are characterized by a dominance of amorphous material in the clay fraction, a low bulk density, a relatively high content of organic matter in the A horizon, a friable consistence, smeariness, and a high sorptive capacity. These soils are widely spread in the volcanic regions throughout the world (see map attached)[1] and occur under a great variety of climatic conditions ranging from humid temperate to humid tropical. They are known under many different names, i.e. Volcanic Ash soils or Trumao soils (South America); Andosols, Black Dust soils or High Mountain soils (Indonesia); Kuroboku, Black Volcanic Ash soils, Kurotsuchi, Andosols, Humic Allophane soils or Brown Forest soils (Japan); Alvisols or Yellow Brown Loams (New Zealand); Talpetate soils (Nicaragua); and Andepts, or Hydrol Humic Latosols (U.S.A.).

The soils covered by these names have many properties in common but the range of variability in their characteristics seems to differ widely as a result of the difference in the criteria which have been used to delimit and subdivide them. Thus the difference in nomenclature often reflects a difference in concept and is not only a matter of different name giving.

In order to establish a correlation between the classification units related to soils derived from volcanic ash, it is necessary to analyse the criteria upon which the separation of these soils is based.

On the general soil map of Japan at the scale of 1:500,000 the soils derived from volcanic ash are called Kuroboku (from Kuro meaning black and Boku meaning friable). These soils are defined (1) as having "... a layer of black colour, both Munsall value and chroma being equal to or less than 2, developed at the upper part of terrestrial soils." The organic matter content in these soils ranges from 15 to 20 percent, and they show a low stickiness and plasticity. It is assumed that a close relationship exists between the occurrence of Kuroboku and grassland vegetation and that the formation of these soils has been influenced by hydromorphic conditions (2). A subdivision of the Kuroboku is made on the basis of the thickness and the colour of the organic matter layer. Soils derived from volcanic ash which have a dark brown surface soil are considered as "light coloured Korobuku". If these light coloured soils occur under forest they are grouped with the Brown Forest soils or Acid Brown Forest soils of the Japanese classification.

In the new classification recently introduced in New Zealand (3) (5), soils derived from volcanic ash are classified as Alvic soils, a name which is a contraction of Amo-Fulvic. The Fulvic soils – a unit of category I in the classification scheme – "occur characteristically in the humid areas; typically they are well drained without spectacular differentiation of horizons." (3) In category II the soils are subdivided according to the "main energy status as indicated approximately by the latitudinal and

---

[1] The areas of major distribution shown on the map include mainly humid regions where the soils derived from volcanic ash show the characteristics mentioned above. Volcanic ash also occurs in arid regions, such as in Chile, Nicaragua and the United States of America, but these areas have not been shown on the map.

altitudinal zones and by soil moisture", which are indicated by the prefixes "pro" (temperate), "ad" (subtropical) and "el" (elevated). 1/

In category III the soils are separated according to the kind of argillization of the soil body. The soils derived from volcanic ash are characterized by a dominance of amorphous clays from which the name Amo-Fulvic or Alvic soils is derived. According to the grade of argillization, which is used as an indicator of the grade of weathering, the Alvic soils are further subdivided into Subalvic (weakly weathered), Alvic (moderately weathered), and Hydrol Alvic soils (strongly weathered). In category IV the soils are subdivided according to the kind and degree of illuviation, gleying, accumulation, etc. Thus the Alvic soils are further split according to the presence of a B horizon, a fragipan or iron accumulation. Separations made in category V are based on the "state of enleaching" (weakly leached, moderately leached, strongly leached) which is indicated by the base saturation or the degree of salinity. Characteristics of the parent material, texture and type of organic matter, are used as classification criteria in categories VI and VII of the system.

In the U.S.D.A. 7th Approximation of a comprehensive soil classification system (6), the soils derived from volcanic ash are included in the suborder of the Andepts of the order of the Inceptisols which have diagnostic horizons that are thought to form rather quickly, and which do not represent significant illuviation or eluviation, or extreme weathering. The Andepts are further characterized by having 60 percent or more of allophane in the clay fraction or 60 percent or more volcanic ash in the silt and sand fraction. At the great group level the Andepts are subdivided into Cryandepts (with low mean annual temperatures and low mean summer temperatures), Durandepts (with a duripan), Ochrandepts (with a light coloured A horizon), Umbrandepts (with a dark coloured A horizon but a low to medium base saturation), Mollandepts (with a dark coloured A horizon and a high base saturation) and Hydrandepts (with clays that dehydrate irreversibly into gravel sized aggregates.

For South America, A.C.S. Wright (7) has used the name "Humic Allophane Soils". Their classification is based on the grade of weathering, the amount and kind of organic matter ("melanization"), and the degree of leaching of the soils in analogy with the work done in New Zealand. Furthermore, strong emphasis is given to the environmental conditions. The Humic Allophane soils of South America are subdivided into soils of high latitude (Chile, Argentina), where maritime and Mediterranean climates prevail, and soils of low latitudes (Ecuador, Colombia) which have an equatorial climate. It is considered that the soils of low latitudes are rarely subjected to sharp or prolonged fluctuations in temperature and enjoy a more even moisture regime, while the soils formed in high latitudes are more likely to be subjected to prolonged periods of more extreme conditions. The available data on soil morphology and physical and chemical characteristics show that the soil properties vary in relation to the different environmental conditions. However, further studies will be required to establish an accurate correlation, and so far the subdivision is mainly based on climatic data. For the soils of the low latitudes a distinction is made according to moisture conditions and elevation; for the soils of high latitudes a distinction is made according to the length of the dray season and the amount of rainfall.

In Indonesia, Tan Kim Hong (4) proposed a tentative subdivision of Andosols into soils with a low humic/fulvic acid ratio ($<0,2$) and a low variable charge ($CEC_v < 30$ me/100 g), which correspond to the lowland Andosols of Sumatra, and soils with a high humic/fulvic acid ratio (>0.5) and a medium to high variable charge ($CEC_v > 30$ me/100 g) which correspond to the highland Andosols of Java.

1/ When a certain class is dominantly represented in one climatic zone, the pertinent prefix may be omitted from the name. This applies to the prefix "pro" in the case of the soils derived from volcanic ash in New Zealand where they occur mainly in temperate climatic conditions.

From a comparative study of different systems of classification, it appears that the central concept of the soils derived from volcanic ash is very similar in the different countries. The characteristic features of the modal soil are: a relatively thick and dark coloured A horizon with a moderately high organic matter content, a dominance of amorphous material in the clay fraction, a low bulk density, a friable and smeary consistence, a high sorptive capacity, a medium base saturation, a moderately acid pH, no evidence of clay movement in the B horizon, generally occurring in humid warm temperate conditions. As a result a high degree of equivalence exists between the modal soils included in Kuroboku (Japan), typic Umbrandept (U.S.A.), Adalvic soils (New Zealand) and the moderately weathered, moderately leached, strongly melanized Humic Allophane soils (South America). However, since the criteria for delimiting and subdividing these units are different, as shown above, a serious overlap may occur between soils which lie outside the central concept. Thus the typic Umbrandept (U.S.A.) may include Kuroboku, light coloured Kuroboku and thick Kuroboku (Japan) since in the U.S.A. classification, the colour and thickness of the A horizon are not the only distinctive criteria at a high level of classification. On the other hand, Kuroboku (Japan) may include Umbrandepts and Cryandepts (U.S.A.) since the climatic criteria which are used in the U.S.D.A. 7th Approximation are not applied in the Japanese soil classification. Soils similar to hydric Umbrandepts (U.S.A.) and hydrous Elalvic soils (New Zealand) are excluded from the Kuroboku group when they occur under forest, in which case they are grouped with the Brown Forest soils of the Japanese classification. The Adalvic soils (New Zealand) may include Umbrandepts and Mollandepts (U.S.A.); the distinction on account of the base saturation is also made in New Zealand but at a much lower level of classification. The distinction which is made in Indonesia between Andosols of low lying tropical areas, and those occurring under warm temperate conditions of high altitudes on the basis of the humic/fulvic acid ratio of the organic matter, is not found in other schemes of classification.

Through field research and a study of the literature it has been possible to draw up a comparative table of classification units used in different countries for soils derived from volcanic ash. It appears that although a certain degree of equivalence exists between these units, a perfectly fitting correlation will be obtained only if adjustments can be made in the various classification systems with regard to the criteria used for the delimitation and subdivision of these soils.

The difference in climatic conditions under which soils derived from volcanic ash occur is not generally reflected by the soil characteristics. It seems that a same morphological end product may be formed by different combinations of soil forming factors. The characterization of the organic matter by its humic/fulvic acid ratio, as done in Indonesia, may provide a means to differentiate these soils in relation to differences in climate, but this approach has to be tested further before it can be more generally applied. However, since temperature and rainfall are of very great importance for agricultural utilization it is felt that these factors of environment should be taken into account for the subdivision of these soils. In the U.S.A. the soils occurring in cold climates are classified separately (Cryandepts) while in New Zealand a distinction is made between soils occurring in temperate (prefix "pro") and subtropical (prefix "ad") climates or at high elevation (prefix "el"). However these subdivisions may not suffice to separate the different climatic phases on a world wide basis. It seems therefore necessary that besides an agreement on a morphological definition of the soils derived from volcanic ash, a more precise and uniform approach is needed for the subdivision of these soils according to the different climatic conditions in which they occur.

References

1. Ministry of Agriculture and Forestry, Japanese Government, Volcanic ash soils in Japan, Tokyo, June 1964.

2. OHMASA, M., Genesis and Morphology of Volcanic Ash Soils in Japan. Report on the "Meeting on the Classification and Correlation of Soils from Volcanic Ash", World Soil Resources Report No. 14, FAO, Rome 1965.

3. POHLEN, I.J., Soil Classification in New Zealand, Transactions International Soil Conference, Wellington, 1962.

4. TAN KIM HONG, The Andosols in Indonesia. Report on the "Meeting on the Classification and Correlation of Soils from Volcanic Ash", World Soil Resources Report No. 14, FAO, Rome 1965.

5. TAYLOR, N., The Classification of Ash derived Soils in New Zealand. Report on the "Meeting on the Classification and Correlation of Soils from Volcanic Ash", World Soil Resources Report No. 14, FAO, Rome 1965.

6. U.S.D.A., Soil Conservation Service, Soil Classification, a comprehensive system, 7th Approximation, Washington, 1960.

7. WRIGHT, A.C.S., The "Andosols" or "Humic Allophane Soils", of South America. Report on the "Meeting on the Classification and Correlation of Soils from Volcanic Ash", World Soil Resources Report No. 14, FAO, Rome 1965.

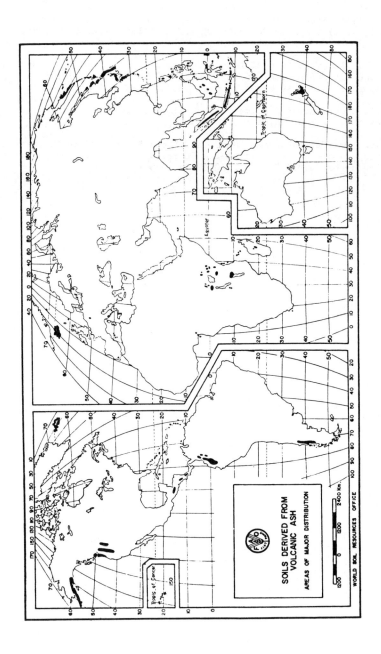

Part II

GEOGRAPHIC DISTRIBUTION OF ANDOSOLS

Editor's Comments
on Papers 4 Through 7

4   **LEAMY et al.**
    The Morphological Characteristics of Andisols

5   **SIMONSON AND RIEGER**
    Soils of the Andept Suborder in Alaska

6   **TAN**
    The Andosols in Indonesia

7   **FREI**
    Andepts in Some High Mountains of East Africa

The geographic distribution of Andosols is associated with the presence of volcanic activity that produces volcanic ash. Major global tectonic and volcanic activity occur mainly in the circum-Pacific zone, but to some extent such activity can also be found in Iceland, Spain, Italy, Rumania, the Canary Island, and East Africa. Localized spots of remnants of tectonic activity also have been reported in West Europe.

Generally Andosols are not found according to major climatic and vegetational zones in the world. They can occur from the arctic to the tropical regions and from subalpine to the warm humid climates in the lowlands of the tropics. Andosols are extensively distributed around the world. The major area where they can be found is Japan, but relatively large areas of these soils have also been reported in Korea, the Philippines, Indonesia, Hawaii, El Salvador, and Chile (Dudal, 1964). With the increased knowledge about Andosols, the soils have currently been identified in the United States (Robertson, 1963), Mexico (Gutierrez Rodriquez, 1980), Costa Rica (Martini, 1976), Colombia (Guerrero, Davila, and Torres, 1979), and Argentina (Inostroza, 1981). Other countries where Andosols have been located are Spain (Rodriguez Pascual et al., 1978), the Canary Islands (Sanchez Dias, Guerra Delgado, and Fernandez Caldos, 1978), Italy (Eschena and Gessa, 1967), France (Duchaufour and Souchier, 1966), Rumania

(Conea, 1972), and Taiwan (Lai and Leung, 1967), and the list continues to grow.

The four Papers in this part have been selected to show the global distribution of Andosols. Papers 4, 5, and 6 discuss the occurrence of Andosols in New Zealand, Alaska, and Indonesia, respectively, covering the circum-Pacific zone, and Paper 7 discusses Andosols on the African continent.

Using the FAO-UNESCO general world soil map, Dudal (1976) estimated the land area covered by Andosols to be approximately 100 million hectares, but Leamy (Paper 4) indicated that, based on more detailed soil maps, the figure should be adjusted to 124 million hectares, or 0.8 percent of the world's land surface. Even the latter is perhaps on the low side since in many countries the areas covered by Andosols have either not been mapped, or completely mapped yet, or yet have still to be discovered. Many are located in heavy mountainous or rugged terrain. As Leamy (Paper 4) noted, the areas with Andosols in the world do not cover extensive acreages at any one place but rather were found as localized spots associated with the presence of volcanoes. On the other hand, Dudal and Supraptohardjo (1957, 1961) and Dudal and Jahja (1957) reported the occurrence of Andosols in all the mountain regions in Indonesia. At altitudes above 600 m above sea level, the soils in Indonesia were classified by Dudal and Supraptohardjo above as Andosols, whereas the soils located at altitudes below 600 m above sea level were called *Latosols*. Although the tendency to find Andosols in the tropics mostly at higher elevation is also illustrated by Paper 7, where Frei noticed that the soils were located in East Africa at altitudes above 2500 m, information provided by other authors indicates that it is equally possible to find these soils at sea level (Wright, 1964; Tan and Van Schuylenborgh, 1961). In Indonesia, it appears that volcanic ash produces not only Andosols but also other soils (Tan and Van Schuylenborgh, 1959), such as Spodosols, Alfisols, Ultisols, and Oxisols. Andosols also can be found in Indonesia at elevations below 600 m above sea level, where conditions are conducive for developing the characteristic features of these soils (Mohr, 1937; Druif, 1939; Tan, 1960). In contrast to the "highland or mountain" Andosols, the "lowland" Andosols are distributed in relatively small localized areas. Paper 6 tries to summarize the distribution of Andosols in Indonesia. A similar situation occurs in South America, where Wright (1964) described the occurrence of Andosols in both the high mountains and the lowland areas. Paper 6 and that by Wright tend to lend some support to Leamy's statement (Paper 4) about the localized distribution of Andosols in the various countries in relatively small areas.

REFERENCES

Conea, A., 1972, Andosols in Rumania, "P"rvi Natsionalen Kongres po Pochvoznanie, Bulgaria, 1969, pp. 465-472.

Druif, J. H., 1939, De Bodem van Deli (Sumatra). III. Toelichting bij de Agrogeologische Kaarten en Beschrijving der Grondsoorten van Deli (Sumarta), *Mededelingen Deli Proefstation* (N. Sumatra) **3-4:**1-74.

Duchaufour, P., and B. Souchier, 1966, Andosolic Soils on Volcanic Rocks of Vosges, *Sci. Terre.* **11:**345-265.

Dudal, R., 1976, Inventory of the Major Soils of the World with Special Reference to Mineral Stress Hazards, in *Plant Adaptation to Mineral Stress in Problem Soils,* M. J. Wright, ed., Proc. Workshop, Cornell Univ. Argic. Expt. Stn., Ithaca, New York, pp. 3-14.

Dudal, R. (ed.), 1964, Meeting on the Classification and Correlation of Soils from Volcanic Ash, Tokyo, Japan, June 11-27, 1964, *Food and Agric. Org., United Nations, World Soil Resources Rept.* **14,** 169p.

Dudal, R., and H. Jahja, 1957, Soil Survey in Indonesia, *Contr. Gen. Agric. Res. Stn. Bogor, Indonesia* **147:**3-8.

Dudal, R., and M. Supraptohardjo, 1957, Soil Classification in Indonesia, *Contr. Gen. Agric. Res. Stn. Bogor, Indonesia,* **148:**1-23.

Dudal, R., and M. Supraptohardjo, 1961, Some Consideration on the Genetic Relationship between Latosols and Andosols in Java, Internat. Congress Soil Sci., 7th Trans. (Madison, Wisconsin) **4:**229-233.

Eschena, T., and C. Gessa, 1967, The Andosols of Sardinia, University Sassari, Sardinia, *Studi Sassar,* ser. 3, **15:**363-386.

Guerrero, R., A. Davila, and C. Torres, 1979, The Effect of Land Use on Some Parameters of Fertility in Two Soil Types (Andept and Tropept) in the Highlands of Pasto, Colombia. I. Organic Matter and Exchangeable Bases, *Anales de Edafologia y Agrobiologia* **38:**173-185.

Gutierrez Rodriguez, E., 1980, *Effect of the Application of Organic and Inorganic Fertilizers on the Chemical and Physical Properties of Andosols,* vol. 3, Chapingo, Colegio de Postgraduados, Secretaria de Agricultura y Recursos Hidraulicos, Mexico, 176p.

Inostroza, O., 1981, Land Preparation of Trumao soils (Andepts) for Rape Seed Sowing, Instituto de Investigaciones Agropecuarias, Santiago, Agric. Tecnica **41(1):**31-40.

Lai, C. Y. and K. W. Leung, 1967, The Physical and Chemical Properties of Ando Soils in Yungmingshan, *Jour. Taiwan Agric. Res.* **16(14):** 1-7.

Martini, J. A., 1976, The Evolution of Soil Properties as It Relates to the Genesis of Volcanic Ash Soils in Costa Rica, *Soil Sci. Soc. America Proc.* **40:**895-900.

Mohr, E. J. C., 1937, *De Bodem der Tropen in het algemeen en die van Nederlands Indie in het bijzonder,* R. L. Pendleton, trans. Ann Arbor, Mich., 1944, 765p.

Robertson, R. H. S., 1963, Allophanic Soil from Trail Bridge, Oregon, with Notes on Mosaic Growth in Clay Minerals, *Clay Minerals* **5:**237-247.

Rodriguez Pascual, C., J. Galvan, M. L. Tejedor Salguero, and E. Fernandez Caldas, 1978, Mineralogical Study of the Clay Fraction of Andepts, in a Cronological Sequence, by Electronic Microscopy, *Anales de Edafologia y Agrobiologia* **37:**989-1001.

Sanchez Dias, J., A. Guerra Delgado, and E. Fernandez Caldas, 1978, The Surorder Andept in Gran Canaria Island, *Anales de Edafologia y Agrobiologia* **37**:387-400.

Tan, K. H., 1960, The Black Dust Soil of Deli (Sumatra) (in Indonesian, with English summary), *Tehnik Pertanian* **9**:77-93.

Tan, K. H., and J. Van Schuylenborgh, 1959, On the Classification and Genesis of Soils Derived from Andesitic Volcanic Materials under a Monsoon Climate, *Netherlands Jour. Agric. Sci.* **7**:1-22.

Tan, K. H., and J. Van Schuylenborgh, 1961, On the Classification and Genesis of Soils Developed over Acid Volcanic Materials under Humid Tropical Conditions. II, *Netherlands Jour. Agric. Sci.* **9**:41-45.

Wright, A. C. S., 1964, The "Andosols" or "Humic Allophane" Soils of South America, Meeting on the Classification and Correlation of Soils from Volcanic Ash, Tokyo, Japan, June 11-27, 1964, *Food and Agric., Org., United Nations, World Soil Resources Rept.* **14**:9-21.

4

Copyright © 1980 by the New Zealand Society of Soil Science
Reprinted from pages 17-34 of *Soils with Variable Charge*, B. K. G. Theng, ed., New Zealand Society of Soil Science, Lower Hutt, N. Z., 1980, 448p.

THE MORPHOLOGICAL CHARACTERISTICS OF ANDISOLS

M. L. LEAMY, Soil Bureau, Department of Scientific and Industrial Research, New Zealand.
G. D. SMITH, Geologisch Instituut, Rijksuniversiteit Gent, Belgium.
F. COLMET-DAAGE, Office de la Recherche Scientifique et Technique Outre-Mer, Martinique.
M. OTOWA, Hokkaido National Agricultural Experiment Station, Japan.

I. INTRODUCTION

Soils formed on materials of volcanic origin have commanded particular attention for many reasons, among which are their specific geographic distribution, their readily identifiable origin and their distinctive properties. Most studies have been devoted to soils formed on volcanic ash (e.g., Ministry of Agriculture and Forestry, Japanese Government, 1964; World Soil Resources Report 14, 1965; Gibbs, 1968; Inter-American Institute of Agricultural Sciences, 1969). In this chapter we attempt to summarize the main morphological characteristics common to most soils which, because of their assemblage of distinctive properties, are recognized as having a common genetic derivation from volcanic materials. We will first outline some of the developments in soil classification which have led to the current proposal to provide an Order of Andisols for at least those soils that are now identified in Soil Taxonomy (Soil Survey Staff, 1975) as Andepts. This is not intended to compromise or restrict ongoing discussion on a proposal which is still being debated and which cannot yet be finalised (International Committee on the Classification of Andisols (ICOMAND), Circular 1, p.2, 1979).

A. Nomenclature and Classification

The name "Ando", which has commonly been associated with volcanic ash soils, was introduced in 1947 during reconnaissance soil surveys in Japan by American soil scientists (Simonson, 1979). The name was connotative of soils formed on volcanic ash which had thick, dark surface horizons and were acid in reaction. It is derived from the Japanese *Anshokudo* which was a descriptive term in common use, meaning dark *(an)*, coloured *(shoku)*, and soil *(do)*.

Part of the name was retained as a formative element in new systems of nomenclature defined in soil classifications developed by F.A.O. (FAO-Unesco, 1974) for the World Soil Map project, and in the U.S.A. as Soil Taxonomy (Soil Survey Staff, 1975).

The legend for the FAO-Unesco World Soil Map identified these soils as Andosols and described them as "soils formed from materials rich in volcanic glass and commonly having a dark surface horizon" (FAO-Unesco 1974).

In Soil Taxonomy (Soil Survey Staff, 1975), the Andept suborder of Inceptisols accommodates soils formed predominantly in volcanic ash. These soils have a low bulk density, an appreciable amount of allophane of high exchange capacity, or consist mostly of pyroclastic materials.

Subsequent testing and use of Soil Taxonomy has revealed a number of serious defects in the classification of Andepts.

1. The definition of the suborder excludes a number of soils that should be included. Andepts are required to have an exchange complex dominated by amorphous materials throughout the upper 35 cm and to have low bulk densities, or to have more than 60% (by weight) vitric volcanic ash, cinders or other pyroclastic material. Some Andepts have a surface horizon of 15 to 25 cm thickness that does not react to NaF as is required by the definition of domination by amorphous materials. Such soils conform to the definition at depth and probably should not be excluded at the highest categoric level. Studies on the interactions between humus and inorganic constituents in volcanic ash soils have indicated that humus evolves from forms with a very low complexing ability for Al and Fe to forms that complex Al and Fe in A1 horizons (Wada & Higashi, 1976). Recent Japanese work has highlighted the importance of "active" Al in volcanic ash soils (Wada, 1977; Wada & Gunjigake, 1979). In some cases (Mizota & Wada, 1980) it is concluded that the NaF reaction is not specific for allophane and a reaction is obtained with any "active" Al even if bound with humus. More data on the fluoride reactivity of the upper horizons of volcanic ash soils are clearly required before clear taxonomic parameters can be derived.

   In addition, the 60% limit of vitreous materials has been interpreted in some cases to mean that any soil formed in ash and having less than 40% clay must be classified as an Andept unless it has a diagnostic horizon prohibited in Inceptisols. Such an interpretation includes soils more properly classified as Entisols.

2. Base saturation by NH_4OAc has been used as a differentia with a limit of 50%, the same limit used for mineral soils that have crystalline clays. The significance of this limit in Andepts is open to very serious question because the clays are mostly amorphous and the cation exchange capacity (CEC) is largely pH dependent.

3. Thixotropy has been used as a differentia, but the decision that a given horizon is, or is not thixotropic, is very subjective and cannot be made uniformly.

4. The soil moisture regime has not been used as a differentia for Andepts as it has for all other soils. Interpretations for a given family cannot be made without the use of climatic phases.

5. The darkness of the epipedon is weighted heavily in subgroup definitions, but in warm intertropical areas there seems to be little or no relation between colour and carbon content, degree of weathering, or any other property.

6. For most mineral soils a fragmental particle size class is provided, but not for Andepts. Coarse pumice falling from the air commonly has a basal layer without appreciable fine earth.

7. Inadequate emphasis was given to the unique moisture retention properties of the Andepts. The irreversible drying effect was used only to define the Hydrandepts. New data require reconsideration of the effects of drying. For instance, Rousseaux & Warkentin (1976) have suggested that for some allophane soils from the West Indies and Japan the chemical composition of the allophane component determines the micropore size distribution, which in turn determines the water retention properties. Colmet-Daage *et al.* (1967) have documented water retention properties and the effects of drying on Andepts of Ecuador. Such properties could possibly be substituted for the present unsatisfactory property of thixotropy.

The identification of these taxonomic defects has resulted in the preparation, by G. D. Smith, of an unpublished document dated 10 April 1978, subsequently amended slightly by letter of 13 February 1979, titled "A Preliminary Proposal for Reclassification of Andepts and some Andic Subgroups". In this document the name Andisol was introduced. This name, rather than Andosols, was proposed because the latter is currently used elsewhere with other definitions (FAO-Unesco 1974), and because the connecting vowel o is supposed to be restricted to Greek formative elements.

This proposal is the basis for the discussions and operations of ICOMAND which currently has about 100 members throughout the world receiving correspondence by way of circulars (ICOMAND Circular No. 1, 3 April 1979; ICOMAND Circular No. 2, 25 January 1980; ICOMAND Circular No. 3, 30 April 1980).

Because it has not been formally published, the proposal is not widely accessible and for this reason as well as its central relevance to a discussion of the morphology of Andisols, that part dealing with the Order definition is reproduced here.

B. Definition

Andisols are defined as mineral soils that do not have an aridic* moisture regime or an argillic, natric, spodic, or oxic horizon unless it is a buried horizon, but have one or more of a histic, mollic, or umbric epipedon, or a cambic horizon, a placic horizon, or a duripan; or, the upper 18 cm, after mixing, have a colour value, moist, of 3 or less and have 3% or more organic carbon in the fine earth; and, in addition, have one or more of the following combinations of characteristics:

1. Have, to a depth of 35 cm or more, or to a lithic or paralithic contact that is shallower than 35 cm but deeper than 18 cm, a bulk density of the fine earth fraction of less than 0.85 g/cm^3 (at 1/3 bar water retention of undried samples) and the exchange complex is dominated by amorphous materials.
2. Have, in the major part of the soil between a depth of 25 cm and 1 m or a duripan, a placic horizon, or a lithic or paralithic contact that is deeper than 35 cm but shallower than 1 m, a bulk density of the fine earth fraction of less than 0.85 g/cm^3 (at 1/3 bar water retention of undried samples) and the exchange complex is dominated by amorphous materials.
3. Have 60% or more, by weight, of noncalcareous vitric volcanic ash, pumice or pumice-like fragments, cinders, lapilli, or other vitric volcaniclastic materials either to a depth of 35 cm or more, or in the major part of the soil between 25 cm and 1 m or a lithic or paralithic contact that is shallower than 1 m, and the pH in the major part of these horizons of 1 g of fine earth in 50 ml of 1N NaF is 9.2 or more after two minutes.
4. Have a weighted average (by thickness of subhorizons) in the major part of the soil between a depth of 25 cm and 1 m or a duripan or paralithic or lithic contact shallower than 1 m, a water retention of undried fine earth at 15 bar pressure of 40% or more; an ustic moisture regime, and a bulk density of the fine earth fraction of less than 0.9; and either a ratio of 15 bar water percentage (undried) to the meq of exchangeable bases is 1.5 or less, or the pH of 1 g of fine earth in 50 ml of 1N NaF is 9.4 or more after two minutes, or both.

Explanation of Definition

The opening paragraph transfers the present requirements of Andepts from Inceptisols to Andisols, and, for the vitric great groups (item 3), presently Vitrandepts, eliminates problems about cambic horizons (which may not have sandy particle size) and

---

* During the 3rd International Soil Classification Workshop, 1980, soils with andic properties were tentatively identified in Syria (H. Eswaran, pers. comm.). Substantiation and testing may lead to the deletion of this requirement.

the thickness of the umbric epipedon. The concept of a sandy particle size does not apply well in Andisols, particularly those in rhyolitic ash because of their high water holding capacity. The carbon limit is intended to eliminate problems with basaltic or andesitic ash, cinders, and lapilli, all of which are nearly black.

Item 1 is intended to bring into Andisols the soils formed in thin deposits of ash. It does not differ significantly from the present definition of Andepts.

Item 2 is intended to bring into Andisols the soils formed in thicker deposits of ash but lacking allophane in the surface horizon. In the West Indies and South America, soils that should be grouped with Andisols commonly do not react to NaF in the upper 15 to 25 cm but react strongly below. The Japanese literature reports the absence of allophane in some surface horizons, although it is identified below (Tokashiki & Wada, 1975). (The ICOMAND deliberations currently include discussion on the role and significance of "active" Al, particularly in upper horizons (ICOMAND Circular No. 2, 1980)).

Item 3 is intended to clarify the classification of soils formed in ash but belonging to other orders. If a soil with a bulk density greater than 0.85 g/cm^3 is classified as an Andisol because it has formed in ash, and glass is thought to be present, it must react to NaF. However, some soils such as the yellow-brown pumice soils of New Zealand have coarse textures, vesicular particles and small surface area (N.Z. Soil Bureau, 1968). Because of this, the pH is required to rise only to 9.2 instead of the 9.4 required for domination by amorphous materials. Not all of the yellow-brown pumice soils reach a pH as high as 9.4 (unpublished data, N.Z. Soil Bureau).

Item 4 is intended to clarify the classification of the present Eutrandepts of Hawaii. Some are calcareous in the lower horizons. Water retention, bulk densities, and CEC are appropriate for Andisols, but not all react adequately to NaF, or they are calcareous and the test is meaningless. Exchangeable bases are very high, so the ratio of 15 bar water to bases is low (unpublished data, U.S. Soil Conservation Service).

Not all soils on volcanic ash are intended to be classified as Andisols. Entisols may form in recent historic ash, and if conditions are suitable, Spodosols and Mollisols occur on volcanic materials. In strongly weathered volcanic ash Ultisols, Alfisols and Oxisols may be developed.

C. Distribution

The distribution of Andisols clearly cannot be outlined beyond general statements while the proposal is still being tested. However, the FAO-Unesco (1974) Soil Map of the World reveals that soil associations dominated by Andosols occupy about one hundred million hectares, or 0.76% of the world land area (Dudal, 1976). However, soils formed on volcanic material are not distributed in wide zonal regions, but are associated with occurrences of volcanoes which, on a global scale, do not occupy large areas in any one place. A more detailed analysis of the sheets of the Soil Map of the World, with the exception of those for South-East Asia and Europe where area measurements are currently not available, shows that more than 124 million hectares, or 0.84% of the world's land surface is occupied by soil associations which have some Andosol content (Fig. 1).

Volcanic ash soils occur primarily in association with the major global tectonic zones. The circum-Pacific zone of volcanic and tectonic activity is an example of this pattern. The distribution and properties of volcanic ash soils in the Pacific region have been documented from Japan (Yamanaka&Yamada, 1964; Yamada, 1977), Korea (Shin, 1965) and Kamchatka (Liverovskii, 1971) in the north-western Pacific; from Alaska (Simonson &Rieger, 1967) the U.S.A. (Flach, 1965) and Mexico (Aguilera, 1969) in North America; in central America (Rico, 1965; McConaghy, 1969; Martini, 1969; Colmet-Daage et al.,

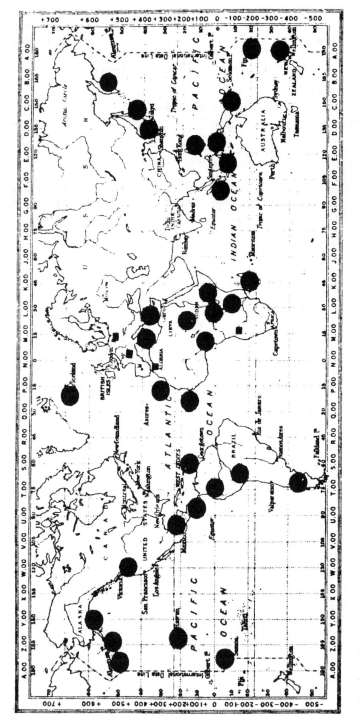

Fig. 1 : Diagrammatic global distribution of Andisols. Areas of major occurrence are indicated by circles, and those of minor occurrence by squares.

1970; Alvarado & Buol 1975); and from Ecuador (Colmet-Daage *et al.*, 1967); Colombia (Luna, 1969); Peru (Zavaleta, 1969); Chile (Valdés, 1969) and Argentina (Wright, 1965) in South America. In the southern and south-western Pacific, occurrences are documented from New Zealand (N.Z. Soil Bureau, 1968), Fiji (Twyford & Wright, 1965), Samoa (Wright, 1963), Vanuatu (Tercinier & Quantin, 1968), and Papua-New Guinea (Haantjens *et al.*, 1967), while in the western and central Pacific the distribution includes Indonesia (Tan, 1965), Taiwan (Lai & Leung, 1967), the Phillipines (Mariano, 1965) and Hawaii (Swindale & Sherman, 1965).

Data from Africa include East Africa (Frei, 1978), Sudan (White, 1967) and Cameroun (Sieffermann, 1969); and from Europe, Italy (e.g. Buondonno, 1964; Lulli, 1971; Violante & Violante, 1973; Eschena & Gessa, 1967), France (e.g., Duchaufour & Souchier, 1966; Hétier, 1973), Spain and the Canary Islands (e.g., Gallardo *et al.*, 1973; Fernandez-Caldas, 1975) and Rumania (Conea, 1972).

II. MORPHOLOGY

The definition of Andisols allows the following diagnostic horizons: umbric, histic, mollic, ochric, cambic, placic and a duripan. The following horizons are not permitted: argillic, natric, spodic, or oxic. There is thus no distinctive, defined diagnostic horizon central to the concept of the Order. And yet, it is not uncommon for these soils to have attracted a local name, such as *Trumao* in Chile, or *Kurobokudo* in Japan – literally meaning "soil from volcanic ash with light stones breaking down in time to a black acid soil". These names often derive from a distinctive feature of soil morphology. The obvious example is "Ando", connotative of a dark colour, but they are also referred to as "soapy hill" in the West Indies which seems to be a clear reference to a characteristic of their morphology and behaviour; and many in New Zealand are known as yellow-brown loams, inferring a distinctive combination of loamy texture and specific colours.

Such names identify properties which distinguish volcanic ash soils from soils with which they are associated in many parts of the world. They do not necessarily identify properties common to all Andisols everywhere, and it is clearly not possible to compile a detailed morphological specification at Order level. However, the following generalizations do capture the essential morphological characteristics of many of these soils as discerned by investigators in Japan, New Zealand, U.S.A. and Latin America (Wright, 1965; Forsythe *et al.*, 1969). They are deep soils, often stratified as a result of periodic accumulation. The upper horizons are in many cases darker coloured and thicker than those of associated soils on non-volcanic materials; subsoils are brown to yellow and have a slippery consistence; textures are predominantly loamy; structures are crumb or granular in topsoils and blocky in subsoils.

Some of the ICOMAND discussions, and particularly submissions from Costa Rica and New Zealand, have strongly advocated the identification of an andic diagnostic horizon or andic materials, which would become the main thrust of the Order definition (ICOMAND, Circular No. 2, 1980).

Clearly, Andisols do have some quite extraordinary morphological characteristics which are most uncommon in other Orders.

A. Colour

Despite the pre-eminence of the formative elements "And-", not all Andisols have deep, dark topsoils. Certainly, the upper horizons often have a very dark colour (very dark grey to black) and great thickness particularly in Japan, especially Hokkaido, and the

Altiplano of the Andes. On very recent volcanic formations, the colour is equally very dark but the thickness can be less. In a tropical climate, the colour appears less dark and more vivid (dark brown to dark red-brown) and the accumulation of humus less important than in a cold climate, although it is clearly greater than is observed in other tropical soils (Quantin, 1972).

In young Andisols with low allophane content, colour tends to be dictated by the parent material. Soils on basic volcanic ash include many dark coloured minerals such as hypersthene, hornblende, augite, olivine, and magnetite. The volcanic ash is thus dominantly black and is low in organic matter (O.M.) or allophanic material. Other volcanic parent materials such as pumice and some dacitic rocks are almost white. A small quantity of O.M. will, however, turn these materials very dark, at least in the surface horizon.

In the hot, isohyperthermic regions, and in the temperate, thermic regions in Central Chile, fulvic acid extracted by pyrophosphate solution clearly dominates. The soils are dark brown on the surface and yellow at depth.

On the other hand, in cool isomesic regions, or in the cooler isothermic regions with an almost constant temperature, humic acid dominates. The soils are black throughout the profile, sometimes up to several metres. There is thus a clear relationship between colour and the nature of the O.M., for there is virtually no difference in soil mineralogy.

Deep, dark topsoils are most commonly and consistently reported from Japan (e.g., Egawa, 1977). Characteristics of the *Kurobokudo*, which are extensive on Hokkaido, Tohoku, Kanto, and Kyushu are that A horizons are blackish in colour, very rich in humus (15-30%) and 30-50 cm in thickness. Occasionally they may be as thick as 100 cm because of redeposition or intermittent thin ash fall. B horizons are brown to yellowish brown. In thick A horizons the average content of humus in virgin soils is 20%. From a global viewpoint, this is the highest content of humus consistently found in terrestrial soils. As in chernozems, herbaceous plants are assumed to be the main source of humus. In herbaceous plants the Gramineae family such as Susuki *(Miscanthus sinensis)* is considered to be the most important source, because a large quantity of plant opal derived from the Gramineae family is found in A horizons. High water-holding capacity may have accelerated the vigorous growth of herbaceous plants from the beginning of pedogenesis. Humus has been preserved in association with amorphous material, especially soluble aluminium, formed in the course of weathering of the mineral constituents (cf. Chapter 6). The accumulated humus has a higher content of humic acid than found in other genetic soil types and the ratio of humic acids to fulvic acids usually exceeds 1. Because of the dominance of humic acid the C/N ratios of A horizons are as high as 15-25 (cf. Chapter 12). This very high content of humus in A horizons is associated with other distinctive properties such as low bulk density, high water holding capacity and very high CEC (Oba, 1976).

The micromorphology of these dark A horizons from Japan was documented by Kawai (1969) who inferred that they were commonly associated with a blocky loose micromorphological fabric or a fine grained porous fabric.

Ortho, macro-sized vugh voids, with numerous channel voids in their vicinity, and many ortho, micro-sized vugh voids in plasma are special characteristics of the blocky, loose fabric, along with loose aggregation of primary structure.

In the fine grained porous fabric, micropeds are dispersed and there is very high porosity. Associated organic carbon contents ranged from 13-21% and soil colours are 7.5YR or 10YR hue with a chroma of 1.

In New Zealand, in a predominantly mesic temperature regime, Andisol topsoils are deep, dark and humic when compared with associated soils, but do not reach the high values for these properties reported from Japan and Chile. For instance, the thickness of black coloured A horizons rarely exceeds 20 cm and organic carbon contents range between 8 and 12% (N.Z. Soil Bureau, 1968, Part 3; Gibbs, 1968).

Marked correlation between native vegetation and colour of the upper horizons has been recorded in soil survey reports. This is particularly so in the pumice soils of the central North Island where, for instance, the change from native podocarp forest to scrub dominated by bracken *(Pteridium aquilinum var. esculentum)* is marked by an abrupt increase in blackness of the topsoil (Vucetich & Wells, 1978).

Properties of A horizons in volcanic ash soils, and particularly the stage of humification and C/N ratio, have been linked to the methoxyl-carbon content by Japanese workers (Kosaka, 1963). A comparative study of New Zealand and Japanese volcanic ash topsoils using methoxyl-carbon contents as a guide (Birrell, 1966) suggested that Japanese soils generally are at a later stage of humification than their New Zealand counterparts.

B. Field Texture

A striking feature of Andisols high in allophane, is that despite apparently high measured clay contents, the soil material is not sticky. Field textures are distinctive and readily identified by experienced pedologists. They are variously described as slippery, greasy, soapy, smeary, or unctuous. This is very different from the clayey feeling typical of soils containing crystalline clays such as kaolinite, montmorillonite, or halloysite.

Although apparently well drained, the soil may have a high water content which can be easily expressed between the fingers. The soil is plastic but it does not adhere to the fingers, as do soils containing well crystallised clays. It is this which imparts the greasy feel in the moist state and the sensation of a silty texture. When dry the soil loses its greasiness and becomes friable and powdery.

Experienced pedologists are able to estimate reasonably accurately the moisture content at about pF 3 on the basis of the combination of field texture and consistence, bearing in mind that this is a function of the amount of allophane present and the degree of moisture held by the allophane, which, in turn, is related to the climatic history of the soil. When moisture content at pF 3 is about 20-50%, the soil does not feel very unctuous and the cohesion, although evident, is weak. When moisture content is about 50-100%, the unctuous feel is usually obvious. When the soil contains more than 100% moisture, it is soapy with strong cohesion. Moisture is expressed on crushing between the fingers. With more than 150 to 200% moisture, it is harder to break down the blocks with the fingers.

This distinctive field texture is correlated with allophane content and with positive reaction to the Fieldes & Perrott (1966) allophane field test. Andisols dominated by vitric volcanic material do not have these properties.

C. Consistence

As with field texture, description of the consistence of Andisols high in allophane has defied the formal terms. It is most commonly described as fluffy, and although this property is probably a reflection of low bulk density and high porosity resulting from high allophane content, it is a striking and detectable morphological characteristic.

The presence of imogolite in Andisols, identified particularly in Japan (Yoshinaga & Aomini, 1962) and Chile (Besoain, 1969), also seems to have an effect on consistence. For instance, in the temperate regions of Chile with annual temperature variations > 5°C, the presence of imogolite is associated with extremely friable soils, in spite of the very high percentages (100%) of water in the freely drained soil. The yellow B horizon is composed of tiny round aggregates and the soil is puffy, loose and very soft. Furthermore, water appears to be absorbed or lost in the same way as in young halloysitic soils. Water retention is high – 100% or more – but after air-drying the soil is able to re-absorb most of the water, whereas where allophane is dominant the water is irreversibly lost.

D. Density

A low bulk density is a required characteristic for all Andisols, except those dominated by vitric volcanic ash. Although the actual value is not measurable in the field, the relative lightness of soil clods is readily identifiable by experienced pedologists. These low bulk densities and associated high porosity seem to be related to the high specific surface area characteristic of allophane.

E. Associated Characteristics

Andisols dominated by allophane have a number of particular characteristics which have no specific morphological expression, but which can be inferred from a combination of properties identifiable in the field. These include:
(a) a critical limit of stability under pressure. Waxy, sensitive clays rich in allophane and common in Andisols on sites with high rainfall exhibit a sudden and large change in strength with deformation under increasing load (N.Z. Soil Bureau, 1968, Part 2). The moisture content is usually too high for satisfactory compaction for foundations.
(b) a high capacity for water retention associated with very high specific surface area. The "field capacity" (measured at pF 3) on a soil kept in its initial moist state is generally greater than 100% of the weight of dry soil (at 105°C). It can reach 300%. Colmet-Daage et al. (1967) have studied this property in detail and have stated that beyond a certain threshold of dehydration (about pF 4.2) there is an irreversible dehydration of the soil.
(c) structural stability. Because of the natural stability of allophane, most Andisols have a very stable structure towards water. This explains their high porosity and their very low susceptibility to erosion (Sieffermann, 1969).
(d) a strong affinity for O.M. which tends to be protected from decomposition (Wada & Higashi, 1976).

Some of these characteristics are important and dominant enough to be expressed in the Andisol proposal at the great group level.

Andisols with unusually specific properties have been identified at high altitude in the Andes (Wright, 1965). The very wet, cold regions of the Andes have low evapotranspiration and are often cloud covered. The soils are black, extremely spongy with high water content (up to 200% at pF 2.5), and react strongly and instantaneously to the Fieldes & Perrott (1966) allophane test. They are very high in O.M. and have a very high CEC which varies with the pH. They have all the characteristics of allophanic soils but with very high percentages of O.M. and they occupy vast areas. French pedologists working in the region have called them organo-Hydrudands. They vary in depth from 60 cm to several metres and were formed partially from very fine volcanic ash erupted from the volcanoes and carried far distances by wind. The fine ash sticks together because of the permanent dampness and is held in place by the vegetation to produce a combination of Histosol and Andisol.

F. Soil Forming Ash Showers

The intermittent accumulation of volcanic ash has considerable impact on both the genesis and morphology of Andisols. In active volcanic regions it is uncommon to find soils which do not have layered ash deposits and buried soils within the control section. Interpretation of the morphology of such profiles is a prime element in unravelling the history of volcanic activity and has given rise to extensive pedological investigations in the fields of tephrachronology and tephrastratigraphy particularly in Japan (e.g., Sasaki, 1974) and New Zealand (Vucetich & Pullar, 1964, 1969, 1973; Pullar & Birrell, 1973).

III. SELECTED EXAMPLES OF PROFILE MORPHOLOGY

Three profile descriptions have been selected to demonstrate part of the range of morphological properties displayed by Andisols (Appendix 1 and Table 1). The pedons described are well-known and comprehensively documented soils from the Pacific Basin and are central examples of the great groups they represent.

The Taupo soil from New Zealand is classified in the N.Z. Genetic Soil Classification as a yellow-brown pumice soil (N.Z. Soil Bureau, 1968). It is pumiceous with high proportions of volcanic glass throughout and the fine earth has a pH in NaF which ranges from 10.2 in the (B) horizon to 8.8 in the C13 horizon. It has a udic moisture regime and 15 bar water values on air dry samples ranging from 12.2% to 2.0%, and on field moist samples ranging from 23.4% to 3.3% (Cotching & Rijkse, 1978). In the Andisol proposal it is a Vitrudand.

Table 1. Morphological comparison of selected pedons.

	Taupo (New Zealand)	Imaichi (Japan)	Hilo (Hawaii, USA)
A horizon thickness	12 cm	82 cm	40 cm
A horizon colour	10YR 3/1 very dark grey	7.5YR 1/1-2/1 black	10YR 3/3 dark brown
Dominant subsoil colour	10YR 5/6 and 2.5Y 6/4 yellow brown	10YR 6.5/7 brownish yellow	2.5YR 3/4 and 5YR 3/4 dark reddish brown
Dominant field texture	sandy loam & loamy sand	silt loam	silty clay loam
Dominant consistence	friable	compact	friable
Range of structure	moderate fine blocky & crumb to single grain	weak fine granular to massive	weak fine granular to strong fine blocky
Tephra layers	Three identified	Two identified	
Buried soils	One identified	One identified	
Ped coatings			thick, translucent gelatinous in subsoil
Other properties	high proportion of vesicular pumice		all horizons dehydrate irreversibly

Salient morphological features of the Taupo soil are the coarse texture throughout; friable consistence becoming loose in some horizons; weakly developed structures; a weakly developed yellowish brown cambic horizon; and the presence of tephra layers and a buried soil within the control section. Bulk density figures range from 0.66 – 0.73 g/cm^3.

The Imaichi soil from Japan is an example of a "Kuroboku" soil (Ministry of Agriculture and Forestry Japanese Government 1964 Chapter VIII). Bulk density figures range from 0.44-0.71 g/cm^3 and figures for phosphate adsorption coefficient are high. It occurs in a udic moisture regime and has an epipedon more than 80 cm thick with moist colour values and chromas 2 or less throughout. Figures for organic carbon range from 20.19% in the A11 horizon to 13.24% in the A13. In the Andisol proposal it is a Melanudand.

The most striking morphological characteristic is the thick, dark epipedon which is described as slightly compact to compact, and apart from the surface horizon, massive throughout. The field texture is dominantly silt loam and one other tephra layer is identified at the base of the profile.

The Hilo soil from Hawaii has been classified as a Hydrol Humic Latosol (Swindale & Sherman, 1965), and as a Typic Hydrandept (Soil Conservation Service, 1976). Dry bulk density figures range from 0.61 to 0.80 g/cm^3. No data were available for pH in NaF or phosphate retention. It occurs in an isohyperthermic temperature regime, and figures for 15 bar water of the field moist samples range from 69.8% in the Ap horizon to 149.5% in the B28 horizon. In the Andisol proposal it is a Hydrotropand.

The most distinctive morphological feature of the Hilo soil is the irreversible drying which occurs in all horizons. Thick, translucent, gelatinous coatings of amorphous colloidal oxides are a prominent feature of lower horizons. Uniform dark reddish brown colours and silty clay loam textures prevail throughout the subsoil. It is probable that most profiles comprise numerous ash bands but the deep, strong weathering tends to obscure them.

Table 1 compares the main morphological features of the Taupo, Imaichi and Hilo profiles.

IV. MORPHOLOGICAL PARAMETERS USED IN THE ANDISOL PROPOSAL

Apart from aquic properties which are employed at the suborder level, salient morphological characteristics are used as differentiae at the great group level. *Aquic* suborders are defined on the presence of an histic epipedon or on the occurrence of low chromas, or mottles or both. A suggested addition to this definition is a placic horizon that rests on a duripan. In Japan some Andisols developing under an aquic regime have very thick, surface horizons with high O.M. contents which are dark enough to obscure low chroma colours or mottles. A great group of Melanaquands has been proposed and an additional aquic criterion based on the identification of active ferrous iron will be required (ICOMAND Circular 1, pp. 10-11, 1979; Circular 2, pp. 14-15, 1980).

Melanic great groups are proposed in the Aquands, Borands and Udands. The definition requires a thick epipedon with low colour value and chroma and high content of organic carbon. Such great groups accommodate the thick, black soils which are well known from Japan (Egawa, 1977) and from the high Andean regions of South America (Wright, 1965). The Imaichi series (Appendix 1) displays melanic properties.

Vitric great groups are provided in all suborders and are defined on the content of 15 bar water. They accommodate coarse textured, commonly weakly developed and in most cases vitreous or pumiceous soils. Vitrudands are extensive in the central North Island of New Zealand on pumice deposits (Taupo soil, Appendix 1).

Placic great groups are provided in three suborders, Borands, Udands and Tropands. Placoborands have not yet been identified (ICOMAND Circular 2, p.9, 1980) and may not exist. Placotropands and Placudands are known to one of the authors (GDS), but descriptive data have not yet been documented. As in other orders placic great groups are defined by the presence of a placic horizon within a specific depth in half or more of each pedon.

Duric great groups are identified in Xerands and Ustands, and tentatively in Aquands. They are defined on the presence of a duripan. Hardpan layers have been recorded in Andosols in South America (Wright, 1965) and in Japan where they are known as "masa" or "kora" cemented horizons and may occur in a udic moisture regime (ICOMAND Circular 2, p. 15; Appendix 1). It is not clear from the available evidence whether these are pedogenic or geologic horizons.

Hydric great groups are provided for the freely drained soils that rarely or never become drier than field capacity in the Tropands and Udands. They are defined on the basis of their undried 15 bar water content. It is not certain that Hydrudands exist, but some data suggest that they might occur in Chile (ICOMAND Circular 1 1979, Appendix 1). Hydrotropands are well known from Hawaii (Swindale & Sherman, 1965) where their most distinctive morphological feature is the irreversible drying they undergo when exposed (Hilo soil, Appendix 1).

Subgroups defined on morphological criteria include hydric, vitric, placic and aquic which have definitions similar to the great group with the same formative element; entic, thapto-histic, lithic, pachic, ruptic and ustollic already used and defined in Soil Taxonomy (Soil Survey Staff, 1975); and psammic which is defined on the basis of particle size and is provided for Andisols that are particularly subject to wind erosion.

V. CONCLUSIONS

This review has not encompassed the total spectrum of Andisols, because data on many important members, particularly those in intertropical regions, very cold climates, and dry areas, are still being generated through the activities of ICOMAND.

However, it is clear that many of the very distinctive and specific properties of these soils are reflected in their morphology. A feature of this brief review has been the discovery that comprehensive and relevant morphological descriptions of Andisols are well documented from only a few countries. This is not altogether surprising in view of the relatively short history of global interest in such soils. Current momentum in the classification field should improve this situation.

VI. ACKNOWLEDGMENT

Grateful acknowledgment is made to Keith Vincent, Pedologist, Soil Bureau for the design and execution of Figure 1, and for the calculation of areas.

VII. REFERENCES

ORSTOM – Office de la Recherche Scientifique et Technique Outre-Mer (Paris).

Aguilera, N. 1969. Geographic distribution and characteristics of volcanic ash soils in Mexico. In: Panel on Volcanic Ash Soils in Latin America, A. 6. Inter-American Inst. Agr. Sci., Turrialba, Costa Rica.

Alvarado, A., Buol, S. W. 1975. Toposequence relationships of Dystrandepts in Costa Rica. Soil Sci. Soc. Am. Proc., 39: 932-7.

Besoain, E. 1969. Imogolite in volcanic soils of Chile. Geoderma, 2: 151-69.

Birrell, K. S. 1966. Methoxyl-carbon content of soils as an index of humification. N.Z. J. Agric. Res., 9: 444-7.

Buondonno, C. 1964. The soils of Napoli province. An agrochemical survey. Ann. Fac. Agr. Protici, 29: 221-51 (in Italian).

Colmet-Daage, F., Cucalon, F., Delaune, M., Gautheyrou, J. et M., Moreau, B. 1967. Caractéristiques de quelques sols d'équateur dérivés de cendres volcaniques. I. Essai de caractérisation des sols des régions tropicales humides. Cah. ORSTOM, Série Pédologie, 5: 3-38.

Colmet-Daage, F., Cucalon, F., Delaune, M., Gautheyrou, J. et M., Moreau, B. 1967. Caractéristiques de quelques sols d'équateur dérivés de cendres volcaniques. II. Conditions de formation et d'évolution. Cah. ORSTOM, Série Pédologie, 5: 353-92.

Colmet-Daage, F., de Kimpe, C., Delaune, M., Gautheyrou, J. et M., Sieffermann, G., Fusil, G. 1970. Caractéristiques de quelques sols dérivés de cendres volcaniques de la côte pacifique du Nicaragua. Cah. ORSTOM, Série Pédologie, 8: 113-72.

Conea, A. 1972. Andosols in Rumania. "P"rvi Natsionalen Kongres po Pochvoznanie," Bulgaria, 1969, pp. 465-72.

Cotching, W. E., Rijkse, W. C. 1978. Proceedings of the twenty-sixth New Zealand Soil Bureau Conference. Soil Bureau, D.S.I.R., N.Z. Restricted publication, 128 pp.

Duchaufour, P., Souchier, B. 1966. Sols Andosoliques et roches volcaniques des Vosges. Science Terre, 11: 345-65.

Dudal, R. 1976. Inventory of the major soils of the world with special reference to mineral stress hazards. In: M. J. Wright (Editor), Plant Adaptation to Mineral Stress in Problem Soils. Proceedings Workshop, Cornell Univ. Agr. Exp. Stat., Ithaca, N.Y., pp. 3-14.

Egawa, T. 1977. Properties of soils derived from volcanic ash. In: Y. Ishizuka and C. A. Black (Editors), Soils derived from volcanic ash in Japan. International Maize and Wheat Improvement Center (CIMMYT), Mexico, pp. 10-63.

Eschena, T., Gessa, C. 1967. The andosols of Sardinia. Studi Sassaresi, 15: 363-86 (in Italian).

FAO/Unesco, 1974. Soil Map of the World. Unesco, Paris.

Fernandez-Caldas, E., Tejedor Salguero, M. L. 1975. Andosoles de las Islas Canarias, Santa Cruz de Tenerife. Caja General de Ahorros, 210 pp.

Flach, K. 1965. Genesis and morphology of ash-derived soils in U.S.A. In: World Soil Resources Report 14, FAO/Unesco, pp. 61-70.

Fieldes, M., Perrott, K. W. 1966. The nature of allophane in soils. Part 3. Rapid field and laboratory test for allophane. N.Z. J. Sci., 9: 623-9.

Forsythe, W., Gavande, S., Gonzalez, M. 1969. Problems related to the physical properties of volcanic ash soils. In: Panel on Volcanic Ash Soils in Latin America, B.3. Inter-American Inst. Agr. Sci., Turrialba, Costa Rica.

Frei, E. 1978. Andepts in some high mountains of East Africa. Geoderma, 21: 119-31.

Gallardo, J. F., Garcia Sanchez, A., Saavedra, J. 1973. Properties of Andosols of the Sierra de Francia (western Spain). Anales Edafol. Agrobiol., 32: 1135-41 (in Spanish).

Gibbs, H. S. 1968. Volcanic-ash soils in New Zealand. N.Z. Department of Scientific and Industrial Research. Information Series No. 65, 39 pp.

Haantjens, H. A., Reynders, J. J., Mouthaan, W. L. P. J. 1967. Major soil groups of New Guinea and their distribution. Commun. Dep. agric. Res. R. trop. Inst. 55, 87 pp.

Hetier, J. M. 1973. Caractères et repartition des sols volcaniques du Massif-Central, Part II. Science du Sol, No. 2: 97-109.

ICOMAND*. 1979-80.
  Circular No. 1, 3 April 1979, 28 pp.
  Circular No. 2, 25 January 1980, 41 pp.
  Circular No. 3, 30 April 1980, 76 pp.

Inter-American Institute of Agricultural Sciences, 1969. Panel on Volcanic Ash soils in Latin America, Turrialba, Costa Rica.

Kawai, K. 1969. Micromorphological studies of Andosols in Japan. Bull. Nat. Inst. Agr. Sci., Series B, 20: 77-154.

Kosaka, J. 1963. Division of the process of humification in upland soils and its application to soil classification. Soil Sci. Plant Nutr., 9: 14-8.

Lai, C. Y., Leung, K. W. 1967. The physical and chemical properties of Ando soils in Yungmingshan. J. Taiwan. Agric. Res., 16: 1-7.

Liverovskii, Yu A. 1971. Volcanic ash soils of Kamchatka. Pochvovedenie, No. 6, 3-11.

Lulli, L. 1971. Volcanic soils around Lake Bracciano (Rome). Annali dell' Istituto Sperimentale per lo Studio e la Difesa del Suolo Firenze, 2: 23-130 (in Italian).

Luna, C. 1969. Genetic aspects of Colombian andosols. In: Panel on Volcanic Ash Soils in Latin America, A.3. Inter-American Inst. Agr. Sci., Turrialba, Costa Rica.

---

* International Committee on the Classification of Andisols, C/- Soil Bureau, D.S.I.R., Lower Hutt, New Zealand.

McConaghy, S. 1969. Geographic distribution and characteristics of volcanic ash soils in the Antilles. In: Panel on Volcanic Ash Soils in Latin America. A.4. Inter-American Inst. Agr. Sci., Turrialba, Costa Rica.

Mariano, J. A. 1965. Volcanic ash soils of the Philippines. In: World Soil Resources Report 14, FAO/Unesco, pp. 53-5.

Martini, J. A. 1969. Geographic distribution and characteristics of volcanic ash soils in Central America. In: Panel on Volcanic Ash Soils in Latin America. A.5. Inter-American Inst. Agr. Sci., Turrialba, Costa Rica.

Ministry of Agriculture and Forestry, Japanese Government 1964. Volcanic Ash Soils in Japan, 295 pp.

Mizota, C., Wada, K. 1980. Implications of clay mineralogy to the weathering and chemistry of Ap horizons of Ando soils. Geoderma, 23: 49-63.

N.Z. Soil Bureau 1968. Soils of New Zealand. Parts 1, 2 and 3. N.Z. Soil Bureau Bulletin 26.

Oba, Y. 1976. Andosols in Japan. Encyclopedia of Plant Nutrition, Soils & Fertilizers, Yokendo, Tokyo, pp. 260-2.

Pullar, W. A., Birrell, K. S. 1973. Age and distribution of Late Quaternary pyroclastic and associated cover deposits of Central North Island, New Zealand. N.Z. Soil Survey Reports 1, 2.

Quantin, P. 1972. Les andosols. Revue bibliographique des connaissances actuelles. Cah. ORSTOM, Série Pédologie, 10: 273-301.

Rico, M. 1965. Report on soils of volcanic ash origin in El Salvador. In: World Soil Resources Report 14, FAO/Unesco, pp. 23-9.

Rousseaux, J. M., Warkentin, B. P. 1976. Surface properties and forces holding water in allophane soils. Soil Sci. Soc. Am. J.,40: 446-51.

Sasaki, T. (Editor) 1974. Distribution of the Late Quaternary pyroclastic deposits in Hokkaido, Japan. Miscellaneous Publication Hokkaido Nat. Agr. Exp. Stat. No. 4.

Shin, Y. H. 1965. Volcanic ash soils of Korea. In: World Soil Resources Report 14, FAO/Unesco, pp. 50-2.

Sieffermann, G. 1969. Les Sols de Quelques Régions Volcaniques du Cameroun. Thèse, Université de Strasbourg, 285 pp.

Simonson, R. W. 1979. Origin of the name "Ando Soils". Geoderma, 22: 333-5.

Simonson, R. W., Rieger, S. 1967. Soils of the Andept suborder in Alaska. Soil Sci. Soc. Am. Proc., 31: 692-9.

Soil Conservation Service (U.S. Department of Agriculture) 1976. Soil Survey Laboratory Data and Descriptions for Some Soils of Hawaii. Soil Survey Investigations Report No. 29, 208 pp.

Soil Survey Staff 1975. Soil Taxonomy. A basic system of soil classification for making and interpreting soil surveys. U.S. Dept. Agr. Handbook 436, 754 pp.

Swindale, L. D., Sherman, G. D. 1965. Hawaiian soils from volcanic ash. In: World Soil Resources Report 14, FAO/Unesco, pp. 36-49.

Tan, K. H. 1965. The andosols in Indonesia. In: World Soil Resources Report 14, FAO/Unesco, pp. 30-5.

Tercinier, G., Quantin, P. 1968. Influence de l'altération de cendres et ponces d'age récent sur la nature, les propriétés et la fertilité des sols aux Nouvelle-Hébrides. Cah. ORSTOM, Série Pédologie, 6: 203-34.

Tokashiki, Y., Wada, K. 1975. Weathering implications of the mineralogy of clay fractions of two Ando soils, Kyushu. Geoderma, 14: 47-62.

Twyford, I. T., Wright, A. C. S. 1965. The soil resources of the Fiji Islands. Government of Fiji, 570 pp.

Valdés, A. 1969. Geographic distribution and characteristics of volcanic ash soils in Peru. In: Panel on Volcanic Ash Soils in Latin America A.1. Inter-American Inst. Agr. Sci., Turrialba, Costa Rica.

Violante, P., Violante, A. 1973. The andosols of Vulture. Annali della Facolta di sciienze Agrarie della Università degli Studi di Napoli Portici, IV 7, 219-38 (in Italian).

Vucetich, C. G., Pullar, W. A. 1964. The stratigraphy of Holocene ash in the Rotorua and Gisborne districts. In: N.Z. Geological Survey Bulletin 73, pp. 43-63.

Vucetich, C. G., Pullar, W. A. 1969. Stratigraphy and chronology of Late Pleistocene volcanic ash beds in Central North Island, N.Z. N.Z. J. Geol. Geophys.,12: 784-837.

Vucetich, C. G., Pullar, W. A. 1973. Holocene tephra formations erupted in the Taupo area, and interbedded tephras from other volcanic sources. N.Z. J. Geol. Geophys.,16: 745-80.

Vucetich, C. G., Wells, N. 1978. Soils, Agriculture and Forestry of Waiotapu Region, Central North Island, N.Z. Soil Bureau Bulletin 31, 100 pp.

Wada, K. 1977. Active aluminum in Kuroboku soils and non- and para-crystalline clay minerals. Nendokagaku, 27: 143-51 (In Japanese).

Wada, K., Higashi, T. 1976. The categories of aluminum- and iron-humus complexes in Ando soils determined by selective dissolution. J. Soil Sci., 27: 357-68.

Wada, K., Gunjigake, N. 1979. Active aluminum and iron and phosphate adsorption in Ando soils. Soil Sci., 128: 331-6.

White, L. P. 1967. Ash soils in Western Sudan. J. Soil Sci., 18: 309-17.

Wright, A. C. S. 1963. Soils and land use of Western Samoa. N.Z. Soil Bureau Bulletin 22, 190 pp.

Wright, A. C. S. 1965. The "Andosols" or "Humic Allophane" soils of South America. In: World Soil Resources Report 14, FAO/Unesco, pp. 9-22.
World Soil Resources Report 14, 1965. Meeting on the classification and correlation of soils from volcanic ash, Tokyo, Japan 1964. FAO/Unesco, 169 pp.
Yamanaka, K., Yamada, Y. 1964. Distribution of volcanic ash soils. In: Volcanic Ash Soils in Japan. Ministry of Agriculture and Forestry, Japanese Government, pp. 4-7.
Yamada, S. 1977. Distribution and morphology of soils derived from volcanic ash in Japan. In: Y. Ishizuka and C. A. Black (Editors), Soils derived from volcanic ash in Japan. International Maize and Wheat Improvement Center (CIMMYT), Mexico pp. 1-9.
Yoshinaga, N., Aomine, S. 1962. Imogolite in some Ando soils. Soil Sci. Plant Nutr., 8: 22-9.
Zavaleta, A. 1969. Geographic distribution and characteristics of volcanic ash soils in Peru. In: Panel on Volcanic Ash Soils in Latin America.A.2. Inter-American Inst. Agr. Sci., Turrialba, Costa Rica.

APPENDIX 1

Profile description data for three examples of Andisols.

TAUPO SANDY LOAM (Reference: Cotching & Rijkse, 1978)

LOCATION:	8 km south of Waiotapu, Wharepaina, west of State Highway (Rotorua-Taupo) back of Mr Bell's property, northern side of silage pit at the end of main race Central North Island, New Zealand.
MAP SHEET:	NZMS 1 N85 Grid reference: 781747
TOPOGRAPHY:	Slope: Flat to easy rolling Landform: Terrace Altitude: 335 m a.s.1.
SOIL DRAINAGE:	Well drained
VEGETATION:	Present: Improved pasture (ryegrass clover) Native: Manuka scrub, and bracken fern The area had light podocarp forest and scrub prior to Polynesian fires
PARENT MATERIAL:	Taupo Pumice, on Whakatane Ash, Mamaku Ash, Rotoma Ash, Waiohau Ash, Rotorua Ash and Rerewhakaaitu Ash
CLIMATE:	Mean annual rainfall 1500-1550 mm Mean annual temperature 12°C
LAND USE:	Dairying, semi-intensive sheep farming, beef and exotic forestry

DESCRIPTION:

A	0-12 cm	very dark grey (10YR 3/1) sandy loam; friable; weakly developed coarse subangular blocky structure breaking to moderately developed fine subangular blocky and crumb structure; abundant roots; few fine lapilli; distinct smooth boundary.
(B)	12-18 cm	yellowish brown (10YR 5/6) gritty sandy loam; friable; moderately developed medium and fine subangular blocky structure breaking to moderately developed fine crumb structure; abundant roots; few coarse and many fine lapilli; distinct smooth boundary.
C_{11}	18-36 cm	light yellowish brown (2.5Y 6/4) loamy sand; friable; massive breaking to weakly developed medium and fine angular blocky structure crushing to fine crumb structure; many fine strong brown (7.5YR 4/6) iron stained root channels; very few black (N2) charcoal concentrations; very few dark greyish brown (2.5Y 4/2) inclusions; horizon becomes greyer with depth; few roots; distinct smooth boundary.
C_{12}	36-44 cm	dark greyish brown (10YR 4/2) and light brownish grey (2.5Y 6/2) coarse sand; friable; single grain; few roots; few fine strong brown (7.5YR 4/6) iron stained root channels; (Rhyolite Block Member); distinct smooth boundary.
C_{13}	44-65 cm	light grey (2.5Y 7/2) pumice gravel; loose; single grain; many roots; distinct strong brown and yellowish brown iron staining in upper 7cm of horizon; sharp smooth boundary.
D_r	65-70 cm	grey (5Y 5/1) loamy sand; friable; massive; few roots; (Rotongaio Ash); distinct wavy boundary.
II uA	70-86 cm	brown (10YR 4/3) greasy sandy loam; friable; weakly developed medium subangular blocky structure crushing to weakly developed crumb structure; few roots; few medium very dark brown (10YR 2/2) organic matter concentrations; many fine lapilli; indistinct irregular boundary.
II C	86-100+ cm	yellowish brown (near 10YR 5/8) greasy loamy sand; very friable; very weakly developed medium angular blocky structure breaking to moderately developed medium crumb structure; very few roots.

IMAICHI (Reference: Ministry of Agriculture & Forestry, Japanese Government, 1964)

LOCATION:	Dozawa, Imaichi-shi, Tochigi Prefecture (Longitude, 139°44', East; Latitude, 36°41', North)
TOPOGRAPHY:	Upland; slope, west 2°, Elevation, 310 m (1030 feet)
CLIMATE:	Mean annual temperature, 12.5°C (55°F); Annual precipitation, 1522 mm (59.9 inches). (Utsunomiya Local Meteorological Observatory)
PARENT MATERIAL:	Wind blown volcanic ash
VEGETATION:	Deciduous forest, Nara (*Quercus serrata*), Kunugi (*Quercus accutissima*) and others.

DESCRIPTION:

A11 0-12 cm (0 to 5 inches)
Black (7.5YR 2/1), very dark brown (10YR 3/1.5) when dry, silt loam; weak, fine granular structure, few fine pores; slightly compact (16mm), slightly sticky, slightly plastic; semi-moist, many roots; smooth, clear boundary.

A11 12-28 cm (5 to 11 inches)
Black (7.5YR 1/1), very dark gray (1.25Y 2.5/1) when dry, silt loam; massive, few fine pores; compact (20mm), slightly sticky, slightly plastic, semi-moist; common roots; smooth, gradual boundary.

A13 28-62 cm (11 to 24 inches)
Black (7.5YR 1/1), very dark gray (2.5Y 3/1) when dry, silt loam, few weathered yellow pumice gravel; massive, few fine pores; compact (20mm), slightly sticky, slightly plastic; semi-moist; common roots; smooth, gradual boundary.

AIIB 62-82 cm (24 to 32 inches)
Black (7.5YR 2/1), dark grayish brown (1.25Y 4/2) when dry, silt loam; many weathered yellow pumice gravel and few half weathered gravel; massive, few fine pores; compact (23mm), slightly sticky, slightly plastic; semi-moist; smooth, gradual boundary.

IIB 82-110 cm (32 to 43 inches)
Brownish yellow (10YR 6.5/7), very pale brown to yellow (10YR 7.5/4.5) when dry, weathered pumice gravel layer (Hichihonzakura); very compact (26mm), non sticky, non plastic; semi-moist to moist.

HILO SILTY CLAY LOAM (Reference: Soil Conservation Service, 1976)

LOCATION:	Island of Hawaii, Hawaii County, Hawaii. Approximately 2.8 km (1.8 miles) north of Hilo Post Office. Sample site is located 30 m (100 feet) south of road at a point 0.5 km (0.3 mile) west of Haaheo School which is at the north end of Wainaku Village.
VEGETATION:	Originally ohia-tree fern vegetation, now cleared and in sugarcane.
CLIMATE:	Average annual precipitation is 438 cm (175 inches). The mean annual temperature is 22.2°C (72°F), the mean January temperature 20.0°C (68°F), and the mean July temperature 23.9°C (75°F).
PARENT MATERIAL:	Volcanic ash.
TOPOGRAPHY:	Undulating to rolling low windward slopes of Mauna Kea, 3 percent slope to east
ELEVATION:	105 m (350 feet)
DRAINAGE:	Well drained; rapid permeability; slow runoff
SOIL MOISTURE:	Moist
REMARKS:	Textures are apparent field textures. Colours are for moist soil. All horizons dehydrate irreversibly to sand and gravel size aggregates

DESCRIPTION:

Ap 0-40 cm (0-16 inches)
Dark brown (10YR 3/) silty clay loam mixed with dark reddish brown (5YR 3/4) by cultivation; weak very fine and fine granular structure; friable, sticky, plastic, moderately smeary; many roots; many very fine and fine interstitial pores; few to common firm ash nodules; abrupt smooth lower boundary.

B21 40-53 cm (16-21 inches)
Dark reddish brown (5YR 3/3) silty clay loam; weak medium subangular blocky structure; friable, sticky, plastic, moderately smeary; many roots; many very fine and fine and common medium and few coarse tubular pores; thick gelatinous coating on ped surfaces, some surfaces appear like clay flow; few firm ash nodules; abrupt smooth boundary.

B22	53-58 cm (21-23 inches)	Dark reddish brown (2.5YR 3/4) silty clay loam, weak fine and very fine subangular blocky structure, friable, sticky, plastic, moderately smeary, common roots; many very fine and fine and common medium and few coarse tubular pores, common firm ash nodules of 13 to 25mm (½ to 1 inch) in diameter, ped surfaces have gelatinous appearance, abrupt smooth boundary.
B23	58-65 cm (23-26 inches)	Dark brown (7.5YR 3/2) silty clay loam; moderate very fine subangular blocky structure; friable, sticky, plastic, moderately smeary, common fine roots; many very fine, fine and medium and few coarse tubular pores; ped surfaces have gelatinous appearance, some surfaces look like clay flows; few firm ash nodules; abrupt smooth boundary.
B24	65-75 cm (26-30 inches)	Dark reddish brown (5YR 3/4) silty clay loam, moderate very fine subangular blocky structure, friable, sticky, plastic, moderately smeary; common fine roots; many very fine, fine and medium and few coarse tubular pores; ped surfaces have translucent gelatinous appearance, some coatings appear like clay flows, few firm ash nodules; abrupt smooth boundary.
B25	75-80 cm (30-32 inches)	Dark reddish brown (2.5YR 3/4) silty clay loam; moderate subangular blocky structure, friable, sticky, plastic, moderately smeary; few fine roots; many very fine, fine and common medium and few coarse tubular pores; ped surfaces have translucent gelatinous appearance, abrupt smooth boundary.
B26	80-83 cm (32-33 inches)	Dark brown (7.5YR 3/2) silty clay loam, moderate very fine subangular blocky structure, friable, sticky, plastic, moderately smeary; few fine roots; many very fine, fine and common medium and few coarse tubular pores; thick translucent gelatinous coatings on ped surfaces that appear like clay flows; few firm ash nodules; abrupt smooth boundary.
B27	83-93 cm (33-37 inches)	Dark reddish brown (5YR 3/4) silty clay loam, moderate very fine subangular blocky structure; friable, sticky, plastic, moderately smeary; few roots; many very fine, fine and medium and few coarse tubular pores; thick gelatinous coating on ped surfaces; few firm ash nodules of dark reddish brown (2.5YR 3/4); clear smooth boundary.
B28	93-123 cm (37-49 inches)	Dark reddish brown (5YR 3/4) silty clay loam; moderate fine and medium subangular blocky structure; friable, sticky, plastic, moderately smeary; few roots; many very fine, fine and medium and few coarse tubular pores; thick gelatinous coating on ped surfaces, clear smooth boundary.
B29	123-128 cm (49-51 inches)	Dark brown (7.5YR 3/3) silty clay loam; strong very fine subangular blocky structure; friable, sticky, plastic, moderately smeary; no roots; many very fine, fine and common medium tubular pores; thick translucent gelatinous coating on ped surface; few firm ash nodules; abrupt smooth boundary.
B210	128-133 cm (51-53 inches)	Dark reddish brown (2.5YR 3/4) silty clay loam; weak very fine and fine subangular blocky structure; friable, sticky, plastic, strongly smeary; no roots; many very fine, fine and medium and few coarse tubular pores; thick gelatinous coating on ped surfaces; common firm ash nodules of 6 to 18mm (¼-¾ inch) in diameter; abrupt smooth boundary.
B211	133-140 cm (53-56 inches)	Dark reddish brown (5YR 3/4) silty clay loam; moderate very fine subangular blocky structure, firm, sticky, plastic, strongly smeary; no roots; many very fine, fine and medium and few coarse tubular pores; translucent gelatinous coating on ped surfaces; few firm ash nodules; abrupt smooth boundary.
B212	140-145 cm (56-58 inches)	Dark reddish brown (2.5YR 3/4) silty clay loam; moderate very fine subangular blocky structure; friable, sticky, plastic, strongly smeary; no roots; many very fine, fine and common medium and few coarse tubular pores; ped surfaces have translucent gelatinous appearance; many firm ash nodules of 6 to 25mm (¼-1 inch) in diameter; abrupt smooth boundary.

B213 145-168 cm Dark reddish brown (5YR 3/4) silty clay loam; strong very fine subangular blocky
(58-67 inches) structure; friable, sticky, plastic, strongly smeary; no roots; many very fine, fine and
common medium and few coarse tubular pores; ped surfaces have gelatinous
appearance; few firm ash nodules.

5

Copyright © 1967 by the Soil Science Society of America
Reprinted by permission from *Soil Sci. Soc. Am. Proc.* **31**:692–699 (1967)

Soils of the Andept Suborder in Alaska[1]

ROY W. SIMONSON AND SAMUEL RIEGER[2]

ABSTRACT

These are the major well-drained soils of nonmountainous areas of Kodiak Island, the Aleutian Islands, and the Alaska peninsula, and they occur as well in southwestern Kenai peninsula. The soils were formerly classified in the Ando group on the basis of similarities to that group as originally proposed in Japan. Three profiles were studied as examples of the group and to illustrate differences in degree of horizonation. Detailed information is given on the morphology and composition of these profiles. The soils have thick dark A horizons, are strongly acid and have low base saturation throughout their profiles, have little textural differentiation in the profile, and have clay fractions dominated by allophane. The soils are now classified as Cryandepts (Inceptisols) in the 7th Approximation and seem closely related to other soils formed in volcanic ash in places as far removed as Japan, New Zealand, and South America. They also share certain characteristics with geographically associated Spodosols (Podzols). Field relationships and the shared characteristics of B horizons suggest that Andepts are readily converted to Orthods (Podzols) following occupation by spruce forest.

Additional Key Words for Indexing: allophane, Ando soils, Cryandepts, Spdosols, volcanic ash, Japan, dark A horizons.

KELLOGG AND NYGARD (18) in their exploratory study of the soils of Alaska identified as "Tundra without permafrost" a group of dark soils formed under tall grass vegetation in the Aleutian Islands, Kodiak Island, the Alaska peninsula, and the southwestern part of the Kenai peninsula. These soils were studied further during soil surveys of parts of Kodiak Island and the Kenai peninsula and, as a result, were reclassified in the great soil group of Ando soils (24). Such soils are now classified in the 7th Approximation (29) as Andepts, a suborder of Inceptisols.

The Andepts in southern Alaska are near the cold end of the range for the suborder. Information on their nature may therefore be of interest. Morphological and analytical data are given in this paper for several profiles selected as examples of the Andepts in Alaska. Before these data are presented, however, the history and general nature of the broad group of Ando soils are described. Lastly, the relationships of the Andepts in southern Alaska to geographically associated Spodosols (Podzols) and to soils derived from volcanic ash in other parts of the world are discussed briefly.

History and General Nature of the Ando Group

American scientists making reconnaissance soil surveys in Japan in 1946 and 1947 recognized that the dark soils formed from volcanic ash did not fit well into any of the great soil groups defined up to that time in the United States. Morphologically, the soils were most like Brunizems (earlier called Prairie soils) and Brown Forest soils in North America.[3] Chemically, however, the dark soils in Japan differed in a number of important respects from the Brunizems and Brown Forest soils of the United States. Consequently, W. S. Ligon[4] proposed in 1947 that these soils be recognized as a separate great soil group, for which he suggested the name, Ando soils.[5] This broad group was described in general terms in the reports of the reconnaissance soil surveys in Japan, the first of which covered the Kanto plain of Honshu (3). Brief notes on the Ando group were also included in a later symposium on soil classification published in the USA in 1949 (34).

The broad group of soils in Japan for which the name, Ando soils, was proposed in 1947 had been identified previously by several names. The group was called Volcanic Ash soils by Seki in 1932 (26), a name used again by Kanno in 1956 (16). Other names proposed after 1932 were Black soils, Brown Forest soils, prairie-like Brown Forest soils, Braunerde, and Onji soils (12, 13, 15). Since 1947, additional names have been suggested, e.g., Humic Allophane soils by Kanno (17) and Kuroboku soils by Matsui et al. (21). A history of the study of the soils from volcanic ash in Japan, including work on their classification and the names proposed for them, is given in a publication by the Ministry of Agriculture and Forestry of Japan in 1964 (14).

Some of the Ando soils in Japan are virtually indistinguishable in profile from many of the Brunizems in the midwestern USA, but others have thicker and darker A horizons. The Ando soils commonly have very dark brown, very dark grayish-brown, or black A horizons of medium texture. These are mostly about 30 cm (1 ft) thick. The dark colors give way gradually with depth to the brown, yellowish brown, or dark yellowish-brown colors of the B horizons, which grade in turn into lighter colored parent materials. On the whole, the Ando soils are rather porous and low in bulk density throughout their profiles. The soils were formed in volcanic ash or in regoliths with high proportions of volcanic ash.

Several distinctions in chemical composition were noted between the Ando soils and the Brunizems and Brown Forest soils (3). General levels of organic matter were higher in Ando soils. Exchangeable bases were lower, on the order of 2 to 9 meq/100 g of soil, and extractable aluminum ranged from about 3 to 8 meq/100 g. The soils had lower pH values, with reaction generally being very strongly acid. Total phosphorus was lower, and the capacity of the soil to fix phosphorus in slowly available form was high. Silica-sesquioxide ratios of the clay fraction were low. Allophane has been found to be the dominant clay mineral in soils formed in

---

[1] Contribution from the Soil Survey, SCS, USDA. Received Feb. 13, 1967. Approved May 16, 1967.
[2] Director, Soil Classification and Correlation, Washington, D. C., and State Soil Scientist, Alaska, respectively.

[3] Names used for great soil groups are from Simonson and Steele (27).
[4] The late Dr. Ligon, on leave from the Department of Agronomy, University of Kentucky, was serving as a soil scientist in the Natural Resources Section, General Headquarters of the Supreme Commander for the Allied Powers.
[5] The word, Ando (pronounced on-dough), was coined from a pair of Japanese words referring to dark soils and consequently is not part of the body of that language.

volcanic ash in Japan (1), including soils of the Ando group. The same clay mineralogy has been observed in similar soils in places as far away as New Zealand (4) and South America (36). Thus, it is evident as the data collected some years ago are examined now that the Ando soils, as the group was first defined, share some important characteristics with Spodosols (Podzols).

Many of the soils which have in the past been classified in the Ando group are believed to be classifiable in the suborder of Andepts in the 7th Approximation. In all probability, not all of these soils are properly classified in this suborder, but which ones do belong cannot yet be determined with certainty. The great soil group of Ando soils was defined in general terms, as were other great soil groups at that time. Definite ranges in characteristics were not specified. Moreover, the information on individual pedons required to classify in the 7th Approximation all of the soils thought at various times to represent the Ando group is not available.

The soils studied in southern Alaska are classified in the suborder of Andepts. That name will be used for those soils in the remainder of this paper.

Environment of Andepts in Alaska

Andepts cover nearly all well-drained nonmountainous areas of the Aleutian Islands, the Alaska peninsula, and the Kodiak Island group, as shown in Fig. 1. Farther north, on the Kenai peninsula, they occur in association with Spodosols. Andepts here occupy areas in which the summer climate is cooler and wetter than the average for the region. These include areas immediately above tree line in the Caribou Hills and the Kenai Mountains, high narrow coastal strips in the southern part of the peninsula, and shallow depressions on level benches in the lowlands bordering Cook Inlet.

The Andepts in Alaska have been formed under a cool, maritime climate. Average temperatures range from 7 to 13C. in July and from −7 to 0C in January. Annual average precipitation ranges from about 600 to 1,600 mm. In winter, precipitation comes as both rain and snow, and the soils are seldom frozen deeply. Because of the low average temperatures and frequent rains, the soils are almost always moist.

The vegetation in most areas of Andepts in Alaska consist of tall grasses, dominantly *Calamagrostis* sp., with associated forbs and shrubs. No forests are found on the Aleutian Islands, but willows (*Salix* sp.), poplar (cottonwood) (*Populus* sp.), and paper birch (*Betula* sp.) occupy some valley bottoms on the Alaska peninsula and Kodiak Island, and large patches of Sitka alder (*Alnus* sp.) occur on the uplands. Forests of Sitka spruce (*Picea* sp.) cover parts of Afognak Island and northern Kodiak Island; clumps of Sitka spruce occur in the Caribou Hills of the southern Kenai peninsula

Active volcanos on the Alaska peninsula and the Aleutian Islands are the source of fairly frequent ash falls in these regions. Layers of recognizable ash are common in many soil profiles (18, 35). The volcanic deposits range in texture from coarse cinders to silt and in thickness from a few centimeters to many meters.

Methods of Study

Description and Sampling—A number of profiles of Andepts were examined during the soil surveys of parts of Kodiak Island and the Kenai peninsula. Other profiles were observed and described in special studies. Some of the profiles described in detail were sampled for laboratory analysis. Full descriptions of three profiles as examples of the suborder are given in this paper. These descriptions follow the standards and terminology of the Soil Survey Manual (28) and the Supplement (30). Samples were collected by horizons from these three profiles for laboratory characterization.

Laboratory Methods—All laboratory analyses[6] were made on material passing a 2-mm sieve. Particle-size distribution was determined by the pipette method (20). Soil pH was measured with a glass electrode in a 1: 1 soil-water suspension, for the most part. Organic carbon was determined by a modification of the Walkley-Black dichromate titration method (22), and nitrogen by a modified AOAC Kjeldahl procedure (2). Free iron oxides were extracted from the soil with sodium hydrosulfite and determined with standard potassium dichromate, following a procedure by Kilmer (19). Cation-exchange capacity was determined by direct distillation of absorbed ammonia, calcium by oxalate precipitation and titration with standard potassium permanganate, and magnesium by the precipitation of magnesium ammonium phosphate and back titration of excess sulfuric acid with standard sodium hydroxide (22). Exchangeable acidity was measured by the barium chloride-triethanolamine method at pH 8.2 (22). Sodium and potassium were determined in the ammonium acetate extract with a Beckman DU flame spectrophotometer. For aluminum, 10 g of soil were allowed to stand overnight in 1N solution of KCl and then leached with additional solution. The amount of aluminum in the extract was determined colorimetrically by an "Aluminon" reagent method (5).

Morphology of Soils

The profiles selected as examples of the Andept suborder in southern Alaska represent the Kodiak series on Kodiak Island and the Kachemak and Island series on the Kenai peninsula.

The Kodiak series was chosen to illustrate the Andepts of Kodiak Island, the Aleutian Islands, and the Alaska peninsula. The profile seems typical for those soils except for a surface layer, about 30 cm (1 ft) thick, of volcanic ash deposited in 1912. A characteristic landscape of Kodiak soils is shown in Fig. 2.

The Kachemak soils occur on the lower margins of the Caribou Hills in the southwestern part of the Kenai peninsula, mostly at elevations of 250 to 450 m (800 to 1,500 ft) above sea level. Although the series is classified as an Andept, the soils have incipient Spodosol horizons in the upper part. These faint horizons are believed to reflect a change in the direction of horizon differentiation from that characteristic of the suborder. A profile of Kachemak silt loam is shown in Fig. 3 and the landscape in which it occurs in Fig. 4.

The Island soils occur in narrow grassed strips bordering or near Cook Inlet and in shallow depressions farther inland. The series illustrates soils with a low degree of horizonation.

Descriptions of the profiles and notes on the sites at which the soils were examined and sampled are given in the remainder of this section. Colors of horizons are for moist conditions.

KODIAK SILT LOAM

The profile site was near Anton Larson Bay Road, about 2.1 km (1.3 miles) north of Naval Ski Chalet, in northeastern Kodiak Island. This was on a slope of 25% in a hilly moraine underlain by slate and graywacke. The vegetation consisted of native grasses and associated forbs and shrubs.

Horizon	Depth in cm (Inches in parentheses)	Description
O11	10–2 (4–1)	Litter of straw and alder leaves.

---

[6] Analyses of the Kachemak and Island soils by the Soil Survey Laboratory, SCS, Lincoln, Neb.; of the Kodiak soil (except for organic carbon) by the Soil Survey Laboratory, SCS, Beltsville, Md.

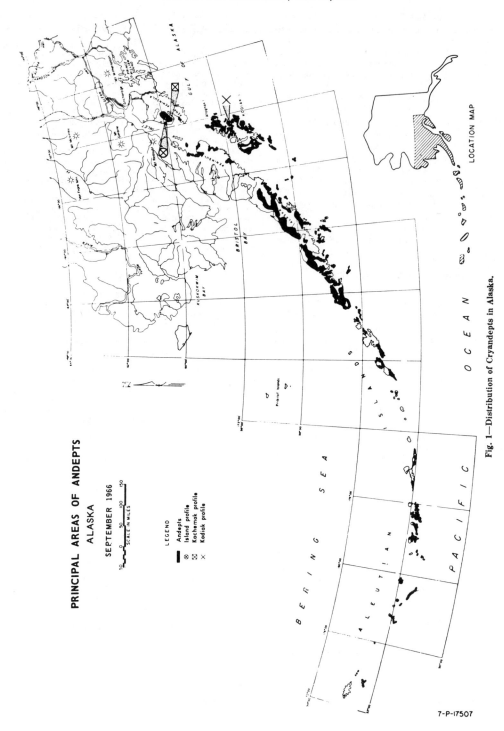

Fig. 1—Distribution of Cryandepts in Alaska.

Horizon	Depth in cm (Inches in parentheses)	Description
O12	2–0 (1–0)	Dark brown (10YR 3/3) mat of partially decomposed organic materials; abrupt smooth boundary.
C1	0–5 (0–2)	Recent volcanic ash. Light gray (10YR 7/1) coarse silt loam; very weak fine platy structure; firm in place, very friable when disturbed; few fine roots; abrupt smooth boundary.
C2	5–15 (2–6)	Recent volcanic ash. Light yellowish-brown (10YR 6/4) coarse silt loam; massive; firm in place, very friable when disturbed; few roots; abrupt smooth boundary.
C3	15–20 (6–8)	Recent volcanic ash. Light yellowish-brown (10YR 6/4) loamy fine sand; loose; few roots; clear wavy boundary.
C4	20–28 (8–11)	Recent volcanic ash. Grayish-brown (10YR 5/2) fine sand; loose; few roots; abrupt smooth boundary.
A11b	28–38 (11–15)	Dark reddish-brown (5YR 2/2) silt loam; weak very fine subangular blocky structure; very friable; many roots; abrupt wavy boundary.
A12b	38–58 (15–23)	Dark reddish-brown (5YR 3/3) silt loam; weak medium subangular blocky structure; friable, sticky; streaks of brown (7.5YR 4/2) silt loam throughout horizon; fewer roots than in horizon above; gradual boundary.
IIAC	58–68 (23–27)	Brown (7.5YR 4/2) gravelly silt loam; many angular rock fragments; abrupt wavy boundary.
IIC	68–88 (27–35)	Olive (5Y 4/3) very gravelly sandy loam mottled with reddish brown.

Fig. 2—Landscape of Kodiak silt loam under native vegetation near head of Kalsin Bay, Kodiak Island. This shows the dominance of grass (*Calamagrostis* sp.) and the presence of some shrubs. The profile in the lower right corner is about 1 m (3+ ft) deep, and the uppermost 30 cm (1 ft), which appears white, is a deposit of ash from an eruption of a volcano on the Alaska peninsula in 1912.

Kachemak Silt Loam

The profile site was in the SE 1/4 SE 1/4 Sec. 7, T. 6 S., R. 13 W., the Kenai peninsula. This was on a slope of 4% in rolling uplands at an elevation of about 315 m. The vegetation consisted of native grasses and associated forbs with patches of Sitka spruce nearby.

Horizon	Depth in cm (Inches in parentheses)	Description
O1	8–0 (3–0)	Mat of roots, leaves, and stems; much charcoal; thin lenses of white sand grains at base of horizon; abrupt smooth boundary.
C1	0–2 (0–1)	Recent volcanic ash. Dark brown (7.5YR 3/3) silt loam; weak very fine granular structure; very friable; roots common; abrupt wavy boundary.
C2	2–5 (1–2)	Recent volcanic ash. Dark reddish-brown (5YR 3/4) silt loam; weak fine granular structure; very friable; roots common; abrupt wavy boundary.
O2b	5–6 (2–2 1/2)	Black (5YR 2/1) mat of burned grass and woody materials; thin lenses of sand, probably ash; roots common; abrupt wavy boundary.
A1b	6–12 (2 1/2–5)	Dark brown (7.5YR 3/2) silt loam; weak very fine granular structure; very friable; releases water and becomes smeary when rubbed; roots common; abrupt wavy boundary.
B21b	12–17 (5–7)	Very dusky red (2.5YR 2/2) silt loam; weak very fine granular structure; very friable; releases water and becomes smeary when rubbed; roots common; clear wavy boundary.

Fig. 3—A profile of Kachemak silt loam, a Typic Cryandept, in the Caribou Hills a few miles northeast of Homer, near the site of the one described and sampled for laboratory characterization. The profile has a darkened surface horizon about 30 cm (1 ft) thick and a gradual change to lighter colors below. Numbers on the scale indicate intervals of 1 foot or approximately 30 cm. The marks on the scale are approximately 2 inches or 5 cm apart.

Horizon		Description
B22b	17-27 (7-11)	Dark reddish-brown (5YR 3/3) silt loam; weak very fine granular structure; very friable; releases water and becomes smeary when rubbed; roots common; gradual boundary.
B3b	27-34 (11-14)	Dark reddish-brown (5YR 3/3) silt loam with patches of dark brown (7.5YR 4/4); weak fine subangular blocky structure; very friable; releases water and becomes smeary when rubbed; roots common; gradual boundary.
C1b	34-42 (14-17)	Dark brown (10YR 4/3) silt loam; weak medium subangular blocky structure; very friable; releases water and becomes smeary when rubbed; fine pores; abrupt wavy boundary.
IIC2	42-57 (17-23)	Olive gray (5Y 5/2) silt loam streaked with dark grayish-brown (2.5Y 4/2); few pebbles; moderate medium platy parting to weak very fine angular blocky structure; friable; fine pores; gradual boundary.
IIIC3	57-72 (23-29)	Olive gray (5Y 5/2) silt loam with common medium faint mottles of olive brown; few pebbles; moderate fine platy structure; friable, fine pores, abrupt smooth boundary over moderately indurated shale.

Fig. 4—Landscape in the Caribou Hills at the site of the Kachemak profile, a few miles northeast of Homer, southwestern part of the Kenai peninsula. This shows the grass vegetation on the smoother and higher sites with spruce in the draws and valleys. The distant peaks with snow fields are in the Kenai Mountains to the east of Kachemak Bay.

Island Very Fine Sandy Loam

The profile site was in the NW 1/4 Sec. 9, T. 2 S., R. 14 W., the Kenai peninsula. This was on a slope of 2% on a broad ridge at an elevation of about 60 m along the coast of Cook Inlet. The vegetation consisted of native grasses and associated forbs.

Horizon	Depth in cm (Inches in parentheses)	Description
O1	5-0 (2-0)	Very dark brown (10YR 2/2) dense mat of finely divided grass roots and stems; abrupt smooth boundary.
A11	0-12 (0-5)	Very dark grayish-brown (10YR 3/2) very fine sandy loam; weak fine granular structure; very friable; dark horizontal streaks which may be the result of fires; many roots; clear wavy boundary.
A12	12-20 (5-8)	Dark brown (10YR 3/3) very fine sandy loam; weak fine subangular blocky structure; very friable; roots common; clear wavy boundary.
B21	20-45 (8-18)	Dark yellowish-brown (10YR 3/4) silt loam; weak prismatic parting to weak coarse subangular blocky structure; friable; roots common; gradual boundary.
B22	45-62 (18-25)	Dark yellowish-brown (10YR 4/4) silt loam, with many poorly defined streaks of very dark brown; weak medium subangular blocky structure; friable; few fine dark concretions; few roots; gradual boundary.
C1	62-92 (27-37)	Olive brown (2.5Y 4/4) silt loam; few faint light olive brown mottles, and many dark brown mottles; massive; friable; few roots; many fine pores; gradual boundary.
C2	92-145 (37-58)	Olive (5Y 4/3) silt loam; streaks of dark brown, apparently old root channels; massive; friable; common pores; abrubt boundary over layered olive very fine sandy loam, and firm fine gravelly silty loam.

Composition of Soils

Data are given in this section on particle size distribution, pH, carbon, nitrogen, cation-exchange relations, extractable aluminum, and free iron oxides.

The data for particle size distribution are given in Table 1. Percentages of clay are believed to be bigger in these soils than indicated by the data. It is known that soils high in allophane are not completely dispersed by standard procedures, and allophane has been found to be the dominant clay mineral in the Island soils.[7] The same mineral is believed to be dominant in clay fractions of the Kachemak and Kodiak soils because of the release of water and the resulting smeariness when soil material from the deeper horizons is rubbed between the fingers.

Data are given in Table 2 for selected chemical characteristics of horizons of three profiles. The data for the Kodiak profile, except those for organic carbon, were taken from the monograph by Kellogg and Nygard (18). They sampled a profile on Kodiak Island which is nearly identical with the one described in this paper. The layer of 1912 ash, however, was sampled as a unit for these earlier analyses. Extractable aluminum was determined only on the samples of Kachemak silt loam, whereas free iron oxides were determined on the samples of Kachemak and Island soils.

The clay fraction was dominated by allophane in samples from the B horizon of the Island profile, as already indicated. The very high cation exchange capacities of the B horizons of the Kachemak soil and of the A12b and ACb horizons of the Kodiak soil reinforce the interpretation that allophane is the dominant clay mineral in those soils as well.

Comparisons of Soils

The presence of the volcanic ash laid down in 1912 as the uppermost layer of the Kodiak profile sets it apart from the

---

[7] We are indebted for this identification to Hsin-Yuan Tu, SCS Soil Survey Laboratory, Beltsville, Maryland. The identification of allophane rests on three lines of evidence. These were the results of differential thermal analysis, the presence of much volcanic glass in the soil, and the presence of aggregates of noncrystalline material.

SIMONSON AND RIEGER: SOILS OF THE ANDEPT SUBORDER IN ALASKA

Table 1—Particle size distribution by horizons in the profiles of three Cryandepts in Alaska

Horizon	Depth, cm	Very coarse sand 2-1 mm	Coarse sand 1-0.5 mm	Medium sand 0.5-0.25 mm	Fine sand 0.25-0.10 mm	Very fine sand 0.10-0.05 mm	Silt 0.05-0.002 mm	Clay < 0.002 mm *	0.2-0.02 mm	0.02-0.002 mm	> 2 mm
\multicolumn{12}{c}{Kodiak silt loam}
C1	0-5	0.2	0.3	1.3	18.0	18.0	60.9	1.3	55.7	37.4	—
C2	5-15	—	0.6	4.9	23.1	11.7	56.8	2.9	46.3	36.7	—
C3	15-20	—	2.2	17.1	47.4	5.1	26.2	2.0	31.8	18.0	—
C4	20-28	—	6.1	26.4	51.4	3.1	11.4	1.6	24.8	8.4	—
A11b	28-38	0.5	2.6	3.0	9.4	15.4	63.8	5.3	44.8	40.0	—
A12b	38-58	0.1	0.2	1.0	15.3	25.4	57.3	0.7	60.4	34.2	—
ACb	58-68										
IIC	68-88	20.5	8.7	3.8	10.5	12.4	37.1	7.0	33.5	22.2	36.6
\multicolumn{12}{c}{Kachemak silt loam}
C1	0-2	<0.1	2.2	6.5	4.4	17.1	65.2	4.6	45.4	39.7	—
C2	2-5	<0.1	<0.1	0.2	3.3	24.1	70.0	2.4	57.0	39.7	—
O2b	5-6	\multicolumn{10}{l}{(No analyses made of this horizon)}
A1b	6-12	0.1†	1.3	4.7	11.4	15.5	55.9	11.1	42.7	34.7	—
B21b	12-17	0.2†	1.1	3.7	9.7	16.2	56.2	12.9	43.6	34.2	Tr
B22b	17-27	0.3‡	0.8	3.0	9.5	16.8	58.4	11.2	44.3	36.4	Tr
B3b	27-34	0.4‡	0.9	3.2	10.1	18.0	59.2	8.2	47.9	35.2	Tr
C1b	34-42	0.3	0.8	3.3	8.9	17.1	60.0	9.6	46.2	36.2	Tr
IIC2	42-57	0.2	0.6	2.1	5.5	12.6	66.6	12.4	43.2	39.3	Tr
IIC3	57-72	<0.1	0.1	0.5	4.8	18.1	61.9	14.6	45.7	37.9	Tr
\multicolumn{12}{c}{Island Very Fine Sandy Loam}
A11	0-12	0.5	1.6	4.0	20.1	25.4	43.8	4.6	66.0	15.1	—
A12	12-20	0.1	2.2	8.0	18.7	21.0	45.0	5.0	58.9	15.9	—
B21	20-45	0.1	1.6	1.9	8.4	23.2	60.6	4.2	63.7	26.1	—
B22	45-62	—	0.6	1.4	8.2	22.1	64.3	3.4	62.8	29.6	—
C1	62-92	0.1	1.8	4.7	11.3	18.9	59.5	3.7	58.3	27.4	—
C2	92-145	—	0.9	2.4	8.1	17.7	68.5	2.4	66.3	25.5	—

* Proportions of clay are believed to be low because of allophane in the clay fraction.
† Includes many organic matter fragments.
‡ Includes few dark aggregates.

Table 2—Chemical characteristics by horizons of the profiles of three Cryandepts in Alaska

Horizon	Depth, cm	pH	Org. C	N, %	C/N ratio	Free Fe_2O_3 %	CEC*	Ca	Mg	Na	K	Mn	H + Al	Al	Sum	Base saturation, %	Ca/Mg ratio
\multicolumn{18}{c}{Kodiak Silt Loam‡}
C	0-25	5.8§						0.2	0.2		0.2	0.01	1.4		2.0	30	1.0
A11b	25-37	4.9	23.+					3.5	1.8		1.2	0.08	95.5		102.1	6	1.9
A12b	37-50	5.1	12.+					1.7	0.8		1.2	0.01	106.6		110.3	3	2.1
ACb	50-65	5.4						0.3	0.6		0.9	0.04	50.1		51.9	4	0.5
IIC	65+	5.3	3.					0.9	0.6		1.1	0.02	91.4		94.0	3	1.5
\multicolumn{18}{c}{Kachemak Silt Loam}
O1	8-0	4.4¶	21.+	0.336	62												
C1	0-2	4.4	8.45	0.247	34	0.4	17.4	1.5	0.7	0.2	0.5		24.6	1.9	29.4	10	2.1
C2	2-5	4.7	5.35	0.163	33	1.0	17.2	1.2	0.4	0.2	0.2		41.1	2.4	45.5	5	3.0
O2b	5-6		\multicolumn{15}{l}{(No analyses made of this horizon.)}
A1b	6-12	4.8	7.09	0.214	33	1.0	22.4	2.7	1.4	0.2	0.1		35.9	3.1	43.4	11	1.9
B21b	12-17	4.7	8.87	0.509	17	2.8	35.4	2.3	0.8	0.2	0.1		76.9	5.1	85.4	4	2.9
B22b	17-27	5.1	5.96	0.376	16	2.3	25.3	1.6	0.3	0.2	<0.1		59.3	3.6	65.0	3	5.3
B3b	27-34	5.1	3.50	0.263	13	1.8	19.9	1.1	0.4	0.1	<0.1		43.3	3.0	47.9	4	2.8
C1b	34-42	5.2	3.20			1.2	18.4	1.0	0.4	0.1	<0.1		38.4	3.2	42.1	4	2.5
IIC2	42-57	5.3	1.69			0.7	14.6	1.1	0.7	0.1	<0.1		26.6	3.7	32.2	7	1.6
IIC3	57-72	5.0	0.48			0.7	17.5	4.9	2.4	0.2	0.1		19.9	6.0	33.3	28	2.0
\multicolumn{18}{c}{Island Very Fine Sandy Loam}
A11	0-12	5.7¶	6.02	0.533	11.3	0.6	23.0	12.8	3.4	0.1	1.2		12.4		29.9	58	3.8
A12	12-20	5.1	4.21	0.374	11.2	1.0	19.2	5.3	1.8	0.1	0.7		19.8		27.7	28	2.9
B21	20-45	5.2	3.66	0.285	12.8	1.2	17.1	3.6	0.8	0.1	0.3		25.0		29.8	16	4.5
B22	45-62	5.2	2.73	0.218	12.5	1.6	14.5	2.4	0.4	0.1	0.3		23.8		27.0	12	6.0
C1	62-92	5.2	2.05	0.162	12.6	1.6	12.3	1.8	0.5	0.1	0.2		19.4		22.0	12	3.6
C2	92-145	5.2	0.93	0.079	11.8	1.1	7.8	1.2	0.4	0.1	0.2		11.2		13.1	14	3.0

*Cation exchange capacity and extractable cations expressed in meq/100 g soil.
†Base saturation = $\frac{\text{Sum extractable bases}}{\text{Sum extractable cations}} \times 100$.
‡Data, except for organic carbon, from Kellogg and Nygard (6) for a profile of Kodiak silt loam nearly identical with the one described in this paper. All layers in the recent ash deposit (0-25 cm) were combined into one sample.
§pH measured at 1:1 soil-water ratio.
¶pH measured at 1:5 soil-water ratio.

others, but that layer has not destroyed the similarities between the remainder of the profile and those of the Kachemak and Island soils. All three have thick, dark A1 horizons. All have transitional horizons between the A and C horizons and lack evidence of clay illuviation. Levels of organic matter are high or very high in the A horizons and decrease gradually with depth, the distribution resembling a diffusion pattern in some ways. All three profiles have deeper horizons that are strongly or very strongly acid, with low base saturation. All have the characteristic smeariness common to soils high in allophane.

The Kodiak profile is appreciably higher in organic matter

in the A1 horizon than are the other two soils. Distinctness of horizons in the Kodiak profile, excluding the recent ash layer, is as great as has been observed in the Andepts of southern Alaska. It is believed that the degree of horizonation is as great as can be found among Andepts in cool climates.

The Island profile, on the other hand, has the lowest degree of horizonation of the three. Differences between the A horizon and deeper ones in amounts of organic matter, in color, and in structure are smaller in the Island profile than in the other two profiles. The A1 horizon is less acid and marked by higher base saturation than those of the other profiles. The Island soil described here occurs along the eastern shore of Cook Inlet and receives sediments blown from the beaches. These periodic additions of new materials are believed to account, at least in part, for the greater depth of the soil, for the low degree of horizonation, and for the higher base saturation, especially of the A1 horizon.

The Kachemak profile has incipient Spodosol horizons immediately below a surface layer of organic matter and thin deposits of recent ash. The amount of free iron oxides is highest in the B horizon, which seems related to the development of the Spodosol horizons. The indications that Kachemak soils are Andepts with some characteristics of Spodosols or are being converted to Spodosols are discussed in a later section of the paper.

The three series studied in southern Alaska are classed in the great group of Cryandepts. The Kachemak and Kodiak series fit well in the subgroup of Typic Cryandepts, whereas the Island series has higher base saturation in the A horizon than characteristic for that subgroup. Presently available evidence indicates that the Island series can also best be classified in the subgroup of Typic Cryandepts.

Andept–Spodosol Relationships

The Andepts in Alaska, typically found under grass, are thought to have been formed under such vegetation. The present distribution patterns of grass and forest indicate, however, that forests are extending their range at the expense of grass in the lands bordering the eastern part of the Gulf of Alaska (10). Furthermore, most of the land below an elevation of about 90 m (300 ft) on Afognak Island and the northern part of Kodiak Island are now covered by forest. Isolated clumps of spruce occur just south of the forested areas on Kodiak Island. In northeastern Kodiak Island, the soils under forest are indistinguishable from those under grass (24), but on Afognak Island, where the forests are thought to have existed longer, the soils have faint A2b horizons immediately beneath A1b horizons.

Further north, in the Caribou Hills of the Kenai peninsula, patches of trees occur as islands in grassland. These hills seem to be a tension zone between regions of forest and grass vegetation. Most parts of the landscape have probably been in trees for a time and in grass for a time in the past. Both kinds of vegetation are thought to have occupied most of the landscape, perhaps more than once, although it does not seem that the whole area was ever under continuous forest.

Even under grass vegetation, incipient Spodosol horizons may be found immediately beneath a surface layer of organic matter in the Kachemak soils. Partly decayed or burned wood fragments are also present in the uppermost horizons in many places. Occupation of these soils by forest for some interval in the past may be responsible for the incipient Spodosol horizons in some or all of these soils. On the other hand, well-developed Spodosols have formed under grasses, forbs, and shrubs on high mountain slopes bordering Prince William Sound and the northern part of the Cook Inlet-Susitna Lowland.

The uplands below elevations of about 240 m (800 ft) adjacent to the Caribou Hills are mostly covered by forest. Soils in these areas are Spodosols with distinct A and B horizons, the latter being high in organic matter. Furthermore, these B horizons tend to have the same or similar consistence as the B horizons of Andepts (23).

Several lines of evidence indicate that Andepts are converted to Spodosols in a relatively short time following their occupation by spruce forest. The relationships between the distribution patterns of soils and vegetation on Kodiak and Afognak Islands support this interpretation. The geographic association of the Island and Kachemak soils with Spodosols and the gradations from one set of soils to the other in the Kenai peninsula (23) are consistent with this interpretation. Moreover, analogous relationships seem to hold for closely related kinds of soils derived from volcanic ash in New Zealand (33). These related soils become podzolized rapidly under forest cover. Whether similar relationship prevail elsewheres is not known, though it seems probable that they could because of the similarities between the B horizons of the Andepts and certain Spodosols.

Relationships to Soils Formed in Volcanic Ash Elsewhere

Soils closely similar to or classifiable in the Ando group have been described in a number of regions with substantial deposits of volcanic ash. Soils which correspond to the buried portion of the Kodiak soils have been described by Johannesson (11) Iceland. "Acid Brown Forest soils," which have properties similar to those of Ando soils of Japan, were recognized on Vancouver Island, British Columbia (6). Forests on these soils are believed to be very recent invaders of "former vegetation which consisted of grass with a canopy of scattered garry oak" (6). Several sequences of soil profiles formed in superimposed deposits of volcanic ash have been described in the Kamchatka peninsula of Siberia (31), which lies in the same latitude as do the Aleutian Islands and the Alaska peninsula. Called Volcanic Forest soils, these seem to be closely similar to the Andepts of Alaska. Andosols were identified by Dudal and Soepraptohardjo (8) and Ando soils by Tan and van Schuylenborgh (32) in several of the islands of Indonesia. Many of these soils seem to belong to the suborder of Andepts. Other soils which seem to be Andepts are the schwartze Andenboeden described by Frei in Ecuador (9). Widespread occurrence of the suborde along the Andes Mountains in South America is indicated by Wright (36). The Eutrophic Brown soils on volcanic ash described by D'Hoore in Africa are believed by him to correspond approximately to the suborder of Andepts (7). Yellow-brown soils from volcanic ash in New Zealand (33) and in New Guinea (25) are other soils that seem to be Andepts.

The Andepts in southern Alaska are comparable in morphology and composition to part of the soils classified in the past in the Ando group. Andepts in general include an important proportion of all soils derived from volcanic ash. The soils in Alaska do not cover the full range of those that were included in the Ando group, however, nor do Andepts cover the full range of soils formed from volcanic ash. Soils representing several orders in the 7th Approximation have been formed from ash. The Andepts of southern Alaska represent a single great group, the Cryandepts, in one order, the Inceptisols.

As Cryandepts, the soils of southern Alaska are distinguished from five other great groups in the suborder. These five are restricted largely to warmer parts of the world. Excluding the low temperatures, however, the Andepts in Alaska are closely similar in morphology and composition to Andepts elsewhere. Further testing of possible similarities and differences among the soils of these widely separated localities will become possible as more data on their morphology and composition are gathered and published.

LITERATURE CITED

1. Aomine, S., and N. Yoshinaga. 1955. Clay minerals of some well-drained volcanic ash soils in Japan. Soil Sci. 79:349-358.
2. Association of Official Agricultural Chemists. 1945. Official and tentative methods of analysis, 6th Ed. Washington.
3. Austin, M. E., et al. 1945. Reconnaissance soil survey of Japan—Kanto Plain Area. General Headquarters of Supreme Commander for Allied Powers. Nat. Resources Sec. Rpt. No. 110-A.
4. Birrell, K. S., and M. Fieldes. 1952. Allophane in volcanic ash soils. J. Soil Sci. 3:156-166.
5. Chenery, E. M. 1948. Thioglycollic acid as an inhibitor for iron in the colorimetric determination of aluminum by means of "Aluminon". The Analyst 73:501-502.
6. Day, J. H., L. Farstad, and D. G. Laird. 1960. Soil survey of southeastern Vancouver Island and Gulf Islands, British Columbia. British Columbia Soil Surv. Rpt. No. 6.
7. D'Hoore, J. L. 1964. Soil map of Africa—Scale 1:5,000,000 —Explanatory monograph. Commission for Technical Cooperation in Africa Publ. No. 93.
8. Dudal, R., and M. Soepraptohardjo. 1957. Soil classification in Indonesia. Indonesian Gen. Agr. Res. Sta. Contrib. No. 148.
9. Frei, E. 1958. Eine Studie uber den Zusammenhang Zwischen Bodentyp, Klima und Vegetation in Ecuador. Plant Soil 9:215-236.
10. Griggs, R. F. 1934. The edge of the forest in Alaska and the reasons for its position. Ecology 15:80-96.
11. Johannesson, Bjorn. 1960. The soils of Iceland. Univ. of Iceland Res. Inst., Dept. Agr. Reports, Series B, No. 13.
12. Harada, M. 1935. Volcanic soils from the province of Tottori in Japan. Soil Research 3:147-168.
13. Hosoda, Katsumi. 1938. Investigation of the black soils in Japan. Memoirs of Tottori Agr. College (Japan) 6:1-238.
14. Japanese Ministry of Agriculture & Forestry. 1964. Volcanic ash soils in Japan. Japanese Ministry of Agriculture and Forestry, Tokyo.
15. Kamoshita, Yukata. 1958. Soils in Japan. Nat. Inst. Agr. Sci. (Japan) Misc. Publ. B, No. 5.
16. Kanno, Ichiro. 1956. A pedological investigation of Japanese volcanic-ash soils. Int. Congr. Soil Sci., Trans. 6th. E:105-109.
17. Kanno, Ichiro. 1961. Genesis and classification of main genetic soil types in Japan. I. Introduction and humic allophane soils. Bul. Kyushu Agr. Exp. Sta. 8:1-185.
18. Kellogg, Charles E., and Iver J. Nygard. 1951. Exploratory study of the principal soil groups of Alaska. US Dept. Agr. Monograph No. 7.
19. Kilmer, V. J. 1960. The estimation of free iron oxides in soils. Soil Sci. Soc. Amer. Proc. 24:420-421.
20. Kilmer, V. J., and L. T. Alexander. 1949. Methods for making mechanical analysis of soils. Scoil Sci. 68:15-24.
21. Matsui, T., T. Kurobe, and Y. Kato. 1963. Pedological problems concerning volcanic ash in Japan. The Qauternary Research (Japan) 3:40-58.
22. Peech, M., L. T. Alexander, L. A. Dean, and J. F. Reed. 1947. Methods of soil analysis for soil fertility investigations. US Dept. Agr. Circ. 757.
23. Rieger, S., and J. A. DeMent. 1965. Cryorthods of the Cook Inlet-Susita Lowland, Alaska. Soil Sci. Soc. Amer. Proc. 29:448-453.
24. Rieger, S., and R. E. Wunderlich. 1960. Soil survey and vegetation, Northeastern Kodiak Island Area, Alaska. US Dept. Agr. Soil Surv. Series No. 17.
25. Rutherford, G. K. 1962. The yellow-brown soils of the highlands of New Guinea. Int. Soc. Soil Sci. Trans. Comm. IV and V (New Zealand), p. 343-349.
26. Seki, Toyotaro. 1932. Volcanic ash loams of Japan proper (their classification, distribution, and characteristics). Int. Congr. Soil Sci., Trans. 2nd (Moscow) 5:141-143.
27. Simonson, Roy W., and J. G. Steele. 1960. Soil (great soil groups). McGraw-Hill Encyclopedia of Sci. & Technology 12:433-441
28. Soil Survey Staff. 1951. Soil Survey Manual. US Dept Agr. Handbook No. 18.
29. Soil Survey Staff. 1960. Soil classification—A comprehensive system. 7th Approximation. US Dep. Agr., Washington.
30. Soil Survey Staff. 1962. Identification and nomenclature of soil horizons. US Dep. Agr., Supplement to Agr. Handbook No. 18.
31. Sokolov, I. A., and Z. S. Karayeva. 1965. Migration of humus and some elements in the profile of volcanic Forest soils of Kamchatka. Soviet Soil Sci. 1965, (No. 5), p. 467-475.
32. Tan, K. H., and J. van Schuylenborgh. 1961. On the classification and genesis of soils developed over acid volcanic material under humid tropical conditions. Netherlands J. Agr. Sci. 9:41-54.
33. Taylor, N. H., and J. E. Cox. 1956. The soil pattern of New Zealand. N. Zeal. Soil Bur. Publ. no. 113.
34. Thorp, James, and Guy D. Smith. 1949. Higher categories of soil classification: Order, suborder, and great soil groups. Soil Sci. 67:117-126.
35. Ulrich, H. P. 1947. Morphology and genesis of the soils of Adak Island, Aleutian Islands. Soil Sci. Soc. Amer. Proc. (1947) 11:438-441.
36. Wright, A. C. S. 1964. The "Andosols" or "Humic Allophane" soils of South America. FAO World Soil Resources Report no. 14.

Copyright © 1965 by the Williams & Wilkins Company
Reprinted by permission from *Soil Sci.* **99**:375-378 (1965)

THE ANDOSOLS IN INDONESIA

K. H. TAN

Bogor Institute of Agricultural Sciences[1]

Received for publication August 10, 1964

In the last few years there have been several studies of the andosols. Since this group of black-colored soils is widely distributed in the volcanic areas around the Pacific basin, it is not surprising that they are gradually receiving more attention.

The unusual behavior of these soils is attributed to their amorphous clay fraction. In Japan, according to the literature, andosols are often characterized by the presence of allophane (1, 6). In New Zealand, however, allophane has been reported (3) as also present in a group of soils derived from volcanic andesitic and rhyolitic ash, and in Hawaii (9) as present in considerable amounts in latosols.

Because of the predominant influence of the allophane content in the soils of Japan, Kanno (6) even proposed to designate Japanese volcanic ash soils as humic allophane soil.

In Indonesia this group of soils is widely spread over the archipelago, from Sumatra in the west (4, 8, 12) over Java (10, 13) to the Lesser Sunda Islands in the east (8).

The possibility that these soils might resemble margalitic soils, as postulated by Gerasimov [see (6), p. 180], is doubtful. The andosols should be clearly separated from margalitic soils, since the andosols possess different morphological, physical, and analytical characteristics.

PARENT MATERIAL

The andosols in Indonesia are developed from a rather wide variety of parent material, but all of them originated from recent Pleistocene eruptions.

In north Sumatra the andosols are formed primarily on andesito-dacitic tuffs and lahars of the Sibajak volcano [for their mineralogical composition, see Tan and Van Schuylenborgh (12)] and are restricted to the lowest parts of the region. In west Java they are found successively from west to east on basalto-andesitic lahar of the Salak volcano (Tjiapus/Bogor), on andesitic tuff of the Tangkuban-prahu volcano (Lembang), and on andesito-basaltic tuff of the Pengalengan Highlands. In east Java the kind of parent material is still under investigation but is expected to be of basaltic type.

These facts indicate that the parent material from which the andosols are developed changes from acid to rather basic types in going from west to east along the volcanic rim of the Indian Ocean. This change in parent material may have some influence in the development of certain differing characteristics among the andosols that will be discussed in this paper.

CLIMATE

In general the climate in Indonesia changes with elevation. Rainfall increases with elevation and probably reaches a maximum at about 1500 to 2000 m. above sea level. Temperature decreases with increasing altitude. Differences in temperature can be calculated by the formula [see reference to Braak, 1923/25 in Tan and Van Schuylenborgh (12)]:

$$t = 26.3 - h \times 0.6° \text{ C.}$$

where h is the elevation in hectometers. Thus the lowlands will have a humid tropical climate, while up in the mountains a warm temperate climate will prevail.

Since the andosols in north Sumatra are restricted to the plains at the foot of the Sibajak volcano, they are formed under the influence of a humid tropical climate (table 1). The andosols in Java are, to the contrary, mostly situated at higher elevations, in what is closer to a warm temperate climate.

Due to these differences in climatic conditions, different humus compositions seemed to be formed in the various andosols of Indonesia. The humus of the Sumatran andosol contains relatively higher amounts of fulvic acids than

[1] Bogor, Indonesia. This article was prepared for the F.A.O. working meeting on volcanic ash soils in Tokyo (Japan) June 11-27, 1964.

the Java andosol (table 2). The humic/fulvic ratio for the Sumatran species is 0.2 for the A horizon, whereas the Java species reveals humic/fulvic ratios of about 0.5 or higher for similar horizons.

These results suggest a preliminary conclusion that there are two kinds of andosols: (a) the Lowland andosols, formed in the plains with a humid tropical climate, and possessing humic/fulvic ratios of about 0.2 or lower, and (b) the Highland andosols, formed at higher elevations under a warm temperate climate, and possessing humic/fulvic ratios of 0.5 or higher.

THE ACTIVE FRACTION

The active fraction of the andosol consists of the organic and inorganic colloidal material. Since the andosols are also characterized by high amounts of organic matter, it seems reasonable to pay the same attention to the organic as to the inorganic material. The combined action of both the organic and inorganic fractions gives rise to the peculiar behavior of these soils.

In addition to our tentative conclusion that there are at least two kinds of andosols, there is another important result of the certain change

TABLE 1
Rainfall in north Sumatra and west Java (2)

Location	Altitude	Rainfall <60 mm.	Rainfall >100 mm.	Type of Climate* Köppen†	Type of Climate* Schmidt/Ferguson‡	Mean Yearly Rainfall
	m.	months	months			mm.
North Sumatra						
Timbanglangkat	29	0.7	9.7	Afa	A	2522
Padangbrahrang	49	0.4	10.9	Afa	A	2847
West Java						
Tjiapus/Bogor	540	0.1	11.8	Afa	A	4880
Lembang	1247	2.6	8.0	Cfhi	B	2915
Malabar	1550	2.7	8.1	Cfi	C	2564

* According to definitions of Köppen and Schmidt/Ferguson as given in (2).

† Mean temperatures: (A) coldest month >18° C. and (C) between 18° and −3° C.; (a) warmest month >22° C. and (h) annually >18° C. (i) Difference between coldest and warmest month is less than 5° C. (f) Humid.

‡ Dividing climate into 8 classes from the formula avg. no. of dry months/avg. no. of wet months × 100%, Class A = 0–14.3%; Class B = 14.3–33.3%; and Class C = 33.3–60.0%.

TABLE 2
Humus composition of the andosols

Profile	Humus (%) Fulvic	Humus (%) Humic	Humic Fulvic
North Sumatra			
A1	81.4	18.6	0.2
A3	100.0	tr	0.0
West Java			
Tjiapus/Bogor			
A11	65.9	34.1	0.5
A12	55.9	44.1	0.8
Lembang			
A11	42.9	57.1	1.3
B1	52.5	47.5	0.9
Pengalengan			
A1	67.7	32.3	0.5
B1	69.9	30.1	0.4

in humus composition, that is an unequal participation of the different humus substances in the migration of sesquioxides. Since fulvic acids may have a considerably higher ability in eluting aluminum than iron,[2] the migration of aluminum in the Sumatran andosol is expected to be relatively higher than in the Java andosol. Investigations into this possibility are still under way and differential thermal analyses (D.T.A.) have already revealed (10) increasing amounts of gibbsite with depth of profile.

The inorganic colloidal fraction consists chiefly of allophane, as is observed with D.T.A. (10, 13). According to Fieldes (5), allophane differs from other clay minerals with respect to its aluminum content, which is held in its structure in tetrahedral coordination. This is supposed to be the main cause of its many unusual chemical activities, for example water retention, base-holding mechanism, phosphate fixation, and formation of very stable mineral-organic colloids.

One of the soil characteristics resulting from the combined effect of the organic and inorganic fractions is the proportion and kind of electric charge which both stimulate in the colloidal

[2] K. H. Tan, A. Satari, and Kuntadi. The relation of soil type to organic matter in the tropics. To be published in Trans. Intern. Congr. Soil Sci., 8th Congr. Bucharest.

material. As is well known, the electrical charge is held to be responsible for the chemical activities observed in the soil. So charge distribution studies were carried out according to Mehlich's method (7), and are being continued with a wide variety of andosols.

Preliminary results (table 3) indicate that the andosols all have a low permanent charge (CEC_p). They seem to be more differentiated by the variable charge (CEC_v). At this time, based on the proportion of the variable charge, three different andosols can be recognized: andosols with low CEC_v ($CEC_v < 30$ me./100 g.); with medium CEC_v ($CEC_v = 30 - 50$ me./100 g.); and with high CEC_v ($CEC_v > 50$ me./100 g.).

Concerning the amount of positive charge, it can be observed tentatively that the andosols with low CEC_v have high anion-exchange capacities (AEC), whereas andosols with medium and high CEC_v show low anion-exchange capacities. In order to come to more definite conclusions, more data are needed.

The results reported here seem to be in accordance with those observed by Fieldes (5) for andosols in New Zealand. Although he conducted his investigation with quite different methods, Fieldes noted that allophane develops an increasing negative charge as pH increases above pH 5, and thus it has a variable negative charge.

TABLE 3
Preliminary results on charge characteristics of andosols

Profile	CEC_p	CEC_v	$CEC_{s.2}$	AEC
	me./100 g.			
North Sumatra				
A1	5.7	29.9	35.6	44.4
A3	1.2	21.1	22.3	34.1
West Java				
Tjiapus/Bogor				
A11	8.1	62.3	70.4	23.9
A12	8.4	64.3	72.7	29.2
Lembang				
A11	4.8	39.0	43.8	24.5
B1	5.5	36.2	41.7	24.5

TENTATIVE CLASSIFICATION

A comparison of tables 2 and 3 shows that a division of andosols based on humic/fulvic ratio may well fit a division of andosols based on amount and kind of electric charge, if the Lembang andosol, since its humic/fulvic ratio is above 1.0, is recognized as a separate group.

Thus it might be tentatively concluded that the andosols in Indonesia may be separated into three groups: those with a humic/fulvic ratio ≤ 0.2 and low CEC_v; those with a humic/fulvic ratio $= 0.5$ to 1.0 and high CEC_v; and those with a humic/fulvic ratio ≥ 1.0 and medium CEC_v.

LAND USE AND EVALUATION

As noted, the andosols are characterized by high organic matter contents. On the whole they are medium-textured and medium-fertile soils. They contain high amounts of nitrogen, have medium levels of potassium, but relatively low contents of phosphate (13). A certain relationship between organic matter and nitrogen content seems to be present. Statistical studies reveal the existence of a positive correlation between organic matter and nitrogen.[3] The

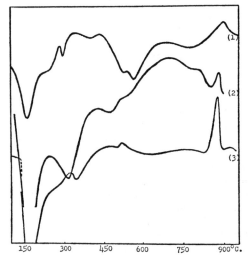

FIG. 1. D.T.A. thermograms of clay fractions (2 µ) [from Tan and Massey (13)]: (*1*) A1/A2 gray-brown podzolic soil (1100 m.); (*2*) A11 andosol (1100 m.); and (*3*) Ap andosol (1100 m.) taken from a nearby cultivated soil similar to (*2*).

[3] *Ibid.*

linear regression shows the following relationship ($r = 0.9770$ with a standard error of 0.0032)

$$N = 0.0702\ C + 0.0760$$

Because of these characteristics, the Sumatran andosols are utilized for industrial crop production. The area of north Sumatra is especially famous for its Deli-tobacco wrappers, which formerly found their market in Holland. Grown on the andosols, the tobacco plant usually receives adequate phosphate fertilizers, little potassium, and less nitrogen. To counteract the adverse effect of bacterial wilt-disease, once every four years the area is cultivated with tobacco.

In Lembang (west Java) the andosols are often considered as being the main factor for a successful horticultural operation. Due to their location in a warm temperate climate, the andosols are favored for growing cabbage, carrots, potatoes, cut flowers, and other European crops. Although high amounts of organic matter are already present in the soil, when growing potatoes the farmers often fertilize it again with tons of organic matter, probably to counteract the toxic effect of the soil's high aluminum content. The average yield of potatoes [6 tons/Ha (11)] would be considered low by European standards.

The more rugged part of the country in Lembang is planted with pine trees. Investigation of the relation of site quality to soil factors suggests a certain superiority of the andosols over podzolic soils (13).

In the Pengalengan highlands, which are covered with these andosols, the best tea plantations in Java are found. The production of approximately 3000 pounds/Ha/year is considered quite high, and the quality of this "Highland Tea" is more favored than the "Lowland Tea" grown at places at lower elevations on different kinds of soils.

SUMMARY

Investigations were carried out on the genesis and characterization of a group of black-colored soils developed from volcanic ash, at present classified as andosols.

Preliminary results on the humus and clay composition, and on the proportion and kind of electric charge, indicated a possible recognition of three kinds of andosols in Indonesia: those with low humic/fulvic ratio and low CEC_v; those with medium humic/fulvic ratio and high CEC_v; and those with high humic/fulvic ratio and medium CEC_v.

Chemical and field studies reveal medium fertility levels. The areas covered with these soils represent productive areas for industrial crops and horticultural operations.

REFERENCES

(1) Aomine, S., and Yoshinaga, N. 1955 Clay minerals in volcanic ash soils in Japan. Soil Sci. 79: 349–358.
(2) Berlage, H. P., Jr. 1949 Dept. Verkeer, Energie en Mÿnbouw. Meteor. en Geophys. Kon. Magn. Obs. Verh. 37.
(3) Birrel, K. S., and Fieldes, M. 1952 Allophane in volcanic ash soils. J. Soil Sci. 3: 156–166.
(4) Druif, J. H. 1939 De bodem van Deli. III. Toelichting bij de agrogeol. kaarten en beschr. der grondsoorten van deli. Med. Deli Proefsta. III/IV: 1–74.
(5) Fieldes, M. 1962 The nature of the active fraction of soils. Trans. Intern. Soil Sci. Conf. New Zealand, pp. 62–78.
(6) Kanno, I. 1961 Genesis and classification of main genetic soil types in Japan: I. Bull. Kyushu Agr. Expt. Sta. VII: 1–185.
(7) Mehlich, A. 1960 Charge characterization of soils. Trans. Intern. Congr. Soil Sci. 7th Congr. Madison II: 292–302.
(8) Mohr, E. C. J., and Van Baren, F. A. 1954 "Tropical soils." Les edition A. Manteau S. A. Bruxelles.
(9) Tamura, T., Jackson, M. L., and Swindale, G. D. 1953 Mineral content of low humic, humic and hydrol humic latosols of Hawaii. Soil Sci. Soc. Am. Proc. 17: 343–346.
(10) Tan, K. H. 1963 Differential thermal analyses of andosols in Indonesia. Research J. Ministry Higher Ed. and Sci. Rept. Indonesia I B, 1: 11–20.
(11) Tan, K. H., and Hutagalung, O. 1960 Penjelidikan sementara mengenai pemupukan tanaman kentang di Indonesia. Tehnik Pertanian IX, 3/4: 94–104.
(12) Tan, K. H., and Van Schuylenborgh, J. 1961 On the classification and genesis of soils developed over acid volcanic material under humid tropical conditions: II. Neth. J. Agr. Sci. 9: 41–54.
(13) Tan, K. H., and Massey, H. F. 1964 Effect of site on the pulpwood productive capacity of *Pinus merkusii*: I. Research J. Ministry Higher Ed. and Sci. Rept. Indonesia I B, 3: 88–101.

7

Copyright © 1978 by Elsevier Scientific Publishing Company
Reprinted from *Geoderma* **21**:119-131 (1978)

ANDEPTS IN SOME HIGH MOUNTAINS OF EAST AFRICA

ERWIN FREI

Swiss Federal Research Station for Agronomy, CH-8046 Zürich (Switzerland)
and
University of Berne, Berne (Switzerland)
(Received August 22, 1977; accepted May 11, 1978)

ABSTRACT

Frei, E., 1978. Andepts in some high mountains of east Africa. Geoderma, 21: 119—131.

Soils at elevations of 2,500 m to 4,000 m above sea level in the Semien Mountains of Ethiopia and on Mount Kenya in east Africa are classified in the current American system as Andepts. The Cryandepts are situated above the forest line in a tussock grass vegetation or in dwarf forests at elevations of 3,000 to 3,800 m a.s.l. In the tall forest above 2,500 m a.s.l. and below 3,000 m there are Dystrandepts with an isothermal soil temperature of more than 8°C. In the legend of the FAO-Unesco soil map of the world the soils could be classified as Humic Andosols.

Histosols may take the place of the Andepts where relative humidity of the atmosphere is high and rainfall is more than 2,000 mm per year.

INTRODUCTION

This preliminary pedological study was part of a research program on the ecology of the Semien Mountains of Ethiopia (Messerli et al., 1974), and of Mount Kenya. The soils of these mountains are not well known, but some indications of the pedological conditions can be gained from reports of geographers, geologists, botanists, biologists and climatologists.

SOIL FORMING FACTORS AND SOIL GENESIS IN MOUNTAINS OF THE TROPICS

Climate

Soil-forming environments are partly determined by the climate. In high mountain regions, one climatic element is a succession of temperature belts, not unlike those extending from the equator to the poles. Progressively colder belts occur at higher elevations (Frei, 1958). Because the air in high mountain regions is rarefied it absorbs little heat and water and is therefore cool and dry. In comparison the soil surface absorbs much heat during day time but loses it rapidly during the night. The isothermal horizon of soils in mountains of the tropics has been observed at a comparatively shallow depth of about 50 cm below the surface (B. Messerli et al., personal communication, 1978).

Andepts between altitudes of 2,500 and 4,000 m a.s.l. in east Africa occur under warm (12°C) to cool temperate (4°C) climates with marked diurnal fluctuations of temperature and humidity at soil surfaces (Jenny, 1930). These conditions are important for the quality and quantity of the organic matter in soils of high mountains. They also cause daily migration of the soil animals from the surface down into the isothermal part of a pedon, producing burrows and reworking organic matter into the subhorizons.

In the area of the Andepts at Mt. Kenya and in the Semien Mountains the total rainfall per year is approximately 1,200—1,800 mm. Under these tropical conditions the amount almost equals the potential evapotranspiration of the vegetation (Kenya Soil Survey, 1975). Because of the seasonal concentration of the rainfall in November to December and March to May the soils dry out periodically and are also leached at intervals. Histosols are distributed on the western slopes of Mount Kenya (Fig. 3) with a total rainfall over 2,000 mm and relatively less sunshine because of a misty veil during the rainy season.

Parent material

The Semien Mountains consist of basalts, whereas Mount Kenya is an extinct volcano. Parent materials around Mt. Kenya consist of porphyritic phonolites and agglomerates. The plug of Mt. Kenya, which forms the highest peaks of the mountain consists of a central core of nepheline syenite (Baker, 1967). The basalts in the Semien Mountains and the rocks of Mt. Kenya are mantled by wind blown sediments and volcanic ash. In contrast to lowlands of the tropics, no old and strongly weathered regoliths occur in these mountains. Parent materials are relatively young, Pliocene, Pleistocene or Recent. Major portions of the soil profiles are formed in silty materials which differ in a number of ways from the residues weathered from the underlying rock (Hurni, 1975). Moreover, markedly contrasting layers (in texture, color, etc.) are common within and below the soil profiles, indicating lithologic discontinuities.

Topography

The strong relief of mountainous areas affects soil formation in several ways. The steep slopes promote downcutting and active removal of weathered materials by landslides, erosion, and soil creep. Consequently, a few sites are virtually devoid of soil and others have especially deep regoliths. Where the transported sediments are deposited, buried soil horizons may be found. Another effect is due to the aspect of direction of exposure of a slope. In mountains, aspect affects the temperature and moisture regimes of soils generally. If differences between two aspects are very large, soils on the respective sites may differ in morphology, as well. In the Semien Mountains and on Mt. Kenya, Andepts in regoliths of various depths occur over a wide range of sites and slopes, being

TABLE I

Sites and depths of investigated soils in Gich (38° 07′ E 13° 17′ N), Semien Mountains

Profile	Landform	Elevation a.s.l. (m)	Aspect or exposure	Slope (%)	Vegetation	Umbric epipedon		Bedrock (cm)	Max. depth of root system (cm)
						(cm)	Munsell moist soil color		
1	ridge, slope	3,640	190° S	15	tussock grassland with shrubs	30	10 YR 2/2	40	40
2	gentle slope	3,634	190° S	20		90	10 YR 2/2	120	120
3	depression slope	3,628	206° SW	5	and *Lobelia rhynchopetalum*	60	10 YR 2/1	150	100
4	plateau	3,622	200° SW	3		70	10 YR 2/2	150	100
5	lower slope	3,535	80° E	30	dwarf mountain forest	150	7,5 YR 3/2 —10 YR 2/2	180	170
6	lower slope	3,535	225° SW	25	*Erica arborea*	80	7,5 YR 3/2	100	100

122 *Frei*

absent only in places subject to seepage of water and on the steepest slopes.

Vegetation

Andepts occur at Mt. Kenya and in Semien under high mountain vegetation. At elevations of 2,500—3,000 m a.s.l. Andepts are developed under a rather tall forest of *Hagenia* and *Podocarpus* trees. Bamboo thickets (*Arundinaria alpina*) appear to grow most vigorously and to form continuous stands where the average annual rainfall exceeds 1250 mm (Trapnell et al., 1976). Andepts are also developed at elevations of 3,000—3,800 m a.s.l. under dwarf forests of *Erica arborea, Philippia keniensis* and *Hypericum revolutum* (Lind et al., 1974; Klötzli, 1975). At 3,500—4,000 m a.s.l. Andepts occur under a dense tussock grass vegetation with scattered *Dendrosenecios* and *Lobelias*.

At an altitude of 4,000 m a.s.l. or higher the degree of freezing at night is intensive enough to cause daily frost heaves in soil surface, harming roots in the frozen layer. Only strong, deep-rooted plants are resistant. The vertical displacement of the ground surface by frost action is followed by solifluction each morning when the soil surface begins to thaw. Because the vegetative cover becomes incomplete at this elevation, increasing areas of bare soil become evident and they show patterns due to frost action (Furrer, 1973; Baker, 1967). At such elevations Andepts are gradually replaced by stony, regosolic types.

SOIL PROFILES STUDIED

Soils in the Semien Mountains

In March 1976 I studied six soil profiles in the area of Gich (Table I). The sites represent the local physiographic units of a high basalt plateau. All investigated profiles were under natural vegetation. Arable soils were not included.

Basalt outcrops appear on slope ridges where denudation is active. At such sites the soil profiles are shallow (40 cm) over underlying consolidated bedrock. On stable, flat plateau sites a layer of brown clay (30 cm), believed to have weathered from basalt, covers the bedrock. On this clay a stony loam has been deposited by old colluvial or solifluction processes. It often forms the cambic horizons (term from Soil Survey Staff, USDA, 1975) of the soil profiles. The umbric epipedon is developed in silty material which is probably airborne and could also be volcanic ash (Messerli, 1975). This wind blown layer of 30—100 cm depth covers the entire land surface. The reason for its variable depth is removal by erosion and, perhaps, differing deposition on leeward and windward slopes.

Some analytical data for a well developed soil profile are given in Table II and in Fig. 1. In the whole pedon containing the root system of the vegetation, there are 120 mm of water absorbed with tensions from 0.1 to 1 bar. The 500 mm of absorbed water between tensions of 0.1 and 15 bar supply sufficient water to the plants for at least 100 days in the dry season. Distributed in two rainy periods per year, there is 1,400 mm precipitation at Gich. In a very deep

TABLE II

Chemical and physical properties of the soil profile 5, Typic Cryandept, Gich

Depth (cm)	Horizon	pH 1)	Exchangeable ions (mequiv./100 g)						CEC	Al 8)	Org. matter o.m. (%)	N (%)	C/N
			H	K	Na 2)	Ca	Mg						
0—10	Histic	5.4	22.4	0.5	0.1	11.8	3.7		38	0	29.6	0.7	24
10—70	Umbric	5.3	22.7	0.2	0.1	11.2	3.2		38	0.6	16.0	0.4	21
70—100	Umbric, org. cutans	5.3	19.8	0.2	0.2	11.8	3.4		35	0.4	9.7	0.3	20
100—150	Cambic	5.4	10.9	0.2	0.1	10.3	3.4		25	—	3.6	0.07	29
150—170	Transition	5.4	8.3	0.1	0.1	9.3	3.5		21	—	1.5	0.05	18

Depth (cm)	Horizon	Clay < 0.002 mm (%)	Silt 0.002—0.05 mm (%)	Pore space (ml/100 ml bulk volume)					Density bulk volume (g/ml) 7)
				total	0—0.1 bar 3)	0.1—1 bar 4)	1—15 bar 5)	>15 bar 6)	
0—10	Histic	30	36						
10—70	Umbric	30	51						
70—100	Umbric, org. cutans	32	48	76	21	10	30	15	0.53
100—150	Cambic	52	35						
150—170	Transition	33	49	72	31	5	17	19	0.66

1) pH in CaCl$_2$, 0.02 N
2) Ion exchange at pH 2 (Hajek et al., 1972)
3) Air-space porosity
4) Capillary pore-space, soil-water easily available to plants
5) Soil-water available to plants
6) Not available water
7) 105°C dry basis, undisturbed sample
8) KCl extract

soil plant growth should not be limited by dryness. Large areas are eroded, however, and in some parts no umbric epipedon can be identified.

The organic matter of the desiccated soil surface is water-repellant in contact with rainwater. There is also much run-off on the hydrophobic, sometimes powdery, dry soil surface. Infiltration of rainwater is therefore limited after a drought (Klötzli, 1975). Under moist conditions the soil profile is permeable, with a water percolation rate between 0.007 and 0.01 cm per second. This is in agreement with the very high non-capillary porosity of 20—30 ml per 100 ml bulk volume at 0.1 bar moisture tension (Fig. 1). Table II also shows the still moderate content of adsorbed metallic cations and almost no mobile aluminium. The chemical conditions for plant growth are thus considered satisfactory.

Soils at Mt. Kenya

Except on the western slope, Andepts cover the mountain between elevations of 2,800 and 4,000 m (Fig. 3). On the western slope between 3,000 and 4,000 m the rainfall is very high. The soil has a lithic contact, which precludes deep

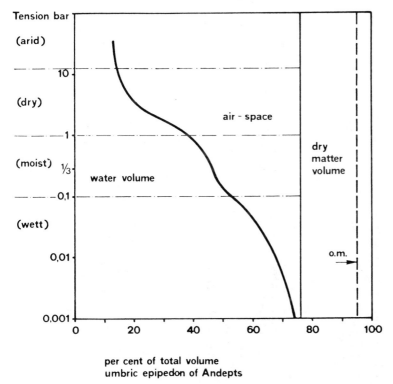

Fig. 1. Tension-moisture curve of the undisturbed umbric epipedon, soil profile 5, Semien Mountains.

TABLE III

Sites and depths of the investigated soils at Mt. Kenya (37° 10'–37° 30'E 00° 00'–20'N)

Profile No.	Landform and place*1	Elevation a.s.l. (m)	Aspect or exposure	Slope (%)	Vegetation	Umbric epipedon (cm)	Munsell, moist soil color	Bedrock (cm)	Max. depth of root system (cm)
1	terrace upper slope, Timau track	3,330	S (local) N (general)	10	tussock grassland	40	7,5 YR 2/2	120	90
2	slope, Naro Moru route	3,000	W	20	Hagenia forest	140	7,5 YR 3/2 4/4	300	150
4	terrace upper slope, Kamweti	2,550	SE	15	Bamboo forest	140	7,5 YR 3/2	—	220
8	lower slope Timau track	3,180	ESE (local) N (general)	35	tussock grassland and Philippia	170	5 YR 2/2	300	220
9	slope Sirimon track	3,240	N (local) NW (general)	20	dwarf mountain forest Hypericum	80		150	

*1 Survey of Kenya (1974).

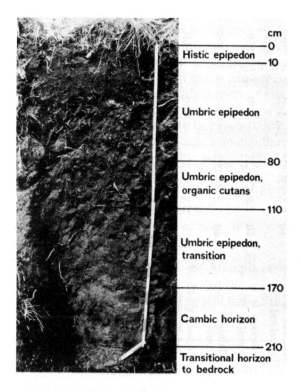

Fig. 2. Soil profile No. 8, Mt. Kenya, 3,180 m a.s.l., Typic Cryandept.

percolation of the water, so that Histosols are very common (Thorp and Bellis, 1960). The umbric epipedon of the Andepts is 40—170 cm thick (Fig. 2 and Table III). The bedrock is of volcanic origin (porphyritic phonolites) but the parent material of the epipedon, as in the Semien Mountains is silty, wind-blown materials. In certain sites solifluction or sheet erosion deposits of the same silt produce very deep soils. Some analytical results for a well developed soil of Mt. Kenya, classified as a Typic Cryandept (Fig. 2) are given in Table IV. The high organic matter content of the soils on Mt. Kenya is very similar to that of the Semien region. The acidity is rather high, but still the content of potassium and sodium ions is relatively high. The distribution of pore sizes, as measured by water retention, indicates extremely high porosity and retention of available water.

Description of profile No. 8 (Fig. 2, Table IV)

Classification:	Typic Cryandept
Location:	Mt. Kenya, N. slope, 3,180 m a.s.l. (Timau Track)
Physiographic position:	lower slope of a small valley
Topography:	35% southeast-facing slope
Drainage:	well drained, rapid permeability (10^{-2} cm/sec)

Vegetation:	shrubs: *Philippia, Hypericum, Erica, Helichrysum*; tussock grasses
Parent Material:	olivine trachyte, basaltic pumice, volcanic ash
Histic epipedon:	0—10 cm; dark reddish brown (5 YR 2/3) moist; partially decomposed organic materials (27% o.m.); very many roots; admixture of silt and very fine crumbs; broken, abrupt boundary. Remark: this horizon is lacking under grassland vegetation.
Umbric epipedon:	10—80 cm; dark reddish brown (5 YR 2/2) moist; silt loam; weak, medium granular structure; very friable, very porous; no coarse fragments, nonplastic; many roots; gradual boundary
Umbric epipedon:	80—110 cm; very dark brown (7,5 YR 2/3 and 2/2) moist; fine sandy loam; very friable; animal borrows (crotavinas); no coarse fragment; medium subangular blocky structure, peds with very dark cutans; gradual boundary
Umbric epipedon:	110—170 cm; dark reddish brown (7,5 YR 3/3) moist; silt loam; medium columnar structure; some peds with very dark cutans; clear boundary
Cambic horizon:	170—220 cm; dark brown to brown (7,5 YR 3/4 and 4/4) moist; silt loam; weak, coarse blocky structure; friable; few roots; gradual boundary
Transitional horizon between the cambic horizon and the bedrock:	220—300 cm; brown (7,5 YR 4/6) moist; sandy loam; very weak coarse blocky structure; firm; abrupt boundary

RESULTS AND CONCLUSIONS

Soil classification

The definition of the Andosols in the Legend of the FAO-Unesco Soil Map of the World (1974) is similar to the requirements for the Andepts of the USA soil classification (Soil Survey Staff, 1975, pp. 230—236). Andepts have an umbric or mollic epipedon of more than 25 cm thickness. The content of organic matter is high, and the soil is very porous and of a low bulk density. Moreover, the mineral exchange complex is dominated by a allophane clay. Like other Inceptisols, the Andepts may also have a cambic horizon beneath an umbric epipedon. These requirements are all met by the black soils of these mountains in east Africa. Two soil profiles representing the typical regional soils of the Semien Mountains and of Mt. Kenya are classified in Table V according to the criteria of the USA classification. Both soil profiles would classify as Typic Cryandepts at the subgroup level of the American classification system because they have relatively low temperatures ($<8°C$).

TABLE IV

Chemical and physical properties of the soil profile 8, Typic Cryandept, north slope of Mt. Kenya

Depth (cm)	Horizon	Exchangeable ions							Org. matter		
		1)				2)			o.m. (%)	N (%)	C/N
		pH	H	K	Na	Ca	Mg	CEC			
			(mequiv./100 g)								
0–10	Histic	5.6	17.6	0.8	0.07	7.6	2.6	28	27	0.6	29
10–80	Umbric	5.5	18.6	0.2	0.07	4.7	1.5	25	15	0.4	24
80–110	Umbric, org. cutans	5.4	14.2	0.1	0.06	3.0	1.0	18	13	0.3	22
110–170	Umbric, transition	5.3	12.5	0.1	0.05	1.4	0.7	15	9	0.2	26
170–210	Cambic	5.2							4	0.1	19
210–230	Transition	5.2	8.1	0.2	0.05	1.2	0.6	10	2	0.1	17

Depth (cm)	Horizon	Clay < 0.002 mm (%)	Silt 0.002– 0.05 mm (%)	Pore space					Density
					3)	4)	5)	6)	7)
				total	0–0.1 bar	0.1–1 bar	1–15 bar	>15 bar	bulk volume (g/ml)
				(ml/100 ml bulk volume)					
0–10	Histic	21.9	52.2						
10–80	Umbric	4.3	70.3	82	42	15	12	13	0.38
80–110	Umbric, org. cutans	2.8	42.6	82	30	11	29	12	0.40
110–170	Umbric, transition	6.8	58.7						
170–210	Cambic	13.2	56.8	68	16	4	32	16	0.77
210–230	Transition	15.6	48.4						

1) pH in $CaCl_2$, 0.02 N
2) Ion exchange at pH 2 (Hajek et al., 1972)
3) Air-space porosity
4) Capillary pore-space, soil-water easily available to plants
5) Soil-water available to plants
6) Not available water
7) 105°C dry basis, undisturbed sample

Other Andepts can be classified as Typic Dystrandepts in the American system. They are usually situated below an elevation of 3,100 m a.s.l. (Fig. 3), and have a warmer soil temperature (> 8°C) in the isothermic layer at approximately 50 cm below the surface. The color of the umbric epipedon of these soils is dark brown or very dark reddish brown. Often the lower part of the umbric epipedon is distinctly darker because of the illuviation cutans, composed of complexes of colloidal organic substances with inorganic colloids e.g. allophanes (Frei, 1964). The reddish brown cambic horizon is well developed and thicker than in the Cryandepts. In the FAO-Unesco Soil Map of the World (1974) all these freely drained dark soils of Mt. Kenya and of the Semien Mountains would classify as Humic Andosols.

TABLE V

Classification of two representative soil profiles of the Semien Mountains and of Mt. Kenya

Criteria (Soil Survey Staff, 1975)	Properties profile 5 Semien	profile 8 Mt. Kenya	Notations
Andepts are Inceptisols that have a depth of > 35 cm	180 cm	230 cm	to a lithic contact
and low bulk density of the fine earth fraction < 0.85 g/ml (1/3 bar)	0.97 g/ml 0.53 g/ml 22 vol%	0.78 g/ml 0.40 g/ml 33 vol%	g undisturbed soil per ml (0.3 bar) g undisturbed soil per ml (105°C) non-capillary porosity (0.1 bar)
Amorphous material is dominant in the exchange complex CEC (pH 8.2) > 150 mequiv./100 g clay,	180 mequiv.	poor dispersion	mequiv./100 g clay
15 bar water: soil/clay = > 1 organic carbon > 0.6%	1 5—17% 90 kg/m2	1.1 7—16% 87 kg/m2	15 bar water, undisturbed sample $K_2Cr_2O_7$ oxidation total organic matter
C/N ratio	20—29	22—29	C% / N%
High anionexchange capacity	38—58 mequiv.	46—60 **mequiv.**	($0.03N$ H_3PO_4) mequiv. PO_4/100 g
	amorphous mat. dominant		thin section examination
	amorphous clay and illites, chlorites, kaolinite		X-ray (Prof. T. Peters, Berne, not published)
Cryandepts have a cryic temperature regime, < 8°C	6—8°C	7—8°C	at a depth of 50 cm
Typic Cryandepts: epipedon has the color and thickness of a mollic epipedon	150 cm	170 cm	dark horizon with organic matter
Munsell value chroma	2—3 2	2 2	Munsell soil colors when moist
Base saturation	39—44%	22—29%	(Ca + Mg + K + Na) · 100/CEC

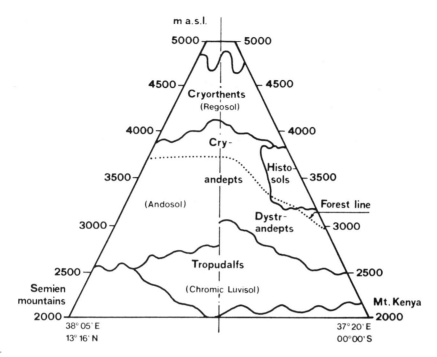

Fig. 3. Schematic distribution of soil types in the Semien Mountains and at Mt. Kenya (altitudes of the highest and the lowest sites).

REFERENCES

Baker, B.H., 1967. Geology of the Mount Kenya area. Ministry of Natural Resources, Geological Survey of Kenya, Report No. 79.
FAO-Unesco, 1974. Soil Map of the World, 1 : 5 000 000. Vol. 1, Legend. UNESCO.
Frei, E., 1958. Eine Studie über den Zusammenhang zwischen Bodentyp, Klima und Vegetation in Ecuador. Plant Soil, IX, 3.
Frei, E., 1964. Micromorphology of Some Tropical Mountain Soils. Elsevier, Amsterdam.
Furrer, G. und Freud, R., 1973. Beobachtungen zum subnivalen Formenschatz am Kilimandjaro. Z. Geomorphol., N.F. Suppl. 16, pp. 180—203.
Hajek, B.F., Adams, F. and Cope, J.T., 1972. Rapid determination of exchangeable bases, acidity and base saturation for soil characterization. Soil Sci. Soc. Am. Proc., 36: 436—438.
Hurni, H., 1975. Bodenerosion in Semien-Aethiopien. Geogr. Helv., 4: 157—168.
Jenny, H., 1930. Hochgebirgsböden. In: E. Blanck (Editor), Handbuch der Bodenlehre. Verlag Julius Springer, Berlin, Vol. 3, pp. 96—118.
Kenya Soil Survey, 1975. Soils of the Kindarum area. Reconnaissance Soil Survey Report R 1. Ministry of Agriculture, Republic of Kenya.
Klötzli, F., 1975. Zur Waldfähigkeit der Gebirgssteppen Hoch-Semiens (Nordäthiopien). Beitr. naturk. Forsch. Südw. Dtl. 34, 131—147.
Lind, E.M. and Morrison, M.E.S., 1974. East African Vegetation. Longman, London.
Messerli, B., Stähli, P. und Zurbuchen, M., 1974. Eine topographische Karte aus den Hochgebirgen Semiens, Aethiopien. Verm. Photogr. Kulturtech. Fachblatt, I - 75.

Messerli, B., 1975. Formen und Formungsprozesse in den Hochgebirgen Aethiopiens. 40. Dtsch. Geographentag Innsbruck, Tagesbericht und wissenschaftliche Abhandlungen.

Soil Survey Staff, 1975. Soil Taxonomy. A Basic System of Soil Classification for Making and Interpreting Soil Surveys. U.S. Govt. Printing Office, Washington, D.C., Agric. Handbook, 436, 754 pp.

Survey of Kenya, 1974. Map and Guide, National Park and Environs, Mount Kenya. Public Map Office, Nairobi.

Thorp, J. and Bellis, E., 1960. Soils of the Kenya highlands in relation to landforms. Trans. 7th ISSS Congr., Madison, Wisc., Vol. V, 329—334.

Trapnell, C.G. and Birch, W.R., 1976. Vegetation 1 : 250,000 map of Kenya, sheet 2, Survey of Kenya, Nairobi, Kenya.

Wada, K. and Higashi, T., 1976. The categories of aluminium- and iron-humus complexes in Ando soils determined by selective dissolution. J. Soil Sci., 27: 357—368.

Part III

GENESIS, MORPHOLOGY, AND CLASSIFICATION

Editor's Comments
on Papers 8, 9, and 10

8 **FLACH et al.**
   Genesis and Classification of Andepts and Spodosols

9 **MOHR, VAN BAREN, and VAN SCHUYLENBORGH**
   Andosols

10 **SMITH**
   A Preliminary Proposal for Reclassification of Andepts and Some Andic Subgroups

Many scientists believe that the formation of Andosols is strongly associated with the formation of amorphous aluminosilicate clay, especially allophane, and the accumulation of humus immobilized in the surface horizon as a result of interactions between the humus and allophane and/or sesquioxides (Fitzpatrick, 1980; Wada and Aomine, 1973). Since the volcanic ash, the parent material, is of Quaternary or recent origin, the weathering processes giving rise to the formation of Andosols must have taken place at a rapid rate. According to Wada and Aomine (1973), Andosols can attain maturity within five thousand years, but Yamada (1977) reported that these soils can be developed between five hundred and fifteen hundred years, depending on the soil formation factors, such as the type of volcanic ash. Although the particle size of volcanic ash may differ considerably among regions and from one to another eruption, the major part of the ash is composed of sand and silt-size particles. Frequently the ash may be mixed with coarser materials, such as pumice, scoria, and lapilli. Nevertheless, such pyroclastic material weathers more rapidly due to its larger surface areas and higher porosities than solid rocks.

Many other soils are derived from volcanic ash. Some of these soils may still contain substantial amounts of unweathered volcanic material in the surface, such as pumice, and others may not exhibit the large accumulation of organic matter and perhaps also lack the dominant presence of the amorphous alumino-silicate clays (Ishizuka and Black, 1977). These kinds of soils are volcanic ash soils by origin but lack the unique features of Andosols; hence they are not Andosols. They are characterized by other properties and may belong to the

Alfisols, Entisols, Oxisols, Spodosols, or Ultisols (Dames, 1955; Van Schuylenborgh and Van Rummelen, 1955; Van Schuylenborgh, 1956; Dudal, 1962.)

Paper 8, by Flach et al., illustrates the concept of formation of Andosols on the basis of the weathering of volcanic ash to allophane and on the formation of humus-allophane and/or humus-aluminum complexes. Paper 9, by Mohr, Van Baren, and Van Schuylenborgh, describes the importance of the soil formation factors—climate, vegetation, and parent materials, for example—involved in the genesis of Andosols. Flach et al. did not use the term Andosols in the text; rather they used *Andepts,* defined as all soils derived from volcanic ash, as well as soils from other parent materials, that contain an abundance of easily weatherable minerals. Nevertheless, all of the soils were described by Flach et al. as being formed by the two dominant processes characteristic for formation of Andosols only. On the other hand, Mohr, Van Baren, and Van Schuylenborgh made a distinction between Andosols and other soils from volcanic ash and indicated that the latter group of soils did not exhibit the features of Andosols.

In both Papers 8 and 9, the authors agree about the wide variations in climate in which Andosols can be formed. The major difference between the two papers is that Flach et al. included as a climatic factor the semiarid climate with a ustic moisture regime, whereas Mohr, Van Baren, and Van Schuylenborgh indicated that volcanic ash in arid and semiarid regions did not develop into Andosols. Wright (1964) also has stated that the unique features of Andosols seldom, if ever, developed under conditions of low soil humidity found in the arid and semiarid areas. Many believe that the presence of adequate soil moisture is one of the requirements for the development of allophane and accumulation of large quantities of organic matter.

Although the distribution of Andosols does not follow major vegetational zones, Mohr, Van Baren, and Van Schuylenborgh noticed certain plant species to be common in regions where Andosols were located. Of the several species studied by these authors, the Nothofagus spp. has been associated more often with Andosols than have other plants in South America, the United States, and New Zealand. The conifers are reported to be the major species in the mountainous regions of Andosols. In Japan, *Miscanthus sinensis, M. sasa,* and bamboo grass are the types of vegetation that have been reported to produce the black humus typical in Andosols (Tokudome and Kanno, 1965, 1968). The annual production of dry matter by the natural grasses *M. sinensis* and *M. sasa* is 3 to 10 tons per hectare (Wada and

Aomine, 1973). A number of Japanese scientists believe that the black-colored humus in Andosols has been the result of anaerobic decomposition of organic matter in poorly drained or meadow-like conditions in the past (Ohmasa, 1964). Others claim that this does not agree with the microorganisms detected in the soil, which reflect the presence of a biologically dry environment.

Although the C/N ratio of the humus can attain values of 20 or higher, the average C/N ratio in the A horizon of Andosols in Japan is 13.8, while in Indonesia values of 8.0 to 12.0 have been noticed for similar horizons. Dudal (Paper 3) suggested that the characterization of the humus in Andosols by its humic acid-fulvic acid ratio, as done in Indonesia by Tan (Paper 6), was perhaps a better method for observing the influence of differences in climate and vegetation on the nature of humus in Andosols.

Most researchers seem to agree that morphologies of Andosols typically exhibit a thick A horizon that is rich in humus and hence dark, if not black, in color. Usually this horizon does not grade into the brown cambic B horizon underneath, but the distinctness of the boundary between A and B horizons, especially in mature Andosols, is frequently sharp and qualifies to be called abrupt or clear. The A horizon is friable. When it is rubbed in a moist condition, it yields water and feels smeary but not sticky. The pedon ends at depth in a C horizon, which is a mixture of unweathered and partially weathered volcanic ash. Figures 1 and 2 show the range from a mature to a recently formed Andosol in Indonesia.

Although some agreement exists on the morphology and main features of Andosols, the classification of these soils has created many problems, especially when the U.S. *Soil Taxonomy* (1975) classified these soils as a suborder of the Inceptisols under the name *Andept*. The *Andepts* were defined as Inceptisols that (1) have to a depth of 35 cm or more, or to a lithic or paralithic contact if one is shallower than 35 cm, one or both of the following: (a) a bulk density (at 1/3 bar retention) in the fine earth fraction of the soil that is <0.85 g per cc, and the exchange complex is dominated by amorphous material, or (b) 60 percent or more of the soil (by weight) is vitric volcanic ash, cinders, or other vitric pyroclastic material, and (2) do not have an aquic moisture regime or do not have the characteristics associated with wetness that are defined for Aquepts (U.S. Soil Survey Staff, 1975). This definition gives room for the inclusion of soils of nonvolcanic origin and many other different interpretations.

One important question is why a soil of such a magnitude of occurrence and importance as the Andosols is placed at the suborder and not at the same taxonomic level (order level) as Oxisols, Ultisols,

Editor's Comments on Papers 8, 9, and 10

Mollisols, and the like. The geographic distribution of Andosols indicates that they occur all over the world, and in many countries they constitute perhaps the major soils. Another equally important problem is the placement of Andosols as young soils. Formation of Andosols is a very rapid process on young Quaternary parent material, while intermittent volcanic eruptions, producing ash fall-outs, may give the appearance to Andosols of being young soils. Therefore a large number of the soils have A/C pedons or soil profiles, but a great number of Andosols are also characterized by A-B-C profiles. These soils, which are taken as the modal soils, are mature soils and are in equilibrium with the prevailing climatic and vegetational conditions in much the same way as the Oxisols, Ultisols, and so on are with their respective environment.

The use of the name *volcanic ash soils* (see Ishizuka and Black, 1977), though attractive, is scientifically untenable because the use of such terminology returns soil science to the past when such names as *granite soils* and *limestone soils* were common. Ishizuka and Black's objection that the name *Andosols* does not cover all soils derived from volcanic ash is somewhat confusing because by using the name *Andosols,* one intends to exclude all other soils that do not exhibit the diagnostic features of Andosols.

With the increased knowledge about Andosols, it became apparent that a reclassification of these soils was urgent. An international effort was started in 1979 under the coordination of ICOMAND (International Committee on the Classification of Andisols) with headquarters located at the Soils Bureau in New Zealand. The term *Andisols* has been proposed to replace all other names previously used for this soil. In the meantime Guy D. Smith, considered the father of the *Soil Taxonomy,* has prepared a detailed reclassification system for the Andepts to conform with world demand on Andosols. His concept, reproduced here as Paper 10, elevates the Andepts from a suborder to the order level under the name *Andisols.* Smith and others consider this name more suitable than *Andosols* because the connecting vowel *o* is supposed to be restricted to Greek formative elements. However, since there are so many inconsistencies in the *Soil Taxonomy* (Tan, Perkins, and McCreery, 1979), I believe that the name *Andosol* does not violate any grammatical principles and has the advantage over *Andisol* of indicating the diagnostic feature of these kinds of soils.

Smith's definition of *Andisol* deviates slightly from that of the Andepts as reported in the *Soil Taxonomy* and covers mainly volcanic ash soils dominated by X-ray amorphous compounds, including humus. The most common diagnostic horizons to be considered for Andisols are Umbric, rarely a Mollic and an Ochric epipedon. Although Leamy et al. (Paper 4) agree with the use of these epipedons, the latter may perhaps become a factor leading to additional arguments. The black A horizon of Andisols or Andosols neither exhibits the exact characteristics of an Umbric, Mollic, or Ochric horizon nor does it behave as an Umbric, Mollic, or Ochric epipedon would, it feels smeary, yields water when squeezed, is not sticky, and so forth. I propose that a new epipedon be created under the name *Andic* epipedon, which applies exclusively to Andosols. The definition can then be altered somewhat to read approximately as follows: Andisols or Andosols are mineral soils that either have an Andic epipedon or have a surface horizon that after mixing to a depth of 18 cm meets all requirements of an andic epipedon. This kind of definition is in conformity with the definition of Mollisols, which includes the requirement of a mollic epipedon (U.S. Soil Survey Staff, 1975).

REFERENCES

Dames, T. W. G., 1955, The Soils of East-Central Java (Indonesia), *Contr. Gen. Agric. Res. Stn., Bogor, Indonesia,* No. 141, 155p.

Dudal, R., 1962, Soil Survey and Its Application in Indonesia, *ETAP (Expanded Program of Technical Assistance) Rept.* **1509,** 71p.

Fitzpatrick, E. A., 1980, *Soils: Their Formation, Classification and Distribution,* Longman, London, 353p.

Ishizuka, Y., and C. A. Black, 1977, *Soils Derived from Volcanic Ash in Japan,* Centro Internacional de Mejoramiento de Maiz y Trigo (CIMMYT), Mexico, 102p.

Ohmasa, M., 1964, Genesis and Morphology of Volcanic Ash Soils, Meeting on the Classification and Correlation of Soils from Volcanic Ash, Tokyo, Japan, June 11-27, 1964, *Food and Agric. Org., United Nations, World Soil Resources Rept.* **14:**56-60.

Tan, K. H., H. F. Perkins, and R. A. McCreery, 1979, Some Observations of the Complexity of Soil Taxonomy, *Jour. Agron. Educ.* **8:**41-44.

Tokudome, S., and I. Kanno, 1965, Nature of the Humus of Humic Allophane Soils in Japan, *Soil Sci. Plant Nutr.* **11**(5):1-8.

Tokudome, S., and I. Kanno, 1968, Nature of the Humus of some Japanese Soils, *Intern. Cong. Soil Sci., 9th, Trans.* (Adelaide, Australia) **3:**163-173.

U.S. Soil Survey Staff, 1975, *Soil Taxonomy: A Basic System of Soil Classification for Making and Interpreting Soil Surveys,* Soil Conservation Service, U.S. Dept. Agric, Agric. Handbook No. 436, 754p.

Van Schuylenborgh, J., 1956, Investigations on the Classification and Genesis of Soils Derived from Acid Tuffs under Humid Tropical Conditions, *Netherlands Jour. Agric. Sci.* **5:**195-210.

Van Schuylenborgh, J., and F. F. F. E. Van Rummelen, 1955, The Genesis and Classification of Mountain Soils Developed on Tuffs in Indonesia, *Netherlands Jour. Agric. Sci.* **3:**192-219.

Wada, K., and S. Aomine, 1973, Soil Development on Volcanic Materials during the Quaternary, *Soil Sci.* **116:**170–177.

Wright, A. C. S., 1964, The "Andosols" or "Humic Allophane" Soils of South America, Meeting on the Classification and Correlation of Soils from Volcanic Ash, Tokyo, Japan, June 11–27, 1964, *Food and Agric. Org., United Nations, World Soil Resources Rept.* **14:**9–21.

Yamada, S., 1977, Distribution and Morphology of Soils Derived from Volcanic Ash in Japan, in *Soils Derived from Volcanic Ash in Japan,* Y. Ishizuka and C. A. Black, eds., Centro Internacional de Mejoramiento de Maiz y Trigo, Mexico, pp. 1–9.

8

Copyright © 1980 by the New Zealand Society of Soil Science
Reprinted from pages 411-426 of *Soils with Variable Charge*, B. K. G. Theng, ed., New Zealand Society of Soil Science, Lower Hutt, N. Z., 1980, 448p.

GENESIS AND CLASSIFICATION OF ANDEPTS AND SPODOSOLS

K. W. FLACH, USDA, SCS, Washington D.C., U.S.A.
C. S. HOLZHEY, USDA, SCS, Lincoln, Nebraska, U.S.A.
F. DE CONINCK, Geologisch Instituut, Rijksuniversiteit Gent, Belgium.
R. J. BARTLETT, University of Vermont, Burlington, Vermont, U.S.A.

I. INTRODUCTION

Concepts of soil genesis are essential for the development of soil classification systems. They allow us to develop taxa of geographically and genetically related soils and to arrange these taxa in a logical framework that improves our perspective about relationships among soils. This perspective, in turn, strengthens concepts of genesis. Because classification systems must be internally consistent, they focus attention toward inconsistencies in the theories on which they are based, thus opening the door for further improvements. The cyclic process of improvement in theory and in classification schemes is operative on an international scale in the improvement of class criteria for Andepts and Spodosols. An understanding of soil genesis is the basis also for the use of classification systems in soil survey. It would be far too expensive to attempt a statistical sample of the soil landscape and the detailed laboratory analysis of all the members of the sample. Rather, the soil surveyors must project the occurrence of soil taxa through genetic models and must verify projections with the identification of a very few, selected pedons.

Chemical and physical properties of both Spodosols and Andepts are determined by chemically and physically highly active *amorphous materials*.[1] The two groups of soils occur predominantly on young parent materials under humid climates. Mostly, they formed under forest vegetation, in many places under conifers, but many Spodosols formed under heath and many Andepts under grassland vegetation.

The primary diagnostic features of Spodosols and Andepts are soil horizons whose active components consist mostly of amorphous materials. The properties of these amorphous materials are incorporated into part of the definition of the spodic horizon, the diagnostic subsurface horizon of Spodosols, and in the "amorphous material dominant in the exchange complex" which is used in the definition of Andepts. A summary of spodic horizons and of "amorphous material dominant in the exchange complex" is given in Tables 1 and 2.

---

[1] The term "amorphous" is used here to indicate absence of long range order that would result in distinct X-ray peaks. Evidence of short range order can, however, be detected by other diagnostic techniques.

Table 1. Diagnostic properties of the spodic horizon.

The spodic horizon has no colour change with increasing depth, or the subhorizon with highest chroma or reddest hue is near the top and the colour change is within 50 cm of the upper boundary.

In addition the spodic horizon meets one or more of the following three requirements. If the temperature regime is frigid or warmer, one of the three must be met at a depth below 12.5 cm. If colder, no depth limit.

1. Has a subhorizon more than 2.5 cm thick that is cemented by organic matter with Fe or Al or both; or
2. Has a sandy or coarse-loamy particle size class and sand grains are covered by cracked coatings or there are distinct dark pellets of coarse-silt size or both; or
3. Meets the following chemical criteria.
   a. If Na pyrophosphate extractable Fe $\geq 0.1\%$,
      then (Na pyrophosphate extractable Fe + Al) $\div$ (measured clay) ≥ 0.2
   b. If Na pyrophosphate extractable Fe $< 0.1\%$,
      then (Na pyrophosphate extractable C + Al) $\div$ (measured clay) ≥ 0.2
   c. (Na pyrophosphate extractable Fe + Al) $\div$ (Na dithionite-citrate extractable Fe + Al) ≥ 0.5
   d. (CEC, pH 8.2) $- \frac{(\text{measured clay} \times \text{thickness, cm})}{2} \geq 65$

Table 2. Diagnostic properties of "Amorphous Material Dominant in the Exchange Complex".

1. CEC, pH 8.2 > 150 meq/100 g measured clay,
   and commonly > 500 meq/100 g measured clay (the higher value because of poor dispersion).
2. If 15-bar water $\geq 20\%$, pH in N NaF > 9.4 at 2 min.
3. (15-bar water)/(measured clay) > 1.
4. Organic C > 0.6%.
5. Differential thermal analysis shows low temperature endotherm.
6. Bulk density of fine earth fraction < 0.85 g/cm^3.

II. GENESIS OF SPODOSOLS

The concept of Spodosols developed from that of "Podzols" of early classification systems. Podzols typically have a light coloured "ashy" eluvial horizon, underlain by a black or reddish illuvial horizon. In contrast to the earlier emphasis on the eluvial horizon, Soil Taxonomy (Soil Survey Staff, 1975) emphasizes the illuvial, spodic horizon (Table 1) as the primary diagnostic feature of Spodosols. Spodosols occur in cold, humid climates usually under conifer or hardwood vegetation on a variety of parent materials. They also occur extensively in warm and tropical humid climates on highly weathered materials, usually quartzitic sands of coastal plains and alluvial lands.

The genesis of spodic horizons may be explained in terms of the formation, migration, and precipitation of organic matter, notably humic substances, complexed with aluminium, and in some cases, with aluminium and iron. The complexing ability and other properties of humic substances, such as pH dependent charge, high specific surface, and water holding capacity are related to the nature and chemical constitution of these materials, discussed more fully in Chapters 12 and 18.

It is generally accepted that humic substances are essentially poly-condensates composed of variable amounts of aromatic, condensed aromatic, and aliphatic structures (e.g. Kononova, 1966; Schnitzer & Khan, 1972; Hayes & Swift, 1978). In humic substances of spodic horizons, there is extensive substitution of these structures by different kinds of functional groups, of which carboxyl (COOH) and phenolic hydroxyl groups are especially important. Thus, 1 gram of fulvic acid contains 4-6 meq COOH, 2-6 meq phenolic OH, about the same amount of alcoholic OH, and 1.5 meq OCH_3. The corresponding values for humic acid are generally slightly lower. Total nitrogen is low and the carbon-to-nitrogen (C/N) ratio is commonly greater than 20 (Righi, 1977; Andreux, 1978; Robin, 1978).

Since, in an aqueous environment, the carboxyl and phenolic hydroxyl groups can dissociate (to give off protons) in the pH range commonly found in soil, and can be solvated by water molecules, humic substances behave as hydrophilic, anionic, colloidal polyelectrolytes, capable of interacting with metal cations.

With some cations (e.g., Na^+, K^+) only weak electrostatic interactions would occur, and a diffuse electrical double layer may form around the polyanions. The resulting double layer repulsion would cause the colloidal particles to disperse and to be transported down the soil profile by percolating wataer. Other cations (e.g. Al^{3+}, Fe^{3+}) tend to form stable coordination/chelation complexes involving the carboxyl and phenolic hydroxyl groups of the polymer.

As the mobile organo-metallic complexes migrate downward, they may become immobilized by one of several mechanisms, (1) complex formation with supplementary cations, (2) arriving at a place where the ionic concentration of the soil solution is higher or its acidity is different as compared with the initial solution, (3) partial desiccation of the system leading to a contraction of diffuse double layers, and (4) inter-molecular coagulation (precipitation) such as may occur when multivalent cations form links between functional groups belonging to different polymeric chains.

When immobilization occurs, part of the hydration water is trapped within the polymer network and the system assumes a gel-like consistence in which van der Waals and hydrogen bonding interactions between polymer molecules can occur. Further loss of water would create voids in the gel structure and this is partly responsible for the high specific surface of the system.

Because the solubility of the mobile organics is reduced as the metal complexes increase (Wright & Schnitzer, 1963; Ong & Bisque, 1968; McKeague et al., 1971), the sesquioxide-depleted eluvial horizon in a Spodosol may be thick, thin, or non-existent, depending on the amounts of sesquioxides encountered by the leaching organic materials. In quartzitic sands the eluvial horizon may be several metres thick while in easily weatherable basic parent materials the spodic horizon may develop in place without an eluvial horizon. If there is no eluvial horizon, the Spodosol may resemble an Andept in morphology.

The illuviated materials in spodic horizons may be in the form of continuous or fractured coatings or of numerous "fluffy" aggregates. The coatings may precipitate with little mixing, forming coatings and bridges of a plasma made up of organic matter and Al, or Al and Fe. These features, covering other soil components and having a polygonal cracking pattern, have been termed "monomorphic coatings" (De Coninck et al., 1974) or "organans" (Brewer, 1974). Monomorphic coatings are always strongly developed in ortstein horizons, suggesting that they are the cementing agents in these horizons. The cracking pattern and the high amount of pores of $< 0.2\ \mu m$ diameter (cf. Chapter 16) indicate that the coatings are deposited in a hydrated or gel state, followed by desiccation and cracking. A high proportion of the organic matter, Al, and Fe are extractable by solutions of sodium pyrophosphate, tetraborate, and dithionite-citrate. The organic matter so extracted is rich in fulvic acids with a high content of COOH and phenolic OH groups (Righi, 1977; Robin, 1978; Higashi et al., 1981). Because the morphology of coatings suggests a direct process of leaching and deposition, their elemental composition should reflect the elemental composition at the time of deposition. Fig. 1 shows the distribution of Fe, Al, and Si in a section of an iron-poor spodic horizon. The aluminium is concentrated in the coatings but the silicon is only in the sand and silt grains. Silicon migrates in acid leaching environments, but it is not precipitated with the organic compounds and aluminium in the coatings (De Coninck et al., 1974; Righi & De Coninck, 1974).

The "fluffy" aggregates, in contrast to the relatively undisturbed coatings, may occur in horizons of greater biological activity in which skeletal components of the soil fabric have been physically forced apart, creating a porous structure of generally low bulk density. The aggregates occur with pedotubules (Brewer, 1964) and pellets. The aggregates may be influenced by or formed during physical mixing, although both they and the pellets resemble faecal matter from small animals (Zachariae, 1965). An alternative explanation of origin invokes precipitation around clay and silt nuclei (Flach, 1960). Clay- and silt-size particles are mixed with the aggregates (De Coninck et al., 1974; Righi, 1977; Robin & De Coninck, 1978). The pedotubules, pellets, and aggregates contain polymorphic organic matter and untransformed plant remains. Determination of organic matter, Al, and Fe extractable by sodium pyrophosphate, tetraborate, and dithionite-citrate, gives lower values for organic matter and often lower for Al than in the monomorphic coatings; the organic matter has a more complicated structure, and a lower COOH and phenolic OH content (Righi, 1977; Robin, 1978; Higashi et al., 1980).

Fig. 2 shows a series of microprobe tracings from a spodic horizon having an aggregate fabric. The Fe and Al are part of the plasma, whereas silica is present only as part of the primary mineral grains. The grains in this case are distributed throughout the fabric. The interpretation is either a cycle of formation and mixing or of precipitation around mineral grains. This fabric can occur beneath an A2 horizon which implies translocation from one horizon to the other. It also occurs very near the surface in materials supplying such abundant aluminium or iron and aluminium (De Coninck et al., 1974; Righi & De Coninck, 1974) that immobilization occurs with very little downward movement. The aggregate fabric contains some fresh plant remains that resist dissolution in complexing agents whereas the fabric containing fractured coatings is almost completely extractable by sodium pyrophosphate.

Using the ^{14}C method, De Coninck (1980) has measured the mean residence time (MRT) of the organic matter in the two kinds of spodic horizon, present in the same horizon sequence, and, in some places, only at a distance of a few cm from each other. The results show that the MRT of polymorphic organic matter is always smaller than the monomorphic type. This indicates a faster turnover and thus a much greater biological activity in horizons with polymorphic organic matter, in line with electron microscope data which suggest that the formation of polymorphic units is biologically mediated. The Cryohumod of Table 3 is likely to have an aggregate fabric whereas the siliceous Typic Haplaquod of Table 6 is likely to have the more uniform fabric in which organics precipitate as a gel and form fractured coatings as the spodic horizon dries. Mean residence time of carbon in the Haplaquod is probably in the thousands or tens of thousands of years (Holzhey et al., 1975).

Spodosols such as Humods with abundant organic matter, easily weatherable minerals, and high biological activity resemble Andepts in physical as well as chemical properties. An example is the Typic Cryohumod (Table 3) described later in this paper.

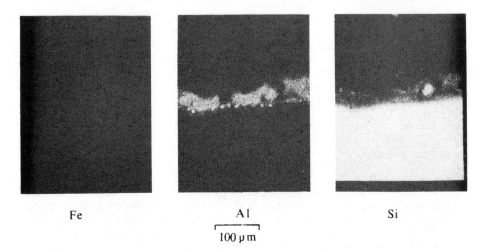

Fe   Al   Si

100 μm

Fig. 1. Microprobe pictures showing Fe, Al, and Si distribution in section cut across a quartz grain coated with a thin clay layer, a small silicate grain and a cracked monomorphic coating of about 10 μm thickness, in a cemented B22h horizon of an Aquod, Antwerp Campine, Belgium.

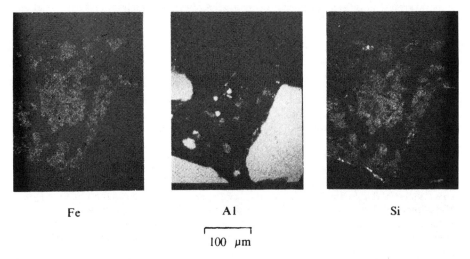

Fe   Al   Si

100 μm

Fig 2. Microprobe pictures showing Fe, Al, and Si distribution in section cut across a polymorphic aggregate in a friable B22ir horizon of a Humod, Antwerp Campine, Belgium.

III. GENESIS OF ANDEPTS

The concepts of Andepts evolved from that of Ando or Kuroboku which are soils with a deep, dark "fluffy" A horizon formed on volcanic ash under a humid climate. The definition of Andepts is much broader than that of Ando and includes soils developed in volcanic ash as well as other parent materials that contain an abundance of easily weatherable materials. Some Andepts in recent volcanic ash have undergone no soil development except organic matter accumulation to form a mollic or umbric epipedon; some are old enough to have a cambic subsurface horizon, low bulk density and characteristics associated with "amorphous material dominant in the exchange complex" (Table 2). Andepts may have formed under a variety of climates ranging from cold humid to warm and tropical humid, and in climates with pronounced dry seasons; the xeric moisture regimes of Mediterranean climates and the ustic moisture regimes of subhumid and semiarid climates. Commensurate with the large range in climates, Andepts occur under many kinds of vegetation and may or may not have a dark surface horizon, and have common boundaries with other soils under a variety of conditions. In general, the Andepts are restricted to the humid, and to the wetter reaches of the xeric and ustic moisture regimes. However, there is not always a sharp boundary between conditions that produce Andepts and conditions that produce other kinds of soils.

Studies on the genesis of Andepts have concentrated on the weathering of volcanic ash to allophane and on the formation of humus-allophane or humus-aluminium complexes.

Given a humid climate, weathering is rapid in volcanic ash. Allophane, a coprecipitate of Al and Si oxides, forms in the B horizons of Andepts, or in buried A horizons where Al-binding humic compounds are less dominant than in the surface horizons (Wada & Higashi, 1976; Wada, 1977; Mizota, 1978). As weathering and soil development proceed, more free silica may be added to allophane, forming halloysite, or possibly other crystalline clay minerals (Wada, 1977). Or, if the environment favours loss of silica, gibbsite may form from silicate minerals. Imogolite is a paracrystalline mineral containing less SiO_2 than allophane. Wada & Harward (1974), suggest that imogolite represents an intermediate phase in the desilication of allophane to form gibbsite in a strongly leaching environment.

Aluminium, released by weathering, forms stable complexes with organic matter near the soil surface. G. G. S. Holmgren (USDA, SCS, Lincoln, Nebraska, pers. comm.) emphasizes that the humic acids are dominantly of larger molecular weight as opposed to the fulvic and small humic acid components of the spodic horizon.

Aluminium released from volcanic ash is retained by humus in the surface horizon where organic residues are most plentiful (Wada & Higashi, 1976). The complexes are analogous to those formed in a spodic horizon, except that in Andepts the availability of humic material may limit the rate of formation while in spodic horizons the availability of sesquioxide may be rate limiting. By suppressing the activity of aluminium the humus makes possible the formation of opaline SiO_2 and retards formation of amorphous alumino-silicate minerals such as allophane (Takashiki & Wada, 1975).

Periodic desiccation may irreversibly affect the chemical and physical properties of the surface horizon of Andepts (Wada, 1977). Drying increases surface acidity and affects bonding between mineral and organic surfaces. The micro-morphology of subsurface horizons of Andepts is similar to that of spodic horizons with a loose polymorphic fabric (Kawai, 1964). This may reflect the activity of soil micro-organisms or local movement of dispersed organic material.

IV. SPODOSOLS AND ANDEPTS WITH SIMILAR MORPHOLOGICAL AND CHEMICAL PROPERTIES

There is little difficulty in distinguishing pedons representing typical Spodosols from those of typical Andepts and in distinguishing the soil material of typical spodic horizons from that of B horizons of Andepts. Salient features of the differences between the two groups of soils have been discussed before and are summarized in Tables 1 and 2. Flach (1972) has enumerated a number of additional differences between spodic horizons and subsurface horizons of soils with "amorphous material dominant in the exchange complex." The latter generally have a pH (1:1, water) above 5.0 even if devoid of exchangeable bases, whereas the spodic horizons (of Spodosols) have lower pH and higher exchangeable Al (KCl extractable). Sodium pyrophosphate at pH 10 extracts a higher proportion of the iron and aluminium in the amorphous materials of Spodosols than in Andepts, and the CEC may increase in the Andepts on sodium dithionite-citrate treatment, whereas it decreases in the Spodosol (Flach, 1972; Franzmeier *et al.*, 1965). Although the Andepts generally have little exchangeable aluminium, they absorb large amounts of fluoride from NaF. Soils of both taxa may have high pH in NaF, although many Spodosols do not. Difficulties arise when one tries to classify soils that formed under very high precipitation on volcanic ash or on parent materials from basic igneous rocks or related sedimentary rocks, where both Spodosols and Andepts may lack a detectable eluvial horizon. The spodic horizon may also meet the chemical criteria for "amorphous material dominant in the exchange complex," and the Andept B horizon may meet the chemical criteria of the spodic horizon. In these environments, the genesis of the two groups of soils is undoubtedly very similar.

Table 3 shows data for two Spodosols of cold and moist climates that were formed on different parent materials but have similar properties. Both resemble Andepts in their physical and chemical properties. They are not Andepts by current criteria because of rather minor differences in colour and bulk density. Undoubtedly pedons that meet all criteria of Andepts could be found nearby. The Humic Cryorthod is formed in volcanic ash beneath a sparse spruce-birch-grass cover. Considerable organic matter has accumulated but the rate of weathering is slow. The soil has a distinct A2-B2 horizon sequence. In the spodic horizon more than half of the sesquioxides extracted by sodium dithionite-citrate are also extracted by sodium pyrophosphate at pH 10. The pH (water) is low and the few bases are mainly in the upper horizons. The pH (NaF) in the B and C horizons is in the same magnitude as in Andepts. Large amounts of silt-size aggregates of complexes contribute to the percent water retained at 15 bar tension. These aggregates resist destruction by oxidizing agents, and resist dispersion, causing clay percentages measured in routine analyses to be lower than percent water retained at 15 bar tension. Below the A2 horizon, the cation exchange capacity decreases much less with depth than does the organic carbon, suggesting a more active kind of organic matter in lower horizons.

The Cryohumod formed under coniferous forests in feldspar-rich glacial till. This soil lacks the A2 horizon of the Alaskan Cryorthod. The A2 horizon may be absent partly because of high biological activity and partly because of the abundance of minerals that readily release aluminium and iron to precipitate organic matter. The iron and aluminium data suggest, however, that some movement of sesquioxides has occurred. The pH (NaF) is greater than 11 in the A and B21h horizons. Sodium pyrophosphate at pH 10 extracts as much Fe and more Al than does sodium dithionite-citrate, which suggests that sesquioxides are associated mostly with the type of organo-sesquioxide complex typical of spodic horizons. We would expect this biologically active soil to have the polymorphic micro-structure already discussed. Morphologically the profile resembles that of an

Table 3. Physical and chemical properties of two Spodosols*.

Horizon	Depth (cm)	Colour (Munsell)	Bulk Density (g/cm³)	Dith.-Cit. Ext. Fe (%)	Dith.-Cit. Ext. Al (%)	Pyro. Ext. Fe (%)	Pyro. Ext. Al (%)	pH NaF	pH H$_2$O	Organic C (%)	Clay (%)	Water Retained 15-bar (%)	Sum Bases	KCl ext. Al	Ext. Acidity	CEC NH$_4$OAc	Sum Cations (Meq/100 g)
\multicolumn{18}{l}{Humic Cryorthod, Medial——Volcanic Ash, Alaska**}
O,1	13-0	5YR 2/2		0.4	0.5	0.3	0.1	8.3	4.7	45		144	16	3.4	113	111	129
A2	0-4	10YR 4/2	0.56	0.8	0.2	0.3	0.1	7.0	4.5	9.3	14	25	3.1	3.2	32	27	35
B21h	4-10	2.5YR 2/2	0.57	2.4	1.5	1.4	1.1	9.5	4.4	10	11	30	1.3	7.6	80	51	81
B22h	10-18	2.5YR 2/4	0.60	2.2		1.3	0.5	11.0	4.4	8.1		31	0.5	3.1	71	42	71
B31	18-28	10YR 4/4	1.16	1.3		0.3	0.4	10.6	5.0	1.6	5	11	0.6	1.4	24	14	25
B32	28-66	10YR 4/3	1.03	1.5	1.2	0.4	0.2	10.7	5.0	1.6		12	0.5	1.6	24	14	24
C	66-91	5Y 5/1	1.30	0.6		0.1	0.3	10.1	5.6	0.50	3	4	0.6	1.2	11	6	11
\multicolumn{18}{l}{Typic Cryohumod, Mixed——Felsic Glacial Till, New York State***}
A1	0-2	5YR 2/1	0.43	0.2	1.3	1.1	1.6	11.2	4.2	19	3	54	51	6.2	73	54	78
B21h	2-13	5YR 2/2	0.46	1.7	1.8	1.7	2.4	11.6	4.1	15	7	36	1.6	6.5	81	40	83
B22h	13-33	5YR 2/2	0.54	1.8	2.2	1.9	3.2		4.5	12	4	29	1.3	5.1	84	39	85
IIB23ir	33-61	7.5YR 3/2	1.53	1.0	0.8	0.5	1.0	11.3	4.8	2.8	2	8	0.4	2.3	26	12	26
IIC1	61-74	10YR 4/2	1.72	0.7	0.6	0.1	0.6		5.2	1.2	1	6	0.2	1.2	16	7	17

* Methods listed, Table 8.
** Cryorthod, S67Ak-58-1, sampled near Nushagak Bay in volcanic ash with possible loess influence. Situated on nearly level portion of a moraine. Veg. sparse spruce, birch forest with grass understorey. Mean annual T. 1°C, mean July T. near 13°C. Mean annual ppt. 650 mm, distributed evenly through the seasons.
*** Cryohumod, S73NY-16-5, sampled in Adirondack Mountains in sandy glacial till from syenite gneiss. Veg. luxuriant pine plantation. Mean annual air T. approx. 5°C; mean July T. approx. 12°C. Mean annual ppt. approx. 980 mm, uniformly distributed.

Andept. The subsurface horizon is a spodic horizon because its colour remains constant with depth. Otherwise the soil meets all the diagnostic criteria of a Cryandept except for the bulk density of the IIB23 horizon.

A brown Podzolic soil with significant accumulation of allophane or allophane-like material is reported by Loveland & Bullock (1976) on basic igneous rocks of Great Britain.

In Table 4, data from two Andepts that share some soil properties with Spodosols are summarized. The Hydrandept from Hawaii is an example of weathering of volcanic ash under an extremely humid, warm climate. Weathering has removed so much silica from the soil that much of the aluminium soluble in 0.5 N NaOH (Hashimoto & Jackson, 1960) must be in forms other than allophane. Sodium hydroxide removed only 7.5% Al_2O_3 and 3.6% SiO_2 in the B22 horizon. Assuming a SiO_2/Al_2O_3 ratio of 2.0 for allophane, only about half of the aluminium can be in the form of allophane. The rest is most likely associated with organic matter. Particle size distribution was not measured in this pedon but similar pedons yield less than 10% clay.

The presence of aluminium that is not in the form of allophane, the large amounts of aluminium recovered in the sodium pyrophosphate extract, and the loss of CEC upon sodium dithionite-citrate treatment (from 114 meq to 70.2 meq/100 g at pH 8.2) (Flach, 1972) suggest similarities to Spodosols. The Ap horizon does, in fact, meet the chemical criteria of the spodic horizon. Yet, the soil resembles other Andepts in containing only little exchangeable (KCl extractable) aluminium in spite of its very low base saturation. In a field test for identifying spodic horizons (Holmgren pers. comm.) this soil behaved in a way that suggests the presence of more fulvic acid than is typical for either Spodosols or Andepts.

The other Andept with properties resembling those of a Spodosol (Table 4) is a Typic Cryandept in Alaska. This soil formed in a cool, humid climate under a northern shrub-grass cover. The profile consists of several slightly weathered ash layers that buried a weakly developed soil profile. The buried A11 horizon (A11b) contains enough iron and aluminium, extractable by sodium pyrophosphate, to meet the criteria of the spodic horizon; also, it has a low pH and contains more exchangeable (KCl extractable) aluminium than is typical of Andepts. It lacks, however, the colour characteristics of a spodic horizon.

V. THE CLASSIFICATION OF ANDEPTS AND SPODOSOLS

A. Spodosols

All soils that have a spodic horizon or a placic horizon are classified in the order Spodosols. At the suborder level (Table 5) the wet Spodosols (having an aquic moisture regime), are recognized as Aquods; the remaining Spodosols are separated into Ferrods, Orthods, and Humods based on the ratio of free iron to total organic carbon in the spodic horizon.

Suborders of Spodosols are further subdivided at the great group level on the basis of soil temperature regimes, the presence or absence of pan horizons, and the ratio of free iron to carbon.

Taxonomic groupings of Spodosols at the suborder and the great group level have proved largely satisfactory, except that Soil Taxonomy does not provide for a clear separation of Aquods with appreciable amounts of aluminium in the spodic horizon from Aquods containing only traces of aluminium. The distinction is important. Spodic horizons containing only traces of aluminium, extractable by sodium pyrophosphate, also contain little extractable by KCl, and do not react strongly with NaF. Similar Aquods that

Table 4. Physical and chemical properties of two Andepts*.

Horizon	Depth (cm)	Colour (Munsell)	Bulk Density (g/cm³)	Dith-Cit. Ext. Fe (%)	Dith-Cit. Ext. Al (%)	Pyro Ext. Fe (%)	Pyro Ext. Al (%)	pH 10 NaF	pH H$_2$O	Organic C (%)	Clay (%)	Water Retained 15-bar (%)	Sum Bases	KCl Ext. Al	Ext. Acidity	CEC NH$_4$OAc	Sum Cations
																(Meq/100 g)	
				Typic Hydrandept, Thixotropic, Isothermic — Volcanic Ash, Hawaii**													
Ap	0-18	7.5YR 3/2	0.51	14	5.6	13	4.3	10.6	5.4	12		102	1.3	0.5	19		53
B21	18-35	7.5YR 3/4	0.33	14				11.6	5.2	6.6		154	1.5	0.6	32		34
B22	35-50	10YR 3/3	0.30	12	7.1	5.9	2.9	11.6	5.4	9.4		167	1.1	0.3	26		39
B23	50-63	10YR 3/3	0.30	12				11.6	5.4	8.5		196	0.4	0.2	23		30
B24	63-70	7.5YR 3/3	0.27	12				11.7	5.5	7.8		193	1.5	0.1			25
B25	70-80	10YR 2/1	0.30					11.7	5.6	7.0		188	0.8	tr			24
IIB26	80-85	10YR 3/2	0.27	9.9								166	0.7				23
IVB28	98-113	2.5YR 3/4	0.24	12 / 17					5.7	5.1		211	1.0	tr			21
				Typic Cryandept, Ashy — Volcanic Ash, Alaska***													
A11	0-13	10YR 3/2	0.76	0.9	0.4	0.4	0.3	10.1	4.6	3.6	4	12	2.6	1.1	19	13	22
A12	13-23	7.5YR 3/2	0.69	1.7				11.2	4.9	3.7		19	2.1	0.8	32	19	34
A11b	23-33	5YR 2/2	0.79	1.8	0.9	0.6	0.5	11.2	5.3	3.3	7	20	1.1	0.3	26	16	27
A12b	33-43	7.5YR 3/2	0.76	1.8				10.9	5.7	2.2		21	2.2	0.2	23	14	25
IIC1	43-56	10YR 2/1		1.3	0.4	0.2	0.2	9.6	5.6	0.80	3	8	2.1	0.1	11	7	13

* Methods listed, Table 8.
** Hydrandept, S62Ha-1-1, sampled on windward slopes of Mauna Kea Mountain in volcanic ash. Pastureland of grasses, forbs, fern and treefern. Mean annual ppt. 3000-3750 mm. Soil T. at 50 cm 19-21°C.
*** Cryandept, S67AK-56-2, sampled on Alaskan Peninsula, on a steep SW slope in volcanic ash over basaltic glacial till. Veg. is cover of grasses, forbs with scattered shrubs (alder). Mean annual air T. 4°C; mean August T. 12°C. Mean annual ppt. 85-140 mm. evenly distributed.

Table 5. Classification of Spodosols.

Aquods – Aquic moisture regime	Fragiaquods	– fragipan
	Cryaquods	– cryic temperature regime
	Duraquods	– cemented albic horizon
	Placaquods	– placic horizon
	Tropaquods	– isomesic or warmer iso regime
	Haplaquods	– Fe/C < 0.2
	Sideraquods	– Fe/C ⩾ 0.2
Ferrods – Fe/C ⩾ 6	Provisional	– No subdivisions
Humods – Fe/C < 0.2	Placohumods	– placic horizon
	Tropohumods	– isomesic or warmer iso regime
	Fragihumods	– fragipan
	Cryohumods	– cryic temperature regime
	Haplohumods	– other
Orthods – other	Placorthods	– placic horizon
	Fragiorthods	– fragipan
	Cryorthods	– cryic or pergelic temperature regime
	Troporthods	– isomesic or warmer iso regime
	Haplorthods	– other

contain little aluminium have been found to have low phosphate fixing capacity (Holmgren pers. comm.). Aquods having little aluminium are probably the dominant form in siliceous coastal plain material in warm temperate and tropical climates, whereas Aquods rich in aluminium are probably the dominant form in less highly weathered materials of cool, humid climates. Aquods in warm and hot, humid climates are probably the most extensive members of the Spodosols. Emphasis on aluminium rather than iron may also be desirable in defining great groups in the suborder Humods.

Table 6 contains data for the two kinds of Aquods. The pedon with mixed mineralogy comes from the extreme northeastern United States. It contains an abundant amount of weatherable minerals and formed in a cool, humid climate. The pedon with siliceous mineralogy comes from the coastal plain of the southeastern United States; it formed under pine and palmetto vegetation on almost pure quartz sands. The spodic horizon in both soils is within the water table for at least part of each year. The theories on the genesis of Spodosols involving interactions between aluminium and soluble organic materials, discussed earlier in this paper, may not apply to Aquods developed in highly quartzitic sands. It is likely that physical processes and soil properties that affect water movement, and the presence of fluctuating water tables, strongly influence the formation of these soils.

B. Andepts and Andaquepts

Andepts and Andaquepts either have properties that reflect "domination by amorphous materials" (Table 2) or they contain 60% or more (by weight) vitric volcanic ash, cinders, or other vitric pyroclastic material. To conform to the classification of other Inceptisols, soils meeting these criteria are recognized at the suborder level as Andepts if they do not have an aquic moisture regime and at the great group level as Andaquepts if they do. Temperature regimes, the presence or absence of pan horizons, CEC, base saturation and changes in the soil fabric upon dehydration are used to define great groups within the suborder Andepts (Table 7).

Table 6. Physical and chemical properties of two Aquods*.

Horizon	Depth (cm)	Colour (Munsell)	Bulk Density (g/cm³)	Dith-Cit. Ext. Fe Al (%)		Pyro. Ext. Fe Al (%)		pH10	pH NaF	pH H₂O	Organic C (%)	Clay (%)	Water Retained 15-bar (%)	Sum Bases	Cation Exchange Properties (Meq/100 g)			
															KCl Ext. Al	Ext. Acidity	CEC NH₄OAc	Sum Cations
\multicolumn{19}{c}{Typic Haplaquod, Sandy, Mixed, Frigid -- Mixed Glacial Till, Maine**}
O21	10-5	2.5YR		0.4	0.3	0.6	0.3	0.5		4.4	37	0.4	119	11	13	95	114	106
O22	5-0	N 2/0		0.3	0.1	0.4	0.3	0.4		4.8	20	1	40	3.5	11	70	55	73
A2	0-10	7.5YR 6/2		tr	0.1	0.1	0.1	0.1	7.9	4.9	1.7	1	4	0.3	2.7	9.1	6.6	9.4
B21hir	10-24	10R 2.5/1		0.2	0.8	0.8	0.7	0.7	11.3	4.9	3.0	0.4	7	0.3	3.8	31	19	31
B22ir	24-30			0.2	0.8	0.6	0.6	0.6	11.3	4.9	2.0	0.4	5	0.3	2.5	23	12	23
B23	30-36	5YR 3/3		0.2	0.6	0.6	0.1	0.6	11.0	5.0	1.2	0.4	4	0.2	1.8	17	8.4	17
B3	36-43	5YR 3/3		0.2	0.5	0.5	—	0.4	11.0	5.1	1.1	0.4	3	0.2	1.4	13	6.3	13
C1	43-70	10YR 5/3		0.1	0.1	0.1	—	0.2	10.0	5.1	0.3	0.4	1	0.2	0.4	2.5	2.0	2.7
C2	70-95	5YR 5/3		0.2	0.1	0.1	—	0.2		5.1	0.2	0.4		0.1	0.3	2.1	1.5	2.2
\multicolumn{19}{c}{Typic Haplaquod, Sandy, Siliceous, Hyperthermic -- Coastal Plain Sand, Florida***}
A11	0-6	N 2/0	1.54	tr	tr	tr	tr	tr	7.3	5.0	2.2	1.1	4	1.5	0.5	5.9	4.4	7.4
A12	6-21	10YR 2/1	1.54	—	tr	—	tr	tr	7.5	4.4	0.7	0.4	1	0.2	0.5	3.5	1.7	3.7
A21	21-35	10YR 5/1	1.66	—	tr	tr	tr	tr	7.4	4.6	0.1	—	0.7	0.1	0.2	0.4	0.5	0.5
A22	35-66	N 7/0	1.68	—	—	—	tr	tr	7.4	5.0	0.1	—	0.7	0.1	0.1	—	0.5	0.1
B21h	66-79	5YR 2/1	1.72	tr	tr	tr	tr	tr	7.3	5.0	1.8	0.4	2	3.8	0.2	7.9	6.4	11.7
B22h	79-99	5YR 3/2	1.64	tr	0.1	tr	0.1	tr	7.7	5.1	1.9	2.8	2	4.3	0.2	8.1	6.3	12.4
B3	99-145	10YR 4/3		0.1	0.2	0.2	0.2	0.1	7.6	5.2	0.4	3.0	1	1.2	0.2	2.1	1.6	3.3
C	145-175	10YR 6/3	1.62	0.1	0.1	0.1	0.2	0.1	7.5	5.6	0.3	5.4	2	1.8	0.1	0.8	1.7	2.6

* Methods listed, Table 8.
** Maine pedon, S78 ME-017-1, formed in sandy glacial outwash with abundant weatherable minerals. Veg. is forest including pine, fir, and maple, with understorey of ferns and sphagnum. Mean ann. air T. approx. 5°C; mean July T. approx. 16°C. Mean ann. ppt. approx. 1000 mm; monthly ppt. greatest in summer.
*** Florida pedon S71F1a-50-1, formed in siliceous coastal plain sands. Veg. is forest including pine, palmetta, shrubs, thin grass understorey. Mean ann. air T. approx. 23°C; mean July T. approx. 28°C. Mean ann. ppt. approx. 1600 mm, monthly ppt. greatest in summer.

Table 7. Classification of Inceptisols dominated by amorphous or vitric pyroclastic materials.

Andepts	Cryandepts	– cryic or pergelic temperature regime
	Durandepts	– duripan
	Hydrandepts	– dehydrate irreversibly
	Placandepts	– placic horizon
	Vitrandepts	– vitric volcanics dominant, low surface area
	Eutrandepts	– high base status
	Dystrandepts	– low base status
Aquepts all Inceptisols with Aquic moisture regime	Andaquepts	– all Aquepts with properties reflecting dominance of amorphous material

Table 8. Methods*.

1. Na dithionite-citrate extraction of Fe and Al––6C2b and 6G7a (Holmgren, 1967)
2. Na pyrophosphate, pH 10, extraction of Fe and Al––6C5a and 6G5a (modified from Bascomb, 1968)
3. pH in NaF, soil: N NaF = 1:50, 2 min stirring––8C1d (Fieldes and Perrott, 1966)
4. pH in water, soil: water = 1:1 except in samples of high water retention, where saturated paste - 8C1a, 8C1b
5. Organic carbon by acid dichromate digestion––6A1 (Peech et al., 1947; Walkley, 1935)
6. Clay by pipette after H_2O_2 and Na hexametaphosphate––3A1 (Kilmer & Alexander, 1949)
7. 15-bar water retention, crushed fabric––8D1
8. Basic cation extraction by NH_4OAc, pH 7–– 6H2, 6O2, 6P2, 6Q2
9. Al extracted by N KCl––6G1
10. Extractable acidity, triethanolamine buffer at pH 8.2––6H2 (Peech, 1947)
11. CEC by NH_4 retained after treatment by NH_4OAc, pH 7––5A8a, 5A6a
12. CEC by sum of cations (bases + acidity)––5A3a (Peech, 1947)
13. Bulk density, moist––4A1

* All methods codes identify standard methods in Soil Survey Investigations Report No 1 (Soil Survey Laboratory Staff, 1972).

In contrast to Spodosols, Andepts of intertropical areas are not separated from those of temperate areas at a high categorical level. This reflects the fact that most Andepts occur in the tropics and in adjacent warm humid climates and that they are less common in temperate areas. The treatment of Andepts and Andaquepts in Soil Taxonomy (Soil Survey Staff, 1975) is not fully satisfactory. Concepts had been developed and tested with a relatively small number of Andepts primarily in Alaska, the extreme western United States, and Hawaii. Also, the placement of Andepts as a suborder of Inceptisols made it difficult to incorporate important features into the framework of Soil Taxonomy. Therefore, G. D. Smith in 1978 made a proposal to restructure the classification of Andepts and Andaquepts based on his experience in Central and South America, and in New Zealand. The proposal is being currently reviewed and tested by an international committee under the chairmanship of M. L. Leamy (DSIR, New Zealand). Major suggestions in the proposal are as follows:

(a) Redefinition of "domination by amorphous material." The proposal would add a measure of phosphate retention and of variable charge; it would drop the measure of dispersibility of clay, and criteria based on differential thermal analysis (DTA) of the current definition,

424 Flach et al.

(b) Create an order of Andisols and of suborders based primarily on moisture and temperature regimes. Subdivisions at the great group level would separate Andisols with relatively small, medium, and large amounts of amorphous material based on their water retention at 15 bar tension; other subdivisions would be based on the presence of pans, and the colour, and organic matter content, of the surface horizon. The proposal is being reviewed thoroughly by soil scientists throughout the areas of occurrence of Andisols and, if accepted, will be incorporated in the next major revision of Soil Taxonomy.

C. Classification at the Family Level

Soil Taxonomy uses soil temperature, mineralogy, and particle size distribution as primary criteria for grouping soils at the family level. In some Spodosols and in most Andepts, however, particle size is not a meaningful concept and apparent differences in important physical and chemical properties are associated with differences in mineralogy. Class criteria that replace particle size classes and mineralogy classes were, therefore, developed for all Andepts (and Andaquepts) and for those Spodosols that have cryic temperature regimes. The classes were designed to reflect differences in the amount and nature of pyroclastic materials in weakly weathered Andepts and differences in tactile characteristics ("feel") and the degree of expression of thixotropic properties in more highly developed Andepts and Spodosols. The proposal for a revised classification of Andisols retains the basic approach that has been used in the past, except that the term "hydrous" would be substituted for thixotropic, and medial and hydrous classes would be defined in terms of 15 bar water retention of fresh and dried soil materials.

VI. SUMMARY

Spodosols and Andepts are important kinds of soils in many parts of the world. Some Spodosols differ drastically from most Andepts but Spodosols formed on basic and easily weatherable parent materials resemble Andepts in many properties. In turn, Andepts formed in extremely humid and in cold climates resemble some Spodosols. Amorphous sesquioxide-organic matter complexes and aluminium-silica material (allophane) largely determine the physical and chemical properties of spodic horizons and of most Andepts, respectively.

Soil forming processes in both groups of soils are similar and are strongly related to kinds of organic matter and to the interactions between organic matter and sesquioxides.

Much is yet to be learned about the genesis of these soils, especially Spodosols of warm, humid climates which have unique properties and have not been studied extensively. Likewise, the genesis, and the physical and chemical properties of Andepts that formed in cold and in extremely wet climates are as yet incompletely understood. Despite their morphological resemblance to other Andepts, these soils may be better classified as Spodosols on the basis of their chemical properties. Much progress has been made in the classification of these soils but new concepts of genesis and a better understanding of important soil properties must be reflected in the revision of the classification system.

VII. ACKNOWLEDGMENTS

The authors gratefully acknowledge major editorial assistance by B. K. G. Theng, Soil Bureau, Lower Hutt, New Zealand.

VIII. REFERENCES

Andreux, F. 1978. Etude des Etapes initiales de la Stabilisation Physicochimique et Biologique d'Acides Humiques Modèles. These de doctorat. Université de Nancy, 174 pp.

Bascomb, C. L. 1968. Distribution of pyrophosphate-extractable iron and organic carbon in soils of various groups. J. Soil Sci., 19: 251-68.

Brewer, R. 1964. Fabric and Mineral Analysis of Soils. John Wiley & Sons, New York-London-Sydney, 470 pp.

Brewer, R. 1974. Some considerations concerning micromorphological terminology. In: G. K. Rutherford (Editor), Soil Microscopy. Limestone Press, Kingston, Ontario, pp. 28-40.

De Coninck, F. 1980. Major mechanisms in formation of spodic horizons. Geoderma, 24: 101-28.

De Coninck, F., Righi, D., Maucorps, J., Robin A. M. 1974. Origin and micro-morphological nomenclature of organic matter in sandy Spodosols. In: G. K. Rutherford (Editor), Soil Microscopy. Limestone Press, Kingston, Ontario, pp. 263-80.

Fieldes, M., Perrott K. W. 1966. The nature of allophane in soils. III: Rapid field and laboratory tests for allophane. N.Z. J. Sci., 9: 623-9.

Flach, K. W. 1960. Sols Bruns Acides in the Northeastern United States. Genesis, Morphology, and Relationships to Associated Soils. Ph.D Thesis. Cornell University, Ithaca, N.Y.

Flach, K. W. 1972. The differentiation of the cambic horizon of Andepts from the spodic horizon. pp. 127-38. Second Panel on Volcanic Ash Soils in Latin American, Pasto, Colombia.

Franzmeier, D. P., Hajek, B. F., Simonson, C. H. 1965. Use of amorphous material to identify spodic horizons. Soil Sci. Soc. Am. Proc., 29: 737-43.

Hashimoto, I., Jackson, M. L. 1960. Rapid dissolution of allophane and kaolinite-halloysite after dehydration. Clays Clay Min., 7: 102-13.

Hayes, M. H. B., Swift, R. S. 1978. The chemistry of soil organic colloids. In: D. J. Greenland and M. H. B. Hayes (Editors), The Chemistry of Soil Constituents. John Wiley & Sons, Chichester, pp. 179-320.

Higashi, T., De Coninck, F., Gelaude, F. 1981. Characterization of some spodic horizons of the Campine (Belgium) with dithionite-citrate, pyrophosphate and sodiumhydroxide-tetraborate. Geoderma, in press.

Holmgren, G. G. S. 1967. A rapid citrate-dithionite extractable iron procedure. Soil Sci. Soc. Am. Proc., 31: 210-1.

Holzhey, C. S., Daniels, R. B., Gamble, E. E. 1975. Thick Bh horizons in the North Carolina Coastal Plain: II: Physical and chemical properties and rates of organic additions from surface sources. Soil Sci. Soc. Am. Proc., 39: 1182-7.

Kawai, K. 1964. Micro soil-structure. In: Volcanic Ash Soils in Japan. Ministry of Agriculture and Forestry, Japan. Chap. IV, p. 74.

Kilmer, V. J., Alexander, L. T. 1949. Methods of making mechanical analyses of soils. Soil Sci., 68: 15-24.

Kononova, M. M. 1966. Soil Organic Matter, 2nd Edition. Pergamon Press, Oxford, 544 pp.

Loveland, P. J., Bullock, P. 1976. Chemical and mineralogical properties of brown podzolic soils in comparison with soils of other groups. J. Soil Sci., 27: 523-40.

McKeague, J. A., Brydon, J. E., Miles, N. M. 1971. Differentiation of forms of extractable iron and aluminium in soils. Soil Sci. Soc. Am. Proc., 35: 33-9.

Mizota, C. 1978. Clay mineralogy of the A horizons of seven Ando soils, central Kyushu. Soil Sci. Plant. Nutr., 24: 63-73.

Ong, H. Ling, Bisque, R. E. 1968. Coagulation of humic colloids by metal ions. Soil Sci., 106: 220-4.

Peech, M. A., Alexander, L. T., Dean, L. A., Reed, J. F. 1947. Methods of Soil Analysis for Soil Fertility Investigations. U. S. Dept. Agr. Circ. 757, 25 pp.

Righi, D. 1977. Genèse et Evolution des Podzols et des Sols Hydromorphes des Landes du Médoc. Thèse de doctorat, Université de Poitiers. 144 pp.

Righi, D., De Coninck, F. 1974. Micromorphological aspects of Humods and Haplaquods of the "Landes du Médoc," France. In: G. K. Rutherford (Editor), Soil Microscopy. Limestone Press, Kingston, Ontario, pp. 567-88.

Robin, A. M. 1978. Genèse et Evolution des Sols Podzoliques sur Affleurements Sableux du Bassin Parisien. Thèse, Université de Nancy, 173 pp.

Robin, A. M., De Coninck, F. 1978. Micromorphological aspects of some Spodosols in the Paris Basin. Proc. Vth Int. Working Meeting on Soil Micromorphology, Granada, Spain, pp. 1019-50.

Schnitzer, M., Khan, S. U. 1972. Humic Substances in the Environment. Marcel Dekker, New York, 327 pp.

Soil Survey Laboratory Staff 1972. Soil Survey Laboratory methods and procedures for collecting soil samples. U. S. Dept. Agr., Soil Survey Investigations Report No. 1, revised, 1972. U. S. Govt. Print. Office, Washington, D. C., 63 pp.

Soil Survey Staff 1974. Soil Taxonomy. A basic system of soil classification for making and interpreting soil surveys. U.S. Dept. Agr., Handbook No. 436, 754 pp.

Tokashiki, Y., Wada, K. 1975. Weathering implications of the mineralogy of clay fractions of two Ando soils, Kyushu. Geoderma,14: 47-62.

Wada, K. 1977. Allophane and imogolite. In: J. B. Dixon and S. B. Weed (Editors), Minerals in Soil Environments. Soil Sci. Soc. Am., Madison, Wisconsin, pp. 603-38.

Wada, K., Harward, M. E. 1974. Amorphous clay constituents of soils. Adv. Agron., 26: 211-60.

Wada, K., Higashi, T. 1976. The categories of aluminium and ironhumus complexes in Ando soils determined by selective dissolution. J. Soil Sci., 27: 357-68.

Walkley, A. 1935. An examination of methods for determining organic carbon and nitrogen in soils. J. Agr. Sci., 25: 598-609.

Wright, J. R., Schnitzer, M. 1963. Metallo-organic interactions associated with podzolization. Soil Sci. Soc. Am. Proc., 27: 171-6.

Zachariae, G., 1965. Spurentierischer Tatigkeit im Boden des Buchenwaldes. Paul Parey, Hamburg and Berlin, 121 pp.

9

Copyright © 1972 by Mouton Publishers
Reprinted from pages 397–418 of *Tropical Soils: A Comprehensive Study of Their Genesis*, 3rd ed., Mouton—Ichtiar Baru—Van Hoeve, The Hague, The Netherlands, 1972, 481p.

ANDOSOLS

E. C. J. Mohr, F. A. Van Baren, and J. Van Schuylenborgh

The Andosols are soils developed on pyroclastic material. They occur in volcanic regions, such as those bordering the Pacific basin: on the west coast of South America, Central America, The Rocky Mountains, Alaska, The Philippine Archipelago, Indonesia, Australian New Guinea, New Zealand, and Japan. The name originated from

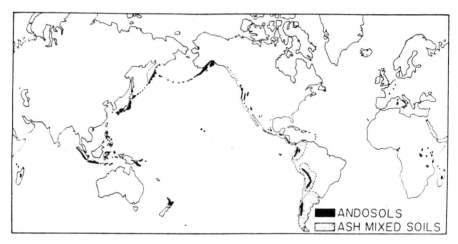

FIG. II.65. *Distribution of Andosols and Ash Mixed Soils (after various sources)*

the Japanese 'ankantsu-shoku-do' meaning dark brown soil (Thorp and Smith, 1949). Other names in literature include the Humic Allophane Soils (Kanno, 1961) and Kuroboku soils (Ohmasa, 1965) of Japan, the Trumao soils of Chile (Wright, 1960), the fresh volcanic ash soils, young volcanic ash soils, 'Brown earth', etc. soils of the Antilles (McConaghy, 1969), the Tapetate, pardo forestal, andosol, and 'latosol' of Central America (Martini, 1969); Black Dust soils, or High Mountains soils, and Andosols of Indonesia (Tan, 1965); the Alvisols or Yellow Brown Loams of New Zealand (Taylor, 1965); the Andepts and Andaquepts of the U.S.D.A. 7th Approximation (1960–1967), and the Andosols FAO/UNESCO Soils Map of the World Project.

The denomination 'Andosols' for soils derived from volcanic ash had become very popular and the Meeting on the Classification and Correlation of Volcanic Ash Soils, in Japan (FAO/Unesco, Report 14, 1965) agreed that the name was suitable.

5.1. SOIL FORMING FACTORS

a) *Climate*

Andosols are found in the temperate dry sub-humid, temperate moist sub-humid, cool temperate humid, and cool superhumid environments of Chile (Wright, 1965); in the cool humid, warm temperate, warm humid, sub-tropical humid, and tropical humid low-lands of Colombia and Ecuador; in the humid sub-tropical and humid tropical undulating to flat territory of Central America; in humid climates in Alaska and the Aleutian chain, along the coast of Washington, Oregon and northern California (Flach, 1965); in the cool semi-arid, tropical semi-arid, and tropical perhumid mountainous territory of Hawaii (Swindale and Sherman, 1965); in the flat or gently undulating land stretching from Hokkaido close to the subarctic region to the extreme southern part of Kyushu in the sub-tropical belt (Ohmasa, 1965).

Soils developed on volcanic ash such as the Regosols and other mineral soils in arid regions and the organic soils in very humid and superhumid environments, do not comply with the modal concept of Andosols.

b) *Vegetation*

A number of plant species are found. Some of the species of plants are probably typical for certain regions or environments i.e. the *Notophagus oblicua* has been associated with the Trumao soils in Chile (Besoain, 1958); *Notophagus* spp. have also been reported on volcanic ash soils in other countries of the world. Ferns are found on acid soils, xerophitic plant species under arid conditions in Peru (Zavaleta, 1969), mosses are common in superhumid climates. Soils on volcanic ash in the humid sub-tropical belt in Colombia provide most suitable conditions for coffee growing. A list of several plant species found on Andosols in several countries follows.

Chile: From south to north in the Trumao zone the plant species include, *Nothofagus oblicua, Nothofagus dombei, Nothofagus antarctica, Nothofagus pumila, Weimania spp., Saxegothea, Laurelia sempervirens, Podocarpus nubigenus* (in mountainous country), *Fittroya cupresoides, Acacia cavenia, Boldea boldus, Aristotelia spp.* (Valdes, 1969).

Colombia: According to Montenegro and Espinal (1963) the following vegetation formations are found in the Andosol zone: humid mountain forest, humid lower mountain forest, humid subtropical forest, and humid tropical forest. The plant species include, grasses, *Quercus spp., Cederela rapanea,* several species of Conifers and *Nothofagus, Ginerum sagittatum, Mimosa pigra, Erythrina glauca, Calliandria spp.*

Central America: Nothofagus spp., and several species of trees of the original vegetation still remain. Coffee, sugar cane are cultivated at below 1500 m and vegetables at higher elevations. Grasses are common in the uplands.

Mexico: Several species of Conifers, *Bromus spp., Muhlenbergia spp., Festuca spp., Agrostis spp.*, etc. (Aguilera, 1969).

U.S.A. (California): The original vegetation includes several species of Conifers, *Nothofagus spp.*, ferns and grasses.

New Zealand: Nothofagus spp., Dracophyllum tarnui, mosses, bracken ferns, grasses *(Dactylis glomerata L.)*

Japan: Several species of Conifers, *Betula spp.*, *Acer pictum*, *Magnolia abovata*, *Salix spp.*, *Quercus serrata*, *Quercus agustissima*, etc. The artificial forest includes *Latrix leptolepis, Quercus dentata, Sasa spp.*, etc.

c) *Parent material*

The parent material of the Andosols is mainly the volcanic debris which remains unconsolidated after it has been deposited. In Japan, the so called 'Humic Allophane Soils', are soils developed on comparatively recent eruptive material from the volcanos (excluding lava and welded tuff, Ohmasa, 1965). In northern Sumatera, Indonesia, soils with similar characteristics to the 'Humic Allophane Soils' are formed on andesito-dacitic tuffs and lahars of the Sibajak volcano (Tan, 1965).

Volcanic tuff consists of rock fragments, single mineral grains, and glass. The glass is the most important of these three components (Pettijohn, 1957) in relation to the formation of Andosols. It forms part of the fine, comminuted material of the formerly liquid magma.

In South America the bulk of the Andosols are developed on finely comminuted ejecta from a magma with a chemical composition between acidic and intermediate. Acidic ejecta mixed with sub-ordinate intermediate magma of Pleistocene times are found in northern South America (Colombia, Ecuador), and intermediate ejecta with sub-ordinate acidic magma from the Quaternary is found in the south (in Chile). In the zone of Los Angeles, Chile, basic material from the Antuco volcano is present (Valdes, 1969).

In Central America, acidic type of rocks are found in El Salvador, Guatemala, and Honduras; basic and intermediate ejecta are present in Panama, Costa Rica, and Nicaragua (Martini, 1969).

In the Antilles, the rock types range from acid dacites to basic lavas but the intermediate andesitic one is the most common (McConaghy, 1969).

The pyroclastic materials in Mexico include basalts rich in olivine (soils of Popocatepetl, Poricutin), andesites (soils of the Itraccihuat), rhyolitic-andesites (soils of the Valle de Puebla), and rhyolites in the northern part of the country (Aguilera, 1969).

The widely distributed deposits in the northern United States are generally of dacitic composition about 6,600 years old (Haward and Borchardt, 1969).

In Japan seven volcanic soil groups have been classified on andesitic, dacitic, or basaltic rock types (Ministry of Agriculture and Forestry, 1964).

The Andosols in Indonesia are developed from a wide variety of parent material. They change from acidic to basic types going from west to east along the Indian Ocean.

Intermediate and rhyolitic rocks are the parent material of the soils in New Zealand (Taylor, 1965).

5.2. ANDOSOL PROFILES

5.2.1. GENERAL CHARACTERISTICS

The Andosols have:
a) AC or ABC profiles, ranging from 30 to 50 cm, and in some cases up to 100 cm depth;

b) dark colours which dominate throughout the profile, although there is a clear difference in colour between the top soil and the subsoil. In humid temperate and in humid tropical uplands (above 2500 m) the colours of the profile are darker than in the humid tropical lowlands (below 1000 m) and in the sub-tropical belt (from about 1000 to 2000 m);
c) very porous, very friable, non-plastic, non-sticky A horizons, merging clearly into the brown B or C horizons;
d) A horizons with a crumb or granular structure;
e) B horizons (if present) with weakly developed blocky structure;
f) high water holding capacity;
g) segregation of aluminium in the form of nodules of gibbsite or segregation of iron oxides in the B or C horizons;
h) 'soapy' feeling of the soil (in the field) when rubbed, becoming almost liquid;
i) sometimes irreversible granulation after air-drying;
j) an amorphous fabric (Luna, 1969), consisting of an isotropic and porous plasm.

The mineralogical properties.
a) Both the silt and fine sand fractions have volcanic glass, the amount of glass varying according to the locality. Some of the grains (particularly hypersthene) appear with a rim of volcanic glass;
b) Ferromagnesian minerals (olivine, pyroxenes, amphiboles), feldspars, and quartz are very common, the amount depending on the origin of the volcanic ash;
c) Allophanes dominate the clay fraction (more than 60 percent) in young soils; allophane and halloysite are found in more developed soils;
d) 'Phytolith' (or 'plant opal') occurs abundantly (Kanno and Arimura, 1958; Luna, 1969).

The physico-chemical properties.
a) Base saturation is low, although the pH is higher than expected.
b) C.e.c. is high when determined by the sodium acetate method at pH 7.
c) Anion exchange capacity is high.
d) The pH of 1 g of soil in 50 ml of 1N NaF exceeds 9.4 after two minutes (Fieldes, 1961).
e) C and N contents are high; the C/N ratio is low.
f) The P content is low (when extracted by 0.5M $NaHCO_3$-solution at pH 8.5) due to strong P fixation (Olsen, 1954).
h) The soils are difficult to peptize.
i) More than 20% moisture is retained at 15 bar pressure.
j) Bulk density is smaller than 0.85.

5.2.2. SOME PROFILES WITH ANALYTICAL DETAILS

Luna (1969) describes a typical profile (profile 39) of the humid subtropical belt in Colombia (S.A.). The profile is situated on the east slope of the Medellin river valley, 9 km from the city of Medellin. Elevation is 2150 m. There is a 40% convex slope. Rainfall is 2600 mm annually. The vegetation consists of grass, ferns and few trees. The parent material is volcanic ash and colluvium of amphibolites. The soil is well-drained. The morphological characteristics are:

Horizon	Depth (cm)	
A1	0–40	Moist, black 10YR2/1 (10YR3/1:d) very humic, silty loam; non-porous aggregates; very friable; non-plastic; poorly rooted; irregular and gradual boundary to
B	40–50	Moist, very dark greyish brown 10YR3/2 (10YR4/2:d) very humic, loam; moderate, medium, prismatic structure, breaking to moderate, medium aggregates; non-plastic, non-sticky; poorly rooted; regular and gradual boundary to
C	+50	Moist, dark brown 7.5YR4/4 (7.5YR5/6:d) humic loam; massive; friable and porous; non-plastic and slightly sticky; no roots.

Analytical details are given in tables II.129, 130, and 131.

TABLE II.129. *Some physical and chemical characteristics of profile 39 (after Luna, 1969)*

Hor.	a^1	Bulk density ($g\,cm^{-3}$)	pH (H_2O)	C (%)	C/N	$\frac{h.a.^2}{f.a.}$	cation exchange characteristics (m.e./100 g of soil)						
							c.e.c.	Ca	Mg	K	Na	Al	H
A1	40.6	0.4	5.1	13.9	15.5	0.5	79	0.2	0.2	0.1	tr	2.5	78
B	46.5	0.5	5.8	7.3	11.0	0.4	60	0.2	0.2	tr	tr	0.4	59
C	34.3	0.4	6.4	1.9	6.9	0.5	49	0.2	0.1	tr	tr	–	48

[1] a = moisture (%) retained at 15 bar pressure. [2] h.a./f.a. = ratio humic acids/fulvic acids.

TABLE II.130. *Chemical composition (partial) of profile 39 (After Luna, 1969; weight percentages)*

Hor.	SiO_2	Al_2O_3	Fe_2O_3	TiO_2	CaO	K_2O	SiO_2/Al_2O_3	SiO_2/Fe_2O_3	Al_2O_3/Fe_2O_3	free Fe_2O_3
A1	36.0	14.4	13.0	0.55	0.5	0.1	4.3	7.5	1.8	4.1
B	39.8	17.0	13.7	0.6	0.6	0.1	4.1	8.3	2.0	4.5
C	39.8	22.4	16.3	0.6	0.6	0.2	3.1	6.6	2.1	4.6

TABLE II.131. *Mineralogical composition of the sand fraction (500–50μ) in percentages*

Hor.	l.m.[1]	h.m.[2]	light minerals							heavy minerals					
			Q	Gl	Ph	Or	Ab	Mi	Misc	Epi	Ho	Oho	Py	Misc	op
A1	56	44	59	2	3	11	21	–	4	2	89	5	4	–	14
B	51	49	48	3	12	12	21	tr	4	–	87	tr	10	3	17
C	62	38	51	3	–	10	31	2	3	–	93	3	4	–	35

[1] l.m. = light minerals; [2] h.m. = heavy minerals. Q = quartz; Gl = volcanic glass; Ph = phytoliths; Or = K-feldspars; Ab = acid plagioclases; Mi = mica; Epi = epidote; Ho = hornblende; Oho = oxyhornblende; Py = pyroxenes; Misc = miscellaneous; op = opaque.

The grain-size distribution in the A_1 horizon is: 21% > 50μ, 55% 50–2μ, and 24% < 2μ. The clay fraction (see fig. II.66) consists of abundant quartz followed by crystoballite, allophane, and gibbsite. Gibbsite increases, and allophane and crystoballite decrease with depth. In the C horizon some halloysite is present.

In spite of the low base saturation, soil pH is not extremely low, a well-known phenomenon in soils with allophane-loaded clay fractions. The pH-dependent charge characteristics, the amphoteric nature of exchangeable aluminium, the complex nature of the amorphous material all play a role. It is believed that some of the amorphous free silica and allophane dissolve in the extracting solution (amorphous silica solubility is constant between pH 3 and 8: see fig. II.22) and the silicic acid is titrated. The silt fraction can also be the site of the exchange reactions, but it is thought that this fraction also contains a considerable amount of amorphous and metastable components; even the sand fraction contains plant opal. Further research into this subject is necessary.

The ratio humic acids/fulvic acids is low, so the content of soluble organic acids is high; this can be expected to have some bearing on sesquioxide translocation. Aluminium hydroxide is extremely mobile (see table II.130 and the X-ray data); this cannot be explained on the basis of its solubility product. It has to be assumed that Al forms soluble complexes with the fulvic acids which are transported downwards with the percolating water and which are then hydrolysed at higher pH values, forming a deposit of aluminium hydroxide (see Part III, Chapter 3). The same process applies, though to a lesser extent, to iron oxide.

It is evident that certain aspects of podzolization, viz. deferritization or cheluviation, are present.

Another selected example (profile 40) of a volcanic ash soil (Luna, 1969) was from Chile (S.A.) from the Centinela Experimental Station of the Ministry of Agriculture (Chile) near Puerto Ocatay. It was situated on a very gently undulating old terrace surface. The parent material is volcanic ash (probably andesitic) overlying alluvial gravels. The elevation is 200 m, the vegetation a grass-forest cover, and the rainfall 2000 mm annually. The profile description is as follows:

Horizon	Depth (cm)	
A_{11}	0–8	Black (10YR2/1:m) to black brown (10YR2/2:d) silt loam; friable; moderately developed fine and very fine granular structure; nonsticky, and very slightly plastic when moist; diffuse boundary to
A_{12}	8–18	Black brown (10YR2/2:m to 10YR3/2:d) silty loam; friable, soft when dry; moderately developed, medium and fine sub-angular blocky structure; very slightly sticky and slightly plastic when moist; diffuse boundary to
B_1	18–43	Dark brown (10YR3/3:m) to grey yellowish brown (10YR4/3:d) silt loam; friable soft loam when dry; strong, medium sub-angular blocky structure, breaking to coarse granules; slightly sticky and slightly plastic when moist; clear boundary to
B_2	43–74	Dark brown (7.5YR3/4:m) to grey yellowish brown (10YR5/4:d) silt loam; moderately developed fine and very fine sub-angular blocky

		structure; friable to firm; slightly sticky and slightly plastic when moist; diffuse boundary to
B₃	74–128	Dark brown (10YR3/4:m) to brown (10YR4/4:d) silt loam; firm to friable; weakly developed fine structure, breaking to very fine granules; slightly sticky and moderately plastic when moist.

The analytical details are given in tables II.132, 133, and 134. As the percentages of the other elements in the clay fraction are in the order of magnitude of 0.1 to 0.5%, the

TABLE II.132. *Some physical and chemical characteristics of profile 40*

Hor.	texture (%)			pH (H_2O)	C (%)	C/N	cation exchange characteristics (m.e./100 g of soil)				
	>50μ	50–2μ	<2μ				c.e.c.	Ca	Mg	K	Na
A_{11}	10	81	9	5.8	18.7	14.4	76	22.5	6.5	1.1	0.6
A_{12}	17	46	37	5.8	7.9	13.3	55	7.0	2.0	0.4	0.3
B_1	12	75	13	5.9	4.3	12.9	34	0.5	1.1	0.2	0.1
B_2	18	70	12	5.9	4.0	10.3	37	0.2	0.2	0.1	0.1
B_3	22	65	13	5.8	4.0	9.9	43	0.2	0.3	0.1	0.1

TABLE II.133. *Partial chemical composition of the clay fractions of profile 40* (weight percentages; after Luna, 1969)

Hor.	SiO_2	Al_2O_3	Fe_2O_3	SiO_2/Al_2O_3	SiO_2/Fe_2O_3	Al_2O_3/Fe_2O_3
A_{11}	26.8	15.4	8.6	3.0	8.3	2.8
A_{12}	23.0	26.0	9.8	1.5	6.3	4.2
B_1	21.9	28.3	9.8	1.3	6.0	4.6
B_2	20.9	30.1	10.2	1.2	5.4	4.5
B_3	19.8	30.8	10.2	1.1	5.2	4.7

data of table II.133 facilitates the calculation of the normative mineralogical composition assuming that allophane (composition $2SiO_2.Al_2O_3.3H_2O$: Wada, 1967) and gibbsite are the most important constituents of the clay. DTA data show that allophane is the dominant mineral and gibbsite increases with depth. The result of the calculation is given in table II.134, which confirms the DTA results (see fig. II.66). As in the previous profile it can be seen that aluminium hydroxide is mobile.

Table II.135 shows that the profile is not as strongly weathered as the previous profile, as the reserve of weatherable minerals is fairly high. The increase in volcanic glass in the surface layers points to a rejuvenation of the profile.

Profile 40 belongs to the 'trumao' type of Chilean soils, to the sub-alvic soils in New Zealand, and probably is the Typic Dystrandept of the 7th Approximation.

TABLE II.134. *Approximate normative mineralogical composition of the clay fractions of profile 40 (equivalent percentages).*

Hor.	Allo	Gibb	Q	Go
A_{11}	70.6	–	17.8	12.6
A_{12}	75.4	12.5	–	12.1
B_1	69.8	18.4	–	11.8
B_2	65.4	22.7	–	11.8
B_3	62.2	25.7	–	12.1

Allo = allophane; Gibb = gibbsite; Q = free silica; Go = goethite.

TABLE II.135. *Mineralogical composition of the sand fraction (500-50µ) in percentages* (after Luna, 1969)

Hor.	l.m.[1]	h.m.[2]	light minerals							heavy minerals					
			Q	Gl	Or	Ab	Lab	An	Misc	Epi	Zo	Ho	Cpy	Opy	op
A_{11}	90	10	20	23	6	7	13	8	23	5	7	8	26	54	2
A_{12}	96	4	18	20	4	1	12	16	29	9	3	7	23	58	2
B_1	90	10	18	18	2	6	12	8	36	2	1	10	19	68	1
B_2	87	13	36	4	8	11	9	17	15	3	tr	7	26	64	9
B_3	89	11	35	2	15	7	16	12	13	5	2	9	17	67	9

[1] l.m. = light mineral fraction; [2] h.m. = heavy mineral fraction. Q = quartz; Gl = volcanic glass; Or = K-feldspar; Ab = acid plagioclases; Lab = intermediate plagioclases; An = basic plagioclases; Misc = miscellaneous; Epi = epidote; Zo = zoisite; Ho = hornblende; Cpy = clino-pyroxenes; Opy = ortho-pyroxenes; op = opaque. Many of the minerals are covered with a rim of volcanic glass.

Profile 40 is only weakly podzolized in comparison with profile 39.

The Yellow Brown Loams from New Zealand are similar to the 'trumao' soils of Chile. One example of these soils was developed on volcanic ash deposits of different ages. C_{14} determinations and the stratigraphy revealed that the age increases from 1,800 and 3,500 years in the top to roughly 10,000 years in the bottom of the profile. The profile (profile 41) is situated at an elevation of 450 m at Powe Hawe's Bay, on a convex slope of 5° in the undulating area of a wide valley. The parent material is rhyolitic volcanic ash over marine sediments. The soil is used as grassland. Rainfall is 1500 mm annually. The profile description is as follows:

Horizon	Depth (cm)	
A_1	2.5–15	Brownish black (5YR2/1) loamy sand with 3% very coarse sand; friable; moderately developed, fine, granular to crumb structure; abundant roots.
B	15–30	Dull yellowish brown (10YR4/3) sand with 3% very coarse sand; very friable; moderately developed, fine crumb structure; many roots.
II A'	30–40	Dull yellow orange (10YR6/3) coarse sand with 8% very coarse sand and gravel, many small pieces of charcoal; slightly compact; single grain structure; few roots.

II B'	40–60	Dull yellowish brown (10YR4/3) sand with 5% very coarse sand; firm; weakly developed medium-sized subangular blocky structure.
II C'	60–70	Dull yellowish brown (10YR4/3) very coarse sand consisting of brownish yellow pumice (85–90%) and grey rhyolite (10–15%); loose and single grain structure.
III B'$_3$	70–90	Yellowish brown (10YR5/6) sandy loam with some very coarse pumice sand in the upper 30 cm; slippery non-sticky consistence; weakly developed coarse subangular blocky structure when dry.
III C'$_1$	105–120+	Light yellow (2.5Y7/4) silt loam derived from Tertiary sandstone.

Classification: Umbric Vitandrept.

TABLE II.136. *Some physical and chemical properties of profile 41*

Hor.	texture (%)			a^1	pH (H$_2$O)	C (%)	C/N	h.a.^{2} / f.a.	cation exchange characteristics (m.e./100 g of soil)				
	>50μ	50–2μ	<2μ						c.e.c.	Ca	Mg	K	Na
A$_1$	31	66	3	11.8	5.5	4.5	12.6	1.0	23.4	2.7	0.6	0.3	0.2
B	41	57	2	6.3	5.7	1.7	10.7	0.3	11.7	0.5	0.1	0.3	0.3
II A	55	44	1	5.4	5.8	1.3	n.d.	1.0	8.6	0.3	tr	0.3	0.3
II B	66	30	4	4.6	5.9	1.1	7.6	n.d.	9.7	0.4	tr	0.2	0.2
II C	82	15	3	4.5	6.0	0.6	7.9	n.d.	6.8	0.2	0.1	0.1	0.1
III B	40	35	25	14.0	5.8	1.7	9.5	n.d.	24.3	0.5	0.7	0.2	0.2
III C$_1$	36	35	29	24.0	6.0	1.2	17.3	n.d.	28.5	0.6	0.5	0.1	0.1

^{1} a = moisture (percentages) retained at 15 atm. pressure. ^{2} h.a./f.a. = humic acid/fulvic acid ratio.

The analytical details are represented in tables II.136 and 137 and in fig. II.66. The discontinuities observed in the profile are fairly well reflected in the analyses, especially the distinct break between deposits II and III. The gradual decrease of the sand fraction and increase of the silt fraction can be attributed to increasing weathering of a homogeneous deposit. The mineralogical analysis (table II.137) also shows a distinct break

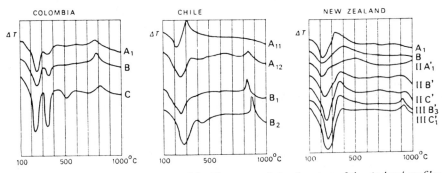

FIG. II.66. *Differential thermal curves of the Na-saturated clay fractions of the Andosol profiles 39, 40, and 41*

between deposit II and III. Apparently, the time lapse between the deposit of materials I and II has been short, and the nature of the material is very similar. This conclusion agrees with the C_{14} datings mentioned above and with the data in fig. II.66. Only the clay fraction of the two lowest horizons have allophane with an exotherm at approximately 900° C (see also section 3, p. 412), indicating that the allophane of the subsoil is better crystallized than that of the upper part of the profile.

The Colombia profile is the oldest one, followed by Chilean profile and finally by the New Zealand profile, with the same age range for the podzolic tendencies in the profiles.

TABLE II.137. *Mineralogical composition of the sand fraction (500–50μ) in percentages*

Hor.	l.m.[1]	h.m.[2]	light minerals							heavy minerals				
			Q	Gl	Or	Ab	Lab	Wp	Misc	Ho	Cpy	Opy	Alt	op
A_1	98	2	2	77	4	2	–	15	–	7	21	61	11	28
B	94	6	tr	97	–	–	–	3	–	4	24	65	7	40
IIA	98	2	–	96	tr	2	–	2	–	5	25	64	6	38
IIB	96	4	tr	89	–	–	–	–	11	–	6	18	61	15
IIC	95	5	tr	67	–	–	–	33	–	tr	6	87	7	55
IIIB	96	4	5	17	–	–	–	70	8	8	26	60	6	32
IIIC	96	4	4	9	–	2	3	78	4	2	21	74	3	17

[1] l.m. = light minerals; [2] h.m. = heavy minerals; Q = quartz; Gl = volcanic glass; Or = K-feldspars; Ab = acid plagioclases; Lab = intermediate plagioclases; Wp = weathering products; Misc = miscellaneous; Ho = hornblende; Cpy = clino-pyroxenes; Opy = ortho-pyroxenes; Alt = alterites; op = opaque. Some of the minerals have a rim of volcanic glass.

An interesting example of the influence of age on clay-mineral formation in volcanic ash soils is found in the study published by the Japanese Ministry of Agriculture and Forestry on this subject (1964). In one profile 15 volcanic deposits occurred and the nature of the clay fractions has been examined by thermal analysis.

From this data it can be seen that the younger deposits are all characterized by the presence of allophane. Kaolinite becomes prevalent from layer IX onwards with the exception of the layers XI and XII in which allophane again occurs in large quantities. One explanation might be that these layers were exposed for too short a period to the atmosphere to allow the transformation of allophane into kaolinite.

Finally two examples from Indonesia will be discussed (Van Schuylenborgh, 1958). The first profile (profile 42) is situated on a practically horizontal part of the slope of Mount Wajang (volcano of Plio-Pleistocene age: Van Bemmelen, 1949, p. 623) South-East of Pengalengan (Preanger, Java). The altitude is 1620 m above sea level. Mean annual rainfall is 2751 mm; the monthly distribution is as follows:

Jan.	Febr.	Mrch	Apr.	Mai	June	July	Aug.	Sept.	Oct.	Nov.	Dec.
332	298	349	300	217	135	88	70	99	210	309	344 mm

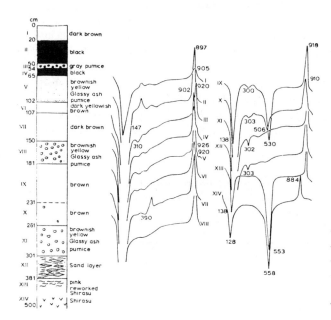

FIG. II.67. *Ash sequences of a Humic Allophane soil of Kanoya, Kagoshima and DTA curves of their clay separates (after Volcanic Ash Soils in Japan, 1964)*

Mean annual temperature is 16.4° C. Vegetation: tropical mountain rainforest with an evapotranspiration of approximately 1000 mm annually. The drainage is excessive and the parent material a basalto-andesitic tuffaceous ash. The profile description reads as follows:

Horizon	Depth (cm)	
$O_1 + O_2$	5–0	Forest litter. Lower part well-decomposed. Mulltype.
A_{11}	0–5	Brownish black (10YR2/2:m) to greyish yellow brown (10YR4/2:d) sandy loam. Well-developed granular to fine subangular blocky structure. Very friable. Abundance of roots. Clear boundary to
A_{12}	5–31	Brownish black (10YR3/2:m) to dull yellowish brown (10YR5/3:d) loam. Well-developed fine subangular blocky structure. Friable. Many roots. Clear boundary to
A_2	31–61	Dark brown (10YR3/4:m) to dull yellow orange (10YR6/4:d) loam. Thin platy structure. Friable. Roots are present. Gradual boundary to
B_1	61–80	Brown (10YR4/4:m) to bright yellowish brown (10YR6/6:d) clay loam. Moderately developed fine subangular blocky structure. Slightly firm. Prominent glassy (clay) coatings on structural units. Few roots. Clear boundary to
IIB'_2	80–95	Brownish black (7.5YR3/2:m) to dull yellowish brown (10YR5/4:d) silty clay loam. Well-developed medium subangular blocky structure. Friable. Very few roots.
$IIIA'_1$	95+	Coaled organic horizon. Old vegetation surface burnt by hot ash deposit.

The soil dehydrates irreversibly into gravel-size aggregates. Therefore it was analysed in the field-wet state. None of the common peptizing agents could be used in the granulometric analysis. However, 10^{-3}M to 10^{-4}M hydrochloric acid produced perfect peptization. Classification: Typic Hydrandept.

The analyses are given in tables II.138 and 139. Table II.139 shows that there is mineralogical difference between horizon IIB₂ and the upper part of the profile. Apparently the last eruption of the volcano consisted of a short first stage in which a basalto-andesitic ash with hypersthene-augite association was deposited, followed by a longer (dying-out) stage in which a similar ash with an olivine association was deposited. Although primary minerals are still present in the solum, glass has already been completely weathered. A considerable amount of fine iron concretions can be observed, indicating that this soil is in a fairly advanced stage of weathering. It is certainly the oldest of the profiles discussed.

Table II.138 shows that there is an accumulation of clay in the B horizon, which was also indicated in the profile description. X-ray analysis showed that the clay was predominantly composed of amorphous material, with only minor admixtures of

TABLE II.138. *Some physical and chemical characteristics of profile 42*

Hor.	texture (%)			C (%)	C/N	pH H_2O	molar ratios of clay fraction		
	>50μ	50–2μ	<2μ				SiO_2/Al_2O_3	SiO_2/Fe_2O_3	Al_2O_3/Fe_2O_3
A_{11}	44.4	48.8	6.8	14.6	12.8	6.3	2.2	9.3	4.1
A_{12}	26.3	55.1	18.6	7.1	9.8	6.0	1.9	8.1	4.3
A_2	31.4	45.2	23.4	4.6	8.1	6.1	1.2	5.6	4.5
B_1	28.1	43.5	28.4	3.3	7.7	6.0	1.2	5.1	4.3
IIB_2	13.2	52.2	34.6	3.5	8.8	5.9	1.1	4.5	4.2

montmorillonite, kaolinite, and gibbsite. The coatings in the B horizon do not show clay orientation, so are not considered coatings in the sense of the definition of the 7th Approximation. Nevertheless, there is movement of clay-sized particles and the B horizon is therefore the horizon of accumulation of clay-sized particles. The glassy or wax-like coatings on the structural units are undoubtedly accumulations of allophane together with gibbsite, as X-ray analysis showed an increase of gibbsite with depth.

TABLE II.139. *Mineralogical composition of the sand fraction of profile 42* (in percentages)

Hor.	total sand fraction											heavy sand fraction					
	Q	Lab	Ho	Hy	Aug	Ol	Gl	op	ic	Z	Misc	op	Ho	Hy	Aug	Ol	Z
A_{11}	–	20	tr	tr	6	1	–	2	23	1	47	9	13	26	33	26	2
A_{12}	2	18	2	1	5	2	tr	1	29	–	40	16	8	34	34	24	1
A_2	2	21	4	1	6	3	1	2	30	1	29	12	17	28	28	17	1
B_1	1	13	7	7	11	3	1	3	25	5	24	29	14	17	33	34	–
IIB_2	tr	6	5	9	3	–	1	1	68	–	7	27	32	51	17	–	2

Q = quartz; Lab = intermediate plagioclases; Ho = green hornblende; Hy = hypersthene; Aug = augites; Ol = olivine; Gl = volcanic glass; op = opaque; ic = iron concretions; Z = zircon; Misc = miscellaneous (rock fragments).

These conclusions are supported by the molar SiO_2/Al_2O_3 ratios of the clay fraction; the decrease of this ratio with depth is caused either by translocation of clay with a low SiO_2/Al_2O_3 (allophane) or by translocation of gibbsite or by both processes. The remainder of the clay in the upper horizons is then residually enriched with components of high SiO_2/Al_2O_3 ratios. This was actually found as X-ray analysis showed that kaolinite increases from the bottom to the top.

This profile represents a fairly strongly podzolized soil; translocation of clay-sized particles took place together with cheluviation of aluminium hydroxide and iron oxides (see SiO_2/Fe_2O_3 ratio in table II.138). It should be emphasized that this clay translocation was only discovered because the samples were not air-dried before analysis and a suitable peptizing agent was used.

The same tendencies can be observed in the second profile, which is slightly younger than the previous one. This profile comes from Mount Tandjungsari of the Salak-complex (Pleistocene age: Hartmann, 1938), near Leuwiliang, Tjanten tea plantation, Western Java (Indonesia). The profile (profile 43) is situated on the flattened top at an elevation of 1100 m above sea level. The mean annual temperature is 19.6° C. Average monthly rainfall and rainfall distribution are:

Total	Jan.	Febr.	Mrch	Apr.	Mai	June	July	Aug.	Sept.	Oct.	Nov.	Dec.
5420	432	420	472	538	497	350	290	338	464	583	576	460

The climate is more humid than in the previous case. The parent material is an andesitic tuffaceous ash with hypersthene-association. The vegetation is a tropical mountain rain forest and the drainage is perfect.

The profile description reads as follows:

Horizon	Depth (cm)	
$O_1 + O_2$	2.5–0	Forest litter, lower part well-decomposed. Mulltype.
A_{11}	0–15	Brownish black (10YR3/2:m) to dark brown (10YR3/3:d) loam with moderately developed granular and subangular blocky structure. Very friable. Abundance of roots. Gradual boundary to
A_{12}	15–28	Dull yellowish brown (10YR4/3:m and d) clay loam. Weakly developed fine subangular blocky structure. Friable. Many roots. Clear boundary to
B_1	28–46	Brown (10YR4/4:m and d) clay. Well-developed medium subangular blocky structure. Friable. Many roots. Gradual boundary to
B_2	46–66	Yellowish brown (10YR5/6:m) to dull yellowish brown (10YR5/4:d) clay. Well-developed fine subangular blocky structure. Pronounced wax-like coatings. Friable. Few roots. Clear boundary to
C	66+	Bright yellowish brown (10YR6/8:m) to yellow orange (10YR7/8:d) sandy clay loam. Very porous. Very friable. No roots.

As the preceding profile, the material dries irreversibly. Classification: Typic Hydrandept.

The mineralogical analysis of the sand fraction (table II.141) shows that the parent material is homogeneous. The amount of iron concretions is smaller than in profile 42, pointing to a less advanced weathering stage.

TABLE II.140. *Some physical and chemical characteristics of profile 43* (after Van Schuylenborgh, 1958)

Hor.	texture (%)			C (%)	C/N	pH (H_2O)	molar ratios clay fraction		
	>50μ	50–2μ	<2μ				SiO_2/Al_2O_3	SiO_2/Fe_2O_3	Al_2O_3/Fe_2O_3
A_{11}	26.2	61.6	12.2	16.6	15.9	4.9	1.8	8.4	4.8
A_{12}	20.5	54.5	25.0	10.1	14.1	5.3	0.9	5.1	5.9
B_1	20.8	46.7	32.5	5.4	14.7	5.4	0.7	5.7	7.5
B_2	28.7	31.0	40.3	2.0	14.6	5.4	0.8	6.6	8.3
C	53.8	17.0	29.2	0.7	8.0	5.6	0.9	10.0	11.5

Translocation of clay-sized particles can also be observed in this profile (see profile description and table II.140). The molar SiO_2/Al_2O_3 ratio again points to movement of allophane and gibbsite. The latter is confirmed by X-ray analysis of the clay fraction, showing an increase of gibbsite with depth. There is an abundance of amorphous material with minor admixtures of montmorillonite, kaolinite, and α-crystoballite. Kaolinite content increases slightly in the upper horizons.

Some volcanic ash soils have horizons with accumulation of clay-sized particles and others not. The allophane is predominantly a weathering product of glass and changes, under conditions of good drainage, its composition from $SiO_2.Al_2O_3.2H_2O$ in the youngest soils to $2SiO_2.Al_2O_3.3H_2O$ in older soils and finally to halloysite and kaolinite in still older soils. This process is accompanied by the separation of gibbsite, which is highly mobile because it complexes with fulvic acids, which are predominant

TABLE II.141. *Mineralogical analysis of the sand fraction of profile 43* (after Van Schuylenborgh, 1958).

Hor.	total sand fraction											heavy sand fraction			
	Q	Plg	Ho	Hy	Aug	Gl	op	ic	Gibb	Ph	Misc	op	Ho	Hy	Aug
A_{11}	2	27	1	12	1	17	4	3	6	9	17	43	8	82	10
A_{12}	3	25	2	11	2	18	3	2	8	7	19	21	8	85	7
B_1	2	29	tr	15	1	21	2	3	6	12	9	30	2	91	7
B_2	2	40	1	13	1	20	5	2	7	–	9	21	4	95	1
C	1	44	tr	9	–	18	4	1	19	–	4	39	1	96	3

Mineral abbreviations as in table II.139, Plg = plagioclases; Gibb = gibbsite, Ph = phytoliths.

in volcanic ash soils. Consequently, some soils show argeluviation and cheluviation, others only cheluviation.

Many volcanic ash soils show irreversible granulation upon air-drying, especially those which are continually moist because of the humid climate. It is therefore useless to analyse air-dried samples (Van Schuylenborgh, 1954), if it is desired to determine the grain size distribution or other physical properties. It is also possible that air-drying effects cation exchange characteristics, although exact data are not available.

5.2.3. WEATHERING AND SOIL FORMING PROCESSES

A great deal of information has been obtained in the last decade on the weathering of volcanic ash, and particularly on the formation of secondary minerals. The weathering process of volcanic ash begins with the leaching of soluble components (H_4SiO_4, Ca^{2+}, Mg^{2+}, Na^+, K^+, etc.) by rainwater (desilication: see Part III, Chapter 1). Carbonic acid accelerates the decomposition of the ash. Sesquioxides will accumulate residually, while aluminium with silicic acid forms secondary minerals. The composition of the ash and the leaching conditions determine the type of the secondary minerals. Commonly, a narrow SiO_2/Al_2O_3 ratio is characteristic for volcanic soils. When allophane predominates in the clay fraction of young volcanic soils, the ratio varies between 1 and 2 (a.o. Birrell, 1965). On increasing age the allophane is transformed into kandites and the ratio is then 2. When drainage is impeded and the parent ash contains a considerable amount of ferromagnesian minerals, smectites can be formed and the ratio is then 3 or more. Sometimes, in intermediate and basic ashes and under excessive drainage, SiO_2/Al_2O_3 ratios are smaller than 1 (see profile 43), due to the presence of free aluminium hydroxide (see also: Tan and Van Schuylenborgh, 1961).

It seems that allophane is formed from volcanic glass and plagioclases (andesine and labradorite) and that ferromagnesian minerals (olivine, pyroxenes, and amphiboles) also participate (a.o. Kanno, 1961). Besoain (1969) has constructed a time scale for the transformation of the primary minerals into the several intermediate products of the formation of kaolinite. Firstly a mixed gel of SiO_2 and Al_2O_3 is formed (called allophane B by Fieldes, 1955), then allophane (allophane A of Fieldes, 1955), then halloysite and finally kaolinite. The time-scale is as follows:

volcanic glass basic feldspars	→	mixed gel of $SiO_2 + Al_2O_3$	→	allophane	→	halloysite	→	kaolinite
time periods	↓	↓ decades centuries		40–80 centuries		thousands of centuries		
	months							

DTA data of a 10,000 years old volcanic ash layer that forms the lower part of a profile in New Zealand (fig. 11.66) indicates that the weathering of the primary minerals has reached the allophane stage. This does not tally with Besoain's scheme. As a matter of fact, such a scheme as Besoain's (1969) has little meaning, as it depends on the environmental conditions. This is clarified by the reaction:

$2NaCaAl_3Si_5O_{16}(s) + 17H_2O + 6H^+ \rightleftharpoons$
(labradorite)
$3Al_2SiO_4(OH)_2 \cdot H_2O(s) + 2Na^+ + 2Ca^{2+} + 7H_4SiO_4$ - - - - - - - - - - (68)
(allophane)
$lgK_{68} = 2lg[Na^+] + 2lg[Ca^{2+}] + 7lg[H_4SiO_4] + 6pH$ - - - - - - - - - (68a)

It depends on the soluble reaction products whether the reaction will proceed to the right or not. If these are removed readily (rapid permeability), the reaction will be more complete than if the leaching process is slow. If leaching is slow it is even possible that allophane is not formed at all; if, for example, there is sufficient Mg^{2+} (set free upon weathering of ferromagnesian minerals), montmorillonite may be formed.

Several other secondary compounds have been found in the clay fraction of volcanic ash soils. They include imogolite (Yoshinaga and Aomine, 1962), palagonite (Birrell, 1965), hisingerite (Sudo and Nakamura, 1952), amorphous silica, and, of course, iron oxides (Matsui, 1959) and titanium oxide (Sherman et al., 1964). However, allophane is the most important product. Its composition (Wada, 1967) varies between $SiO_2.Al_2O_3.2H_2O$ (comparable with Fieldes' allophane B) and $2SiO_2.Al_2O_3.3H_2O$ (comparable with Fieldes' allophane A). The structure proposed by Wada for allophane with the Si/Al ratio of 1, is composed of a silica tetrahedral chain and an alumina octahedral chain joined at one corner. The addition of another alumina octahedral chain to the silica tetrahedral chain results in the formation of allophane with the Si/Al ratio of 1/2.

The DTA curves of fig. 11.66 (after Luna, 1969) indicate the presence of two types of allophane. One of them gives an endothermic reaction in the region 150° to 200° C (present in the upper horizons); the other one (present in the lower parts of the profiles) gives two thermic reactions, viz. an endotherm at 150° to 200° C and an exotherm at 850° to 900° C. The type of allophane present in the upper horizons is apparently a less organized form, the one found in the lower horizons a more ordered type. The last type could well be an intermediate stage in the formation of crystalline minerals (hydrated halloysite, kaolinite) and is possibly comparable with Wada's allophane, composition $2SiO_2.Al_2O_3.3H_2O$. Another intermediate phase to halloysite formation may be imogolite (Wada, 1967), which has an endotherm at 410° to 430° C and appears as thread-like particles of 100 to 200 Å diameter in electron micrographs (Yoshinaga and Aomine, 1962).

The most frequently cited formation mechanism of allophane is isoelectric precipitation of silica and alumina (Egawa, 1965). The mechanism of isoelectric precipitation of gels (mutual precipitation of positively and negatively charged colloids), which is inapplicable for a number of minerals (Van Schuylenborgh and Sänger, 1954), seems to be valid in the formation of allophane. It can occur if amorphous aluminium hydroxide is formed during the weathering of glass; this may be in the positively charged stage and form mixed gels with the electro-negative silica colloids.

Andosols have a much higher organic-matter content than non-volcanic soils under similar circumstances. It is believed that the decomposition of organic matter is hindered by the amorphous aluminium hydroxide in Andosols (Kosaka et al., 1962; see Part I, Chapter 4, p. 180).

Humic and fulvic acids are the most important components of the organic fraction. In Japan the humic acid/fulvic acid ratio of the Humic Allophane Soils ranges from 1 to 2; in the Red Yellow Soils the ratio is approximately 1. Similar values apply to the Andosols of Colombia (Luna, 1969). This means that the fraction of organic acids with low molecular weights is very important. As this fraction is also the most soluble, it can be expected that sesquioxides are mobile because of the formation of soluble complexes. This seems to be especially true for alumina translocation, and less so for the iron oxide. For particulars on complex formation refer to Part III, Chapter 3.

It is difficult to understand the occurrence of clay translocation, as the climate is not favourable for this process. Possibly it can be explained by dissolving in the surface layers and reprecipitation in sub-surface horizons. Coatings of clay-sized materials occur on the peds in the B horizon, so this horizon should be called an argillic horizon, although not exactly in the sense of the definition of the 7th Approximation.

When examining the soil formation on andesitic volcanic material in a continually humid climate as in the case of profiles 42 and 43, Van Schuylenborgh (1958) found the following sequence of soils, their characteristics strongly dependent on the altitude:

> 1100 m altitude: Humic Allophane Soils with accumulation of clay-sized minerals in the B horizon (profiles 42 and 43). The soils show irreversible granulation upon air-drying. Classification according to the 7th Approximation: Hydrandepts.

1100–600 m altitude: Humic Allophane Soils with no accumulation of clay-sized minerals in the B horizon, but with irreversible granulation upon air-drying. Very acid. Aluminium is mobile, iron not. Classification according to the 7th Approximation: Hydrandepts or Dystrandepts.

600–300 m altitude: Brown clay soils with no irreversible granulation upon air-drying. The amount of amorphous material is much smaller than in the soils of the higher altitudes, and is replaced by kaolinite. Argillic B horizon is present. Classification according to the 7th Approximation: Tropudalf or Tropudult.

300–0 m altitude: Reddish-brown and Red clay soils with up to 90% of clay. The clay is predominantly kaolinitic. Classification: Haplorthox.

Under very similar climatic conditions, but on rhyolitic volcanic material, the sequence was (Tan and Van Schuylenborgh, 1961):

2000–1500 m: Soils with a spodic horizon. No irreversible granulation upon air-drying. Classification: Tropohumods.

1500–500 m: Soils with a cambic horizon. Moderately acid (pH 5 to 6.5). There is no irreversible granulation. Classification: Eutropepts or Dystropepts.

500–0 m: Soils with an argillic horizon and low base saturation. Classification: Tropudults.

Under a monsoon climate (2 to 5 months dry; 10–7 months wet) and on andesitic volcanic material, the sequence was (Tan and Van Schuylenborgh, 1959):

> 2500 m: Soils with a cambic horizon, a mollic epipedon and very high contents in organic matter. There is no irreversible granulation upon air-drying. Podzolic tendencies in soil formation are fairly strong. Considerable amounts of volcanic glass and pumice particles are present. Classification: Eutrandepts.

2500–1400 m: Soils similar to the preceding ones, but with weaker podzolic tendencies; the biological activity in the soil is considerably higher. Classification: Eutrandepts.

1400–1000 m: Soils similar to the preceding ones. Podzolic tendencies are absent and biological activity is very high. Classification: Vitrandepts.

1000–300 m: Strongly weathered soils with clay percentage up to 75%. The material is not yet oxic and the B horizon is argillic. Base saturation is higher than 35%. Classification: Haploxeralfs.

5.3. THE CLASSIFICATION OF ANDOSOLS

The classification of Andosols is generally based on their morphological, physical and chemical properties, while the mineralogical composition of the sand, silt, and clay fractions is also taken into account. In some of the classification systems stress has been laid on profile characteristics, whereas in others the effect of the soil forming factors and processes such as climate, location, degree of weathering, composition of organic matter, have been considered.

Typical Andosols are classified in the order of Intrazonal soils because of the domination of the parent material (volcanic ash) over other zonal, pedogenic factors (climate, vegetation). Forms which can be classified into Zonal soils may occur upon ageing of the soil; the parent material no longer has weatherable minerals and the amorphous clays have been converted into crystalline clays.

A preliminary classification of Andosols or Humic Allophane soils has been made in South America (Wright, 1965). They are grouped into soils of high latitudes (Argentina and Chile), and soils of low latitudes (Colombia and Ecuador). The classification is based mainly on the degree of weathering and leaching and on organic matter content. The soils of the high latitudes include seven different environments, ranging from the temperate dry sub-humid (mesomediterranean) conditions to the cold, humid to superhumid environments. In the soils of the low latitudes, five different environments have been (provisionally) recognized, viz. the cool, temperate, tropical, humid, and equatorial lowlands.

The soils on volcanic ash in Central America have not been classified. They are known as: Andosol, suelo de Talpetate, pardo forestal, and latosol (Martini, 1969). Andosols and latosols are the most important of the group that can be used for classification purposes. The soils have mostly forest vegetation covers.

In the U.S.D.A. 7th Approximation, the group of soils developed on volcanic ash are classified in the order of the Inceptisols (the Inceptisols have diagnostic horizons which have been formed relatively recently; they show moderate weathering of the primary minerals and there is little or no eluviation of clay), in the sub-order of the Andepts and in the Great Group of the Andaquepts. The Andepts are classified as having 60% or more of vitric volcanic ash, cinders or other vitric pyroclastic materials in the silt, sand, and gravel fractions or a bulk density of the fine earth fraction of the soil of less than 0.85 per cc in the upper part of the solum, and an exchange complex that is dominated by amorphous material. The Andepts are further sub-divided into the following Great Groups:

Cryandepts (with a mean annual soil temperature of less than 8.3° C and mean summer temperature of less than 15° C), Durandepts (with a duripan within 1 m of the surface), Hydrandepts (with clays that dehydrate irreversibly), Eutrandepts (soils with high base saturation), Dystrandepts (soils with low base saturation) and Vitrandepts.

The Andaquepts have similar characteristics as Andosols, except that these soils also have hydromorphic characteristics.

Very young ash soils with no or only slight profile development are classified in the Entisol order. Ash-derived soils in arid or semi-arid conditions, showing accumulation of organic matter and having clay minerals of the smectite group are grouped in the Aridisol order. Extremely weathered ash-derived soils fall into the Order of the Oxisols (see Part II, Chapter 1). Those with a spodic horizon are grouped in the Spodosol order and those with an argillic horizon in an andic subgroup of the Hapludalfs or Hapludults.

The profiles 42 and 43 of section 3, however, are difficult to classify. They have a horizon with accumulation of clay-sized material which dries irreversibly upon air-drying. It is therefore thought that an alfic and ultic sub-group should be made in the Hydrandepts, depending on base saturation; or an andic sub-group in the Tropudalfs or Tropudults. As the argillic horizon of the soils mentioned is not argillic in the sense of the 7th Approximation, the first suggestion is probably most logical.

In theory the separation of Andosols from Ultisols, Spodosols and Oxisols is simple, following the key to the U.S.A. classification system. However, the value of this separation from the genetic point of view depends on several factors including the composition of the pyroclastic material, time, and environmental conditions of the volcanic ash layers. Organic matter, if present in a large quantity in the profile, can influence the water holding capacity of the soil, bulk density, and can even mask some of the pedogenic processes such as eluviation and cheluviation.

In Japan the 'Kuroboku' are the typical soils derived of volcanic ash (Ohmasa, 1965) and are characterized by a dark-coloured A horizon which has 15 to 30 per cent humus content, and low stickiness and plasticity. It is suggested that the formation of the 'Kuroboku' soils is closely connected with grassland vegetation and humid climatic conditions. The soils are classified on the bases of thickness and colour of the A horizon; therefore, soils with light-coloured A horizon are known as 'light-coloured Kuroboku' (Volcanic Ash Soils in Japan, 1964). The light-coloured soils which occur under forest are grouped as Brown Forest Soils or Acid Brown Forest Soils in the Japanese classification.

In New Zealand (Taylor, 1965), the soils developed on volcanic ash are divided into three main categories, named 'skeletiform' (with regosolic profile), 'fulviform' (with profiles similar to those of the brown soils of the humid temperate regions), and 'podiform', with profiles similar to the podsols. The 'fulviform' and 'podiform' soil categories are further subdivided into 'alvic' and 'sub-alvic' according to the state of weathering of the primary minerals. A third sub-division is made on the basis of certain morphological features such as the formation of a B horizon, fragipans, etc. Other sub-divisions are concerned with the degree of leaching of the soils and the 'mellanized' A horizon.

In Indonesia (Tan, 1965), a tentative classification of Andosols into soils with low humic/fulvic acid ratio (0.2) and low variable charge (CEC_v < 30 m.e./100 g), and soils with high humic/fulvic acid ratio (0.5) and medium to high variable charge (CEC_v > 30 m.e./100 g) is proposed.

The central concept of Andosols is similar in all countries of the world. The soils are dark in colour, very porous, organic and with clay, dominantly the amorphous type (allophane, silica, alumina, iron hydroxides, etc.). However, the correlation of the soils is difficult because of the different criteria used in the several classification systems. The characterization of the organic matter by its humic acid/fulvic acid ratio, as has been done in Indonesia (Tan, 1965), is suggested by Dudal (1965) to be a satisfactory approach to the classification of the soils with respect to differences in climate, but further investigations are needed before this suggestion can be generally applied.

The study of the micromorphology of Andosols may also contribute to a better understanding of the classification problems which are still to be solved. In a recent paper Kawai (1969) provides an addition of 4 sub-groups to those of the 7th Approximation based on the specific nature of the micromorphological fabric of a

TABLE II.142. *Classification of Andosols based on micromorphological characteristics, with special reference to the 7th Approximation* (after Kawai, 1969)

Great Groups	Sub-groups	Author's sub-group	Example
Dystrandepts	Typic Cryandepts	*Entic Cryandepts*	Tokotan
	Typic Dystrandepts	Typic Dystrandepts	Kanuma
		Humic Dystrandepts	Imaichi, Kuju
			Kuroishibaru
			Hirusen
		Umbreptic Dystrandepts	Shinshiro
Eutrandepts	Typic Eutrandepts	Typic Eutrandepts	Miyagasaki
			Yatsugatake
			Miura
Durandepts	Typic Durandepts		
		Humic Durandepts	Imazato
Vitrandepts	Umbric Vitrandepts	Umbric Vitrandepts	Jōnouchi

Subgroups in italics are proposed by the author.

number of Japanese Andosols (table II.142). It is indicated that micromorphological investigations may lead to a more detailed classification of Andosols.

LITERATURE CITED

AGUILERA, N. 1969 Distribución y características de los suelos derivados de ceniza volcánica de Mexico. *Suelos derivados de cenizas volcánicas de America Latina. Centro de Ensenanza e Investigación del IICC, Turrialba, Costa Rica*.

BESOAIN, E. 1958 Mineralogia de las arcillas de algunos suelos volcánicos de Chile. *Agr. Tec. Santiago*, 18: 110–167.

— 1969 Mineralogia de las argillas de suelos derivados de cenizas volcánicas de Chile. *Suelos derivados de cenizas volcánicas de America Latina. Centro de Ensenanza e Investigación del IICC, Turrialba, Costa Rica*.

BIRRELL, K.S. 1965 Some properties of volcanic ash soils. *Meeting on the classification and correlation of soils from volcanic ash. FAO Report 14*, Rome: 74–81.

EGAWA, T. 1965 Mineralogical properties of volcanic ash soils in Japan. *Meeting on the classification and correlation of soils from volcanic ash. FAO Report 14*, Roma: 89–91.

FLACH, K. W. 1965 Genesis and morphology of ash derived soils in the United States. *Meeting on the classification and correlation of soils from volcanic ash. FAO Report 14*, Roma: 111–114.

FIELDES, M. 1955 Clay mineralogy of New Zealand soils. 2. Allophane and related mineral colloids. *N.Z.J. Sci. Tech.*, B. 37: 336–350.

HARTMAN, M.A. 1938 Die Vulkangruppe im Südwesten des Salak-Vulkans in West Java. *Natuurk. Tijdschr. Ned. Ind.*, 98: 216–249.

HAWARD, M.E. and BORCHARDT, G.A. 1969 Mineralogy and trace element composition of ash and pumice soils in the Pacific northwest of the United States. *Suelos derivados de cenizas volcánicas de America Latina. Centro de Ensenanza e Investigación del IICA, Turrialba, Costa Rica.*

KANNO, I. 1961 Genesis and classification of main genetic soil types in Japan. *Bull. Kyushu Agr. Exp. Sta.*, 7: 1–185.

— and ARIMURA, S. 1958 Plant opal in Japanese soils. *Soils and Plant Food*, 4, No. 2.

—, KUMANO, Y., HONJO, I. and ARIMURA, S. 1964 Characteristics and classification of an unirrigated anthropogenic alluvial soil found in the Kumamoto Plain. *Bull. Kyushu Agr. Exp. Sta.*, 10: 11–122.

KAWAI, K. 1969 Micromorphological studies of Andosols in Japan. *Bull. Nat. Inst. Agric.*, Japan: 145–154.

KOSAKA, J., HONDA, CH. and IZEKI, A. 1962 Transformation of humus in upland soils, Japan. *Soil Sci. & Pl. Nutr.*, 8: 191–197.

LUNA, C. 1962 Minerales amorfos en suelos del Depto. de Antioquia. *Trabajo presentado al III Congreso Nacional de Químicos e Ing. Químicos, Baranquilla, Colombia.*

— 1969 Aspectos genéticos de andosoles en Colombia. *Suelos derivados de cenizasvolcánicas de America Latina. Centro de Ensenanza e Investigación del IICA, Turrialba, Costa Rica.*

MARTINI, J.A. 1969 Distribución geográfica y características de los suelos derivados de cenizas volcánicas de Centro America. *Suelos derivados de cenizas volcánicas de America Latina. Centro de Ensenanza e Investigación del IICA, Turrialba, Costa Rica.*

MATSUI, T. 1959 Some characteristics of Japanese soil clays. *Adv. Clay Sci.*, Tokyo, 1: 244–259.

MCCONAGHY, S. 1969 Geographic distribution and characteristics of vocanic ash soils in the Antilles. *Suelos derivados de cenizas volcánicas de America Latina. Centro de Ensenanza e Investigación del IICA, Turrialba, Costa Rica.*

MINISTRY OF AGRICULTURE AND FORESTRY. 1964 Volcanic Ash Soils in Japan. *Jap. Govern. Tokyo*, pp. 211.

MONTENEGRO, E. and ESPINAL, S. 1963 Formaciones vegetales de Colombia Depto. Agrológico, Inst. Geogr. *Agustin Codazzi, Bogotá, Colombia.*

OHMASA, M. 1965 Scope of volcanic ash soils, their extent and distribution. *Meeting on the classification and correlation of soils from volcanic ash. FAO Report 14*, Rome: 56–60.

OLSEN, S.R. 1954 Estimation of available phosphorus in soils by extraction with sodium bicarbonate. *U.S.D.A. circ. 939*, pp. 18.

PETTIJOHN, F.J. 1957 Sedimentary rocks. 2nd. Ed. *New York-London.*

SHERMAN, G.D., MATSUAKA, Y., IKAWA, H. and HAHASA, G. 1964 The role of amorphous fraction in the properties of tropical soils. *Agrochemia*, 7: 146–162.

SUDO, T. and NAKAMURA, T. 1952 Hinsingerite from Japan. *Am. Min.*, 37: 618–621.

TAN, K.H. 1965 The andosols in Indonesia. *Meeting on the classification and correlation of soils from volcanic ash. FAO Report 14*: 30–35.

— and VAN SCHUYLENBORGH, J. 1959 On the classification and genesis of soils derived

from andesitic volcanic material under a monsoon climate. *Neth. J. agric. Sci.*, 7: 1–21.
— and — 1961 On the classification and genesis of soils developed over acid vocanic material under humid tropical conditions. II. *Neth. J. agric. Sci.*, 9: 41–54.

TAYLOR, N. H. 1965 The classification of volcanic ash soils in New Zealand. *Meeting on the classification and correlation of soils derived from volcanic ash. FAO Report 14:* 101–106.

THORP, J. and SMITH, G. D. 1949 Higher categories of soil classification: Order, Suborder, and Great Soil Groups. *Soil Sci.*, 67: 117–126.

VALDES, A. 1969 Distribución geográfica y características de los suelos derivados de cenizas volcánicas de Chile. *Suelos derivados de cenizas volcánicas de America Latina. Centro de Ensenanza e Investigacion del IICC, Turrialba, Costa Rica.*

VAN BEMMELEN, J. M. 1949 The Geology of Indonesia. Vol. IA. *The Hague.*

VAN SCHUYLENBORGH, J. 1954 The effect of air-drying of soil samples upon some physical properties. *Neth. J. agric. Sci.*, 2: No. 1.

— 1958 On the genesis and classification of soils derived from andesitic tuffs under humid tropical conditions. *Neth. J. agric. Sci.*, 6: 99–123.

— and SÄNGER, A. M. H. 1949 The electrokinetic behaviour of iron- and aluminium-hydroxides and -oxides. *Rec. trav. chim. Pays-Bas*, 68 : 999–1010.

WADA, K. 1967 A structural scheme of soil allophane. *Am. Min.*, 52: 690–708.

WRIGHT, A. C. S. 1960 Observaciones sobre los suelos de la zona central de Chile. *Agr. Tec.*, Santiago.

— 1965 The 'Andosols' or Humic Allophane Soils of South America. *Meeting on the classification and correlation of soils derived from volcanic ash. FAO Report 14:* 9–20.

YOSHINAGA, N. and AOMINE, S. 1962 Imogolite in some Ando soils. *Soil Sci. & Pl. Nutr.*, 8: 22–29.

ZAVALETA, A. 1969 Distribución geográfica y características de los suelos derivados de cenizas volcánicas del Peru. *Suelos derivados de cenizas volcánicas de America Latina. Centro de Ensenanza e Investigación del IICC, Turrialba, Costa Rica.*

10

Copyright © 1978 by the New Zealand Soil Bureau, Department of Scientific and Industrial Research

Reprinted from an unpublished paper, 1978, 20p.

A PRELIMINARY PROPOSAL FOR RECLASSIFICATION OF ANDEPTS AND SOME ANDIC SUBGROUPS

Guy D. Smith

The classification of Andepts presented in Soil Taxonomy has a number of serious defects. First, the definition of the suborder clearly excludes a number of soils that should be included if we consider all of their properties. They were required to have their exchange complex dominated by amorphous materials throughout the upper 35 cm and to have low bulk densities, or to have more than 60 percent vitreous ash, etc. Many Andepts have a surface horizon of 15 to 25 cm thickness that does not react to NaF as is required by the definition of domination by amorphous materials. The 60 percent limit of vitreous materials has been interpreted by some to mean that any soil formed in ash and having less than 40 percent clay must be classified as an Andept unless it has a diagnostic horizon prohibited in Inceptisols.

Second, base saturation by NH_4OAc has been used as a differentia with a limit of 50%, the same limit used for mineral soils that have crystalline clays. The significance of this limit in Andepts is open to very serious question because the clays are mostly amorphous and the CEC is largely pH dependent. Few Andepts have more than traces of KCl extractable Al^{3+}, and pH values less than 5.2 are rare, even though amounts of extractable bases are very small, less than 1 meq per 100 g of soil, and base saturation (NH_4OAc) is well below 10%. The Dannevirke silt loam B horizon, for example, has a CEC of 10.7, 0.3 meq of bases, and only 0.3 meq of Al^{3+}. The clay mineralogy is a mixture of chlorite, vermiculite, and allophane, qualifying for domination by amorphous materials. The pH (water) is 5.3. The Papakauri clay loam B_{22} has a CEC of 31 meq, 8% base saturation, 0.00 meq Al^{3+}, and a pH of 6.5. It has no layer lattice clays.

Third, thixotropy has been used as a differentia, but the decision that a given horizon is or is not thixotropic is very subjective and cannot be made uniformly. Thixotropy is partly a function of the water content and also parly a function of the stress applied. During the 1960 Chilean earthquake, road fills made from Andepts failed as a result of thixotropy, though no pedologist would have considered the soil to be thixotropic.

Fourth, the soil moisture regime has not been used as a differentia for Andepts as it has for all other soils. Interpretations for a given family cannot be made without the use of climatic phases.

Fifth, the darkness of the epipedon is weighted heavily in subgroup definitions, but in warm intertropical areas there seems to be little or no relation between color and carbon content, degree of weathering, or any other property. Many volcaniclastic materials are black when deposited and become lighter in color with weathering.

Sixth, for most mineral soils a fragmental particle-size class is provided, but not for Andepts. Coarse pumice falling from the air commonly has a basal layer without appreciable fine earth.

Seventh, inadequate emphasis was given to the unique moisture retention properties of the Andepts. The irreversible effect of drying was used only to define the Hydrandepts. New data require reconsideration of the effects of drying. Rousseaux and Warkentin (SSSAJ, 1976; V.40; 446-456) have studied the effects of drying on pore size and CEC. F.Colmet-Daage has studied the effects of drying on Andepts of Ecuador (Cah.ORSTOM Serie Pedologie, 1967; V.5, 3-38, and 353-392) and of the Antilles (Ref. unavailable). Consideration needs to be given to the possibility of substituting these properties for the presently unsatisfactory property of thixotropy as well as the definitions of ashy and medial.

The moisture retention seems to be a function of two variables, the amount of drying that has occurred in nature in the soil during its formation, and the amount of amorphous clay produced by weathering. In a soil formed in a single tephra, the 15-bar water content of dried or undried samples can vary appreciably, but the effect of drying tends to be relatively constant. The water contents of the Tipoka loam, Egmont County, N.Z. (unpublished data, N.Z. Soil Bureau) can be used to illustrate this. They are as follows:

Depth cm	15-bar water undried	15-bar water air dry	Δ	% decrease of fresh 15-bar water
17-35	62	21	41	66
44-51	60	19	41	68
57-62	66	19	47	71
71-90	43	13	30	70

The differences here are presumably due to the degree of weathering.

The Inglewood series, also in ash from Mt. Egmont, with a mesic temperature regime and a perudic moisture regime, can illustrate the effects of weathering in a soil formed in two tephra (unpublished data, Massey Univ.). The sample at the 80-90 cm depth is from a buried soil in an older tephra, but the moisture regimes under which the two soils formed are believed to be similar.

Depth cm	15-bar water undried	15-bar water air dry	Δ	% decrease of fresh 15-bar water
30-40	43	14	29	68
45-55	47	12	35	74
65-75	20	8	12	60
80-90	67	16	51	76

A Hydrandept, Ecuador 115 (Colmet-Daage, op. cit.) represents a soil in a strongly weathered tephra with an isohyperthermic temperature regime and a perudic moisture regime.

Depth cm	15-bar water undried	15-bar water air dry	Δ	% decrease of fresh 15-bar water
20-40	148	36	112	76
50-80	170	35	134	79
90-120	192	33	159	83

Guy D. Smith

Compare the buried A of the Inglewood, 80-90 cm, weathering at depth in a continuously moist environment, and the Hydrandept, weathering in a similarly moist environment. The 15-bar water contents vary widely, 67 vs 148 to 192, but the change on drying, expressed as the percentage of the moist retention, is virtually the same. The percentage loss on drying seems to be a function of the environment. The 15-bar water of the air dry samples seems to be a function of the amount of amorphous clay. Data for a Hydrandept from Hawaii, the Akaka series, show a loss of 80 to 83 percent of the undried 15-bar water. The data are virtually identical with those of the Ecuadorian Hydrandept.

The new data also require a re-examination of the present definitions of the classes of combinations of particle size and mineralogy, as mentioned earlier. To a considerable extent, they are all subjective. Ashy, for example, "feels like a sand or a loamy sand after prolonged rubbing". Medial "feels loamy . . . after prolonged rubbing. The fine earth fraction is not thixotropic".

In the discussion that follows, geologic terms follow the apparent intent of the AGI Geological Glossary except for terms that specify particle size. Because the Wentworth scale is used by geologists, particle size limits have been modified to fit the more normal particle size classes used in soil laboratories. Thus, ash in the glossary is less than 4 mm, but here it is <2 mm. The Glossary lists more than one set of size limits for some terms, and only relative terms for others. The terms are used as defined below.

Ash: Fine pyroclastic material less than 2 mm in all dimensions.
Cinders: Uncemented juvenile, vitric and vesicular pyroclastic material, more than 2 mm in at least one dimension, with an apparent specific gravity (including vesicles) of more than one, and less than 2.0.
Clastic: A rock or sediment composed of transported fragments of pre-existing rock.
Lapilli: Non or slightly vesicular pyroclastics, 2 to 76 mm in at least one dimension, with an apparent specific gravity of 2.0 or more.
Pumice: Light colored vesicular pyroclastic material with a composition approaching rhyolite and an apparent specific gravity (including vesicles) of less than one.
Pumice-like: Vesicular pyroclastic materials other than pumice but having an apparent specific gravity (including vesicles) of less than one.
Pyroclastic: Clastic rock material formed by volcanic explosion or aerial expulsion from a volcanic vent.
Vitric: Pyroclastic material that contains more than 75% glass.
Volcaniclastic: A clastic rock containing volcanic material in whatever proportion and without regard to its origin or environment.

NEW PROPOSALS FOR DEFINITIONS OF CLASSES OF COMBINATIONS OF PARTICLE SIZE AND MINERALOGY

The following classes are proposed:

Pumiceous: More than 60% of the whole soil is composed of pumice or pumice-like fragments coarser than 2 mm, with insufficient fine earth (or volcaniclastic materials) to fill interstices coarser than 1 mm in at least 10% of the volume of the soil; pumiceous fragments are two-thirds or more of the fragments coarser than 2 mm (by volume).

Cindery: Sixty percent or more of the whole soil (by weight) composed of volcanic ash, cinders, lapilli, and pumiceous fragments; one-third or more (by volume) is cinders and/or lapilli.

Ashy: More than 60% of the whole soil (by weight) volcanic ash, cinders, pumice, or other vitric volcaniclastics; less than 35% (by volume) is 2 mm in diameter or larger; less than 30% water retention at 15-bars on undried samples of fine earth, *and* less than 12% on air dried samples.

Ashy-pumiceous: Thirty five percent or more by volume is greater than 2 mm; pumice or pumice-like fragments larger than 2 mm are two-thirds or more (by volume) of the fraction greater than 2 mm; fine earth is otherwise ashy.

Ashy-skeletal: Thirty five percent or more by volume is greater than 2 mm; pumice and pumice-like fragments are less than two thirds of the fraction greater than 2 mm; fine earth fraction is otherwise ashy.

Medial: Less than 35% (by volume) is greater than 2 mm; water retention at 15-bars is 12% or more on previously dried samples; *or* water retention at 15-bars of undried samples is between 30 and 100%; the exchange complex is dominated by amorphous materials.

Medial-pumiceous: Thirty-five percent or more (by volume) is greater than 2 mm; pumice or pumice-like fragments larger than 2 mm are two-thirds or more (by volume) of the fraction greater than 2 mm; fine earth is otherwise medial.

Medial-skeletal: Thirty-five percent or more (by volume) is greater than 2 mm; pumice and pumice-like fragments are less than two-thirds (by volume) of the fraction greater than 2 mm; fine earth fraction is otherwise medial.

Hydrous: Less than 35% (by volume) is greater than 2 mm; water retention at 15-bars is 100% or more on undried samples of the fine earth; the exchange complex is dominated by amorphous materials.

Hydrous-skeletal: Thirty five percent or more (by volume) is greater than 2 mm; pumice and pumice-like fragments are less than two-thirds of the fraction greater than 2 mm; fine earth fraction is otherwise hydrous. (*Note:* Hydrous-pumiceous is not presently known to occur but should be recognized if found).

Explanation of Proposed Changes

The classes of combinations of particle-size and mineralogy differ significantly from those given in Soil Taxonomy. They have been defined at this point because they appear in the definitions of taxa that follow.

It has been pointed out that the presently defined terms are subjective in application. Thixotropy is a function of the stress applied ant the moisture content of the sample. Ashy, it is said, feels like a sand or loamy sand, and medial feels loamy but is not thixotropic. This proposal attempts to

Guy D. Smith

define ashy and medial in terms of measurable properties, and substitutes hydrous for thixotropic.

Skeletal classes are distinguished from pumiceous classes because of the vastly different properties of pumice, on the one hand, and cinders, lapilli, and lava on the other. The available water retention of ashy-pumiceous soils in New Zealand is as much as 16 to 22% on a volume basis. This is 10 times or more that of sandy-skeletal soils. Engineering properties of pumice are also very different from fragments that are not vesicular. Farm experience in New Zealand shows that a pumiceous soil becomes ashy in the surface horizons just from trampling by livestock. Highway construction problems are also very different.

A class of pumice-like fragments has been added because the AGI Glossary restricts pumice to a rhyolitic composition, but similar dark-colored andesitic fragments occur in New Zealand.

The limits between ashy, medial, and hydrous are subject to modification but were selected as follows:

The limit of 100% 15-bar water for hydrous was suggested first by F. Colmet-Daage. Inspection of all available data showed that few samples were close to this limit. They were either considerably higher in Hydrandepts, or appreciably lower. Only a few that are less than 100% approach this value. It seems to be a very good natural limit.

The limit between ashy and medial was selected in part by deciding, after field discussion, whether a given horizon should be considered ashy or medial, and then determining the 15-bar water on the dried and undried samples (Massey University unpublished data). The definitions were fitted to the data. An attempt was made to follow Soil Taxonomy "definitions" for ashy and medial. The data show that the field estimate is influenced both by the water content, reflected in the figures for undried samples, and by the amount of amorphous materials reflected in the figures for dried samples. The definitions therefore involve both values. It should be noted that if the 15-bar water of undried samples exceeded 30%, the dried samples exceeded 12%, but samples having more than 40%, undried, approached the 12% limit when dried. Some New Zealand reference samples having up to 70% moist, had less than 12% when dried.

PROPOSAL FOR THE RECLASSIFICATION OF ANDEPTS

The proposal that follows elevates the suborder of Andepts to an order, introduces soil moisture and temperature regimes to define suborders, divides some present great groups (but not all) among the suborders, and proposes definitions of typic subgroups for the great groups. In a later section, proposals are made for modifications of Entisols and Spodosols. The proposed definitions are given first and are followed by explanations of the reasons for making them.

Andisols: This name is proposed rather than Andosols because the latter name is currently used elsewhere with other definitions and because the connecting vowel o is supposed to be restricted to Greek formative elements.

The central concept of Andisols is that of a soil developing in volcanic ash, pumice, cinders, and other volcanic ejecta and in volcaniclastic materials, with an exchange complex that is dominated by x-ray amorphous compounds of Al, Si, and humus, or a matrix dominated by glass, and having one or more diagnostic horizons other than an ochric epipedon. Bulk densities are always comparatively low in most horizons, though the absolute values vary with the degree of weathering, the humidity of the soil climate, and in a very few with the degree of cementation by silica or other cements. The most common diagnostic horizons are an umbric, or rarely a mollic epipedon, and a cambic horizon, or an ochric epipedon and a cambic horizon. In the driest climates, there may also be a duripan, and in the wettest climates, a placic horizon is not uncommon.

Andisols have several important properties in common. The amorphous clays have a low permanent charge and a high pH dependent charge. Aluminum toxicity is rare. Phosphate fixation and water retention are high relative to most other soils with comparable textures. Precentages of carbon tend to be high relative to other mineral soils, but bulk densities tend to be low, and the weight of carbon per unit volume does not differ so greatly as the percentages. Andisols pose unique engineering problems particularly because of the fragility of pumice and because it is not uncommon that the liquid limit may be reached before the plastic limit. They can occur with any moisture regime except aridic and any temperature regime.

While Andisols are related to parent materials that have or had appreciable amounts of glass, many other kinds of soil may form in similar materials. Spodosols, Mollisols, Ultisols, Alfisols, and Oxisols may also form in ash if conditions are suitable. Entisols may be in recent historic ash, and Andisols in ash falls that have been dated up to about 15,000 years BP.

Definition

Andisols are mineral soils that do not have an aridic moisture regime or an argillic, natric, spodic, or oxic horizon unless it is a buried horizon, but have one or more of a histic, mollic, or umbric epipedon, or a cambic horizon, a placic horizon, or a duripan; or, the upper 18 cm, after mixing, have a color value, moist, of 3 or less and have three percent or more organic carbon in the fine earth; and, in addition, have one or more of the following combinations of characteristics:

1. Have, to a depth of 35 cm or more, or to a lithic or paralithic contact that is shallower than 35 cm but deeper than 18 cm, a bulk density of the fine earth fraction of less than 0.85 g/cc (0.9?) (at 1/3 bar water retention of undried samples) and the exchange complex is dominated by amorphous materials.
2. Have, in the major part of the soil between a depth of 25 cm and 1 m or a duripan, a placic horizon, or a lithic or paralithic contact that is deeper than 35 cm but shallower than 1 m, a bulk density of the fine earth fraction of less than 0.85 g/cc (at 1/3 bar water (0.9?) retention of undried samples) and the exchange complex is dominated by amorphous materials.

3. Have 60% or more, by weight, of noncalcareous vitric volcanic ash, pumice or pumice-like fragments, cinders, lapilli, or other vitric volcaniclastic materials either to a depth of 35 cm or more, or in the major part of the soil between 25 cm and 1 m or a lithic or paralithic contact that is shallower than 1 m, and the pH in the major part of these horizons of 1 g of fine earth in 50 ml of 1N NaF is 9.2 or more after two minutes.
4. Have a weighted average (by thickness of subhorizons) in the major part of the soil between a depth of 25 cm and 1 m or a duripan or paralithic or lithic contact shallower than 1 m, a water retention of undried fine earth at 15-bars pressure of 40% or more; an ustic moisture regime and a bulk density of the fine earth fraction of 0.9 or less; and in addition either a ratio of 15-bar water percentage (undried) to the meq of exchangeable bases is 1.5 or less, or the pH of 1 g of fine earth in 50 ml of 1N NaF is 9.4 or more after two minutes, or both.

Explanation of Definition

The opening paragraph transfers the present requirements of Andepts from Inceptisols to Andisols, and, for the vitric great groups (item 3), presently Vitrandepts, eliminates problems about cambic horizons (which may not have sandy particle size) and the thickness of the umbric epipedon. The concept of a sandy particle size does not apply well in Andisols, particularly those in rhyolitic ash because of their high water holding capacity. The carbon limit is intended to eliminate problems with basaltic or andesitic ash, cinders, and lapilli, all of which are nearly black.

Item 1 is intended to bring into Andisols the soils formed in thin deposits of ash. It does not differ significantly from the present definition of Andepts.

Item 2 is intended to bring into Andisols the soils formed in thicker deposits of ash but lacking allophane in the surface horizon. In the West Indies and South America, soils that should be grouped with Andisols commonly do not react to NaF in the upper 15 to 25 cm but react strongly below. The Japanese literature reports the absence of allophane in some surface horizons, although it is the only clay identified below.

Item 3 is intended to clarify the classification of soils formed in ash but belonging to other orders. If a soil with a bulk density greater than 0.85 g/cc is classified as an Andisol because it has formed in ash, and glass is thought to be present, it must react to NaF. However, because surface area may be small, as in yellow-brown pumice soils of New Zealand, the pH is required to rise only to 9.2 instead of the 9.4 required for domination by amorphous materials. Not all of the yellow-brown pumice soils reach a pH as high as 9.4 (unpublished data, N.Z. Soil Bureau).

Item 4 is intended to clarify the classification of the present Eurtrandepts of Hawaii. Some are calcareous in the lower horizons. Water retention, bulk densities, and CEC are appropriate for Andisols, but not all react adequately to NaF, or they are calcareous and the test is meaningless. Exchangeable bases are very high, so the ratio of 15-bar water to bases is low (unpublished SCS data).

It should be noted that the proposed definition transfers the Andaquepts to Andisols along with the Andepts.

KEY TO SUBORDERS

A. Andisols that have artificial drainage or an aquic moisture regime and have one or more of the following:
   a. A histic epipedon
   b. At a depth of less than 50 cm or immediately below an epipedon that has color values, moist, of 3 or less, dominant colors, moist, on ped faces or in the matrix, if peds are absent, as follows:
      (1) If there is mottling, chroma of 2 or less
      (2) If there is no mottling, chroma of 1 or less
      (3) Distinct or prominent, coarse or medium mottles due to segregation of iron within or immediately below 18 cm of the surface or any Ap deeper than 18 cm [irrespective of] chroma
   c. A placic horizon that rests on a duripan
   AQUANDS

B. Other Andisols that have a frigid or cryic temperature regime
   BORANDS

C. Other Andisols that have a xeric moisture regime
   XERANDS

D. Other Andisols that have an ustic moisture regime or a duripan, or both
   USTANDS

E. Other Andisols that have an isomesic or warmer iso-temperature regime or a hyperthermic temperature regime
   TROPANDS

F. Other Andisols (that have a udic or perudic moisture regime)
   UDANDS

Explanation of Key to Suborders

The suborders proposed are more or less parallel to suborders of Alfisols and several other orders. However, a suborder of Tropands is added, parallel to Tropepts, because of the traditional emphasis on the color of the Andepts. In warm, humid, intertropical areas, colors of the A_1 horizons seem poorly related to carbon, CEC, or any other property, and somewhat different differentiae seem to be needed. While the Andisols of Hawaii have dark colors, it is almost universally true that Hawaiian surface soils and subsoils have moist values of 3 or less, and the darkness seems partly due to iron rather than carbon.

Guy D. Smith

Duripans are known to occur in both Xerands and Ustands, but some West Indian soils that have duripans seem to have moisture regimes marginal to udic. It seems better to keep these soils together in the taxonomy.

KEYS TO GREAT GROUPS

AQUANDS

 AA. Aquands that have a duripan or a placic horizon that rests on a duripan

    DURIAQUANDS

 AB. Other Aquands that have 15-bar water retention of previously dried samples of less than 15% on the weighted average of all horizons between 25 cm and 1 m and have less than 30% 15-bar water on undried samples of the same horizons

    VITRAQUANDS

 AC. Other Aquands that do not have a placic horizon

    HAPLAQUANDS

BORANDS

 BA. Borands that have an epipedon 30 cm or more thick with color values, moist, of 2 or less and chromas of less than 2 throughout, or have a subsurface horizon (a buried A_1) that meets these requirements and has an upper boundary within 30 cm of the surface if it is 30 cm thick, or has an upper boundary within 50 cm of the surface if it is 50 cm or more thick; and has 8% or more organic carbon throughout these thicknesses

    MELANOBORANDS

 BB. Other Borands that have a cryic or pergelic soil temperature regime

    CRYOBORANDS

 BC. Other Borands that have a placic horizon within 1 m of the surface in half or more of each pedon

    PLACOBORANDS

 BD. Other Borands that have 15-bar water retention of previously dried samples of less than 15% on the weighted average of all horizons between 25 cm and 1 m or a lithic or paralithic contact shallower than 1 m *and* have less than 30% 15-bar water on undried samples of the same horizons

    VITRIBORANDS

 BE. Other Borands.

    HAPLOBORANDS

XERANDS
- CA. Xerands that have a duripan or a placic horizon that rests on a duripan

  DURIXERANDS

- CB. Other Xerands that have 15-bar water retention of previously dried samples of less than 15% on the weighted average of all horizons between 25 cm and 1 m or a lithic or paralithic contact shallower than 1 m *and* have less than 30% 15-bar water on undried samples of the same horizons

  VITRIXERANDS

- CC. Other Xerands

  HAPLOXERANDS

USTANDS
- DA. Ustands that have a duripan

  DURUSTANDS

- DB. Other Ustands that have 15-bar water retention of previously dried samples of less than 15% on the weighted average of all horizons between 25 cm and 1 m or a lithic or paralithic contact shallower than 1 m *and* have less than 30% 15-bar water on undried samples of the same horizons

  VITRUSTANDS

- DC. Other Ustands

  HAPLUSTANDS

TROPANDS
- EA. Tropands that have a placic horizon within 1 m of the soil surface in half or more of each pedon

  PLACOTROPANDS

- EB. Other Tropands that have 15-bar water retention of undried samples of 100% or more on the weighted average of all horizons between 25 cm and 1 m or a lithic or paralithic contact shallower than 1 m

  HYDROTOPANDS

- EC. Other Tropands that have a 15-bar water retention of previously dried samples of less than 15% on the weighted average of all horizons between 25 cm and 1 m or a lithic or paralithic contact that is shallower than 1 m and have less than 30% 15-bar water on undried samples of the same horizons

  VITRITROPANDS

Guy D. Smith

    ED. Other Tropands

                                                 HAPLOTROPANDS

UDANDS
    FA. Udands that have a placic horizon within 1 m in half or more of each pedon

                                                 PLACUDANDS

    FB. Other Udands that have a 15-bar water retention of undried samples of 100% or more on the weighted average of all horizons between 25 cm and 1 m or a lithic or paralithic contact that is shallower than 1 m

                                               HYDRUDANDS

    FC. Other Udands that have an epipedon 30 cm or more thick with color values, moist, of 2 or less and chromas of less than 2 throughout, or have a subsurface horizon (a buried A_1) that meets these requirements and has an upper boundary within 30 cm of the surface if it is 30 cm thick, or has an upper boundary within 50 cm of the surface if it is 50 cm or more thick; and has 8% or more organic carbon throughout these thicknesses

                                               MELANUDANDS

    FD. Other Udands that have a 15-bar water retention of previously dried samples that is less than 15% on the weighted average of all horizons between 25 cm and 1 m or a lithic or paralithic contact that is shallower than 1 m *and* have less than 30% 15-bar water on undried samples of the same horizons

                                                 VITRUDANDS

    FE. Other Udands

                                                 HAPLUDANDS

Explanation of the Great Groups

The proposal provides duric great groups only in the soils that become seasonally dry, the Xerands and Ustands.

Hydric great groups are provided for the freely drained soils that rarely or never become drier than field capacity. The limit of 100% 15-bar water is the same as that proposed for the hydrous combination of particle size and mineralogy. Most Hydrandepts are far above this limit. It is not certain that Hydrudands exist, but they are provided temporarily because the literature suggests that they occur in Chile. None have been found in New Zealand, perhaps because the ash is too young in the most humid soils.

Placic great groups are provided in three suborders, Borands, Udands, and Tropands. The writer has not seen any Borands, but it seems probable that Placoborands exist. Placic horizons are known to the writer in both Tropands and Udands. They could occur in Aquands but have not been seen by the writer in mottled soils.

Vitric great groups are provided in all suborders. While the ashy class is proposed as having less than 12% 15-bar water after drying, a limit of 15% is proposed tentatively for the great groups to permit vitric great groups to have some medial horizons. The limit of 30% 15-bar water is the same as that of the ashy classes because few medial samples seem to approach this limit. Values of 40 to 70% are common in New Zealand soils even though the dried samples have less than 12% 15-bar water.

Two melanic great groups are proposed tentatively. The significance of the black color is not entirely clear, but the black soils have been distinguished from other Andisols in the past, and certainly the morphology is vastly different. Umbric epipedons from one to two meters thick seem to be very common. Most Melanoborands seem to have cryic temperature regimes, particularly in the Andes of South America, but they bear almost no resemblance to the Cryoborands of Alaska, particularly in appearance.

SUBGROUPS

There follows a list of subgroups that seem necessary or potential for the various great groups that have been suggested earlier. Following this list, proposed wordings are given for the items in the typic subgroup definitions to provide for the suggested subgroups. For the definition of the typic subgroup of a given great group, the reader must note which subgroups are proposed for that great group and then refer to the wording suggested to provide for each of the proposed subgroups.

Proposed Subgroups in Addition to Typic

AQUANDS
  Haplaquands
    Allic
    Entic
    Hydric
    Tropic
    Ustic
    Vitric
    Xeric
    Thapto-Histic

  Vitraquands
    Allic
    Entic
    Tropic
    Ustic
    Xeric
    Thapto-Histic

BORANDS
  Cryoborands
    Allic
    Placic
    Pergelic
    Tropic
    Vitric

  Haploborands
    Allic
    Aquic
    Entic
    Placic
    Vitric

Melanoborands
  Acric
  Allic
  Aquic
  Cryic
  Lithic
  Pachic
  Vitric

Placoborands
  Ruptic

Vitriborands
  Entic
  Placic

Proposed Subgroups in Addition to Typic

TROPANDS
  Haplotropands
    Acric
    Allic
    Aquic
    Entic
    Hydric
    Lithic
    Oxic
    Placic
    Vitric

  Hydrotropands
    Altic
    Lithic
    Placic

  Placotropands
    Ruptic

  Vitritropands
    Aquic
    Allic
    Entic
    Lithic
    Placic
    Psammic

UDANDS
  Hapludands
    Allic
    Aquic
    Entic
    Hydric
    Lithic
    Placic
    Vitric

  Hydrudands
    Altic
    Lithic
    Placic

  Melanudands
    Acric
    Allic
    Aquic
    Hydric
    Lithic
    Pachic
    Vitric

  Placudands
    Ruptic

  Vitrudands
    Allic
    Aquic
    Entic
    Lithic
    Placic
    Psammic

USTANDS
  Durustands
    Vitric

  Haplustands
    Aquic
    Entic
    Lithic
    Ustolic
    Vitric

  Vitrustands
    Entic
    Lithic
    Psammic

XERANDS
  Durixerands
    Vitric

  Haploxerands
    Aquic
    Entic
    Lithic
    Vitric

  Vitrixerands
    Entic
    Lithic
    Psammic

Subgroup Definitions

The suggested wording of items in the definitions of typic subgroups to provide for the subgroups proposed in the various great groups are as follows:

ACRIC: Have, in all subhorizons between 25 cm and 1 m, extractable bases plus KCl extractable Al, expressed as Al^{3+}, that is 1.5 meq per 100 g fine earth or more when the sum of bases plus Al^{3+} is divided by $(2.5 \times \%$ 15-bar water [air dried])/100.

Explanation: Extractable bases and KCl extractable Al are commonly very low in Andisols of humid regions because of the absence of any permanent charge in the amorphous colloids. The range in extractable cations in Andisols is from less than 0.2 meq to more than 50 meq in Ustands, and some provision is needed for the soils with extremely low amounts of bases where Ca deficiencies create problems with root growth. Potential Al toxicities are also present, but quantities of cations can be so low that calculations of Al saturation are unreliable. The use of the 15-bar water percentage of dried samples multiplied by 2.5 is an attempt to adjust values for the amounts of clay present. This relationship was discussed earlier.

ALLIC: Have, in all subhorizons between a depth of 25 cm and 1 m, or, if an Ap is present, between a depth of 25 cm below the base of the Ap and 1 m, or between these depths and a lithic or paralithic contact shallower than 1 m, KCL extractable Al, expressed as Al^{3+}, that is less than one-third of the sum of extractable bases, if the sum of bases plus Al^{3+} is 1.5 meq or more when divided by $(2.5 \times \%\ 15\text{-bar water [air dried]})/100$.

Explanation: Most Andisols have only traces or no KCl extractable Al. The most common reasons for its presence are the presence of some crystalline clay and the use of acid forming fertilizers. Because the fertilizers can affect horizons just below the depth of fertilizer application, the presence of KCl extractable Al must be tolerated in cultivated soils below the Ap. If the sum of cations is very low, as in the acric subgroups, the ratio between bases and Al cannot be measured accurately. Therefore, Al saturation is proposed for use only if there are some tenths of a meq of cations present. The use of the 15-bar water percentage of air dried soil is an attempt to adjust absolute values for the amounts of clay present. The ratio proposed is for an Al saturation of 25%. Some will consider this too low. Some will consider that the Al saturation is not mappable. An alternative is to use the pH in water or KCl. A pH of 5.1 in water should be excluded from the typic subgroups. In the West Indies, a few unquestionable Andisols have pH values in water of as low as 4.7. These must be distinguished from typical Andisols by some means. Pedon 6 of Soil Taxonomy, with a pH of about 5.1 is about 80% Al saturated.

ALTIC: Have a negative Δ pH of 0.3 or less (pH KCl-pH H_2O) in some sub-horizon within 1 m of the soil surface.

Explanation: This subgroup is proposed only for hydric great groups. Most soils presently classified as Hydrandepts or as hydric subgroups of Dystrandepts have a net positive charge or no charge in deeper horizons. Negative charges in surface horizons seem to be due more to organic matter than mineral colloids, and decrease with depth as the organic matter diminishes. At the same time, anion retention increases with depth, particularly of sulfates. This is considered typical. The altic subgroup (L altus, high, for high negative charge) is an extragrade provided for the hydric great group soils that have an appreciable negative charge at depth.

AQUIC: Do not have distinct or prominent medium or coarse mottles due to segregation of iron, and do not have mottles that have chromas of 2 or less, within 1 m of the soil surface if the mottled horizon is saturated with water at some season of the year or the soil is artificially drained.

Explanation: Aquic subgroups are provided for the Andisols that are mottled at depth and that either have ground water in the mottled horizons or have been artificially drained. The loss of the 15-bar water on drying tends to be high if expressed as the percentage of the 15-bar water of undried samples. Hence, the provision for hydric subgroups needs to be waived in aquic subgroups. It should be noted that high chroma mottles may be found at the contact between strongly contrasting particle size classes, but should not place soils in aquic subgroups because the contact will not meet the specifications for saturation with water.

CRYIC: Do not have a cryic temperature regime.

Explanation: A cryic subgroup is provided only for Melanoborands, many but not all of which are cryic. The frigid is called typic so that the term cryic will appear in the name of all cryic soils. Most of the Cryic Melanoborands will probably be in a Cryic Tropic subgroup.

ENTIC: Have 5% or more organic carbon throughout the upper 25 cm, or have a subsurface horizon that has an upper boundary within 30 cm of the surface (a buried A_1) that meets this requirement.

Explanation: Entic subgroups are suggested because they are in Soil Taxonomy, but with very serious reservations. Color value is useless in Tropands and in soils formed in black cinders, lapilli, and some ashes, as a diagnostic, so carbon contents are substituted. It seems possible that the vitric subgroups can be substituted for the entic, and the latter dropped. The utility of this subgroup needs discussion. West Indian Tropands are commonly eroded and subsoils exposed by the practice of hoeing down slope. The entic subgroups would require different series for eroded and uneroded soils.

HYDRIC: Lose less than 75% of the 15-bar water of undried samples by air drying on the weighted average (by thickness) of all horizons between 25 cm and 1 m or a lithic or paralithic contact shallower than 1 m, and have less than 70% 15-bar water before drying.

Explanation: Some soils in Haplic great groups approach the hydric great groups both in 15-bar water in undried samples and in the loss of 15-bar water on drying, if that is expressed as the percentage of the 15-bar water in the undried samples. The Patua loam of New Zealand would be an example. It has a mesic temperature and about 4 m of precipitation. The 15-bar water of undried samples is about 75%, and that of air dried samples is 11%, a loss of 85% on drying. Hydrotropands lose about 80% of their 15-bar water on drying. The requirement of 70% 15-bar water in fresh samples introduced for the hydric subgroups by this item may not be needed, but is suggested to eliminate the very slightly weathered ash. This provision should be waived in aquic subgroups, which also show a high loss of 15-bar water on drying.

LITHIC: Do not have a lithic contact within 50 cm of the surface.

Explanation: This item has been used throughout Soil Taxonomy except in Oxisols.

OXIC: Have in some subhorizon between 25 cm and 1 m, less than 30% 15-bar water in air dried samples if the sum of bases plus KCl extractable al, expressed as Al^{3+} is less than 2.5 meq per 100 g fine earth when divided by $(2.5 \times \%$ 15-bar water [air dried])/100, and if there is less than 10% weatherable minerals in the 0.2 to 0.02 mm fraction.

Explanation: In warm humid climates, some Andisols in old tephra seem to grade into Oxisols with a loss of weatherable minerals, extractable cations, and of allophane or allophane-like clays. The latter are largely changing to halloysite and kaolin. Because these more completely weathered volcaniclastics lack much silt or sand, the 15-bar water in dried samples begins to approach 40%. Hence, the soils that have a high 15-bar water content after drying and that lack appreciable amounts of weatherable minerals and extractable cations are placed in oxic subgroups..The oxic subgroup is presently suggested only for Haplotropands.

PACHIC: Have an umbric epipedon that is less than 1 m thick.

Explanation: The umbric epipedon of soils in Melanic great groups may vary from a minimum of 30 cm to a maximum of 2 m or more. The thick epipedons commonly cover the entire landscape irrespective of the position on the landscape. They could represent slow accumulation of ash, but the evidence in the Andes is against this hypothesis. There, closely associated Mollisols formed in presumably the same ash may have a well developed argillic horizon in the upper third of a very thick mollic epipedon. This suggests stability rather than accumulation. No method of distinguishing pachic and cumulic subgroups seems practical, so no cumulic subgroups are suggested. There is a question about the thickness limit of the pachic subgroup. Perhaps 75 cm would be better than 1 m.

PERGELIC: Have a mean annual soil temperature higher than 0° Celcius.

Explanation: This definition is the same as that in other kinds of soil in Soil Taxonomy.

RUPTIC-PLACIC: Do not have an intermittent placic horizon within 1 m of the surface in more than one-fifth of the area of each pedon.

Explanation: This subgroup is provided for soils that have an intermittent placic horizon in less than half of each pedon (the limit for placic great groups) but in more than one-fifth of the pedon. It is proposed to tolerate small areas of intermittent placic horizons in typic subgroups because they probably are not significant barriers to water movement or root growth. It is also proposed to permit, in typic subgroups, incipient accumulations of iron in thin horizons if they are soft and plastic and do not interfere with roots.

RUPTIC SUBGROUPS OF PLACIC GREAT GROUPS: Have a placic horizon that is continuous throughout each pedon, or is present in 90% or more of each pedon.

Explanation: While there are large areas of placic great groups in which the placic horizon is continuous, the presence of small areas within a placic horizon must be tolerated in typic subgroups. If the areas without a placic horizon become significant, ruptic subgroups are proposed, following the term ruptic with the name of the great group whose definition fits the soil where the placic horizon is absent. Thus, subgroup definitions would read "like the typic except for (the above item), and . . . followed by identification of the appropriate great group".

PSAMMIC: Have less than 70% fine to coarse sand (0.1 to 2 mm), or more than 35%, in volume, greater than 2 mm in some subhorizon within 1 m of the surface.

Explanation: This extragrade subgroup is provided for Andisols that are particularly subject to blowing and drifting. The definition is as nearly comparable to that of Psamments as is possible, but because of dispersion difficulties with Andisols and the vesicular nature of some of the sands, it is necessary to rely on sieving rather than sedimentation. Many pumiceous sands float in water. Pumice particles of gravel size, up to about 20 mm in the largest dimension, are found in coppice dunes mixed with andesitic ash fine and medium sands. However, in the least dimension, these gravels approach 2 mm in thickness. For the present, it seems adequate to sieve the less than 2 mm fraction.

TROPIC: Do not have an iso-temperature regime.

Explanation: Tropic subgroups are proposed for Aquands and Cruyoborands. The usage in Aquands is parallel to that in Aquepts. In Cryoborands it seems essential because temperature is not specified in families of cryic great groups, but the potential uses of a cryic soil in mid- or high-latitudes are very different from those in low latitudes. In intertropical regions, cryic soils have frost every night, but in higher latitudes there is a frost free growing season. Melanoborands that have an isofrigid temperature should be in a Cryic Tropic subgroup.

USTIC: If not irrigated, are not dry in some or all parts of the moisture control section for as long as 90 cumulative days in most years.

Explanation: Ustic subgroups are provided for Aquands as a temporary expedient pending revision of the definition of the ustic moisture regime and the possible definition of an additional moisture regime for the wet and dry soils of intertropical regions. These are soils that must be drained during the rainy season and, if perennial crops are to be grown, irrigated during the dry season. The writer has previously proposed similar ustic subgroups for aquic great groups in several orders for soils in Guyana and Venezuela where dry seasons are very long, but rainy seasons very wet.

USTOLLIC: Do not have a subhorizon within 1.5 m of the surface that contains soft, powdery secondary lime.

Explanation: The ustollic subgroup is provided in Soil Taxonomy for Eutrandepts. This merely continues the present subgroup, but because they would be restricted to Ustands, the subgroup name might better be "mollic" to prevent repetition of the formative element "ust."

VITRIC: Have 12% or more 15-bar water after air drying on the weighted average of all subhorizons between 25 cm and 1 m or a lithic or paralithic contact shallower.

Explanation: Melanudands and Cryoborands include some soils that, in other suborders, would meet the definitions of vitric great groups. The nature of the epipedon of Melanudands, and the temperature of Cryoborands are considered more important than the ashy nature of the soil. In haplic great groups, the 15-bar water content of the fresh samples may range from 30 to 100%, but the 15-bar water of air dried samples may be less than 12%, one of the limits of the ashy class. The vitric soils that are ashy are identified at the family level. The medial soils are not distinguished according to the dry 15-bar water contents at the great group or family levels. This is considered the best available measure of the amount of amorphous clay present, and may range in haplic great groups from less than 10 to more than 30%.

Vitric subgroups should not have different definitions in different great groups, or the system becomes overcomplicated. Hence, only the 15-bar water content of dried samples is used in the definition, and the limit is one of the limits of the ashy class.

XERIC: Unless irrigated, are not dry in all parts of the moisture control section for as long as 45 consecutive days during the 6 months following the winter solstice, in 6 or more years out of 10.

Explanation: The xeric subgroup definition is more or less parallel to that of the Xeric Albolls, but the period following the winter solstice has been extended to permit exhaustion of the ground water. These soils, like the ustic subgroups, require both drainage and irrigation for perennial crops or summer crops.

THAPTO-HISTIC: Do not have a buried Histosol with an upper boundary within 1 m of the surface.

Explanation: This subgroup definition is parallel to others in Soil Taxonomy.

Family Differentiae

Proposed redefinitions of combinations of particle size and mineralogy have already been given. The strongly contrasting classes require some attention. Medial horizons over ashy horizons in New Zealand show some mottling at the contact, and should be considered strongly contrasting. Medial and hydrous contacts have not been observed. It is proposed that medial and ashy be considered strongly contrasting, and that *any* contact with pumiceous, which is similar to fragmental, be considered strongly contrasting. Note that this does not include ashy-pumiceous or medial-pumiceous.

Extension of Proposals to Spodosols and Entisols

Soil Taxonomy provides for the use of combinations of particle-size and mineralogy in selected taxa of Spodosols and Entisols. Some changes seem

Guy D. Smith

needed. The spodosols of New Zealand that are developed in ash and pumice have much higher contents of carbon and 15-bar water (fresh) than those developed in loess or till. It is proposed that ashy, medial, and hydrous, and their combinations with pumiceous and skeletal, be used in andic subgroups of Spodosols. It is proposed that these be defined by adding the item, where needed, to the typic subgroup definition as follows: "Do not have, below the spodic horizon, horizons that meet requirements for ashy, medial, or hydrous fine earth." In New Zealand, soils that would not meet this specification, i.e. those that are ashy or medial below the spodic horizon, have mesic or frigid temperature regimes and resemble the Andisols in all properties below the spodic horizon. Pedon 27, Soil Taxonomy, p.538, is an example of a hydrous, cryic soil that should be an Andic Spodosol.

Very recent volcaniclastics, less than 50 to 100 years old, have fallen on pre-existing soils having almost any slope. The definitions of Soil Taxonomy create the ridiculous classification of Fluvents on the ridge crests where the slope is less than 25%, and Orthents on the slopes below where the slopes are greater than 25%. The definition of Fluvents should be modified to exclude soils in which the horizons that have a higher carbon content than the overlying horizon meet the requirements for Andisols.

Part IV
PHYSICAL CHARACTERISTICS

Editor's Comments
on Papers 11 and 12

11   MAEDA, TAKENAKA, and WARKENTIN
Physical Properties of Allophane Soils

12   SHOJI and ONO
Physical and Chemical Properties and Clay Mineralogy of Andosols from Kitakami, Japan

Andosols exhibit physical characteristics that differ considerably from those of other mineral soils. Unique among the characteristics are the low bulk density, problems in dispersion, high water-holding capacity, and irreversible physical changes on drying. Paper 11 summarizes most of the major physical characteristics of Andosols. Although Warkentin and Maeda (1980) have recently published another article on the physical properties of Andosols, the older publication, reproduced in this volume, gives a better coverage of the subject material than the new article. Paper 12 provides a limited discussion on selected physical properties of Andosols in Japan and is included here for comparison of data in Paper 11. Some selected physical data are also discussed in Paper 17. Because few scientists have investigated the physical properties of Andosols, my selection of articles for reproduction was limited here.

Although many scientists believe that the strange physical properties of Andosols are attributed to the presence of high amounts of organic matter and allophane and/or other amorphous clay, Warkentin and Maeda (1974) and Maeda and Warkentin (1975) reported that allophane was primarily responsible for the unusual physical behavior of these soils. The fact that organic matter also plays an important role in the development of the physical characteristics in Andosols can be illustrated by some examples.

The high levels of organic matter tend to impart to the soils a dark color and are also the reason for the presence of extremely low bulk density values, high water-holding capacity, and formation of very stable granular structures. Bulk density values ranging from 0.3 to 0.8

g/cc are common in A horizons of Andosols. A dry bulk density of < 0.85 tons/m^3 has been employed for characterization of the soils in the U.S. *Soil Taxonomy* (U.S. Soil Survey Staff, 1975), while Maeda, Soma, and Warkentin (1983) indicated that the value of 0.85 tons/m^3 perhaps can be used as a reference point dividing Andosols from other mineral soils of nonvolcanic origin.

Of the amount of water retained by the soil, the portion available for plant growth, corresponding to field capacity, has been reported to occur at 0.1 bar instead of at 0.3 bars suction as in other soils. Extremely high amounts of water are also present at 15 bars suction, which is usually considered the wilting point (Gradwell, 1976). In this respect, Rousseaux and Warkentin (1976) noted that this unique problem in water retention was perhaps due to the proportion of micropores and the structure of allophane. From their studies on surface properties and forces holding water in Andosols, these authors found a high proportion of micropores possessing a radius $\leq$ 20 Å. Since in addition most of the hydroxyl groups of allophane were believed to occur at the surfaces of the mineral (Wada, 1966, 1967) giving to allophane the strong affinity for water, Rousseaux and Warkentin (1976) concluded that pore geometry and nature of active surfaces were the reasons for this exceptionally high water retention capacity of Andosols. Maeda, Takenaka, and Warkentin (Paper 11), however, noted that a decrease in organic matter content could bring about a large decrease in water content and cautioned about adhering to the idea that water retention in Andosols is ascribed totally to allophane.

Another important physical property of Andosols that is less frequently mentioned or investigated is the structure. The soil structure is composed of a macrostructure and a microstructure component. In terms of macrostructure, the A horizons are generally characterized by a peculiar granular structure formed by a process Mohr (1937) called mountain granulation. This structure differs from the ordinary granular structures of other soils by the fact that the structural units are very resistant to the impact of falling raindrops. Because of this resistance and since in dry condition they feel gritty, the units have been referred to as "pseudo sand" (Mohr, 1937). Most probably peptized aluminum and iron oxide play an important role as binding agents. Warkentin and Maeda (1980) believe that gel masses serve as coatings in the aggregation of the clay particles. The size, shape, and arrangements of particles and voids, known as microstructure or soil fabric (Fitzpatrick, 1980; Brewer, 1964), can be investigated by using soil thin sections; however, not much information in this respect is available in the literature. Figures 3 and 4 show thin sections of an

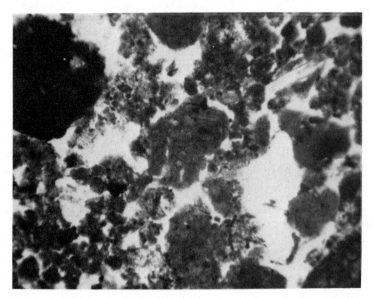

Figure 3. Thin section of the A horizon of an Andosol in the mountain regions north of Bandung (West Java, Indonesia; Obj. = 10×, Occ. = 10×).

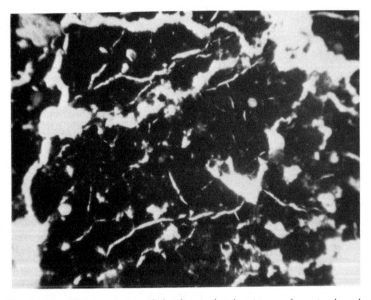

Figure 4. Thin section of the buried A horizon of an Andosol in the mountain regions north of Bandung (West Java, Indonesia; Obj. = 10×, Occ. = 10×).

Andosol in Indonesia. The samples were prepared from the Andosol pedon shown in Figure 2. The yellowish-brown (10 YR 5/6-4/4) soil matrix of the A horizon (Figure 3) reveals a porous, granular, and loose appearance. Amorphous oxide starts to flocculate, giving to the soil mass a brown coloration. According to Kubiena (1938, 1970), such a soil fabric is called an "Earthy Braunlehm" or an agglomeratic fabric; Brewer (1964) would call it an agglomeroplasmic fabric. Egawa (Paper 17) noted that a fine-grained, loose fabric, as shown in Figure 3, was the most characteristic microstructure of Andosols. On the other hand, the buried A horizon of the Andosol pedon does not possess such characteristics. As Figure 4 shows, the soil fabric of the buried A horizon exhibits an angular blocky structure in terms of the classification of Fitzpatrick (1980). The structural units seem to be built up mostly of peptized black colloidal organic matter, forming a rather massive, smeary appearance with little or no micropores. The black to very dark gray (7.5 YR 3/2-3/3) dense fabric is dotted by reddish-yellow (7.5 YR 6/6) iron oxide particles. Using the table for microstructures provided in Paper 17, this type of fabric fits the description of a blocky, loose fabric, while in Kubiena's terms (1938) it qualifies for an agglomeroplasmic fabric. Using a more recent description provided by Kubiena (1970), however, the soil fabric has some similarities to an "Anmoor-like" fabric, which finds support with the descriptions presented by Fitzpatrick (1980). Whatever the name, such a soil fabric dominated by large amounts of black humus tends to lend support to the idea of an internal peat formation in the buried A horizon as indicated by Kubiena's term *Anmoor-like*.

REFERENCES

Brewer, R., 1964, *Fabric and Mineral Analysis of Soils,* Robert E. Krieger Publ. Co., New York, 482p.

Fitzpatrick, E. A., 1980, *The Micromorphology of Soils. A Manual for the Preparation and Description of Thin Sections,* Univ. Aberdeen, Dept. Soil Sci., 186p.

Gradwell, M. W., 1976, Available Water Capacities of Some Intrazonal Soils of New Zealand, *New Zealand Jour. Agric. Res.* **19:**69-79.

Kubiena, W. L., 1938, *Micropedology,* Collegiate Press, Ames, Iowa, 243p.

Kubiena, W. L., 1970, *Micromorphological Features of Soil Geography,* Rutgers Univ. Press, New Brunswick, New Jersey, 254p.

Maeda, T., K. Soma, and B. P. Warkentin, 1983, Physical and Engineering Characteristics of Volcanic Ash Soils in Japan Compared with Those in Other Countries, *Irrigation Eng. and Rural Planning* **3:**16-31.

Maeda, T., and B. P. Warkentin, 1975, Void Changes in Allophane Soils Determining Water Retention and Transmission, *Soil Sci. Soc. Am. Jour.* **39:**398-403.

Mohr, E. C. J., 1937, *De Bodem der Tropen in het Algemeen en die van Nederlands Indie in het Bijzonder,* trans. R. L. Pendleton as *Soils of Equatorial Regions,* John Edwards, Ann Arbor, 1944, 765p.

Rousseaux, J. M., and B. P. Warkentin, 1976, Surface Properties and Forces Holding Water in Allophane Soils, *Soil Sci. Soc. Am. Jour.* **40:**446-451.

U.S. Soil Survey Staff, 1975, *Soil Taxonomy. A Basic System of Soil Classification for Making and Interpreting Soil Surveys,* Soil Conservation Service, U.S. Dept. Agric., Agric. Handbook 436, 754p.

Wada, K., 1966, Deuterium Exchange of Hydroxyl Groups in Allophane, *Soil Sci. Plant Nutr.* (Japan) **2:**176-182.

Wada, K., 1967, A Structure Scheme of Soil Allophane, *Am. Minerlogist* **52:**690-708.

Warkentin, B. P., and T. Maeda, 1974, Physical Properties of Allophane Soils from the West Indies and Japan, *Soil Sci. Soc. Am. Proc.* **38:**372-377.

Warkentin, B. P., and T. Maeda, 1980, Physical and Mechanical Characteristics of Andisols, in *Soils with Variable Charge,* B. K. G. Theng, ed., New Zealand Soc. Soil Sci., Lower Hutt, New Zealand, pp. 281-301.

11

Copyright © 1977 by Academic Press, Inc.
Reprinted from *Advances Agronomy* **29**:229-264 (1977)

PHYSICAL PROPERTIES OF ALLOPHANE SOILS

T. Maeda,[1] H. Takenaka,[2] and B. P. Warkentin[3]

```
 I. Introduction .................................................... 229
    A. Allophane Soils ............................................. 229
    B. General Nature of Physical Properties of Allophane .......... 231
 II. Index Properties ............................................... 232
    A. Grain Size Distribution ..................................... 232
    B. Plasticity .................................................. 234
    C. Surface Area and Heat of Wetting ............................ 237
    D. Mineral Density ............................................. 238
    E. Thermal Conductivity ........................................ 238
III. Structure of Allophane Soils ................................... 241
    A. Description of Structure .................................... 241
    B. Model for Physical Properties of Allophane .................. 244
 IV. Physical Characteristics of Allophane Soils .................... 246
    A. Volume Change ............................................... 246
    B. Water Retention ............................................. 247
    C. Water Transmission .......................................... 250
    D. Field Studies on Infiltration and Evaporation ............... 252
    E. Water Available for Plant Use ............................... 252
  V. Soil Engineering .............................................. 253
    A. Compaction .................................................. 254
    B. Strength .................................................... 256
    C. Consolidation ............................................... 259
    D. Soil Stabilization .......................................... 260
    E. Adhesion and Cohesion ....................................... 260
    References .................................................... 261
```

I. Introduction

A. ALLOPHANE SOILS

The term "allophane soils" used in the title is not uniquely defined. It is used here to describe soils having observed properties common to soil materials which arise from weathering of pyroclastics. Volcanic ash soils is another term which

---

[1] Department of Agricultural Engineering, Hokkaido University, Sapporo, Japan.
[2] Department of Agricultural Engineering, University of Tokyo, Tokyo, Japan.
[3] Department of Renewable Resources, McGill University, Montreal, Canada.

could have been used. The reader should interpret "allophane" in this less specific sense.

The concept of allophane is changing as more becomes known about the material. Imogolite is now distinguished from allophane, but is included in our use of the term "allophane soils." Imogolite, with its thread-shaped particles, has long-range order in one direction. It has a characteristic form in electron micrographs and has an identifiable X-ray diffraction pattern. Allophane, as the term is now used, has only short-range crystalline order, but does have an identifiable spherule form in electron micrographs. There are "amorphous materials" in soils which do not have this spherule form. It is likely that some of these materials are of pyroclastic origin.

Possibly the term "andept" or "andosol" would have been a better choice to describe these soils. However, for the nonspecialist, the term "allophane soil" will probably best bring to mind the soils which we will be describing here. While these soils are dominated by allophanic properties, other minerals are often present in addition to allophane and imogolite. There are soils which contain small amounts of imogolite and/or allophane which would not be considered allophane soils here because they do not possess the physical properties associated with allophane.

Allophane soils are widely distributed. They occur frequently in the Caribbean and Andean lands, as well as in the Pacific areas of Indonesia, Japan, New Zealand, and the United States. More studies on physical properties have been carried out in Japan than in any other country. One of the purposes of this review is to make the results of the published Japanese studies more readily available to readers of the English language. The references cited are mostly in Japanese, but usually have summaries in English. The more recent articles often have legends for figures and tables in English. Some of the papers are written in English.

The justification for a review on physical properties of allophane soils is that they have distinctive properties which distinguish them from other soils. Soils can be divided into three groups on the basis of physical properties. In the first group, void characteristics determine physical properties, and void volume changes little with changes in water content. These soils, with sands as the example, can be treated as rigid, porous media. In the second group, the nature and extent of surfaces determines physical properties. Volume changes accompany water content changes; these changes are reversible even though they show hysteresis. Physical–chemical descriptions of behavior are often more useful than mechanical descriptions. Swelling clays are examples of soils in this group. In the allophane soils, of the third group, void characteristics rather than surface area determine physical properties. There are volume changes accompanying water content changes, but the effects are largely irreversible. The matrix changes on drying, and the dried soil can be considered a different material.

Much of this review will, therefore, be in the form of comparing physical properties of allophane soils with the properties measured for soils with crystalline clay minerals. It is assumed that most readers will be familiar with the latter.

B. GENERAL NATURE OF PHYSICAL PROPERTIES OF ALLOPHANE

The measured values of physical properties of allophane soils, in summary, show: they have low natural bulk density, high 15-bar water content, and high natural water content; that medium to low amounts of water are available to plants; they have high liquid limit and low plasticity index; they are difficult to disperse; and that there are irreversible changes in all these properties on drying.

Several good general summary descriptions of physical properties of allophane soils have been published. Swindale (1964) lists the following features:

> ... deep soil profiles, usually with distinct depositional stratification, and normally friable in the upper part; topsoils as thick as one metre, and dark brown to black in color, containing humic compounds which are comparatively resistant to microbial decomposition; prominent yellowish brown to reddish brown subsoil colors with a smeary feel when the soil is wet; very light and porous profiles with a low bulk density and high water-holding capacity; rather weak structural aggregation, with easily destroyed porous peds lacking in cutans, and lack of horizontal differentiation in the subsoil except for the occurrence of duripans in some soils; .... Smeary consistencies are marked only in soils in very humid or per-humid climates. The soils which form in per-humid climates tend to dry irreversibly when they are allowed to dry out in road cuts or banks. This feature of irreversible drying is a useful classification criteria, although the soils in the field never become dry enough to exhibit the property to any significant extent.

Fieldes and Claridge (1975) have summarized the early studies by Fieldes and his co-workers on New Zealand allophane soils:

> ... allophane in its early stages of formation could be visualized as gel-like fragments of random aluminosilicate held together by cross-linking at a relatively small number of sites. The fragments have an open internal structure, which originally, in the hydrogel state, enclosed much water. Until the water is removed, rearrangements of materials with more ordered structure cannot occur, and the moist clay has a weak "waxy" consistence. When the water is removed by drying, the structure collapses and further cross-linking takes place so that the process cannot be reversed; and the resultant material has considerable mechanical strength. Thus, it is not possible to reconstitute the moist hydrogel structure by rewetting, although the more compact xerogel is still open enough to have a strong affinity for water. ...

Allophane soils generally have a friable surface soil and massive structure in the subsoil, which however has a relatively high permeability. The friable structure of the surface soil is partly due to effects of drying. Often allophane soils have several layers with very different physical properties which affect water movement and water available for plant use.

Many properties, such as volume or water retention which decrease on drying, show an irreversible decrease beyond 10- to 15-bar suction.

Physical properties of allophane soils do not show the dependence upon exchangeable cation which is prominent in soils with crystalline minerals. For example, Kubota (1971) showed that there was no difference in glycerol adsorbed on allophane with different exchangeable cations Mg, Ca, Sr, and Ba, and only about 4% lower adsorption with K as compared with Li. For bentonite there is a 20% difference for the divalent ions and nearly a 100% difference for the monovalent ions. Water vapor adsorption showed a similar pattern.

II. Index Properties

A. GRAIN SIZE DISTRIBUTION

The grain size distribution, also called particle size distribution or mechanical analysis, of a soil is the most widely used index property for physical properties of soils. Much effort has been spent in soil science on grain size measurements. For soils with crystalline clay minerals, especially in glaciated areas, and for clay contents less than 30%, one can predict many soil properties from the grain size distribution (Warkentin, 1972). However, for allophane soils the grain size is not an adequate index property.

The index properties used for allophane soils include water retention (Flach, 1964; Colmet-Daage et al., 1967) and plasticity (Warkentin, 1972). Packard (1957) used surface area as a measure of clay content, as did Birrell (1966). Flach (1964) recommends using the 15-bar water retention to estimate clay content.

The difficulty in obtaining dispersion, against both chemical and physical forces, and the uncertainty of what is the unit particle of an allophane soil, are the reasons for the limited usefulness of grain size in predicting physical properties of allophane soils. Chemical defloccuation is the problem with wet allophane subsoils, while in surface soils which have been dried the problem is cementing to form larger particles. These cementing bonds can be broken to different degrees. There is a considerable literature on the problems of dispersion of allophane soils (Gautheyrou et al., 1976). It is not possible from most of the studies to separate the effects of chemical deflocculation from physical dispersion. Therefore, the general term "dispersion" is used here.

The difficulty of dispersing allophane soils has been noted by many people. Davies (1933) was one of the first to study the problem and recommended using 0.002 N HCl for dispersion. Kanno (1961) used 0.002 N HCl as a dispersant for Japanese allophane soils. Optimum pH for good dispersion of Kanto loam, both surface and subsoil, is in the pH range 2.5–3.5 (Tada and Yamazaki, 1963).

Allophane soils flocculate in sodium silicate solutions, but polyphosphates can sometimes be used as dispersants. Sodium pyrophosphate was a better dispersant than sodium metaphosphate for Indonesian soils (van Schuylenborgh, 1953). Low pH or calgon were used for Japanese soils (Kobo and Oba, 1964); the former works better for subsoils and the latter for surface soils with organic matter. Sherman et al. (1964) emphasized that the usual dispersing agents could not be used for allophane soils. Ultrasonic vibration was found to increase dispersion (Kobo and Oba, 1964), and is now generally used for allophane soils. Oba and Kobo (1965) found that ultrasonic dispersion released clay size grains from aggregates. A number of papers describe the use of this method (Gautheyrou et al., 1976). Espinoza et al. (1975) found that ultrasonic treatment still gave much lower values for clay content than did the estimate from the 15-bar water content, i.e., clay = FBP × 2.5.

Ahmad and Prashad (1970) have taken a different approach to dispersion of allophane soils. They found good dispersion by reversing the charge with zirconium to get a positively charged particle.

Drying the sample decreases the measured clay content. This can be attributed to the cementing on drying. The phenomenon has been described by many workers, e.g., Sherman (1957), Birrell (1966), Wesley (1973). The magnitude of the effect varies with the particular allophane soil. Kubota (1972) measured the approximate soil suction at which irreversible bonding of clay into sand-size grains occurs on drying. The clay and silt contents began to decrease when the pF exceeded 3.5; fine sand-size grains were formed. Increases in coarse sand-size grains were not measured until about pF 5, at which time the clay and silt-size grains were at a minimum. The fine sand grains were then being bonded to coarse sand-size grains. No further changes in grain size distribution occurred at suctions above pF 5.5.

Particles of pumice break down on stirring, and sand-size particles settle more slowly than expected because of internal pores (Youngberg and Dyrness, 1964). They also have a long wetting time because of entrapped air.

Kobo (1964) has summarized the Japanese studies on dispersion of allophane soils, and Colmet-Daage et al. (1972) report on an extensive series of tests of dispersion of allophane soils of the Antilles and Latin America. Their results are summarized as follows. There is no one best method which can be recommended. Surface and subsoils react differently, as do allophane soils containing different components such as gibbsite or halloysite. Both flocculation and incomplete physical dispersion occur. Undried samples always disperse more completely than air-dried or oven-dried samples, the difference being much larger for subsoils than for surface soils. Surface soil samples generally disperse better at high pH of 10 or 11 with ammonium or sodium hydroxide (the Kanto loam is an exception), while subsoils generally disperse better at pH 3 with HCl. Subsoils generally flocculate at high pH. Sodium pyrophosphate is an effective dispersant for surface soils, but metaphosphate is not. Allophane soils containing gibbsite are difficult to disperse in acid or basic suspensions. With small amounts of gibbsite, dispersion appears to be better at pH 3; with larger amounts the best dispersion is obtained at high pH. Soils with halloysite disperse at high pH, but not at low pH. Ultrasonic vibration is recommended for dispersion. Birrell and Fieldes (1952) had also noted that the presence of gibbsite makes allophane soils more difficult to disperse.

Dispersion of surface soils at low pH might be attempted when solubilization or organic matter at high pH would interfere with subsequent measurements. The use of pyrophosphates might also interfere with other measurements on separated soil fractions.

Control of pH is critical for dispersion at low pH, but not at high pH.

The results given by Kubota (1972) are representative of the effect of different treatments on hydrandepts. On a B horizon sample the clay contents measured were as follows: 10kc–300W sonic dispersion, 56%; standard shaking on moist sample, 31%; on air-dry sample, 5%; and on an oven-dry sample, 1%.

Baba (1971) was able to disperse allophane soils in an alkaline medium only when ultrasonic treatment was used. Under these conditions sodium silicate was a more effective dispersant than sodium polymetaphosphate.

While it is generally preferable to work with undried samples, this is not always possible. Undried samples may not be desirable if the soil contains predominantly sand and gravel because of the difficulty in obtaining a representative subsample of a wet soil.

Grain size analysis, therefore, has a limited usefulness in characterizing allophane soils. The measurement should be done on field-moist samples (e.g., Schalscha et al., 1965) and the details of the method should be given. Since different allophanes react differently to dispersion treatments, some experimentation is necessary to obtain maximum dispersion (Colmet-Daage et al., 1972). The method most generally used for dispersion is ultrasonic vibration and low pH.

B. PLASTICITY

The plasticity is one of the physical properties which distinguish allophane from crystalline materials. The name "allophane" comes from the striking change on drying of allophane clays. Glassy when wet, the allophanes become earthy on drying (Grim, 1953). The wet material is plastic, and the dry earthy material is nonplastic.

Many papers have documented the plasticity limits, or Atterberg limits, and their change on drying (Birrell, 1951; Gradwell and Birrell, 1954; van Schuylenborgh, 1953; Yamazaki and Takenaka, 1965; Wesley, 1973; Warkentin and

Maeda, 1974; Kodani et al., 1976; and others). Wet allophane soils have a high liquid limit, but also a high plastic limit, and hence a low range of water content over which they are plastic. As the samples are gradually dried, the liquid limit decreases more rapidly than the plastic limit. Highly allophanic soils become nonplastic before they reach the air-dry water content. The nonplastic state is where the plastic limit cannot be measured, or where its measured value equals or exceeds the liquid limit.

Since the samples are completely rewetted during the determination of plasticity limits, the decreases in liquid and plastic limits on drying indicate an irreversible decrease in hydration of the allophane surfaces. The nature of these irreversible changes is discussed in Section III, A.

The values of plasticity limits are shown on a Cassagrande plot in Fig. 1. Crystalline clays have plasticity values which fall near the "A" line. The allophane samples fall far from the line, with the most allophanic samples having the highest liquid limit and the lowest plasticity index. This has suggested the use of plasticity values in classification of allophane soils (van Schuylenborgh, 1953; Gradwell and Birrell, 1954; Warkentin, 1972; Warkentin and Maeda, 1974). The measurement of plasticity is readily made, while other measures of allophane are difficult to make. The intensity of allophanic characteristics would be highest for samples with high liquid limit and low plasticity index, and lowest for values approaching the A line. Samples which remain plastic on air-drying or oven-drying would have low allophanic characteristics.

The difficulty in using plasticity values for classification is that the measured values depend upon degree of previous drying and upon content of organic matter. Since the drying history of surface soil samples is usually not known, air-dry samples may have to be used. The method might be more suitable for subsoils.

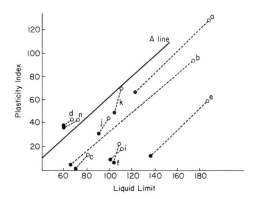

FIG. 1. Cassagrande plasticity chart with values for Japanese allophane soils (from Yamazaki and Takenaka, 1965). ○, Fresh soil; ●, air-dried soil.

Organic matter contributes to the water held at the liquid limit (Kodani et al., 1976), an effect which decreases irreversibly on drying. Maeda et al. (1976) have shown the separate contributions of organic matter and mineral matter to the liquid limit of allophane soils (Table I). The liquid limit is increased from 1.5 to 3% for each 1% organic matter in the samples they used. Their results indicate that not all the organic matter in a soil sample contributed to the liquid limit. Bonfils and Moinereau (1971) found that both liquid limit (L.L.) and plastic limit (P.L.) were strongly related to organic matter content (O.M.); the equations were: L.L. = 2.7 O.M. + 41 and P.L. = 2.7 O.M. + 34. The plasticity index was not correlated with organic content.

The activity values (ratio of plasticity index to clay content) measured for allophanes are variable. Northey (1966) gives values of 1.2 to 1.5, Wesley (1973) gives value below 0.6. The meaning of these values is uncertain because the measurement of clay content is difficult.

The measurement of plasticity is more difficult for allophane samples than for crystalline clays, and the precision is lower. This results from the low range of water content over which the samples are plastic. The slope of the liquid limit determination (water content versus log number of blows) is also more variable for allophane clays than for crystalline clays. The one-point liquid limit method (Sowers, 1965) cannot be recommended for allophanes. The degree of remolding and working of the soil with water affects the liquid limit, especially of subsoils (Ikegami and Tachiiri, 1966).

Measurements of liquid limit by Soma and Maeda (1974) on soils which were gradually dried show a sharp break at a specific water content where irreversible changes occur (Fig. 2). Drying does not produce irreversible change for the soil shown until the water content falls below 100%. The shrinkage limit also occurs at this water content.

TABLE I
Effect of Organic Matter on Liquid Limit of Allophane Soil[a]

Soil	Organic content (%)	Natural water content (%)	Water content at liquid limit		Decrease in liquid limit on air-drying	
			Whole soil (%)	Organic matter removed (%)	Due to organic matter (%)	Due to allophane (%)
A	29	142	180	83	33	36
B	20	116	164	72	28	27
C	20	119	147	72	36	15
D	17	133	151	107	8	40
E	23	108	172	107	52	5
F	11	44	62	42	2	3

[a]From Maeda et al. (1976).

PHYSICAL PROPERTIES OF ALLOPHANE SOILS

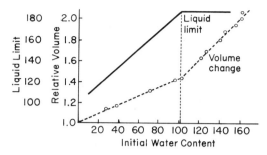

FIG. 2. Decrease of liquid limit and volume of allophane soil on gradual drying (from Soma and Maeda, 1974).

Freeze-drying produces less particle bonding than air-drying, and results in higher measured plasticity (Warkentin and Maeda, 1974). Ultrasonic vibration increases the measured liquid limit (Soma and Maeda, 1974) because it causes some dispersion.

There is little change in liquid limit with different exchangeable cations. For some samples the Na-soil has a slightly higher liquid limit than the Ca-soil, for others it is reversed (Yazawa, 1976). The flow index (slope of the water content versus number of blows curve in the determination of liquid limit) is also not consistently different for Na or Ca allophane soils (Yazawa, 1976).

C. SURFACE AREA AND HEAT OF WETTING

Measurements of surface area for allophane soils have been reviewed recently by Wada and Harward (1974). This literature will not be reviewed in detail here.

The surface area values are high, in the range of 300–600 $m^2 g^{-1}$, but some of this surface is not accessible to large molecules. This is not equivalent to the internal and external surfaces of swelling clays, but is due to small size of voids and small necks leading to voids. The heats of adsorption indicate that physical adsorption rather than chemical adsorption is dominant (Fieldes and Claridge, 1975).

Total surface area is often calculated from the amount of ethylene glycol adsorbed, and external surface is measured from nitrogen adsorption; internal surface area is estimated from the difference. Aomine and Egashira (1970) found ratios of total to internal surface from 2.3 to 3.0, while Fieldes and Claridge (1975) report ratios of 1.8 to 2.6 when the nitrogen surface was measured after heating to 600°C. Egashira and Aomine (1974) found that total surface area measured on samples vacuum-dried over P_2O_5 was higher than for oven-dried samples.

Aomine and Egashira (1970) measured the heat of immersion of allophane soils in comparison with soils containing crystalline minerals. They found that for equal surface areas, allophane soils had heats of immersion about twice as large as montmorillonite soils. The ratio of heat of immersion to surface area ranged from 0.047 to 0.056 cal m^{-2} for allophane and from 0.023 to 0.029 for montmorillonite soils. The exchangeable cation had only a small effect on heat of immersion of allophane soils. For montmorillonite soils, hydration of exchangeable cations is a more important part of total heat of immersion. Water molecules are bonded more strongly on allophane surfaces than on montmorillonite surfaces. When the heat of immersion was plotted against initial water content of the allophane soils, the oven-dry samples fell below the smooth curve joining the vacuum-dried and moist samples. This indicated to the authors that oven drying had altered the nature of the allophane surface.

Maeda *et al.* (1976) measured heat of wetting values for an allophane soil ranging from 7.3 cal g^{-1} at 4% organic matter to 11.2 cal g^{-1} at 26% organic matter.

D. MINERAL DENSITY

Some measured values of mineral density, or specific gravity, of wet allophanes are low, in the range of 1.8–1.9 g cm^{-3} (Fieldes and Claridge, 1975). These values had been accepted in earlier studies. However, other measurements show values of 2.7 or higher. Forsythe *et al.* (1964) quote values of 2.7–2.9 g cm^{-3}.

Wada and Wada (1975) measured values of 2.72 to 2.78. They took special precautions to remove entrapped air from the samples. If the unit particle of allophane is accepted to be a hollow spherule of 50 A outside diameter and about 30 A inside diameter, water movement into and out of this sphere would be difficult and could account for the low mineral density sometimes measured.

Bonfils and Moinereau (1971) measured values of 2.32 to 2.70, the lower values being for horizons with large amounts of organic matter—about 25%.

E. THERMAL CONDUCTIVITY

The thermal conductivity of a soil depends upon the conductivities of the components—mineral, organic, water, and air. Because the path of heat transfer from one component to another cannot be easily specified, the conductivity of a soil cannot be readily calculated from the conductivities of the components. Models exist for some heat flow paths, but in general the preceding statement is true. However, soil thermal conductivities vary in a predictable way with properties of the components. On this basis allophane soils would be expected to have thermal conductivity and thermal diffusivity values which are lower than the corresponding values for soils with crystalline clay minerals. Diffusivity is the ratio of conductivity to heat capacity and is the constant which relates temperature changes in the soil to the temperature gradient.

The conductivity of glass is lower than that of clay minerals or quartz (Cochran et al., 1967). Allophane soils have a lower bulk density than soils with crystalline minerals; this should result in lower thermal conductivity. The higher water content would give a higher heat capacity and hence lower thermal diffusivity.

Some of the results obtained are summarized in Table II.

Yakuwa (1943) made extensive measurements on soil temperature and thermal properties of different soils in Japan, including allophane soils.

Higashi (1951) measured a value of 0.32 cal g^{-1} $°C^{-1}$ for specific heat of dry allophane soil. Thermal diffusivities were calculated from amplitude ratios and phase differences of temperature waves; the phase differences gave more consistent results. While the dense packing seems to give higher values of diffusivity, the scatter was such that no conclusions could be drawn on effect of bulk density (Table II). The maximum in the diffusivity curve occurred at 50% water on a weight basis or about 0.25 on a volumetric basis. Thermal conductivity, calculated from specific heat and diffusivity, was higher at high bulk density. The loose packing had a bulk density of around 0.5 g cm^{-3}; the dense packing varied from 1.0 at 0% water to 0.7 g cm^{-3} at 50% water. Higashi (1952) also measured thermal properties of frozen allophane soils. The thermal diffusivity is similar for frozen and unfrozen soils below a water content of about 30%, but at higher water contents the diffusivity of frozen soils increases very quickly.

Cochran et al. (1967), using a line heat source probe, found that thermal conductivity of a pumice material was very low, only slightly higher than for a peat soil (Table II). They used this to explain low night temperature and the high incidence of frost in pumice soil areas of Oregon. The maximum in the diffusivity—water content curve occurred at a low volumetric water content of 0.04 for the C horizons of the soil.

Maeda (1968) measured temperature gradients in soils near Sapporo, Japan. He calculated thermal diffusivities from amplitude ratios and phase differences. The diffusivity of fine pumice was higher than for an alluvial soil, and the allophane soil was the lowest (Table II).

Kasubuchi (1975a) found that the specific heat of allophane clay (0.229 cal g^{-1} $°C^{-1}$) was higher than bentonite (0.209), kaolin (0.201), or quartz (0.170). Kasubuchi (1975b) measured thermal conductivity of different soils using a line heat source probe. The values again show conductivity and diffusivity values for allophane which are 20 to 35% of those for soils with crystalline minerals. The higher heat capacity and lower heat conductivity of allophane soils results in slower temperature changes.

TABLE II
Measured Thermal Properties of Allophane Soils

Material	Bulk density (g cm^{-3})	Water content (w, %)	Heat capacity (cal cm^{-3} °C^{-1})	Thermal conductivity (mcal cm^{-1} sec^{-1} °C^{-1})	Thermal diffusivity [cm^2 sec^{-1} ($\times 10^{-3}$)]	Reference
Pumice soil	1.15	11	0.32	1.1	3.4	Yakuwa (1943)
Allophane soil	Loose	0	0.20	0.22	1.13	Higashi (1951)
	Loose	50	0.39	0.81	—	Higashi (1951)
	Dense	0	0.29	0.32	2.34	Higashi (1951)
	Dense	50	0.53	1.23	—	Higashi (1951)
Crown glass	—	—	—	2.3	—	Cochran et al. (1967)
Pumice	0.76	0	0.15	0.37	2.6	Cochran et al. (1967)
		40%	0.55	1.25	2.1	Cochran et al. (1967)
Clay soil	—	0	0.30	0.60	—	Cochran et al. (1967)
		40	0.70	3.80	—	Cochran et al. (1967)
Allophane soil	—	—	—	—	2.4	Maeda (1968)
Alluvial soil	—	—	—	—	3.5	Maeda (1968)
Fine pumice	—	—	—	—	6.6	Maeda (1968)
Allophane	0.2–0.3	40	—	0.5–0.7	1.5–1.7	Kasubuchi (1975b)
Alluvial soil	0.5	40	—	2.0–2.5	3.7–4.0	Kasubuchi (1975b)
Diluvial soil	0.5	40	—	3–4	4.5–5.5	Kasubuchi (1975b)
Black andosol	0.77	40	—	0.6	1.5	Miyazawa and Konno (1976)
Brown andosol	0.72	40	—	0.8	1.7	Miyazawa and Konno (1976)

The values reported by Miyazawa and Konno (1976) confirm the other measurements (Table II). They found a maximum thermal diffusivity of 1.8 to 2.0 cm^2 sec^{-1} at a volumetric water content of 0.50. They used these values to describe the thermal regime of the soils.

III. Structure of Allophane Soils

A. DESCRIPTION OF STRUCTURE

A major difficulty in describing physical properties of allophane soils is a lack of understanding of the structure (size, shape, and arrangement of particles and voids) at different levels of observation. This is compounded by the apparent change in nature as well as amount of surface exposed as the soil dries. These changes are irreversible when the sample is dried at high suction, although they are usually reversible at low suction.

The structure at the lowest level, 10–100 Å, is being defined in recent studies, and structure at the highest level, 1–10 mm, is also known. But little information exists at intermediate levels, for example over the range of void sizes retaining water for plant use. The differences in physical properties among allophane soils cannot, as yet, be related to differences in structure; often the differing physical properties are attributed to different amounts of allophane present. This is reminiscent of the situation 70 years ago when different soil properties were explained by different contents of "kaolin" in the soil.

An exception to this lack of understanding is pumice, where the physical properties can be related to size, shape, continuity, and strength of voids (Sasaki et al., 1969).

The terminology used to describe structure in the 0.01 to 10 μm range varies with the discipline, or even with the author. No attempt will be made in this review to suggest a preferred terminology. Structure will be used as defined by Brewer (1964) to include size, shape, and arrangement of particles and voids; fabric is the component of structure which describes arrangement.

The structure of allophanes in the 1- to 10-mm range is given in many field descriptions. The total porosity is high, with typical bulk density values from 0.3 to 0.8 g cm^{-3}. Some of the low-silica allophane soils in Hawaii have bulk density values around 0.1 g cm^{-3} (Sherman et al., 1964). There is a marked difference between surface soil and subsoil, due mostly to effects of drying. The topsoil usually shows good aggregation, with well-defined and stable interaggregate voids. The subsoils show the properties of undried allophane, with restricted profile development. The porosity is high and tubular pores are present, although these pores have restrictions that limit percolation of water (Tabuchi et al., 1963).

The structure of imogolite in the 10- to 100-A range is known from electron micrographs supplemented by X-ray diffraction, electron diffraction, and salt absorption measurements (Wada et al., 1970; Wada and Henmi, 1972). The unit is a hollow tube with inside and outside diameters of 7–10 and 17–21 A, respectively. A number of tubes lying parallel form threads of 100 to 300 A in diameter. Three different classes of pores would be present in such a material: intraunit pores of about 10 A in diameter, intrathread (interunit) pores of the same or slightly smaller diameter, and interthread pores of several hundred Angstrom (A) units diameter.

The "unit particle" of allophane has been defined (Kitagawa, 1971) as a spherical particle with a diameter of about 55 A (Birrell and Fieldes, 1952). The density of this particle is about 1.9 g cm^{-3}. Kitagawa (1971) has assumed that these unit particle spheres exist in close packing in the air-dry state. This forms the "microaggregates" commonly found. On heating, the layer of adsorbed water is lost, and the spheres come closer together. Grinding distorts the unit particle spheres allowing closer packing, which, in the electron micrographs, gives the appearance of sheets rather than the individual spherical particles of the dried sample.

The close packing of these 55-A diameter units would again result in voids about 10 A in diameter, so the water absorption properties would be similar to imogolite. Irregular packing would produce some larger pores.

These models, therefore, explain many observed properties. The decrease in surface area on grinding is due to collapse of some of the small pores. Kitagawa (1971) has shown that calculation of particle size from surface area and density also gives values of about 55 A diameter. He found that phosphate adsorption was not decreased by grinding on drying, which argues against any chemical change in the surface, specifically in the number of hydroxyls at the surface. Interparticle bonding is then by physical forces that would be weaker than if chemical bonding were involved.

These measurements were made on a series of allophane soils from Japan. It is not known whether all allophanes have this same "unit particle." Measurements by Rousseaux and Warkentin (1976) of water vapor absorption show a maximum in pore volume at diameters of 7–10 A for allophanes from the Caribbean and from Japan. The Caribbean samples show a narrower size distribution that could result from closer and more regular packing of unit particles. Fujiwara and Baba (1973) calculated effective pore size from nitrogen absorption measurements, and found a maximum at 25 A. The volume of these 25-A pores was about one-twentieth of the volume of 10-A pores found from water absorption by Rousseaux and Warkentin (1976). These measurements confirm the unit particle in allophane soils.

Measurements of water vapor absorption are of necessity made on dried samples. It is reasonable to assume that undried samples consist of the same unit particles, but that they are further apart and more irregular. Such a model would explain water retention, plasticity, etc. It is more difficult to relate differences in these properties among soils, wet or dry, to different structures in the model. Some allophane soils become nonplastic on air drying, others do not. This could be due to differences in arrangement of unit particles, or to different nonallophane components in the soils. Also, it is not known how differences in chemical composition such as the Al/Si ratio affect the structure, although these differences are correlated to differences in physical properties (Rousseaux and Warkentin, 1976). The Caribbean samples, for example, have lower SiO_2/Al_2O_3 ratios than the Japanese samples.

Ito (1964) postulated that allophane structure consisted of individual particles and massive particles (structure units), with only the latter affected by drying. The individual particles retain their undried properties, and different phases of shrinkage are due to different effects of individual particles and structure units.

The irreversible changes in physical properties of allophane soils on drying set them apart as a separate group of soils. In the extreme, allophane soils change from highly plastic when wet, to sandy when dry. This has been described by many workers, e.g., Sherman *et al.* (1964), who suggest research on changing the colloidal properties of allophane soils by promoting dehydration. The particles become cemented together to form units of sand size that are sufficiently strong to withstand the usual manipulation in field or laboratory. Neither the forces involved in this cementing nor the fabric of the units have been adequately described. A tentative model for structure is described in Section III, B.

The observed pore structure of allophane soils has been described by a number of workers in relation to permeability. Tabuchi *et al.* (1963) described the channels for water flow that they observed in thin sections of surface and subsoils. Interaggregate pores conducted water in surface soils, tubelike pores were found to conduct water in subsoils. Takenaka *et al.* (1963) report similar results. Nagata (1963) found that plots of log air permeability against air porosity were linear functions which could be related to the kind of structure in the soil sample. For some soils log K_a versus η_a consisted of two straight-line portions, depending upon changes in structure with void size.

A more detailed description is available for the pores in pumice (Borchardt *et al.*, 1968; Maeda *et al.*, 1970; Tsujinaka *et al.*, 1970). On the basis of water-retention measurements carried out in different ways, Maeda *et al.* (1970) distinguished dead, active, semiactive, secondary active, and semidead pores. The active and semiactive pores usually occupy the largest volume, but semidead and dead pore space may be large in some samples of pumice. The pumice materials have been described by Sasaki (1957).

Some information on structure, especially on the fabric component, can be obtained from measurements of rheotropy or thixotropy of allophane soils. The behavior of allophane on mechanical manipulation is usually not true thixotropy, i.e., a reversible sol/gel transformation. The changes on remolding described by Takenaka and Yasutomi (1965) are discussed in Section V, B.

A number of papers have reported experiments with soil conditioning chemicals; the general impression is that these chemicals are only marginally effective in promoting aggregation in allophane soils. This may be partly due to mixing the chemicals with soils at water contents which are not optimum for aggregation. Sudo and Suzuki (1963) found that sodium alginate increased stable aggregates in an allophane soil but the polyelectrolyte CMC did not. They report, in their literature review, that additives are generally considered to have little effect on aggregation of allophane soils. Kawaguchi *et al.* (1963) report that bentonite added to polyvinyl alcohol is effective in aggregation. Fujioka *et al.* (1965) found an effect of soil conditioners on allophane soils, but Terasawa (1967) found very little effect of synthetic polymers on aggregate formation.

B. MODEL FOR PHYSICAL PROPERTIES OF ALLOPHANE

What is required is a model for structure, including fabric, of allophane soils. This model must explain the known physical properties of allophane and their changes on drying. The model should predict other physical properties. It should also be related to important differences in chemical composition and properties, for example the Si/Al ratio. The model must describe structure in the 0.01- to 100-μm range, where physical properties can be explained.

Measurements to date on allophane soils allow only a general specification of this model. The unit particle of about 50 A is well established and can be taken as the starting point. This particle has an "internal" water content. The unit particle has been established in dried samples and is assumed to be present also in wet samples. These unit particles are then weakly bonded together to form domains in the diameter range of 0.01 to 1 μm. The modal size may be 0.05 to 0.1 μm. The voids between particles in these domains account for the large water retention above 15-bar suction. Drying brings the particles closer together, increasing the bulk density of the domains and decreasing water retention above 15 bar. Irreversible changes occur when the unit particles come sufficiently close together to allow strong bonding between unit particles. These domain units may be the clay-size grains measured in a grain size determination. They are not broken up on remolding and account for the high water content at the plastic limit. They are the units that move on plastic readjustment. Organic molecules are held within the domains.

The domains are arranged in clusters in the size range of 1–100 μm. These units have weaker bonding than the domains. Remolding breaks up the clusters, releasing water held within the cluster. Drying causes shrinkage of the domains and rearrangement into clusters of highei bulk density. In some allophane soils the bonds holding the clusters together after drying are strong, and the clusters become the units of dried soils. Bonding within clusters would then also contribute to irreversibility. This may be true only in soils that are composed entirely of allophane and do not have crystalline minerals or oxides mixed in. Water in the plant-available range is held within the clusters.

The water-retention curves for allophane soils show an approximately linear change of water content between 0.03 and 1 bar, with decreasing amounts of water retained below and above these values. The break at 1 bar indicates a change at void diameters of 30 μm; the upper boundary of cluster size could be drawn there. This would put much of the plant available water between clusters.

Kubota (1971) has studied the formation of units in the sand- and silt-size range. Drying is necessary for pedogenic formation of these units. These units fall within the range defined here as "clusters." These units are stable against dispersion if the allophane soil has been dried. Dehydration alone can cause irreversible binding. The formation of these clusters is best studied on subsoil samples of allophane soils that have been continually moist. Presumably the clusters are already present in surface soil horizons that have been previously dried.

Kubota (1971) divides the clay-size grains into three types on the basis of their potential for forming sand- and silt-size grains on drying. The clay content was measured on sonic-dispersed samples of subsoils of wet allophanes. Type I is active free clay which can form large grains on air drying. This is measured as the difference between clay content of moist and air-dry soils on shaking. Type II is the aggregated clay that is released from moist soils by sonic dispersion. These are pedogenically formed aggregates. Type III is inactive free clay not affected by air-drying. This is the measured clay content of an air-dry soil. Type I clay content increases from the Ap to the B_2 horizon, while Type III decreases. Type II remains approximately constant with depth.

Monolayer adsorption of water vapor on allophane is strong but multilayer adsorption of more than two water layers is weak compared with layer silicate minerals. Therefore unit particles of allophane can come in close contact. Kubota (1971) also states that hydroxyaluminum groups are the site for adsorption of water molecules; they are also the sites for bonding of unit particles in an irreversible aggregation.

The details remain to be filled into this model. The size ranges of the fabric units, domains, and clusters, have been chosen mostly for convenience of description. They will undoubtedly need refinement.

Neither the size boundaries nor the terminology of domains and clusters are suggested as being definitive. They are both used here for convenience. The clusters may be better described as microaggregates. However, Kitagawa (1971) refers to the unit particles of 55 A as microaggregates, and Kubota (1971) refers to sand- and silt-size units as aggregates.

IV. Physical Characteristics of Allophane Soils

A. VOLUME CHANGE

Wet allophane soils show a large volume decrease on air-drying, and a limited volume increase on rewetting. Most of the volume change is irreversible. This distinguishes allophane soils from swelling mineral soils, where the volume changes are more nearly reversible. The amount of shrinkage of an allophane soil depends upon the initial water content and the fabric changes during drying, both of which depend upon the allophane content. Many physical properties change in association with this volume decrease, e.g., permeability increases. However, volume change in the field cannot be predicted quantitatively from shrinkage measured on small samples because the boundary conditions are different.

Wet allophane samples dried to intermediate water contents, suctions below 10-20 bars, will regain most of the volume on rewetting. Volume change is approximately reversible over this range (Takenaka, 1961).

Volume change curves can be plotted as changes in measured volume, linear dimensions, bulk density, or void ratio with change in water content. The measurements are not precise because it is difficult to measure volume repeatedly on a sample as the water content changes. The precision is sufficient to characterize shrinkage on drying, but greater precision is desirable in describing volume change for the water-retention curve (Section IV, B).

The general nature of the shrinkage curves is shown in Fig. 3 for a highly allophanic soil (Cl) from Dominica, West Indies, and for a soil (Nl) from Hokkaido, Japan, with low allophanic properties (Warkentin and Maeda, 1974). There is a break in the volume change curve which can be called a shrinkage limit, but the volume change at higher water contents is not "normal" shrinkage where volume decrease equals water content decrease (Takenaka, 1961; Ito, 1964).

The shrinkage limit is less pronounced and occurs at a higher water content as allophane content increases. The shrinkage limit occurs between 50 and 100% water for allophane soils, a much higher value than for crystalline clays (Takenaka, 1965; Warkentin and Maeda, 1974; Soma and Maeda, 1974). Takenaka (1965) found that the shrinkage limit occurred near pF 6 for a sample of Kanto loam. He related shrinkage to soil suction values and showed that the amount of shrinkage depended upon the rate of drying. The measured water content at the shrinkage limit also increases with increasing organic matter content of the soil (Takenaka, 1973). He measured a 15% increase in shrinkage limit for 10% increase in organic matter up to 25%. Remolding does not change the value of the shrinkage limit (Takenaka, 1965), but drying increases the shrinkage limit.

The usefulness of the shrinkage limit is in the information it gives about fabric and structure of allophane soils. The high water content at which the slope of

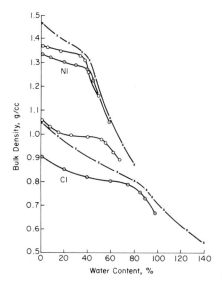

FIG. 3. Shrinkage curves for two allophane soils. Reproduced from *Soil Science Society of America Proceedings,* Volume 38, page 375, 1974 by permission of the Soil Science Society of America. ×, Field moisture; ⊙, air dry; ○, oven dry.

the shrinkage curve changes and the high value of residual shrinkage (between the shrinkage limit and zero water content) indicates a random arrangement of units. Measured shrinkage is isotropic, again indicating random arrangement. For crystalline minerals, the shrinkage limit is lowest for high swelling soils. The shrinkage limit is, therefore, diagnostic for allophane soils (Warkentin and Maeda, 1974).

Despite the large amount of shrinkage on drying, one would predict little visible cracking for allophane soils in the field because the shrinkage is taken up in small spaces between clusters (Section III, B). Cohesion of allophane is low and it decreases if the samples are dried (Soma and Maeda, 1974). Allophane soils do not form dry clods with dimensions of tenths of a meter.

The effect of remolding a soil depends upon its degree of consolidation (Croney and Coleman, 1954). Remolding an overconsolidated soil exposes new surfaces for water retention. Allophane soils are underconsolidated. Remolding breaks some of the fabric bonds, decreases the soil suction, and increases the amount of shrinkage (Takenaka, 1965).

B. WATER RETENTION

This section will deal with soil water characteristics measured on allophane soil samples in the laboratory; the mechanisms of water retention will be discussed. The field water regime is described in Section IV, D.

It has been difficult to relate soil suction measured in the laboratory to the water regime of allophane soils in the field. Only two variables—water content and suction—are usually measured; while the complete description of the water-retention curve of an allophane soil requires the specification of four variables— water content, suction, sample volume or percent water saturation, and initial degree on drying of the sample. Only for coarse-grained soils, where suction and water content determine the system, have water-retention curves been useful in predicting the field water regime. Measurements in the laboratory on fine-grained soils, especially swelling clays, have not been useful in predicting soil water behavior in the field. The water-retention curve for swelling crystalline clay soils requires specification of three variables—water content, suction, and volume. If the structure is disturbed on sampling this can be considered a fourth variable. Initial bulk density is assumed to be the same as that in the field, even though this is not always assured.

Another difficulty in applying laboratory measurements to the field is the unreliability of water-retention measurements during wetting of fine-grained soils. The measurements are almost always made only on the drying part of the cycle.

Therefore, the main use for water-retention measurements for allophane soils is in characterizing soil surfaces or void-size properties. In addition, Colmet-Daage and his colleagues (e.g., Colmet-Daage et al., 1967) have used pF values for wet and dry allophane soils as an index property (see Section II, A). Hughes and Foster (1970) have suggested that the water content at a specific suction can be used to rank the degree of disorder, or the amount of extractable materials, in halloysite and allophane. Galindo-Griffith (1974) showed that the 15-bar percentage was not correlated with surface area, but was correlated with reactive alumina and with the point of zero charge. Many physical properties of allophane soils are interrelated, as has been shown in a number of papers (e.g., Warkentin and Maeda, 1974).

Water retention by allophane soils is determined by the size distribution of voids, not by the amount of surface area (e.g., Fujiwara and Baba, 1973). This fact differentiates allophane soils from soils with swelling clay minerals.

The amount of water held by wet allophane samples is high (Table III); the very high water contents at 15-bar suction are striking. This results from the large volume of small voids. Drying the samples decreases water retention at any suction value. These measurements have been made by a number of people. Misono et al. (1953) made a detailed study of water-retention characteristics of a number of Japanese allophane soils. Colmet-Daage and his co-workers (1967, 1970) have published measurements for a wide range of allophane soils from the Caribbean, Central America, and South America. These reports provide very valuable source material on properties of allophane soils. A complete list of papers is available in the bibliography by Gautheyrou et al. (1976).

TABLE III

Measured Water Content at Different Suctions for Allophane Soils

Sample	Water content at suction			Reference
	0.001 bar	0.3 or 0.5 bar	15 bar	
Kuriyagawa				
B1–Wet	275	160	80	Misono et al. (1953)
–Air-dried	170	80	50	Misono et al. (1953)
Dominica				
–Wet	145	90	70	Maeda and Warkentin (1975)
–Air-dried	130	70	50	Maeda and Warkentin (1975)
Hawaii				
Hydrandept				
–Apl	–	138	100	Flach (1964)
–B24	–	235	174	
Martinique				
–Fresh	–	207	83	Colmet-Daage et al. (1972)
–Dry	–	143	42	Colmet-Daage et al. (1972)
Ecuador				
–Fresh	–	255	192	Colmet-Daage et al. (1967)
–Dry	–	36	33	Colmet-Daage et al. (1967)
Ecuador				
0–110 cm, weakly allophanic				
–Fresh	–	65	45	Colmet-Daage et al. (1967)
–Dry	–	40	30	Colmet-Daage et al. (1967)
50–300 cm, strongly allophanic				
–Fresh	–	150	115	Colmet-Daage et al. (1967)
–Dry	–	40	35	Colmet-Daage et al. (1967)

The measurements by Colmet-Daage and Cucalon (1965) illustrate the large decreases on drying in water content at pF 4.2 and 2.8. The amount of water held between these two suction values, an estimate of plant-available water, also decreases on drying. In extreme cases for certain horizons this decrease was from 45% to 3% and from 68% to 5% by weight.

Soils with a high content of allophane have an S-shaped water retention curve, similar to the shape for a coarse-grained soil. The approximately linear portion on the water-content–log-suction plot is between 0.01 and 1-bar suction (Maeda and Warkentin, 1975). For swelling crystalline clays this linear portion extends from less than 0.01 bar to around 100 bar.

Forsythe (1972) states that the loss in water retention on drying is greater at lower suctions than at high suctions. Maeda and Warkentin (1975) confirmed this for samples with moderate allophanic properties; for highly allophanic soils the effect appeared to be reversed. Forsythe (1972) found that volumetric water content increased on air-drying; Maeda and Warkentin (1975) found a lower volumetric water content and lower percent saturation of samples after drying.

Parfitt and Scotter (1972) report on an allophane soil from Papua, New Guinea, with a low 15-bar percentage of 36% and a 0.1-bar percentage of 120%. This results in a high content of plant-available water.

Colmet-Daage et al. (1967) checked the influence of organic matter on water retention by allophanic soils. They realized that treatment with peroxide could aid in dispersion, so that the measured difference could not be attributed solely to organic matter. They found that treatment with hydrogen peroxide to remove organic matter had no influence on water retention at pF 2.5. At pF 4.2 some of the treated samples had lower and some had higher water retention than the control samples. They concluded that organic matter had only a small influence on water retention for the soils which they studied. However, in humic allophane soils, the organic matter content is important in water retention. Swindale (1964) reports that for some Hawaiian soils, percent field water content (w) increases with percent organic matter (O.M.): w = 4.2 + 34.7 O.M. Takenaka (1973) and Maeda et al. (1976) pointed out that water retention by allophane soils high in organic matter content was especially decreased on drying.

A number of authors have divided the water-retention curve into three or more classes of water (e.g., Misono et al., 1953). The breaks in the water-retention curves indicate some validity in this approach, but there is no evidence that forces of water retention are different and can be separated in this way. The retention appears to be due to voids of different sizes except at high suction values, larger than 100 bar, where surface adsorption is involved. The water-retention curve at suctions below 100 bar can then be used to obtain a qualitative determination of void size distribution (Misono et al., 1953; Maeda and Warkentin, 1975).

Masujima (1962) found that different exchangeable cations did not change the water retained between 10 and 100 bar.

C. WATER TRANSMISSION

Rate of water transmission through allophane soils is high, due to the low bulk density and the granular structure of surface horizons (Table IV). At the same void ratio, allophane soils have a higher saturated hydraulic conductivity than soils with montmorillonite (Maeda and Warkentin, 1975).

Drying increases the saturated hydraulic conductivity at constant bulk density. The increase is at least two orders of magnitude for highly allophanic soils but

TABLE IV
Saturated Hydraulic Conductivity of Allophane Soils

Sample	D_b (g cm^{-3})	Saturated K (cm sec^{-1})	Reference
Onuma			
—Wet	—	6×10^{-4}	Kubota (1972)
Dominica			
—Wet	1.0	3×10^{-8}	Maeda and Warkentin (1975)
—Air-dry	1.0	1×10^{-5}	Maeda and Warkentin (1975)
Ecuador			
—Dry	—	2×10^{-3}	Colmet-Daage et al. (1967)
Costa Rica			
—Field	—	3×10^{-3}	Forsythe (1975)
Kanto loam			
—Surface	—	10^{-2}	Tabuchi (1963)
—Subsoil	—	10^{-2}	Tabuchi (1963)
Averages	—	2×10^{-2} to 2×10^{-4}	Yamanaka (1964)

becomes smaller when the content of allophane decreases (Maeda and Warkentin, 1975). This indicates that dried samples have a larger proportion of large voids, i.e., that the small voids are lost preferentially on drying (Misono et al., 1953). The common observation that dried allophane soils resemble sands in their physical properties is due to formation of aggregates with small internal porosity and hence large interaggregate void volume.

Tada (1965) described non-Darcy flow of water in fresh allophane soil. The conductivity was constant to hydraulic gradients of about 15, then increased about 20 times to gradients of 50, after which the conductivity again decreased. Dried allophane soil did not show this effect.

Iwata (1963, 1966) published a thorough study of water movement in unsaturated soils to clarify water redistribution and field capacity concepts. He found that allophane soils had a much higher unsaturated hydraulic conductivity than soil with crystalline clay minerals when compared at the same suction. For this reason the field capacity occurred at a higher suction in allophane soils. Iwata (1963) measured unsaturated hydraulic conductivity values of 0.1 cm day^{-1} at 0.1 bar and 1 cm day^{-1} at 0.05 bar for an allophane soil. These values were 3 to 4 times higher than measurements on an alluvial soil. Maeda and Warkentin (1975) measured values of 0.01 cm day^{-1} at 1 bar and 10^{-4} cm day^{-1} at 7 bar. The unsaturated conductivity of allophane soils was higher than that of crystalline clay soils at suctions below 2 bar.

The advance of the wet front during infiltration also increases from wet to dry allophane samples (Maeda and Warkentin, 1975). This is opposite to the effect for crystalline clay minerals. The diffusivity was four orders of magnitude larger for the dry sample. The change in soil water diffusivity with change in bulk density was about the same as for soils with crystalline clay minerals.

El-Swaify and Swindale (1968) found that relatively high levels of salinity and sodium in irrigation water could be used on allophane soils and still maintain adequate permeability. The sodium caused only slight changes in structure of the soil. El-Swaify (1973) found anion effects on hydraulic conductivity for allophane soils.

D. FIELD STUDIES ON INFILTRATION AND EVAPORATION

High percolation rates in the surface and subsoil present a problem in managing rice paddies in Japan. A percolation rate of 3–4 cm day^{-1} is desired. Often bentonite is used to decrease percolation rates because soil compaction by rolling is not sufficient. Another difficulty is the variability of percolation rates within a field; percolation does not follow a normal distribution and a few high values can result in ratios of the mode to the mean of 0.05 to 0.3 (Ishikawa *et al.*, 1963). This makes effective bentonite dressings difficult to achieve. Birrell (1952) comments that the natural variation in water content and compaction characteristics of allophane soils in the field has made engineering investigations difficult.

The variability of physical properties of allophane soils in the field has received considerable attention. A number of Japanese research workers cooperated in a study organized by the Japan Society for Irrigation, Drainage, and Reclamation Engineering on field variability of physical properties (e.g., Kuroda, 1971, Tokunaga and Sato, 1975).

Forsythe (1975) has reviewed measured infiltration rates for allophane soils in Central America. The values are generally high, initial infiltration rates of 20–70 cm hour^{-1} and 2-hour rates of 5–20 cm hour^{-1} are reported. He points out that these high rates make the soils unsuitable for furrow or flood irrigation.

Nakano *et al.* (1970), in a series of three papers, reported laboratory measurements of infiltration and evaporation from columns of layered volcanic ash soils of different grain size, and field measurements of water movement and evaporation. Internal soil drainage can often be increased by mixing the layers to overcome boundary effects (Kon, 1967).

E. WATER AVAILABLE FOR PLANT USE

Relatively little information has been published on field studies of water available to plants in allophane soils. Many numbers for available water are based on water held between 15-bar and 0.5- or 0.3-bar suction. There is evidence that

15 bar is too high for the upper limit; that 5- to 8-bar suction is observed in the field (e.g., Misono and Terasawa, 1957; Masujima and Mori, 1962). It is also to be expected that the lower limit may be closer to 0.1 bar for allophane soils with relatively high permeability. Kira *et al.* (1963) found the field capacity to be at 0.08 to 0.1 bar, and commented on the difficulty of estimating field water regime from water-retention curves. Chichester *et al.* (1969) found the field capacity occurred at suctions of less than 0.05 bar for a pumice soil in Oregon. Youngberg and Dyrness (1964) found the value was below 0.1 bar.

Shiina and Takenaka (1961) used the water content after 24 to 48 hours drainage as the upper limit. They found that crop growth decreased markedly when the suction exceeded 1.5 bars in the root zone. They also found considerable water movement up from wet subsoil layers, enough to supply 50% of evapotranspiration in their experiment. Masujima and Kon (1963) also found that water movement up from the subsoil contributed to available water.

Shiina (1963) rejected the use of 0.5- and 15-bar suctions to define the available water. He said this must be based upon growth period and weather conditions as well as soil suction. Plants use water held at suctions as low as 0.03 bar, and initial wilting can occur at 2 bar. Masujima and Mori (1962) noted that water was not equally available over the range of available water, and that availability depended upon plant and soil factors.

Masujima and Mori (1962) studied physical properties and plant growth; increasing noncapillary porosity was associated with decreasing water availability. Optimum porosity for growth of beans was 30% at 0.5 bar and 20% at 0.03 bar. They suggested that land improvement could be achieved by mixing in different particle sizes to achieve an optimum void size distribution.

Forsythe *et al.* (1964) contains a good summary review of measured physical properties such as bulk density and available water content. They rate available water in many allophane soils as average to low, although some allophane soils have a high available water content. On a volume basis the available water in allophane soils is not markedly different from soils with crystalline minerals (Swindale, 1964). On a weight basis, the numbers are very high because of the low bulk density of allophane soils.

Bonfils and Moinereau (1971) report relatively high values of available water (15 to 21%), with the available water increasing with increase in organic matter content.

V. Soil Engineering

The term "cohesive volcanic ash soils" is often used in the soil engineering literature. Cohesion indicates clay properties. The term allophane soils is used in this review in the same sense, and will, therefore, be used in this section as well.

Many of the engineering properties of allophane soils in Japan have been studied on "Kanto loam," an allophane soil found on the Kanto plain north of Tokyo. The subsoil, especially, of the Kanto loam shows typical allophane properties.

A. COMPACTION

Soils are compacted when large earth-moving equipment is employed in land reclamation. From the viewpoint of engineering, it is important to know the compaction characteristics of soils. The compaction curve for a soil is a plot of water content versus the bulk density which can be achieved by a specified compactive effort at that water content. This curve shows a maximum bulk density at a water content called the "optimum water content" for compaction.

Compaction produces a moderate increase in density and a large increase in strength of allophane soils. However, on resaturation the strength is again decreased (Northey, 1966).

Allophane soils have a remarkably high natural water content compared with soils containing crystalline clay minerals, and moreover water-holding characteristics change during drying. These differences are reflected in differences in compaction characteristics. Allophane soils have low maximum bulk densities, in the range 0.8–1.3 g cm^{-3} and relatively high optimum water contents (Table V). The optimum water content is much below the natural water content.

The compaction curve of a soil with crystalline clay minerals is the same regardless of initial water content at the beginning of the compaction test; however, allophane soils show different curves depending upon the initial water content and the amount of remolding produced during testing (Birrell, 1951). An undried allophane soil does not show a distinct maximum in bulk density,

TABLE V
Compaction Characteristics of Two Typical Allophane Soils

Sample	Maximum dry density (g cm^{-3})	Optimum water content (w, %)	Reference
Kanto loam			
–Natural	0.69	105	Kuno and Mogami (1949)
–Air-dry	0.79	86	Kuno and Mogami (1949)
–Oven-dry	0.86	74	Kuno and Mogami (1949)
Java andosol			
–Natural	0.55	127	Wesley (1973)
–Air-dry	0.69	95	Wesley (1973)
–Oven-dry	0.79	80	Wesley (1973)

and hence no identifiable optimum water content. Since the natural water content exceeds the optimum water content, this curve can be measured only by gradually drying samples for compaction. The bulk density increases only gradually as the water content is decreased. Once the soils are dried, they show typical compaction curves. This behavior is illustrated in Fig. 4, which is typical of the results obtained. These characteristics are described by Kuno and Mogami (1949), Birrell (1951), Tada (1965), Tokunaga (1965), Northey (1966), Frost (1967), Takenaka (1973), Wesley (1973), Adachi and Takenaka (1973), and others.

The compaction curve obtained by first drying and then rewetting a sample, therefore, forms a loop with only the drying portion showing a maximum density (Takenaka, 1973). This is due to the reduction in the amount of water retained at any suction value. This loop forms when the soil is dried below a critical water content, which Tada (1965) found to be at a suction of 15 bar for the Kanto loam subsoil. This is the highest suction which the subsoil would be subjected to under natural soil conditions with drying only by plant roots.

Since it is often difficult to find the optimum water content and the maximum dry density on the compaction curve of fresh allophane soil, the criterion of maximum dry density cannot be applied for embankment work. The degree of saturation is used in Japan in place of the maximum dry density to follow the degree of compaction of allophane soil (Kuno and Yabe, 1960, 1962).

The permeability of compacted soils is an important engineering consideration. When allophane soils are compacted, with a decrease in water content, the dry density increases slightly, but the permeability also increases. According to Tada's (1965) experiments, coefficient of permeability reaches a minimum near the natural water content, and tends to gradually increase on lowering of water content. Tokunaga's (1969) experiments studied these phenomena in relation to

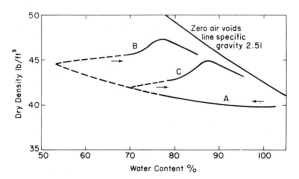

FIG. 4. Compaction curve for allophane soil on gradual drying (from Northey, 1966). A, Dried from natural water content (both samples, 7592B and 7592C); B, dried to 53% and rewet (7592B); C, dried to 70% and rewet (7592C).

TABLE VI
Permeability of Compacted Allophane Soils[a]

Water content (%)	Permeability (cm sec^{-1})	Bulk density (g cm^{-3})
150	7×10^{-6}	0.5
100	10^{-5}	0.65
50	10^{-2}	0.75

[a]From Tokunaga (1969).

consistency and structure of the soil. Microscopic observations revealed that in the range of low water content the soil has a granular structure with large voids. Therefore, it has a high permeability coefficient in spite of the higher bulk density. As the water content increases near the plastic limit, aggregated blocks are developed causing hindrance to water passage and lowering permeability. Further increase of water content up to around the natural water content causes the aggregated blocks to flow into a pastelike structure. In this range of water contents the coefficient of permeability is a minimum, and the dry density is low because of high water content (Table VI).

Allophane soils rich in humus show a different behavior. The organic matter forms stable humus which combines with soil particles forming an aggregated structure. Permeability increases slightly at high water content for compacted soil.

B. STRENGTH

Allophane soils can be stable in the undisturbed state, often occurring in relatively steep banks (Northey, 1966; Wesley, 1973). They are also relatively resistant to erosion, although landslides and water erosion occur on steep slopes. The strength of allophane soils disturbed by excavation and embankment is remarkably lower than that of the undisturbed soils. Once disturbed, allophane soils are too weak to ensure trafficability for construction equipment (Highway Research Board, 1973).

The Kanto loam has the bearing capacity to support buildings of four or five stories in spite of its high natural water content. The unconfined compressive strength of Japanese allophane soils is in the range of 1.0–2.3 kg cm^{-2} (Highway Research Board, 1973). This is more than five times the strength of alluvial nonvolcanic soils at the same water content. It is also recognized that the bearing capacity, estimated from the N-value obtained from the standard penetration test, is higher than that of the nonvolcanic clays. For allophane soils in the undisturbed condition, the experimental equation between the N-value of the standard penetration test and the bearing capacity (q_a) is given as follows:

$$q_a = (2 \text{ to } 2.5)N$$

For alluvial clay soils the equation is:

$$q_a = (1 \text{ to } 1.3)N$$

Birrell (1951) found small friction angles, 0–8°, and low shear strength, 0.2–0.4 kg cm^{-2}, from triaxial test results. Pope and Anderson (1960) measured values as low as 1° for friction angle and 0.5 kg cm^{-2} for the cohesion parameter. Wesley (1973) found undrained shear strength values of 1.0 to 1.2 kg cm^{-2}, and *in situ* vane shear results of 0.7 to 1.0 kg cm^{-2}. The unconfined compressive strength of allophane soils decreases with increasing organic matter content (Yamanouchi and Yasuhara, 1972).

Most authors comment on the large variability over short distances in engineering properties of allophane soils (Pope and Anderson, 1960; Northey, 1966; Wesley, 1973). This high heterogeneity in the field makes it difficult to use measured results in design of earth structures.

Another feature of allophane soils is that strength does not increase with depth, i.e., with increasing overburden pressure. Since it is not possible to estimate strength from the natural water content, no practical indices exist to estimate the strength.

The strain at failure is only 2–3% for allophane soil, against 2–6% for other clay soils. When allophane soils are disturbed, the compressive strain increases (Gradwell and Birrell, 1954; Takenaka, 1965). Gradwell and Birrell (1954) and Komamura and Takenaka (1973) show the Mohr circles for shearing resistance as a function of stress for undisturbed and remolded allophane soils.

Allophane soils have moderate measured sensitivity, the ratio of undisturbed to remolded shear strength. Birrell (1951) gives values of 6–12, Wesley (1973) measured values between 1 and 3. Wells and Furkert (1972) showed that water in undisturbed allophanes was held in hydrogen-bonded clusters. Remolding breaks up these clusters, and the water becomes distributed as single-linked molecules.

Allophane soils have high water contents and a well-developed soil structure. The lowering of the strength on remolding is due to change in water-holding characteristics and structure peculiar to allophane soils. On disturbance, water which was held firmly in the voids is released and free to flow. The measured soil suction has decreased; the decrease is largest at low suction values. In consequence, the soils soften and lose strength. The soils least resistant to this softening are buried soils with high organic matter, especially from old volcanic ash layers (Takenaka, 1966; Komamura and Takenaka, 1973; Takenaka and

Yasutomi, 1965). The changes in stress and pore pressure with time of remolding or kneading are shown in Table VII. The data are taken from Takenaka and Yasutomi (1965). They review the changes in soils on remolding, which can lead to either softening or hardening: (1) the structure units are forced together, leading to decrease in suction; (2) the units are separated, leading to increased suction; (3) water in voids is released, giving decreased suction; (4) the structure unit is broken, exposing new surfaces which absorb water, leading to increase in suction; (5) absorbed water is released, leading to decreased suction. Depending upon the net result of these changes, either hardening or softening can occur. The usual case for allophane soils is softening on remolding.

Adachi and Takenaka (1973) measured the increase in unconfined compressive strength with increasing soil suction. The relationship was linear on a log–log plot for samples in which the suction was increased by gradually decreasing the water content during remolding. However, when samples were dried to different water contents before remolding, strength did not increase beyond a suction of about 1 bar.

If disturbed soils whose strength has been decreased by remolding are kept at the same water content, the strength is gradually recovered with time. This strength regain is one of the features of allophane soils; it is often called "thixotropy" (Takenaka and Yasutomi, 1965; Komamura and Takenaka, 1973; Yasutomi, 1974). Table VIII shows typical results for strength regain. The soils which show little strength regain are thought to soften by structure units being forced together and releasing water. No subsequent movement apart of units takes place. The large strength regain is shown by soils in which water is released from within the structure on remolding. This water can be reabsorbed by the units on standing, and hence the soil hardens.

Wells and Furkert (1972) describe the sensitivity test used to recognize

TABLE VII
Strength and Pore Pressure Changes due to Remolding for Two Allophane Soils[a]

Remolding time (minutes)	Stress (kg cm^{-2})		Suction (millibar)	
	A	B	A	B
0	1.1	0.6	150	90
2	–	–	10	110
10	0.6	0.8	10	50
20	0.2	0.75	–	–
40	0.15	0.6	–	–

[a]From Takenaka and Yasutomi (1965).

TABLE VIII
Strength Regain After Remolding for Two Allophane Soils[a]

Time (days)	Cohesion (kg cm^{-2})		Friction angle (degrees)	
	A	B	A	B
1	0.15	0.29	1.8	12.0
10	0.19	0.30	1.8	12.8
100	0.20	0.41	2.0	13.3

[a] From Komamura and Takenaka (1973).

allophane soils in the field in New Zealand. When pressed between thumb and fingers under increasing pressure the soil suddenly shears, releasing free water.

Fujioka *et al.* (1965) report that polyvinyl alcohol soil stabilizers increase the strength of allophane soils, decrease swelling, and increase the maximum dry density.

Umeda and Nagasawa (1974) found a decreased shear strength and decreased soil suction after freezing an allophane soil.

C. CONSOLIDATION

The e-log p curves of allophane soils have been obtained by a number of workers (e.g., Gradwell and Birrell, 1954; Suzuki, 1973). Allophane soils are fairly compressible, once the preconsolidation pressure has been exceeded. Fieldes and Claridge (1975) quote values, from New Zealand studies, of 2.4 to 3.1 m^2 yr^{-1} for coefficient of consolidation. This high value is due to the aggregation of the clay-size grains.

The initial void ratio of allophane soils is much larger than for soils with crystalline clay minerals, from 2 to 3.5 for low humic allophane and 6 to 7 for allophane soils with high humus content. The e-log p curve of undisturbed allophane soils shows a clear flex point, so it is easy to find the preconsolidation pressure. For the Kanto loam in Japan and for New Zealand soils, the values range from 1 to 3 kg cm^{-2}, but are not correlated with depth. The curve for disturbed samples shows a smooth decrease in void ratio, and it is difficult to obtain a preconsolidation load (Highway Research Board, 1973). The preconsolidation pressure often exceeds the overburden pressure due to the strength of the aggregates (Gradwell and Birrell, 1954).

Allophane soils show a large secondary consolidation, which decreases as the applied load increases (Birrell, 1951; Northey, 1966). This is due to breakdown of the structure units. Secondary consolidation increases with increasing organic matter content of the soil (Yamanouchi, 1965).

The observed rebound for undisturbed and for remolded allophane soils on removing the load is remarkably small. Thus the elastic compression is very small compared with the total consolidation. Initial settlement, secondary consolidation, and coefficient of compressibility are decreased by remolding allophane soils.

D. SOIL STABILIZATION

Soil stabilization is necessary in order to use allophane soils, especially with high organic content, for highways or earth dams. Results on stabilization of highly organic allophane soil obtained by Yamanouchi (1963) were as follows:

Portland cement is not useful for stabilization because calcium is absorbed by the organic matter, and hardening of cement is hindered. Addition of calcium chloride, quick hardening cement, or calcium oxide is useful for stabilization. Arizumi (1962) found that hydrated gehlenite ($2CaO, Al_2O_3, SiO_2 \cdot nH_2O$) was formed by mixing Kanto loam with calcium hydroxide or calcium sulfate. The effect of asphalt emulsion is not dependable nor is a high density obtained, because of the high amount of water added to facilitate mixing. The combined use of asphalt emulsion and Portland cement gave better results than asphalt emulsion alone (Yamanouchi, 1963). Dispersing agents, such as calcium lignosulfonate, and aggregating agents, such as copolymerized vinylacetate and maleic acid, increased soil strength. Soil stabilization of organic allophane soil is best achieved with lignin materials such as spent sulfite liquor or its extract, with potassium dichromate, aluminum sulfate, or ferric chloride as the auxiliary agent.

Northey and Schafer (1974) developed methods of testing the effect of adding lime to wet allophane soils, and found increased soil strength additions of about 5% lime.

E. ADHESION AND COHESION

Adhesion of allophane soil to a metal surface is about one-quarter of that measured for a soil with crystalline clay minerals (Yamanaka, 1964). Adhesion is low at low water content, then quickly increases to a maximum as water content is increased.

Cohesion of allophane soil is low at zero water content, then increases to a maximum before decreasing again. Soils with crystalline clay minerals have a high cohesion at zero water content, which decreases rapidly as water content increases (Yamanaka, 1964). The cohesion of an air-dried allophane soil is lower than that of an undried soil (Yamanaka, 1964; Maeda and Soma, 1974).

REFERENCES

Adachi, T., and Takenaka, H. 1973. *Trans. Jpn. Soc. Irrig. Drain. Reclam. Eng. (Nogyo Doboku Gakkai Ronbunshu)* **43**, 26–32.
Ahmad, N., and Prashad, S. 1970. *J. Soil Sci.* **21**, 63–71.
Aomine, S., and Egashira, K. 1970. *Soil Sci. Plant. Nutr. (Tokyo)* **16**, 204–211.
Arizumi, A. 1962. *Rep. Jpn. Exp. Stn., Civ. Eng.* **110**.
Baba, H. 1971. *J. Fac. Agric. Iwate Univ. (Iwate Daigaku Nogakubu Hokoku Univ.)* **10**, 283–297.
Birrell, K. S. 1951. *Proc. Congr. R. Soc. N.Z., 7th* pp. 208–216.
Birrell, K. S. 1952. *Proc. Aust. N.Z. Conf. Soil Mech. Found. Eng., 1st* pp. 30–34.
Birrell, K. S. 1966. *N.Z. J. Agric. Res.* **9**, 554–564.
Birrell, K. S., and Fieldes, M. 1952. *J. Soil Sci.* **3**, 156–166.
Bonfils, P., and Moinereau, J. 1971. *Cah. ORSTOM, Ser. Pedol.* **9**, 345–363.
Borchardt, C. A., Theisen, A. A., and Harward, M. E. 1968. *Soil Sci. Soc. Am., Proc.* **32**, 735–737.
Brewer, R. 1964. "Fabric and Mineral Analysis of Soils." Wiley, New York.
Chichester, F. W., Youngberg, C. T., and Harward, M. E. 1969. *Soil Sci. Soc. Am., Proc.* **33**, 115–120.
Cochran, P. H., Boersma, L., and Youngberg, C. T. 1967. *Soil Sci. Soc. Am., Proc.* **31**, 454–459.
Colmet-Daage, F., and Cucalon, F. 1965. *Fruits* **20**, 19–23.
Colmet-Daage, F., Cucalon, F., Delaune, M., Gautheyrou, J., Gautheyrou, M., and Moreau, B. 1967. *Cah. ORSTOM, Ser. Pedol.* **5**, 1–38.
Colmet-Daage, F., Gautheyrou, J., Gautheyrou, M., de Kimpe, C., Sieffermann, G., Delaune, M., and Fusil, G. 1970. *Cah. ORSTOM, Ser. Pedol.* **8**, 113–172.
Colmet-Daage, F., Gautheyrou, J., Gautheyrou, M., de Kimpe, C., and Fusil, G. 1972. *Cah. ORSTOM, Ser. Pedol.* **10**, 169–191.
Croney, D., and Coleman, J. D. 1954. *J. Soil Sci.* **5**, 75–84.
Davies, E. B. 1933. *N.Z. J. Sci. Technol.* **14**, 228–232.
Egashira, K., and Aomine, S. 1974. *Clay Sci.* **4**, 231–242.
El-Swaify, S. A. 1973. *Soil Sci.* **115**, 64–72.
El-Swaify, S. A., and Swindale, L. D. 1968. *Trans. Int. Congr. Soil Sci., 9th* **1**, 381–389.
Espinoza, W., Rust, R. H., and Adams, R. S., Jr. 1975. *Soil Sci. Soc. Am., Proc.* **39**, 556–561.
Fieldes, M., and Claridge, G. G. C. 1975. *In* "Inorganic Components" (J. E. Gieseking, ed.), Soil Components, Vol. 2, pp. 351–393. Springer-Verlag, Berlin and New York.
Flach, K. W. 1964. *Proc. Panel Volcanic Ash Soils Latin Am., Turrialba, Costa Rica* Pap. A.7.
Forsythe, W. M. 1972. *Panel Volcanic Ash Soils Am., Pasto, Colombia* pp. 481–495.
Forsythe, W. M. 1975. *Proc. Soil Manage. Trop. Am., North Carolina State Univ.* pp. 155–167.
Forsythe, W. M., Gavande, S. A., and González, M. A. 1964. *Proc. Panel Volcanic Ash Soils Latin Am., Turrialba, Costa Rica* Pap. B.3.
Frost, R. J. 1967. *Proc. Southeast Asian Reg. Conf. Soil Eng., 1st, Bangkok* pp. 44–53.

Fujioka, Y., Nagahori, K., and Sato, K. 1965. *Trans Jpn. Soc. Irrig. Drain. Reclam. Eng. (Nogyo Doboku Gakkai Ronbunshu)* **10,** 1–6.
Fujiwara, H., and Baba, N. 1973. *Trans. Jpn. Soc. Irrig. Drain. Reclam. Eng. (Nogyo Doboku Gakkai Ronbunshu* **48,** 29–33.
Galindo-Griffith, G. G. 1974. Ph.D. Thesis, Univ. of California, Riverside.
Gautheyrou, J., Gautheyrou, M., and Colmet-Daage, F. 1976. "Chronobibliographie des Sols à Allophane." ORSTOM, Cent. Antilles.
Gradwell, M., and Birrell, K. S. 1954. *N.Z. J. Sci. Technol., Sect. B* **36,** 108–122.
Grim, R. E. 1953. "Clay Mineralogy." McGraw-Hill, New York.
Higashi, A. 1951. *J. Fac. Sci., Hokkaido Univ., Ser. 2* **4,** 21–29.
Higashi, A. 1952. *J. Fac. Sci., Hokkaido Univ., Ser. 2,* **4,** 95–107.
Highway Research Board. 1973. "On the Earthworks of Kanto Loam," pp. 78–114. Kyorits Shyuppan Press, Tokyo.
Hughes, I. R., and Foster, P. K. 1970. *N.Z. J. Sci.* **13,** 89–107.
Ikegami, M., and Tachiiri, M. 1966. *Rec. Land Reclam. Res.* **15,** 37–41.
Ishikawa, T., Tokunaga, K., and Tsukidate, K. 1963. *J. Agric. Eng. Soc. Jpn.* **31,** 22–30.
Ito, M. 1964. *Trans. Jpn. Soc. Irrig. Drain. Reclam. Eng. (Nogyo Doboku Gakkai Ronbunshu)* **9,** 1–4.
Iwata, S. 1963. *J. Agric. Eng. Soc. Jpn.* **30,** 3845–394.
Iwata, S. 1966. *Bull. Natl. Inst. Agric. Sci., Ser. 13* **16,** 149–176.
Kanno, I. 1961. *Bull. Kyushu Agric. Expt. Sta. (Kyushu Nogyo Shikenjo Iho)* **7,** 67–73.
Kasubuchi, T., 1975a. *Soil Sci. Plant Nutr. (Tokyo)* **21,** 73–77.
Kasubuchi, T. 1975b. *Soil Sci. Plant Nutr. (Tokyo)* **21,** 107–112.
Kawaguchi, K., Kita, D., and Mori, H. 1963. *J. Sci. Soil Manure, Jpn.* **34,** 7–12.
Kira, Y., Ambo, F., Sōma, K., and Ito, K. 1963. *Trans. Jpn. Soc. Irrig. Drain. Reclam. Eng. (Nogyo Doboku Gakkai Ronbunshu)* **7,** 76–80.
Kitagawa, Y. 1971. *Am. Mineral.* **56,** 465–475.
Kobo, K. 1964. *FAO World Soil Res. Rep.* **14,** pp. 71–73.
Kobo, K., and Oba, Y. 1964. *In* "Volcanic Ash Soils in Japan," pp. 30–31. Min. Agric. For., Japan.
Kodani, Y., Kono, H., and Uchida, K. 1976. *Trans. Jpn. Soc. Irrig. Drain. Reclam. Eng. (Nogyo Doboku Gakkai Ronbunshu* **60,** 7–13.
Komamura, M., and Takenaka, H. 1973. *J. Agric. Sci. Tokyo Univ.* **17,** 331–341.
Kon, T. 1967. *Res. Corresp. Soil Plant Growth Hokkaido* **58,** 28–50.
Kubota, T. 1971. *J. Clay Sci. Soc. Jpn.* **11,** 73–84.
Kubota, T. 1972. *Soil Sci. Plant Nutr. (Tokyo)* **18,** 79–87.
Kuno, G., and Mogami, T. 1949. *Technol. Eng. Rep., Tokyo Univ.* **3,** 7, 8.
Kuno, G., and Yabe, M. 1960. *Tech. Rep. Civ. Eng. Jpn.* **4,** 33–41.
Kuno, G., and Yabe, M. 1962. *Tech. Rep. Civ. Eng. Jpn.* **6,** 15–24.
Kuroda, M. 1971. *Trans. Jpn. Soc. Irrig. Drain. Reclam. Eng. (Nogyo Doboku Gakkai Ronbunshu)* **36,** 14–20.
Maeda, T. 1968. *J. Hokkaido Branch Agric. Meterol. Jpn.* **19,** 67–72.
Maeda, T., and Soma, K. 1974. *Soil Phys. Cond. Plant Growth, Jpn.* **30,** 15–22.
Maeda, T., and Warkentin, B. P. 1975. *Soil Sci. Soc. Am., Proc.* **39,** 398–403.
Maeda, T., Sasaki, S., and Sasaki, T. 1970. *Trans. Jpn. Soc. Irrig. Drain. Reclam. Eng. (Nogyo Doboku Gakkai Ronbunshu)* **31,** 25–28.
Maeda, T., Soma, K., and Sasaki, S. 1976. *Trans. Jpn. Soc. Irrig. Drain Reclam. Eng.* **61,** 9–17.
Masujima, H. 1962. *Res. Bull. Hokkaido Natl. Agric. Exp. Stn.* **77,** 40–47.
Masujima, H., and Kon, T. 1963. *Res. Bull Hokkaido Natl. Agric. Exp. Stn.* **80,** 70–76.
Masujima, H., and Mori, T. 1962. *Res. Bull. Hokkaido Natl. Agric. Exp. Stn.* **79,** 1–35.

Misono, S., and Terasawa, S. 1957. *Bull. Natl. Inst. Agric. Sci., Ser. B* **7**, 77–103.
Misono, S., Terasawa, S., Kishita, A., and Sudo, S. 1953. *Bull. Natl. Inst. Agric. Sci., Ser. B* **2**, 95–124.
Miyazawa, K., and Konno, T. 1976. *Res. Bull. Hokkaido Natl. Agr. Exp. Stn.* **114**, 89–118.
Nagata, N. 1963. *Trans. Agric. Eng. Soc. Jpn.* **7**, 37–42.
Nakano, M., Tabuchi, T., and Yawata, T. 1970. *Trans. Jpn. Soc. Irrig. Drain Reclam. Eng. (Nogyo Doboku Gakkai Ronbunshu)* **31**, 17–24.
Northey, R. D. 1966. *N.Z. J. Sci.* **9**, 809–832.
Northey, R. D., and Schafer, G. J. 1974. *N.Z. J. Sci.* **17**, 131–150.
Oba, Y., and Kobo, K. 1965. *J. Sci. Soil Manure, Jpn.* **36**, 203–206.
Packard, R. Q. 1957. *Soil Sci.* **83**, 273–289.
Parfitt, R. L., and Scotter, D. R. 1972. *Papua New Guinea Agric. J.* **23**, 9–11.
Pope, R. J., and Anderson, M. W. 1960. *Am. Soc. Civ. Eng. Res. Conf. Shear Strength Cohesive Soils, Univ. Colorado, Boulder* pp. 315–340.
Rousseaux, J. M., and Warkentin, B. P. 1976. *Soil Sci. Soc. Am., J.* **49**, 446–451.
Sasaki, S. 1957. *J. Sci. Soil Manure, Jpn.* **28**, 17–21.
Sasaki, T., Maeda, T., and Sasaki, S. 1969. *Trans. Jpn. Soc. Irrig. Drain. Reclam. Eng. (Nogyo Doboku Gakkai Ronbunshu)* **27**, 57–60.
Schalscha, E. B., Gonzalez, C., Vergara, I., Galindo, G., and Schatz, A. 1965. *Soil Sci. Soc. Am., Proc.* **29**, 481–482.
Sherman, G. D. 1957. *Science* **125**, 1243.
Sherman, G. D., Matsusaka, Y., Ikawa, H., and Uehara, G. 1964. *Agrochimica* **8**, 146–163.
Shiina, K. 1963. *Bull. Agric. Eng. Res. Stn.* **1**, 83–156.
Shiina, K., and Takenaka, H. 1961. *Trans Jpn. Soc. Irrig. Drain. Reclam. Eng. (Nogyo Doboku Gakkai Ronbunshu)* **2**, 49–55.
Soma, K., and Maeda, T. 1974. *Trans. Jpn. Soc. Irrig. Drain. Reclam. Eng. (Nogyo Doboku Gakkai Ronbunshu)* **49**, 27–34.
Sowers, G. F. 1965. *In* "Methods of Soil Analysis, Part I" (C. A. Black, D. D. Evans, J. L. White, L. E. Ensminger, and F. E. Clark, eds.), Monogr. No. 9, pp. 391–399. Am. Soc. Agron., Madison, Wisconsin.
Sudo, S., and Suzuki, T. 1963. *Trans. Agric. Eng. Soc. Jpn.* **7**, 104–108.
Suzuki, A. 1973. *Bull. Fac. Eng. Kumamoto Univ.* **100**, 1–33.
Swindale, L. D. 1964. *Proc. Panel Volcanic Ash Soils Latin Am., Turrialba, Costa Rica* Pap. B.10.
Tabuchi, T. 1963. *Trans. Agric. Eng. Soc. Jpn.* **7**, 32–37.
Tabuchi, T., Tabuchi, K., and Nagata, N. 1963. *Trans. Agric. Eng. Soc. Jpn.* **7**, 53–60.
Tada, A. 1965. *Trans Jpn. Soc. Irrig. Drain. Reclam. Eng. (Nogyo Doboku Gakkai Ronbunshu)* **14**, 41–45.
Tada, A., and Yamazaki, F. 1963. *Trans. Jpn. Soc. Irrig. Drain. Reclam. Eng. (Nogyo Doboku Gakkai Ronbunshu)* **5**, 17–23.
Takenaka, H. 1961. *Rec. Land Reclam. Res.* **12**, 23–27. Fac. Agric., Univ. Tokyo.
Takenaka, H. 1965. *Trans. Agric. Eng. Soc. Jpn.* **14**, 32–35.
Takenaka, H. 1966. *Soil Phys. Cond. Plant Growth, Jpn.* **14**, 21–25.
Takenaka, H. 1973. *Trans. Jpn. Soil Mech. Found. Eng. No. 180* **21**, 13–19.
Takenaka, H., and Yasutomi, R. 1965. *Trans. Agric. Eng. Soc. Jpn.* **14**, 54–59.
Takenaka, H., Tabuchi, T., Tabuchi, K., and Tada, A. 1963. *Trans. Agric. Eng. Soc. Jpn.* **7**, 61–67.
Terasawa, S. 1967. *Bull. Natl. Inst. Agric. Sci., Ser. B* **19**, 197–228.
Tokunaga, K. 1965. *In* "Soil Physics" (F. Yamazaki, ed.), pp. 248–260. Yokendo Press, Tokyo.

Tokunaga, K. 1969. 'Soil Physics." Yokendo Press, Tokyo.
Tokunaga, K., and Sato, T. 1975. *Trans. Jpn. Soc. Irrig. Drain. Reclam. Eng. (Nogyo Doboku Gakkai Ronbunshu)* **55**, 1–8.
Tsujinaka, N., Sasaki, T., Maeda, T., and Sasaki, S. 1970. *J. Fac. Agric., Hokkaido Univ.* **56**, 267–291.
Umeda, Y., and Nagasawa, T. 1974. *Trans. Jpn. Soc. Irrig. Drain Reclam. Eng. (Nogyo Doboku Gakkai Ronbunshu)* **54**, 6–10.
van Schuylenborgh, J. 1953. *Neth. J. Agric. Sci.* **1**, 50–57.
Wada, K., and Harward, M. E. 1974. *Adv. Agron.* **26**, 211–260.
Wada, K., and Henmi, T. 1972. *Clay Sci.* **4**, 127–136.
Wada, K., Yoshinaga, N., Yotsumoto, H., Ibe, K., and Aida, S. 1970. *Clay Miner.* **8**, 487–489.
Wada, S., and Wada, K. 1975. *Annu. Meet. Clay Sci. Soc. Jpn.*
Warkentin, B. P. 1972. *Can. J. Soil Sci.* **52**, 457–464.
Warkentin, B. P., and Maeda, T. 1974. *Soil Sci. Soc. Am., Proc.* **38**, 372–377.
Wells, N., and Furkert, R. J. 1972. *Soil Sci.* **113**, 110–115.
Wesley, L. D. 1973. *Geotechnique* **23**, 471–494.
Yakuwa, R. 1943. *Mem. Sapporo Meterol. Observ.* **2**(2), 41–104.
Yamanaka, K. 1964. *In* "Volcanic Ash Soils in Japan," pp. 69–75. Min. Agric., Japan.
Yamanouchi, T. 1963. *Proc. Asian Reg. Conf. Soil Mech. Found. Eng., 2nd, Tokyo* **1**, 359–363.
Yamanouchi, T. 1965. *Annu. Rep., Jpn. Road Assoc.* pp. 1–21.
Yamanouchi, T., and Yasuhara, K. 1972. *Trans. Jpn. Soc. Civ. Eng.* **13**, 23–29.
Yamazaki, F., and Takenaka, H. 1965. *Trans. Agric. Eng. Soc. Jpn.* **14**, 46–48.
Yasutomi, R. 1974. *J. Soc. Rheol. Jpn.* **2**, 53–57.
Yazawa, M. 1976. *Trans. Jpn. Soc. Irrig. Drain. Reclam. Eng. (Nogyo Doboku Gakkai Ronbunshu* **65**, 8–14.
Youngberg, C. T., and Dyrness, C. T. 1964. *Soil Sci.* **97**, 391–399.

12

Copyright © 1978 by The Williams & Wilkins Co.
Reprinted from *Soil Sci.* **126**:297-312 (1978)

PHYSICAL AND CHEMICAL PROPERTIES AND CLAY MINERALOGY OF ANDOSOLS FROM KITAKAMI, JAPAN

SADAO SHOJI AND TSUYOSHI ONO

Faculty of Agriculture, Tohoku University, 1-1, Amamiyamachi-Tsutsumidori, Sendai, Japan

ABSTRACT

Physical and chemical properties and the clay mineralogy of Normal and Light-Colored Andosols from Kitakami, Japan were studied. The profiles of both Andosols had features reflecting the repeated falls of tephras with different ages.

The Normal Andosol, which is a common Andosol in Japan, showed unique properties, such as large total porosity, high water retention, low bulk density, remarkable accumulation of organic matter, high fluoride pH value, large phosphate absorption, etc. The Light-Colored Andosol, which was formed by truncation, also had some of these unique properties. It was noted that chloritized 2:1 minerals were dominant, and allophane and imogolite were absent in the clay fractions of all the soil samples derived from volcanic ash, whereas allophane and imogolite were dominant in the clay fractions of all the soil samples from pumice.

It had been assumed that the unique physical and chemical properties of a common Andosol were largely attributable to the high content of allophane. The results obtained in this study show, however, that most of the unique properties of the Andosols from Kitakami are closely related to the dithionite-citrate-soluble constituents, especially alumina, to a great extent.

INTRODUCTION

Andosols show unique physical and chemical properties, such as low bulk density, high water retention, striking accumulation of organic matter, high fluoride pH value, weak cation retention, high phosphate absorbing capacity, etc. Most of these properties have been assumed to be largely attributable to the high allophane content of Andosols. Recently, however, Shoji and Masui (1972), Shoji and Saigusa (1977), and Tokashiki and Wada (1975), have observed that the formation of allophane is inhibited by organic matter in the A1 horizons of some Japanese Andosols. These observations strongly suggest the need for careful reexamination of the existence and the role of allophane as related to the unique properties of Andosols described above.

The Kitakami area, Iwate prefecture, has a wide distribution of Andosols derived from tephras with different ages (Shoji and Ono 1978; Ono and Shoji 1978). The preliminary study of the clay fractions of these soils, made by the present authors, has indicated the absence or virtual absence of allophane. Therefore, it is interesting to study the genesis, properties, and clay mineralogy of such Andosols.

The purposes of the present paper are to show physical and chemical properties and clay mineralogy of selected Andosols occurring on the well-drained sites at Kitakami, Iwate prefecture, and to discuss relationships between the central characteristics and the clay mineralogy of the Andosols.

SOIL AND ENVIRONMENTAL FACTORS

The study area is located at Kitakami, Iwate prefecture (lat. 39°15′ N, long. 141°5′ E). As shown in Fig. 1, the main Andosols are Normal Andosol, Light-Colored Andosol, and Aquic Andosol.[1] The genesis of these soils has been inves-

---

[1] Andosols (Kuroboku soils) occurring in the study area are divided into three subgroups according to the soil classification system proposed by the Land Survey Section of the Economic Planning Agency, of the Japanese government (1969). Andosols are defined in this system as soils that formed from air-borne pyroclastic materials or from parent materials containing large amounts of air-borne pyroclastic materials. *Normal Andosol* is Andosol whose surface soil is black

183

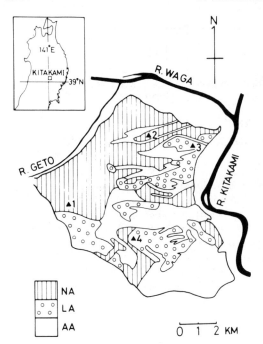

FIG. 1. Map showing the distribution of Andosols and sampling sites. NA = Normal Andosol; LA = Light-Colored Andosol; AA = Aquic Andosol.

tigated in detail (Shoji and Ono 1978; Ono and Shoji 1978). These studies showed that the topography of the study area, consisting of three Pleistocene terraces, is closely related to the genesis of Andosols to a large extent. The Kanagasaki terrace, the youngest one, is an alluvial fan formed by the Geto river (about a 190-m elevation on the top of the alluvial fan and a 70-m elevation on the margin of the alluvial fan). The Murasakino terrace (about a 110-m elevation), the next youngest, and the Nishine terrace (about a 120-m elevation) occur mainly on the margin of the alluvial fan. The major soils were found to be Normal Andosol and Aquic Andosol on the Kanagasaki terrace, Normal Andosol and Light-Colored Andosol on the Murasakino terrace, and Light-Colored Andosol and Aquic Andosol on the Nishine terrace.

(both value and chroma are two or less than two in Munsell notation) and greater than 25 cm in depth. *Light-Colored Andosol* is Andosol whose surface soil is black but less than 25 cm in depth, or is low in humus content and light-colored. *Aquic Andosol* is Andosol whose black surface soil is approximately 50 cm or more in depth, and whose subsoil shows gray color and contains iron mottlings, or has a gleyed layer.

Most of the parent materials originated from Mt. Yakeishi, which is located about 30 km west of the study area. Ono and Shoji (1978) indicated that the parent materials consist of tephras with different ages and textures, showing the intermittent activity of Mt. Yakeishi (Table 1). The rock type of volcanic ash,[2] which is contained in the A1 horizons of Normal Andosols, was determined to be dacitic, and that of all the other tephras, andesitic. The Light-Colored Andosols on the Murasakino and Nishine terraces are truncated soils that lack the dacitic ash contained in the A1 horizons of the Normal Andosols.

The mean annual temperature is 10.6°C, and the mean annual precipitation is 1360 mm in the study area. Though the original vegetation was Sasa-Fagetum crenatae, the present vegetation is strongly influenced by human activities, as shown in Table 1.

MATERIALS AND METHODS

Two Normal Andosols and two Light-Colored Andosols were selected for the present study (Fig. 1). These soils are representative Andosols occurring on well-drained sites in the study area. Morphological characteristics are given in Table 1.

The soil profiles were described in the field and sampled for physical, chemical, and clay mineralogical studies.

The physical methods used were: analysis of solid, liquid, and gas percentages by the volumanometric method, using the core soil samples taken in the field (Misono 1958); bulk density, by using the fine-earth fraction (Soil Survey Staff 1975); and soil texture, by the pipet method (Day 1965). The pretreatment and dispersing agents for the pipet analysis will be shown in the clay mineralogical analysis.

The chemical methods used were: pH, potentiometrically in water (1:2.5) and in N NaF (1:50) (Soil Survey Staff 1975); exchange acidity (Y_1), by the titration of 125 ml of N KCl soil extract (100:250) with 0.1 N NaOH; CEC by the procedure of Wada and Harada (1969); exchangeable bases with N NH_4OAC; phosphate absorption with 2.5 percent $(NH_4)_2HPO_4$ (1:10) (Kobo and

[2] Volcanic ash is defined in this paper as air-borne pyroclastic material consisting mainly of fragments less than 4 mm in size. Volcanic pumice is defined as air-borne pyroclastic material with vesicular structure mostly greater than 4 mm.

TABLE 1

Morphological properties of four soil profiles

Horizon	Depth, cm	Color, moist	Texture[a]	Humus, %	Structure	Consistency, moist	Hardness, kg/cm²[b]	P.M.[c]	Age, years B.P.
Normal Andosol No. 1 (Kanagasaki terrace)									
A11	0–15	7.5YR 1.7/1	LiC	18.5	3fgr	fr	8	VA	1000
A12	15–24	7.5YR 2/1	LiC	15.6	2fgr	fr	11	VA	to
A13	24–35	10YR 2/2	CL	12.0	2msbk	fr	24	VA	5000
IIBb	35–65	7.5YR 5/6	LiC	0.8	2msbk	fi	37	RD	
IICb	65+	10YR 6/8	CL	0		fi	37	RD	
Normal Andosol No. 2 (Murasakino terrace)									
A11	0–11	7.5YR 2/2	LiC	9.5	3fgr	fr	28	VA	1000
A12	11–20	7.5YR 2/2	LiC	8.6	3fgr	fr	28	VA	to
A13	20–33	7.5YR 2/2	LiC	7.4	2msbk	fr	28	VA	5000
IIB1b	33–45	5YR 4/6	HC	2.0	1msbk	fi	94	VA	10,000
IIB3b	45–77	10YR 5/6	HC	1.3	1msbk	fi	94	VA	to 20,000
IIIAb	77–104	5YR 4.5/6	HC	2.0	1mbk	fi	68	VA	>20,000
IVBb	104–120	7.5YR 5/6	HC	2.9	0	fr	nd	VP	>20,000
IVCb	120+	10YR 5/8	SCL	1.6	0	fr	nd	VP	>20,000
Light-Colored Andosol No. 1 (Murasakino terrace)									
A11	0–10	7.5YR 3/2	LiC	9.6	2fgr	fr	68	VA	10,000
A12	10–20	7.5YR 3/4	LiC	5.5	2msbk	fr	94	VA	to
IIB1b	20–34	5YR 4/6	HC	1.8	1msbk	fi	112	VA	
IIB3b	34–56	7.5YR 4.5/6	HC	1.1	1msbk	fi	112	VA	20,000
IIIAb	56–86	5YR 4/6	HC	1.2	1msbk	fi	68	VA	>20,000
IVBb	86–100	7.5YR 5/6	LiC	2.9	0	fr	50	VP	>20,000
IVCb	100+	10YR 5/8	CL	0	0	fr	nd	VP	>20,000
Light-Colored Andosol No. 2 (Nishine terrace)									
A1	0–14	7.5YR 2/3	HC	8.2	2fgr	fr	44	VA	10,000
IIBb	14–35	5YR 4/6	HC	1.2	2msbk	fi	68	VA	to 20,000
IIIAb	35–64	7.5YR 4.5/6	HC		1msbk	fi	80	VA	>20,000
IVBb	64–78	5YR 4/6	HC		1msbk	fr	80	VA	>20,000
IVCb	78–106	10YR 6/8	SC		0	fr	37	VP	>20,000

[a] Determined by the international classification of soil texture. HC: heavy clay; LiC: light clay; CL: clay loam; SC: sandy clay; SCL: sandy clay loam.

[b] Determined by the method of K. Yamanaka (1962).

[c] Parent material: VA, volcanic ash; VP, volcanic pumice; RD, river deposit of volcanic material.

The present vegetation of sampling sites is as follows: Normal Andosol No. 1, *Cryptomeria japonica, Magnolia obovata*; Normal Andosol No. 2, *Cryptomeria japonica, Quercus serrata*; Light-Colored Andosol No. 1, *Pinus densiflora, Sasa*; Light-Colored Andosol No. 2, *Fagus crenatae, Quercus serrata, Sasa*.

Oba 1973); total C, by the dry combustion method; dithionite-citrate soluble silica, alumina, and iron by the method of Mehra and Jackson (1960).

Clay mineralogical analysis

The fractions less than 2 μ, used for the clay mineralogical study, were separated from the field-moist soils by the following procedure (Shoji and Saigusa 1977).

In order to remove soil organic matter, all the soil samples of the A1 horizons were first subjected to the 6 percent H_2O_2 treatment. The clay fractions were dispersed in deionized water by sonication at 20 KHz for 15 min. To assist dispersion, the pH values of the suspensions were adjusted by addition of HCl or NaOH. The clay fractions were dispersed at about pH 10 for the soils from volcanic ash and at about 4 for the soils from pumice.

The mineralogical composition of each clay specimen was determined by chemical, infrared, and x-ray analyses, and by electron microscopic observation.

Total silica, alumina, and iron of desalted air-dried clay specimens (original clays) were determined by the colorimetric methods after Na_2CO_3 fusion. The compositions of amorphous clay constituents were determined by the successive differential dissolution methods as follows: the original clays were subjected to the dithionite-citrate treatment of Mehra and Jackson (1960), and the residue received the 0.15 M acid oxalate treatment (pH 3.5), as described by Higashi and Ikeda (1974). The infrared spectra of the materials removed by the dissolution treatments were obtained by the method described by Wada and Greenland (1970) and Wada and Tokashiki (1972).

The x-ray diffraction analysis was conducted using the oriented clay specimens before and after the dissolution treatments. For determining clay mineralogical composition, the following treatments were used: Mg saturation and glycerol solvation, K saturation and heating at 300°C and 550°C for 1 h, and hydrazine solvation. The relative abundance of crystalline minerals was estimated by determining the peak intensity of the x-ray diffractograms. Electron microscopic observation of the dithionite-citrate treated clays was carried out using a JEM 100 B electron microscope. For the observation, clay suspensions were spotted onto microgrids or carbon-coated collodion films.

RESULTS AND DISCUSSION

Morphology of selected soil profiles

Important morphological features of the selected soil profile are summarized in Table 1. The sampling sites are shown in Fig. 1. A Normal Andosol, No. 1, on the Kanagasaki terrace has developed from younger dacitic volcanic ash (1000 to 5000 years old) overlying the buried soil (IIBb and IICb horizons) formed from the river deposit of volcanic material (Ono and Shoji 1978). The thick A1 horizon, showing a rejuvenation type of humus horizon, has properties characteristic of most Andosols in Japan.

Another Normal Andosol, No. 2, occurring on the Murasakino terrace, has a soil profile consisting of multistoried soils derived from repeated ash and pumice falls. The parent materials of the A1 horizon include the same volcanic ash (1000 to 5000 years old) as observed in the Normal Andosol No. 1. It was noted that these parent materials are well mixed by human activity (Ono and Shoji 1978). This A1 horizon also has unique properties similar to those shown by the same horizon of the Normal Andosol No. 1. The IIBb horizon, which has developed from the old andesitic parent ash (10,000 to 20,000 years old), is heavy in texture and compact, reflecting strong surface weathering before the overburden of the younger dacitic ash (Ono and Shoji 1978). The B horizons of all the soil profiles have weak, subangular, blocky structure, and correspond to a color B horizon. The color, texture, and compactness of the IIIAb horizon, formed from andesitic ash (20,000 or more years old), also indicate that this horizon was subjected to strong surface weathering prior to burial by other volcanic ash. The IVBb and IVCb horizons have derived from andesitic pumice. As described above, the soils beneath the modern A1 horizon are old, buried soils, and so their properties are considerably different from those of the subsoils of most modern Andosols.

Light-Colored Andosols Nos. 1 and 2 are truncated soils, as already described. The A1 horizons of these Andosols have developed from the materials corresponding to the IIBb horizon of the Normal Andosol No. 2. Therefore, it is obvious that there are considerable differences in the properties of A1 horizons between the Normal and Light-Colored Andosols. Nevertheless, the horizons from IIBb down to IVCb in the Light-Colored Andosols show properties similar to those of the corresponding horizons of the Normal Andosol No. 2.

PHYSICAL PROPERTIES

The physical properties of the soil samples are given in Table 2. The results of mechanical analysis indicate the stratified nature of the soil profiles, resulting from the repeated falls of volcanic ash and pumice. For example, the clay content of the soil samples of the Normal Andosol No. 2 ranges from 33 to 41 percent in the A1 horizon, 52 to 56 percent in the IIBb horizon, and 68 percent in the IIIAb horizon.

The solid phase or total porosity (gas phase and liquid phase) of an average soil varies in the neighborhood of 50 percent (Baver, Gardner, and Gardner 1972). The solid phase is much smaller than 50 percent in most soil samples, however, especially the soils belonging to the A1 horizons of the Normal Andosols and the soils

TABLE 2

Physical properties of soil samples

Horizon	Mechanical composition				Proportion			Total porosity	Bulk density
	Coarse sand	Fine sand	Silt	Clay	Solid	Liquid	Gas		
	Wt, %				Vol, %				g/ml
Normal Andosol No. 1									
A11	7.4	28.1	28.1	36.4	22.8	51.3	25.9	77.2	0.64
A12	8.0	28.4	30.0	33.6	24.3	51.7	24.0	75.7	0.66
A13	8.5	25.7	40.6	25.2	21.4	50.7	27.9	78.6	0.75
IIBb	3.1	33.9	29.4	33.6	26.1	51.7	22.2	73.9	0.93
IICb	5.3	41.3	29.5	23.8	35.9	49.5	14.6	64.1	1.04
Normal Andosol No. 2									
A11	7.8	15.9	42.2	34.1	26.1	35.0	38.9	73.9	0.76
A12	7.6	15.3	44.5	32.7	25.6	37.3	37.1	74.8	0.81
A13	3.9	12.4	43.1	40.7	24.0	49.7	26.3	76.0	0.77
IIB1b	4.7	7.0	36.5	51.8	34.1	43.8	22.1	65.9	0.94
IIB3b	4.3	6.3	33.6	55.8	37.8	56.1	6.1	62.2	0.97
IIIAb	4.2	4.7	22.8	68.3	32.6	62.9	4.5	67.4	0.82
IVBb	5.7	10.8	24.5	59.0	20.9	67.2	11.9	79.1	0.66
IVCb	46.4	11.7	16.5	25.4	22.8	44.4	32.8	77.2	0.62
Light-Colored Andosol No. 1									
A11	6.6	18.6	37.2	37.7	35.3	39.4	25.3	64.7	0.86
A12	5.6	18.9	36.2	39.2	41.2	44.9	13.9	58.8	1.01
IIB1b	3.5	8.2	33.8	54.6	42.3	49.6	8.1	57.7	1.14
IIB3b	4.4	7.4	32.7	55.5	41.8	50.7	7.5	58.2	1.07
IIIAb	4.0	5.1	22.6	68.3	31.0	61.3	7.7	69.0	0.98
IVBb	16.1	23.4	20.2	40.4	22.3	68.9	8.8	77.7	0.79
IVCb	38.8	13.1	27.8	20.3	24.4	40.6	35.0	75.6	0.62
Light-Colored Andosol No. 2									
A1	4.6	14.4	35.8	45.2	32.0	46.1	21.9	68.0	0.82
IIBb	5.2	13.2	35.8	45.8	38.8	49.9	11.3	61.2	0.95
IIIAb	5.0	9.5	32.5	53.0	40.1	53.6	6.3	59.9	0.92
IVBb	3.0	9.2	22.7	65.1	31.6	59.3	9.1	68.4	0.93
IVCb	51.4	12.1	4.3	32.2	23.5	55.4	21.1	76.5	0.62

derived from pumice. The liquid percentage is remarkably large for all the soils, especially the soils derived from pumice, suggesting the plentiful existence of amorphous clay materials and vesicular pores of pumice. The gas percentage is medial to high in almost all the soils of the A1 horizons and the soils derived from pumice, and low for the old soils derived from volcanic ash. Thus, the proportions of gas, liquid, and solid of the soil samples reflect various factors, such as texture, soil structure, content of organic matter, and clay mineralogical composition, as shown later.

It is well known in Japan that one of the most remarkable features of Andosols is very low bulk density, values ranging mostly from 0.45 to 0.75 (Ministry of Agriculture and Forestry, Japan 1964). The Soil Survey Staff of USDA (1975) also showed that a bulk density less than 0.85 in the fine earth fraction is one of the important criteria to define Andepts. Bulk density values are low for almost all the soil samples, especially the A1 horizon soils of Normal Andosols and the soils formed from pumice.

CHEMICAL PROPERTIES

Table 3 shows the chemical properties of the soil samples. Total carbon is high for the A1 horizons of all the soil profiles, especially Normal Andosol No. 1, and it decreases with depth. It should be noted that the total carbon of the A1 horizons is much less in the Normal Andosol No. 2 than in No. 1, suggesting that human activity has influenced remarkably the accumulation of

TABLE 3
Chemical properties of soil samples

Horizon	T-C	pH H$_2$O (1:2.5)	pH NaF (1:50)	Y$_1$[a]	CEC	Exchange cations Ca	Mg	K	Na	Base saturation degree	Phosphate absorption	Dithionite-citrate-soluble SiO$_2$	Al$_2$O$_3$	Fe$_2$O$_3$	Si/Al$_2$
	%[b]			ml/100g[b]		me/100g[b]				%	mg P$_2$O$_5$/100g[b]	%[b]			
Normal Andosol No. 1															
A11	10.7	4.4	10.7	23.5	25.4	2.1	1.5	0.7	0.2	17.7	2400	0.5	3.2	3.4	0.26
A12	9.0	4.3	11.1	21.0	20.8	0.4	0.5	0.5	0.2	7.7	2700	0.4	3.5	3.8	0.20
A13	7.0	4.6	11.3	19.7	19.4	0.2	0.2	0.4	0.2	5.2	3000	0.4	3.9	4.1	0.17
IIBb	0.5	4.8		34.1	14.6	0.1	0.5	0.2	0.2	6.8	1900				
IICb	tr	5.1		15.8	15.3	0.2	1.2	0.1	0.2	11.1	1700				
Normal Andosol No. 2															
A11	5.5	4.9	10.8	22.3	21.4	0.6	0.3	0.3	0.2	6.5	1700	0.5	3.0	4.4	0.28
A12	5.0	5.2	10.9	17.3	21.4	0.3	0.3	0.3	0.3	5.6	2100	0.5	3.1	4.1	0.27
A13	4.6	5.3	10.8	19.9	23.1	0.3	0.2	0.2	0.4	4.8	2100	0.4	2.8	4.0	0.24
IIB1b	1.1	5.5	10.0	29.1	16.5	0.9	0.7	0.2	0.5	13.9	1300	0.5	2.1	4.4	0.40
IIB3b	0.8	5.5	9.9	26.7	17.0	1.6	1.2	0.1	0.5	20.0	1500	0.4	2.0	4.8	0.34
IIIAb	1.1	5.7	10.9	7.3	19.6	1.8	1.6	0.2	0.5	20.9	2900	0.6	3.4	5.8	0.28
IVBb	1.7	5.8	11.5	1.6	15.0	0.3	0.3	0.1	0.2	6.0	4500	1.1	5.6	3.8	0.33
IVCb	1.0	5.6	11.1	0.9	12.5	0.3	0.1	0.1	0.1	4.8	4500	1.2	4.8	2.2	0.43
Light-Colored Andosol No. 1															
A11	5.6	5.0	9.7	14.2	22.1	2.4	1.2	0.6	0.1	19.5	1600				
A12	3.2	5.1	10.0	11.5	19.2	1.3	1.0	0.3	0.1	14.1	1700				
IIB1b	1.1	5.5	10.2	17.6	17.0	2.1	1.9	0.3	0.3	27.1	1500				
IIB3b	0.6	5.5	9.8	18.7	17.2	1.9	2.1	0.2	0.3	26.2	1600				
IIIAb	0.7	5.6	10.5	19.1	16.1	2.0	2.0	0.2	0.5	29.2	2400				
IVBb	1.7	5.6	11.6	0.5	14.6	0.4	0.6	0.1	0.2	8.9	4800				
IVCb	tr	5.8	11.5	tr											
Light-Colored Andosol No. 2															
A1	5.2	4.8	10.5	21.9	21.3	0.2	0.3	0.4	0.1	4.7	1800	0.3	2.1	3.4	0.24
IIBb	1.2	5.1	9.9	37.6	17.8	0.3	1.1	0.2	0.3	10.7	1600	0.5	2.3	5.5	0.37
IIIAb	0.9	5.5	10.2	22.4	16.3	0.6	1.7	0.1	0.4	17.2	1700	0.5	2.3	5.4	0.37
IVBb	1.1	5.4	10.9	5.3	14.4	0.4	1.1	0.1	0.3	13.2	2400	1.1	3.5	5.5	0.53
IVCb	tr	5.7	11.1	0.3	9.2	0.1	0.1	0.1	0.1	4.3	4200	2.0	5.9	2.3	0.58

[a] Exchange acidity by titration of 125 ml of N KCl soil extract (soil:N KCl = 100:250) with 0.1 N NaOH.
[b] Oven-dry basis.

soil organic matter. Evidence of human activity was found within the profile of Normal Andosol No. 2 (Ono and Shoji 1978).

Normal Andosols, Nos. 1 and 2, and Light-Colored Andosol No. 2 were selected for the determination of dithionite-citrate-soluble SiO_2, Al_2O_3, and Fe_2O_3.

The dithionite-citrate-soluble fractions of all the soil samples are dominated by Al_2O_3 and Fe_2O_3 with small amounts of SiO_2, so the SiO_2/Al_2O_3 molar ratios are very low. The dithionite-citrate-soluble Al_2O_3, which may be closely related to the phosphate absorption, is greater in amount in the soils formed from pumice than in the other soils.

The pH (H_2O) values of the soil samples range from 4.3 to 5.8, and are not correlated with base saturation. For example, among two groups of soils whose base saturation is less than 10 percent, the one formed from volcanic ash and the one from pumice have pH (H_2O) values of 4.3 to 5.3 and 5.6 to 5.8, respectively.

The soil samples show a wide variation in exchange acidity (Y_1). The acidity (Y_1) is greater than 15 for almost all the soils from volcanic ash and less than 2 for the soils from pumice.

Allophane and imogolite have weaker acid properties than montmorillonite, halloysite, and kaolinite (Henmi and Wada 1974; Iimura 1966; Yoshida 1970, 1971, 1976; Wada and Ataka 1958). Exchange acidity (Y_1) indicates the amount of KCl-extractable Al^{3+} ions that is absorbed by strong acid sites of the soil; it is greater for montmorillonite and nil for allophane and imogolite (Iimura 1966; Yoshida 1970, 1971, 1976). Therefore, both pH (H_2O) and exchange acidity (Y_1) values strongly suggest that the clay fraction is dominated by 2:1 layer silicates in the soils formed from volcanic ash and by allophane, imogolite, or both in the soils formed from pumice.

In contrast to the pH (H_2O) and exchange acidity (Y_1) values described above, all the soil samples have pH (NaF) values greater than 9.4, which is one of the characteristics of Andepts whose exchange complex is dominated by "amorphous material"[3] (Soil Survey Staff 1975). For example, Calhoun, Carlisle, and Luna (1972) assumed that the exchange complex of the selected Colombian Andosols was dominated by

---

[3] "Amorphous material" is colloidal material that includes allophane and has all, or most of, the properties of allophane (*Soil Taxonomy*, p. 47).

allophane, using a pH (NaF) value greater than 9.4.

Examining the various factors related to pH (NaF) values of the soil samples shows that the pH (NaF) value is closely correlated with the amount of dithionite-citrate-soluble alumina, but not with any other component. The following regression equation is obtained by using the data of Table 3

pH (NaF) = 9.3 + 2.93
× log (dithionite-citrate-soluble $Al_2O_3\%$)
($r = 0.871^{***}$)

Perrott, Smith, and Michell (1976) reported that the release of hydroxyl ions from silica and ferric oxide gels, on treatment with NaF solution, is very small, above pH values of 7.6 and 9, respectively. On the other hand, hydroxyl release from alumina gel and poorly ordered aluminosilicates is appreciable at pH 9. Therefore, even if the soil does not contain allophane at all, but contains an appreciable amount of dithionite-citrate-soluble alumina, the soil is to show a high pH (NaF) value.

The CEC values are in the range of 9.2 to 25.4 me/100 g, and are greater for all the A1 horizon soils, indicating the important role of soil organic matter. The amounts of exchangeable bases are very small for all the soil samples, and so the degrees of the base saturation are mostly less than 20 percent. Kobo and Oba (1973) also showed that the base saturation of most surface soils of Andosols from 46 localities in Japan is less than 20 percent. Therefore, it is most likely that the low base saturation of the soil samples is mainly due to the leaching of bases under a humid condition.

Phosphate absorption is high for all the soil samples. The average phosphate absorption of 3620 mg P_2O_5/100 g, for the surface soils of the Andosols described above, was obtained using the data reported by Kobo and Oba (1973). The phosphate absorption of the soils formed from volcanic ash is smaller than the average value, but that of the soils formed from pumice is mostly greater than the average value.

Rajan, Perrott, and Saunders (1974) reported that phosphate at higher concentrations could be absorbed on the various types of absorption sites of amorphous aluminosilicates. On the other hand, investigating the components relating to phosphate absorption of Andosols, Kato (1970) and Yoshida and Kanaya (1975) showed that phosphate absorption of Andosols is more closely correlated with the content of dithionite-citrate-soluble alumina and iron than that of allophane.

Statistical analysis of the relationship between phosphate absorption and various components of the soil samples shows that phosphate absorption is closely correlated with the dithionite-citrate-soluble alumina, as given by the following equation

Phosphate absorption (mg P_2O_5/100 g)
$= -278 + 838 \times$ (dithionite-citrate-soluble Al_2O_3 %)   ($r = 0.955^{***}$)

It is not, however, correlated with any other component.

MINERALOGY OF CLAY FRACTIONS

The results of selective dissolution of clay fractions are given in Table 4. The chemical composition of the original clays is considerably different with different clay specimens. For example, the SiO_2/Al_2O_3 molar ratios of the original clays are in the range of 1.16 to 2.85, and are low for the clays of the soils formed from pumice. Therefore, as already mentioned, it is highly possible that there are distinct differences in clay mineralogical composition between soils formed from volcanic ash and soils formed from pumice.

Original clays were fractionated by the successive dissolution treatment into the dithionite-citrate-soluble, acid oxalate-soluble, and residual fractions. The acid oxalate treatment was used to estimate the content of allophane and imogolite (Higashi and Ikeda 1974; Wada and Wada 1976). Each fraction was expressed on the basis of the sum of SiO_2 and Al_2O_3 of original clay. The clay specimens can be divided into two groups on the basis of the results of selective dissolution analysis. Group 1 includes the clay samples separated from the soils formed from volcanic ash. The clay specimens obtained from the soils formed from pumice belong to Group 2 (soil samples of IVBb and IVCb horizons of Normal Andosol No. 2 and IVCb horizon of Light-Colored Andosol No. 2).

The content of dithionite-citrate-soluble fractions ranges from 5.2 to 10.1 percent in the clays of Group 1, and from 19.7 to 20.4 percent in those of Group 2. Compared to dithionite-citrate-soluble Al_2O_3 and Fe_2O_3, the dithionite-citrate-soluble SiO_2 is very small in the clays of Group 1 and small in the clays of Group 2. Therefore, the SiO_2/Al_2O_3 molar ratios of the dithionite-citrate-soluble fractions are very low, ranging from 0.26 to 0.48 for Group 1, and 0.50 to 0.62 Group 2. Figure 2 shows the infrared absorption spectra of clays in the 800 to 1200 cm^{-1} region. The dithionite-citrate-soluble fraction of the clays of Group 2 indicates the presence of allophanelike constituents having absorption at 945 to 955 cm^{-1} (Wada and Tokashiki 1972; Henmi and Wada 1976). The similar spectra were obtained with difficulty for the clays of Group 1, however, whose dithionite-citrate-soluble SiO_2 is very small in amount.

It is notable that the acid oxalate-soluble fraction is very small in amount and very low in SiO_2/Al_2O_3 molar ratio for the clays of Group 1. This fact strongly suggests that allophane and imogolite are absent or virtually absent from these clays. In contrast, the same fraction of the clays belonging to Group 2 ranges from 54 to 68 percent in amount, and 1.23 to 1.42 in SiO_2/Al_2O_3 molar ratio. As seen in Fig. 2, the acid oxalate-soluble fraction of the clays of Group 2 has infrared absorption at 960 to 970 cm^{-1}, which is characteristic of allophane or imogolite (Wada et al., 1970). The residual fraction of these clays gives strong absorption at about 1020 cm^{-1}, indicating the abundant presence of layer silicates. High-resolution electron micrographs of dithionite-citrate-treated clays of Group 2, as given in Fig. 3-3 and 3-6, indicate that there are large amounts of allophane spherules (outside diameter about 50 A) and imogolite strands (outside diameter about 20 A) having characteristics as described by Henmi and Wada (1976) and Wada et al. (1970). Therefore, the acid oxalate-soluble fraction of the clays of Group 2 is determined to consist mainly of allophane and imogolite. On the other hand, allophane and imogolite are not observed in the high-resolution electron micrographs of clays of Group 1, as seen in Fig. 3-1, 3-2, 3-4, and 3-5. These clays are overwhelmingly dominated by thin, irregular sheets with comparatively clear edges. Trace amounts of tubular halloysite are found in the clays of the buried soils belonging to Group 1, as seen in Fig. 3-2 and 3-4.

The x-ray diffractograms of dithionite-citrate-treated clays, as given in Fig. 4, indicate that the clays of Group 1 are dominated by 14 A minerals, but the clays of Group 2 are dominated by amorphous clay materials. The composition of

TABLE 4

Analytical data for selective dissolution of clay specimens

Horizon	Original clay				Dithionite-citrate-soluble fraction					Acid oxalate-soluble fraction					Residual fraction					Remarks[c]	
	Si/Al_2	SiO_2	Al_2O_3[a]	Fe_2O_3	Content [b] %	Si/Al_2	SiO_2	Al_2O_3[a]	Fe_2O_3	Content [b] %	Si/Al_2	SiO_2	Al_2O_3[a]	Content [b] %	Si/Al_2	Si/Al_2 (quartz-free)	SiO_2	Al_2O_3[a]	Quartz % [a]	Gibbsite % [a]
										Normal Andosol No. 1											
A11	2.84	37.25	22.27	11.03	8.4	0.35	0.84	4.13	8.43	3.7	0.18	0.21	2.00	87.9	3.81	3.27	36.20	16.14	5.2	tr
A12	2.75	38.88	24.03	11.41	8.4	0.41	1.02	4.25	8.83	3.0	0.22	0.21	1.66	88.6	3.53	3.10	37.65	18.12	4.6	1
A13	2.33	35.89	26.14	9.61	9.8	0.28	0.86	5.23	8.05	3.4	0.20	0.22	1.86	86.8	3.11	2.70	34.81	19.05	4.5	1
										Normal Andosol No. 2											
A11	2.67	38.61	24.54	9.85	8.3	0.27	0.71	4.55	7.27	3.5	0.25	0.28	1.90	88.2	3.54	2.87	37.62	18.09	7.1	tr
A12	2.67	39.02	24.81	9.42	8.0	0.26	0.68	4.40	7.26	3.5	0.37	0.40	1.84	88.5	3.47	2.79	37.94	18.57	7.5	tr
A13	2.61	39.43	25.66	9.14	7.6	0.26	0.65	4.31	7.06	2.7	0.41	0.34	1.41	89.7	3.28	2.60	38.44	19.94	7.9	tr
IIB1b	2.85	40.79	24.37	8.05	5.3	0.45	0.72	2.74	6.07	1.4	0.25	0.12	0.81	93.3	3.26	2.63	39.95	20.82	7.8	none
IIB3b	2.79	41.33	25.17	8.24	5.2	0.47	0.75	2.69	5.85	1.5	0.26	0.13	0.85	93.3	3.18	2.62	40.45	21.63	7.1	tr
IIIAb	2.10	33.99	27.49	9.23	9.7	0.41	1.15	4.82	6.76	6.4	0.61	1.03	2.88	83.9	2.73	2.36	31.81	19.79	4.3	3
IVBb	1.52	26.10	29.19	5.59	20.1	0.51	2.57	8.56	4.16	54.2	1.36	13.34	16.65	25.6	4.35	3.71	10.19	3.98	1.5	tr
IVCb	1.40	25.15	30.61	3.50	19.7	0.62	2.93	8.03	2.96	66.8	1.42	16.93	20.31	13.6	3.96	3.59	5.29	2.27	0.5	none
										Light-Colored Andosol No. 2											
A1	2.73	41.60	25.93	8.76	6.0	0.41	0.78	3.25	7.04	2.4	0.35	0.28	1.35	91.6	3.23	2.64	40.54	21.33	7.4	tr
IIBb	2.56	39.56	26.24	8.28	5.9	0.48	0.85	3.02	6.62	1.4	0.22	0.11	0.84	92.7	2.93	2.51	38.60	22.38	5.5	2
IIIAb	2.42	38.74	27.21	8.57	6.4	0.48	0.93	3.31	5.65	1.9	0.29	0.18	1.06	91.7	2.80	2.41	37.63	22.84	5.2	2
IVBb	1.39	27.87	34.01	7.91	10.1	0.48	1.37	4.85	5.97	3.8	0.48	0.52	1.84	86.1	1.62	1.46	25.98	27.32	2.5	16
IVCb	1.16	22.50	33.02	3.60	20.4	0.50	2.58	8.74	3.53	67.8	1.23	15.77	21.85	11.8	2.90	2.90	4.15	2.43	tr	none

[a] On the basis of original clay dried at relative humidity of 50%.
[b] On the basis of the sum of SiO_2 and Al_2O_3 of original clay.
[c] The amounts of quartz and gibbsite in clay specimens were determined by x-ray and differential thermal analyses, respectively.

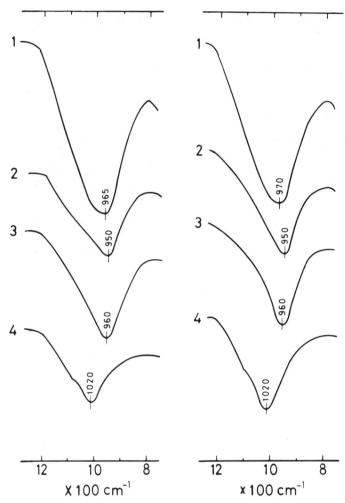

FIG. 2. Infrared spectra of selected clay specimens in the 800–1200 cm^{-1} region (left = IVCb horizon of Normal Andosol No. 2; right = IVCb horizon of Light-Colored Andosol No. 2). 1 = original clay; 2 = dithionite-citrate-soluble fraction; 3 = acid oxalate-soluble fraction; and 4 = residual fraction remaining after the successive treatment.

crystalline clay minerals was studied in detail by x-ray analysis after the various treatments. As shown in the summary of x-ray analysis in Table 5, chloritized 2:1 minerals are the most abundant among the crystalline clay minerals. The chloritized 2:1 minerals of the clay specimens of the A1 horizon soils of Normal Andosol No. 1 are partially expandable, whereas those of all the other clay specimens are nonexpandable. Abundant humus in the A1 horizon of Normal Andosol No. 1 may inhibit the Al-interlaying of expandable 2:1 layer silicates to some extent by the formation of an Al-organic matter complex as Lietzke, Mortland, and Whiteside (1975) mentioned. For example, CEC values of dithionite-citrate-treated clays are 41.5 and 48.4 me/100 g for the A11 and A12 horizons of Normal Andosol No. 1, respectively, and 30.2 and 29.1 me/100 g

FIG. 3. Electron micrographs of selected clay specimens. 1. A11 horizon of Normal Andosol No. 2. 2. IIB3b horizon of Normal Andosol No. 2. 3. IVCb horizon of Normal Andosol No. 2. 4. A1 horizon of Light-Colored Andosol No. 2. 5. IIBb horizon of Light-Colored Andosol No. 2. 6. IVCb horizon of Light-Colored Andosol No. 2: a = allophane; h = halloysite; i = imogolite; scale marker = 500 Å.

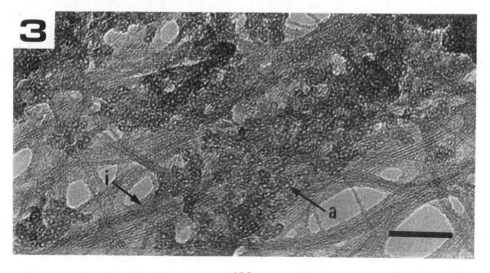

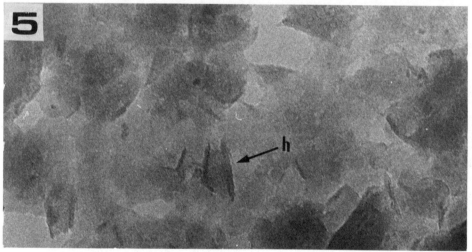

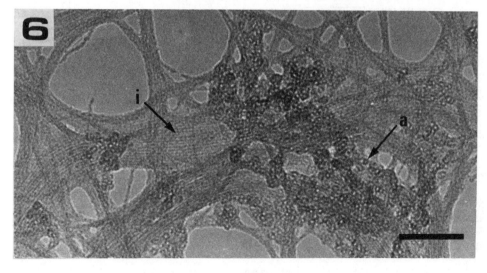

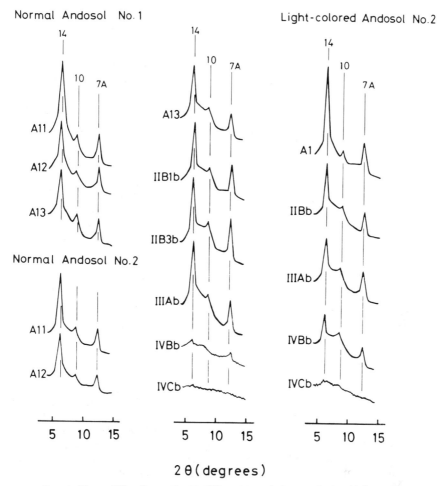

FIG. 4. X-ray diffraction patterns of Ca-saturated clay specimens (deferrated).

for the same horizons of Normal Andosol No. 2, respectively. This fact also indicates the difference in chloritization of 2:1 minerals in these soils. Small amounts of an unknown mineral, which was called a halloysitelike mineral by Mizota and Aomine (1975), were also found to exist in the fine clays of these soils.

As seen in Table 4, the amounts of the residual fraction remaining after the successive dissolution treatment are very large for the clays of Group 1, but small for the clays of Group 2. The SiO_2/Al_2O_3 molar ratio of most clays ranges from 3 to 4, but from 2 to 3 after subtracting the SiO_2 content due to quartz from total SiO_2 of the residual fraction. Some amount of gibbsite appears to be related to the relatively low SiO_2/Al_2O_3 molar ratio of the residual fraction of clay specimens separated from the IVBb horizon soil of Light-Colored Andosol No. 2.

From the results mentioned so far, it is evident that chloritized 2:1 minerals are dominant, and allophane and imogolite are absent from the clay fractions of all the soils formed from volcanic ash. On the other hand, the clay fractions of the soils formed from andesitic pumice are dominated by allophane and imogolite. The genetic problems of 14 A minerals, allophane, and imogolite in Andosols have been discussed in detail by Mizota (1976), Shoji and Saigusa (1977), Yamada, Saigusa, and Shoji (1978) and Wada and Matsubara (1968).

RELATIONSHIP BETWEEN PHYSICAL AND CHEMICAL PROPERTIES AND CLAY MINERALOGY

From the results described so far, it is evident that the unique physical properties of Andosols are not necessarily attributable to the leading role of allophane.

TABLE 5

Composition of crystalline minerals of clay fractions

Horizon	Major constituents	Minor constituents
		Normal Andosol No. 1
A11	chloritized(I)[a]	chlorite, illite, kaolin, quartz
A12	chloritized(I)	chlorite, illite, kaolin, quartz
A13	chloritized(I)	chlorite, illite, kaolin, quartz
		Normal Andosol No. 2
A11	chloritized(II)[b]	chlorite, illite, kaolin, quartz
A12	chloritized(II)	chlorite, illite, kaolin, quartz
A13	chloritized(II)	chlorite, illite, kaolin, quartz
IIB1b	chloritized(II)	chlorite, illite, kaolin, quartz
IIB3b	chloritized(II)	chlorite, illite, kaolin, quartz
IIIAb	chloritized(II)	chlorite, illite, kaolin, quartz, gibbsite
IVBb	chloritized(II)	chlorite, illite, kaolin, quartz
IVCb	chloritized(II)	chlorite, illite, kaolin, quartz
		Light-Colored Andosol No. 2
A1	chloritized(II)	chlorite, illite, kaolin, quartz
IIBb	chloritized(II)	chlorite, illite, kaolin, quartz, gibbsite
IIIAb	chloritized(II)	chlorite, illite, kaolin, quartz
IVBb	chloritized(II)	chlorite, illite, kaolin, quartz
IVCb	chloritized(II)	chlorite, illite, kaolin, quartz

[a] Chloritized(I) = partially expandable, chloritized 2:1 minerals.
[b] Chloritized(II) = nonexpandable, chloritized 2:1 minerals.

Ferruginous aluminous soils in Hawaii have an average bulk density of 0.3 to 0.5, while the ferruginous soils have a bulk density of 0.8 to 1.0 (Wada and Harward 1974). For some Hawaiian humic soils, percent field water content increases with percent organic matter (Maeda, Takenaka, and Warkentin 1977). On the other hand, large total porosity, high liquid percentage, and low bulk density are also common to organic soils (Davis and Lucas 1959). Therefore, unique physical properties can be imparted not only by allophane, but also by sesquioxidic constituents and organic matter. Furthermore, the strong development of granular structure in the A1 horizons (Table 1) appears to be related to these physical properties to some extent.

Unique chemical properties, such as accumulation of humus, high fluoride pH, and high phosphate absorption, are the most striking features of common Andosols. It has also implicitly been assumed that these properties are due to the presence of allophane as the dominant material in the clay fraction.

Total carbon contents of the modern humus horizons are in the range of 10.7 to 4.6 percent, for the Normal Andosols, and 5.6 to 3.2 percent, for the Light-Colored Andosols (Table 3). Since these horizons showed the absence of allophane and imogolite, it is highly probable that dithionite-citrate-soluble alumina and iron play an important role in the accumulation of organic matter. Wada and Harward (1974) stated that the importance of sesquioxidic constituents, rather than allophane, in accumulation of organic matter in some Andosols has been increasingly recognized.

According to the Soil Survey Staff (1975), a fluoride pH value greater than 9.4 is one of the important conditions for Andepts whose exchange complex is dominated by "amorphous material." It was shown, however, that all the soil samples used for the present study have fluoride pH values greater than 9.4, irrespective of differences in the clay mineralogical composition (Table 3), and that there is a close correlation between the fluoride pH value and content of dithionite-citrate-soluble alumina of the soil samples. This fact indicates that the dithionite-citrate-soluble alumina, as well as "amorphous material," can contribute to high fluoride pH value of Andosols, as expected from the experimental result of Perrott, Smith, and Michell (1976), who used different synthetic gels. Therefore, there is a limitation for using fluoride pH measurement to determine if "amorphous material" dominates the exchange complex of Andosols.

It was shown that all the soil samples have high phosphate absorption (Table 3), and that there is a close linear correlation between phosphate absorption and dithionite-citrate-soluble alumina. According to the regression coefficient given above, the molar ratio of phosphate to alumina is approximately 1. Nevertheless, the exact reactions between phosphate and components of the soil samples need further investigation.

From the foregoing discussion, it is obvious that the unique chemical properties of Andosols lacking allophane and imogolite in the clay fraction are largely attributable to the dithionite-citrate-soluble components, mainly alumina.

Andosols, which are widely distributed in Japan, are correlated mainly with Dystrandepts (Matsuzaka 1977). According to the recent soil classification systems proposed by Matsuzaka (1977) and Otowa (1977), however, the predominance of "amorphous material" in the exchange complex is not included in the criteria to define the Andosols.[4] On the other hand, there are many Andosols whose modern humus horizons show an absence or virtual absence of allophane and imogolite from the clay fraction, although they have the unique properties common to most Andosols. Among these are the soils from Towada (Shoji and Saigusa 1977), Hijiori (Yamada, Saigusa, and Shoji 1978), Naruko (Masui et al. 1973; Mizota and Aomine 1975), and Numazawa tephras (Yamada, Saigusa, and Shoji 1978) in northeastern Honshu. Finally, it is evident that sesquioxidic constituents, especially alumina, impart unique physical and chemical properties to some Andosols to a great extent.

REFERENCES

Baver, L. D., W. H. Gardner, and W. R. Gardner. 1972. Soil physics. Wiley, New York, p. 185.

Calhoun, F. G., V. W. Carlisle, and C. Z. Luna. 1972. Properties and genesis of selected Colombian Andosols. Soil Sci. Soc. Am. Proc. 36: 480–485.

Davis, J. F., and R. Lucus. 1959. Organic soils. Michigan State University Agric. Exp. Stn. Spec. Bull. 425.

Day, P.R. 1965. Pipette method of particle-size analysis. In Methods of soil analysis. C. A. Black (ed.), Am. Soc. Agronomy, Madison, Wis., pp. 552–562.

Henmi, T., and K. Wada. 1974. Surface acidity of imogolite and allophane. Clay Miner. 10: 231–245.

Henmi, T., and K. Wada. 1976. Morphology and composition of allophane. Am. Mineral. 61: 379–390.

Higashi, T., and H. Ikeda. 1974. Dissolution of allophane by acid oxalate solution. Clay Sci. 4:205–211.

Iimura, K. 1966. Acidic properties and cation exchange of allophane and volcanic ash soils. Bull. Nat. Inst. Agric. Sci. Japan, B 17: 101–157.

Kato, Y. 1970. Change in phosphorous absorptive coefficient of "Kuroboku" soils through successive H_2O_2-Defferation-Tamm's treatments. J. Sci. Soil Manure Japan 41: 218–224.

Kobo, K., and Y. Oba. 1973. Genesis and characteristics of volcanic ash soil in Japan. 3. The relation between the degree of weathering and parent material to some chemical properties of volcanic ash soil. J. Sci. Soil Manure Japan 44: 126–132.

Land Survey Section, Economic Planning Agency. Japan. 1969. Data of land survey.

Lietzke, D. A., M. M. Mortland, and E. P. Whiteside. 1975. Relationship of geomorphology to origin and distribution of high charge vermiculitic soil clay. Soil Sci. Soc. Am. Proc. 39: 1169–1177.

Maeda, T., H. Takenaka, and B. P. Warkentin. 1977. Physical properties of allophane soils. Adv. Agron. 29: 250.

Masui, J., S. Shoji, M. Saigusa, H. Ando, S. Kobayashi, I. Yamada, and K. Saito. 1973. Mineralogical and agrochemical properties of Kawatabi volcanic ash soil. Tohoku J. Agric. Res. 24: 166–174.

Matsuzaka, Y. 1977. Major soil groups in Japan. Proc. Intern. Seminar on Soil Environment and Fertility Management in Intensive Agriculture. Tokyo, pp. 89–95.

Mehra, O. P., and M. L. Jackson. 1960. Iron oxide removal from soils and clays by a dithionite-citrate system buffered with sodium bicarbonate. Clays Clay Miner. 7: 317–327.

Ministry of Agriculture and Forestry, Japan. 1964. Volcanic ash soils in Japan. Sakurai-Kosaido Printing Co. Ltd., Tokyo.

Mizota, C. 1976. Relationships between the primary mineral and the clay mineral compositions of some recent Andosols. Soil Sci. Plant Nutr. Tokyo 22: 257–268.

Mizota, C., and S. Aomine. 1975. Relationships between the inorganic colloid and the parent material of some volcanic ash soils in Miyagi and Iwate prefectures, Japan. Soil Sci. Plant Nutr. Tokyo 21: 201–214.

Misono, S. 1958. On the measurement of the physical properties of soil by a volumanometric method. J. Sci. Soil Manure, Japan 29: 67–70.

Ono, T., and S. Shoji. 1978. Genesis of Andosols at Kitakami, Iwate perfecture, northeast Japan. 2. Parent materials and soil formation. Quat. Res. Tokyo (Daiyonki-Kenkyu) 17: 15–23.

---

[4] Otowa (1977) defined Andosols as Inceptisols whose parent material is tephra.

Otowa, M. 1977. A proposal of a soil classification system for soil survey of Japan. 1. The introduction and structure of the system. J. Sci. Soil Manure Japan 48: 201–206.

Perrott, K. W., B. F. L. Smith, and B. D. Michell. 1976. Effect of pH on the reaction of sodium fluoride with hydrous oxides of silicon, aluminum, and iron, and poorly ordered aluminosilicates J. Soil Sci. 27: 348–356.

Rajan, S. S. S., K. W. Perrott, and W. M. H. Saunders. 1974. Identification of phosphate-reactive sites of hydrous alumina from proton consumption during phosphate adsorption at constant pH values. J. Soil Sci. 25: 438–447.

Shoji, S., and J. Masui. 1972. Amorphous clay minerals of recent volcanic soils. 3. Mineral composition of fine clay fractions. J. Sci. Soil Manure Japan 43: 187–193.

Shoji, S., and T. Ono. 1978. Genesis of Andosols at Kitakami, Iwate prefecture, northeast Japan. 1. Relationship between topography and soil formation. Quat. Res. Tokyo (Daiyonki-Kenkyu) 16: 247–254.

Shoji, S., and M. Saigusa. 1977. Amorphous clay materials of Towada Ando soils. Soil Sci. Plant Nutr. Tokyo 23: 437–455.

Soil Survey Staff. 1975. Soil taxonomy, a basic system of soil classification for making and interpreting soil surveys. Soil Conservation Service, USDA, AH 436.

Tokashiki, Y., and K. Wada. 1975. Weathering implications of the mineralogy of clay fractions of two Ando soils, Kyushu. Geoderma 14: 47–62.

Wada, K., and H. Ataka. 1958. The ion uptake mechanism of allophane. Soil Plant Food Tokyo 4: 12–18.

Wada, K., and D. J. Greenland. 1970. Selective dissolution and differential infrared spectroscopy for characterization of 'amorphous' constituents in soil clays. Clay Miner. 8: 241–254.

Wada, K., and Y. Harada. 1969. Effects of salt concentration and cation species on the measured cation exchange capacity of soils and clays. Proc. Intern. Clay Conf. Tokyo 1: 561–572.

Wada, K., and M. E. Harward. 1974. Amorphous clay constituents of soil. Adv. Agron. 26: 211–260.

Wada, K., T. Henmi, N. Yoshinaga, and S. H. Patterson. 1972. Imogolite and allophane formed in saprolite of basalt on Maui, Hawaii. Clays Clay Miner. 20: 375–380.

Wada, K., and I. Matsubara. 1968. Differential formation of allophane, "imogolite" and gibbsite in the Kitakami pumic bed. Trans. Intern. Congr. Soil Sci., 9th Congr. Adelaide 3: 123–131.

Wada, K., and Y. Tokashiki. 1972. Selective dissolution and difference infrared spectroscopy in quantitative mineralogical analysis of volcanic ash soil clays. Geoderma 7: 199–213.

Wada, K., N. Yoshinaga, H. Yotsumoto, K. Ibe, and S. Aida. 1970. High resolution electron micrographs of imogolite. Clay Miner. 8: 487–489.

Wada, K., and S. Wada. 1976. Clay mineralogy of the B horizons of two Hydrandepts, a Torrox and a Humutropept, in Hawaii. Geoderma 16: 139–157.

Yamada, I., M. Saigusa, and S. Shoji. 1978. Clay mineralogy of Hijiori and Numazawa Ando soils. Soil Sci. Plant Nutr. Tokyo 24: 75–90.

Yamanaka, K., and K. Matsuno. 1962. Studies on soil hardness. 1. On the soil hardness tester. J. Sci. Soil Manure Japan 33: 343–347.

Yoshida, M. 1970. Acid properties of montmorillonite and halloysite. J. Sci. Soil Manure Japan 41: 483–486.

Yoshida, M. 1971. Acid properties of kaolinite, allophane and imogolite. J. Sci. Soil Manure Japan 42: 329–332.

Yoshida, M. 1976. Acidic properties of soil colloids with special reference to allophane and imogolite. U.S.-Japan seminar on amorphous and poorly crystalline clays, Extended Abstracts, pp. 47–49.

Yoshida, M., and H. Kanaya. 1975. Forms of aluminum compounds in a volcanic ash soil in relation to phosphate fixation capacity. J. Sci. Soil Manure Japan 46: 143–145.

Part V

CHEMICAL CHARACTERISTICS

Editor's Comments
on Papers 13 Through 17

13 **ESPINOZO, GAST, and ADAMS**
Charge Characteristics and Nitrate Retention by Two Andepts from South-Central Chile

14 **BIRRELL and GRADWELL**
Ion-Exchange Phenomena in Some Soils Containing Amorphous Mineral Constituents

15A **GEBHARDT and COLEMAN**
Anion Adsorption by Allophanic Tropical Soils: I. Chloride Adsorption

15B **GEBHARDT and COLEMAN**
Anion Adsorption by Allophanic Tropical Soils: II. Sulfate Adsorption

15C **GEBHARDT and COLEMAN**
Anion Adsorption by Allophanic Tropical Soils: III. Phosphate Adsorption

16 **AMANO**
Phosphorus Status of Some Andosols in Japan

17 **EGAWA**
Properties of Soils Derived from Volcanic Ash

The chemical characteristics of Andosols have been the subject of many investigations, and a large number of articles are available in the literature, most dealing with cation and anion adsorption and phosphate retention. The soils are usually characterized by a sandy loam to loam texture and have acid to moderately acid reactions (Martini, 1976; Peralta, Bornemisza, and Alvarado, 1981; Iniguez and Val, 1982). Most probably the high acidity is the result of the organic fraction since allophane is generally considered weakly acidic in nature. It is assumed that in allophane, Al occurring in tetrahedral sheets is coordinating two O atoms, one OH and one H_2O molecule.

This water molecule is suspected to behave as a Brønsted acid and, depending on pH, may dissociate its proton (Wada, 1980; Henmi, 1977; Henmi and Wada, 1974), which is a very weak reaction for producing very low pH values. The percentage base saturation is usually below 50 percent, and it has been noticed in New Zealand that when base saturations are very low, soil pH in Andosols remains disproportionately high (Birrell, 1964).

One of the most important reasons for the chemical reactions in Andosols is believed to be the variable charge. Many researchers are of the opinion that this charge finds its origin mainly in the clay fraction. Although organic matter may contribute to some degree to increasing the negative charges in Andosols, it has usually been given less credit than allophane. The negative charge in Andosols may vary with changes in pH, ionic strength, and composition of the soil solution—hence, the term *variable* or *pH-dependent* charge. Depending on pH, the charge can even change in sign from negative, through zero, to positive. Although such a variable charge is prominent in Andosols, it is not unique to Andosols since Spodosols, Oxisols, Alfisols, and Ultisols may exhibit varying degrees of variable charges (Theng, 1980).

The amount and sign of charge govern many of the chemical reactions in Andosols. The presence of negative charges is responsible for cation exchange reactions, whereas the positive charges are the reason for the anion exchange capacity of the soils. Since the magnitude of cation and anion exchange may vary with changes in the nature and magnitude of the charges, the latter imposes different problems on the management of these soils, as opposed to that of temperate region soils in the United States and Europe, where soils with a relatively constant charge system are prevalent.

Paper 13 illustrates the variable charge characteristics of Andosols. Paper 14 shows ion exchange and problems in their determination. Papers 15A through 15C are examples of anion adsorption. Paper 16 is included to provide an extension on phosphate retention in Andosols. For a basic theory on development of charges and ion adsorption in general, reference is made to Bowden, Posner, and Quirk (1980) and Tan (1982).

The values for variable charges of Andosols apparently vary considerably from author to author. Parfitt (1980) quoted Espinoza, Gast, and Adams (Paper 13) that the negative charge of an Andosol is 100 me/100 g (pH 8.0), while the positive charge (pH 3.4) is 4 me/100 g. Considering the maximum negative charge of allophane (32-35 me/100g) and the texture of most Andosols (15 to 20 percent clay in A horizons), a negative charge of 100 me/100 g is extremely high when only the inorganic fraction is considered, as indicated by Parfitt (1980). Paper 13 reported in fact that the Andosols investigated exhibited a maximum pH dependent charge of 15 to 20 me/100 g only. Espinoza, Gast, and Adams pointed out the occurrence of a substantial amount of permanent negative charges of 45 to 49 me/100 g, attributed to isomorphous substitution and lattice vacancies in X-ray amorphous clays. In this respect, my analysis of Andosols in Indonesia showed the presence of cation exchange capacities due to permanent charges of 5 to 10 me/100 g (Paper 6). The occurrence of permanent negative charges in Andosols has also been observed recently in Japan. Egawa (Paper 17) quoted several investigators for the presence of permanent charges in Andosols. He noted that some Japanese scientists made a distinction between inside and outside negative charges. The inside charges were attributed to isomorphous substitution inside the crystal, whereas the outside charges were to be the result of mainly organic matter and amorphous minerals; however, no indications were provided as to the nature of the crystalline clays in the Andosols under study. Some Andosols may also contain imogolite, halloysite, kaolinite, montmorillonite, or 14 Å minerals, but in most Andosols these crystalline minerals are accessory minerals compared to allophane and, except perhaps imogolite, are not expected to increase the variable charge to such an extent as reported by Parfitt (1980).

The discovery of permanent negative charges sheds another light on the electrochemical properties of Andosols. It not only contradicts the general opinion that Andosols are dominated by variable charges, but it necessitates shifting the emphasis from allophane to organic matter as the main source of variable charges. Traditionally, variable charges in Andosols have been ascribed mainly to the presence of allophane and/or other amorphous materials in the clay fraction; little or no credit has been given to soil humus. The latter may be valid for most mineral soils, where organic matter constitutes only a very small portion of the soil solid fraction, but in the case of Andosols, in which organic matter content is unusually high in A horizons, most of the variable charges must find their origin in the organic fraction. I have noted that in a number of cases, organic matter content equals or exceeds the clay content in A horizons of Andosols. Andosols with humus contents of 15 to 30 percent and clay contents of 10 to 20 percent in A horizons are not a rare occurrence in Indonesia. Since, in addition, the organic matter in Andosols is expected to be present in intimate relation to allophane and/or other inorganic soil constituents, it is important to emphasize that the active soil fraction is present in the form of organo-mineral complexes or chelates. Such an interaction may bring about changes in electrochemical properties of the individual soil components. In many cases, the complexes or chelates can acquire charges that are lower than the sum of those contributed by the participating constituents. In analogy with terms used in plant nutrition and plant physiology, this effect of interaction causing a reduction in charges can be called antagonism. In many other instances, however, such an interaction can also yield organo-mineral complexes or chelates that have a higher charge than the sum contributed by the individual components, in these cases, the effect can then be considered synergistic. It is well known in basic soil chemistry that an interaction between clay minerals and humic acids adds to the clay surface an acidic functional group that contributes a strong negative charge to the clay (Tan, 1982).

The uniqueness in electrochemical properties of Andosols is not only in a ready change in negative charges with changing soil conditions but also in the capacity of the soils to acquire positive charges with variations in soil pH. That the positive charges can be substantial at ordinary pH levels suitable for plant growth (pH 5.0-6.0) is illustrated by Papers 15A through 15C. Adsorption of Cl^- and $SO_4^=$ was reported to be around 10 me/100 g at pH 6.0, but the value can be doubled or even quadrupled at pH 3.8. The difference between anion exchange with Cl^- or $SO_4^=$ is in the adsorption reaction. Chloride is explained to be adsorbed nonspecifically, whereas $SO_4^=$ is reported to be adsorbed by ligand exchange. Paper 15B added that the affinity of $SO_4^=$ for adsorption by soil was at least ten times greater than that of the nonspecifically adsorbed anions Cl^- and NO_3^-. Paper 15C reported that adsorption of phosphate was even stronger that that of $SO_4^=$, which is not surprising since Andosols have long been known in Japan for their high phosphate retention capacity. A summary on the phosphorus problem in Andosols of Japan is given in Papers 16 and 17. Values for P retention in the range of 400 to 2500 mg P_2O_5/100g are reported in Paper 17. These values are relatively small if compared with values obtained by Rajan (1975), who mentioned values for P retention by allophane of 620 to 1860 mg P/100g.

The presence of allophane is considered essential in the development of a significantly large AEC (anion exchange capacity) value. From his investigations, Wada (1980) found that the AEC in Andosols became substantially large (approaches the CEC value) only if allophane or imogolite was present. Soils containing gibbsite, iron oxides, and 2:1 layer silicates exhibited significantly smaller AEC values. The positive charge was reported not to develop in soils possessing aluminum-humus complexes unless appreciable amounts of allophane and imogolite were also present. From his analysis using

NH_4^+ and Cl^- ions, Wada (1980) also concluded that the following relationship existed between CEC, AEC, pH, and ion concentration:

$$\text{Log CEC} = c + b \text{ Log } C + a \text{ pH}$$

$$\text{Log AEC} = c' + b' \text{Log } C + a'\text{pH}$$

where a, b, c, a', b', and c' are constants and C = ion concentration (0.1 - 0.005 $\underline{N}$ NH_4Cl).

REFERENCES

Birrell, K. S., 1964, Some Properties of Volcanic Ash Soils, Meeting on the Classification and Correlation of Soils from Volcanic Ash, Tokyo, Japan, June 11-27, 1964, *Food and Agric. Org., United Nations, World Soil Resources Rept.* **14:**74-81.

Bowden, J. W., A. M. Posner, and J. P. Quirk, 1980, Adsorption and Charging Phenomena in Variable Charge Soils, in *Soils with Variable Charge*, B. K. G. Theng, ed., New Zealand Soc. Soil Sci., Lower Hutt, New Zealand, pp. 147-166.

Henmi, T., 1977, The Dependence of Surface Acidity on Chemical Composition (SiO_2/Al_2O_3 Molar Ratio) of Allophanes, *Clay Minerals* **7:**356-358.

Henmi, T., and K. Wada, 1974, Surface Acidity of Imogolite and Allophane, *Clay Minerals* **10:**231-245.

Iniguez, J., and R. M. Val, 1982, Variable Charge Characteristics of Andosols from Navarre, Spain, *Soil Sci.* **133:**390-396.

Martini, J. A., 1976, The Evolution of Soil Properties as It Relates to the Genesis of Volcanic Ash Soils in Costa Rica, *Soil Sci. Soc. Am. Proc.* **40:**895-900.

Parfitt, R. L., 1980, Chemical Properties of Variable Charge Soils, in *Soils with Variable Charge*, B. K. G. Theng, ed., New Zealand Soc. Soil Sci., Lower Hutt, New Zealand, pp. 167-194.

Peralta, F., E. Bornemisza, and A. Alvarado, 1981, Zinc Adsorption by Andepts from the Central Plateau of Costa Rica, *Commun. Soil Sci. Plant Anal.* **12:**669-682.

Rajan, S. S. S., 1975, Mechanism of Phosphate Adsorption by Allophanic Clays, *New Zealand Jour. Sci.* **18:**93-101.

Tan, K. H., 1982, *Principles of Soil Chemistry*, Marcel Dekker, New York, 267p.

Theng, B. K. G., ed., 1980, *Soils with Variable Charge*, New Zealand Soc. Soil Sci., Lower Hutt, New Zealand, 448p.

Wada, K., 1977, Allophane and Imogolite, in *Minerals in Soil Environments*, J. B. Dixon and S. B. Weed, eds., Soil Sci. Soc. Amer., Madison, Wisc., pp. 603-638.

Wada, K., 1980, Mineralogical Characteristics of Andisols, in *Soils with Variable Charge*, B. K. G. Theng, ed., New Zealand Soc. Soil Sci., Lower Hutt, New Zealand, pp. 87-107.

13

Copyright © 1975 by the Soil Science Society of America
Reprinted by permission from *Soil Sci. Soc. Am. Proc.* 39(5):842–846 (1975)

Charge Characteristics and Nitrate Retention By Two Andepts from South-Central Chile[1]

W. Espinoza, R. G. Gast and R. S. Adams, Jr.[2]

ABSTRACT

Charge characteristics of two Andepts, or volcanic-ash-derived soils, from south-central Chile were studied using a combination of nitrate retention, pH dependent cation exchange capacity and pH titration curves. Nitrate retention was determined by washing the soil with KNO_3 solutions having the appropriate concentration and pH and determining the amount of nitrate adsorbed. Net surface charge was determined from potentiometric titration curves conducted using a batch technique.

Results show that both soils have a complex mixture of pH dependent and permanent charge surfaces. The maximum pH dependent charge is in the range of 15 to 20 meq/100g at either very low or very high pH and high electrolyte concentration. The permanent negative charge is in the range of 45 meq/100g and is apparently the result of ion substitutions and lattice vacancies in X-ray amorphous oxides and hydrous oxides rather than 2:1 type clay minerals.

Additional Index Words: volcanic ash, permanent charge, pH dependent charge, cation exchange capacity, anion exchange capacity.

SOIL COLLOIDS may be divided into two general categories, those with constant surface charge and variable surface potential vs. those with constant surface potential and variable or pH dependent surface charge (Van Olphen, 1963). Constant charge behavior is generally attributed to the crystalline clay minerals with the charge originating from isomorphous substitution within the clay structure. In the case of the 2:1 type clay minerals, which are composed largely of Si, Al, Fe, or Mg in octahedral or tetrahedral coordination, ion size limitations generally result in a substitution of cations of lower valence for those of higher valence, and hence there is a permanent negative charge on the clay structure (Pauling, 1930; Marshall, 1949). The extent of this negative charge can be determined by cation exchange capacity measurements (Jackson, 1969; Rich, 1961).

The pH-dependent charge on inorganic soil components is due largely to oxides and hydrous oxides (hereafter referred to as oxides) of Fe, Al, and Si. Surfaces of such oxides contain ions not fully coordinated and hence the surfaces are electrically charged (Parks, 1967). Placed in an aqueous environment, these charges are balanced by the chemisorption of water; i.e. H_2O is split into H^+ and OH^- during adsorption to form hydroxylated surfaces (Parks, 1967; Blyholder and Richardson, 1962). Charge can then develop on these hydroxylated surfaces through amphoteric dissociation of the surface hydroxyl groups or adsorption of H^+ or OH^- ions. The extent of the resulting net surface charge is pH dependent and can be determined from pH titration curves.

Since H^+ and OH^- ions, which establish the surface charge, are also potential determining ions, the surface electric potential, Ψ_o, will also vary with pH of the equilibrium solution according to the Nernst relationship (Adamson, 1967; Van Raij and Peech, 1972):

$$\Psi_o = \frac{RT}{F} \ln \left[\frac{(H^+)}{(H^+)_{ZPC}} \right] \quad [1]$$

Here the zero point of charge (ZPC) is the pH at which the net surface charge density is zero. The net surface charge density, σ, on pH dependent charge surfaces varies with electrolyte concentration as well as pH (or Ψ_o) as shown by the Gouy-Chapman equation for the electric double layer (Babcock, 1963):

$$\sigma = \left(\frac{2n_o DkT}{\pi} \right)^{1/2} \sinh \left(\frac{ze\Psi_o}{2kT} \right). \quad [2]$$

According to Eq. [1] and [2], when pH = ZPC, the surface potential and net surface charge density are zero regardless of electrolyte concentration; i.e., this is the point at which pH titration curves for different electrolyte concentrations intersect.

If in conducting pH titration curves, electrolyte free oxide samples are initially placed in indifferent electrolyte solutions (such as NaCl) previously adjusted to a pH equal to the ZPC of the solid, there will be no shift in solution pH; i.e. the zero point of titration (ZPT) will be equal to the ZPC. If the pH of the electrolyte solutions is not at the ZPC, the oxide will adsorb H^+ or OH^-, shifting the solution pH. The direction and extent of the pH shift depends on the electrolyte concentration and the initial solution pH relative to the ZPC of the solid. However, the difference between ZPT and ZPC should be minimal for strongly amphoteric oxides such as Fe_2O_3 and Al_2O_3 which have ZPC's near neutral pH (de Bruyn and Agar, 1962; Parks, 1962, 1965, 1967).

Soils often contain a mixture of crystalline clay minerals having a constant negative charge and oxides having a pH-dependent charge. When such materials are initially placed in solutions of indifferent electrolyte, the ZPT should be relatively near the ZPC for the oxide fraction. Since the clay fraction will adsorb relatively little OH^-, the titration curves above the ZPT will approximate the charge characteristics of the oxides. However, H^+ added to such a mixture during acid titration will be adsorbed by the clay fraction by ion exchange as well as being adsorbed on the oxide surfaces. As a result, the ZPC *appears* to be shifted to a much lower pH than the ZPT. The ZPC observed in such cases is not the ZPC of the oxide surfaces, however, and should be clearly labeled as an "apparent" ZPC of the mixture. Under such circumstances, the ZPT is a better indication of the

---

[1] Paper no. 8920, Scientific Journal Series, Minnesota Agric. Exp. Sta., Univ. of Minn., St. Paul, Minn. 55101. Received 2 Dec. 1974. Approved 22 April 1975.
[2] Former Research Assistant and Professors of Soils, respectively. Department of Soil Science, Univ. of Minnesota, St. Paul, Minn. 55101.

ZPC of the oxide than the "apparent" ZPC observed from the titration curves.

Since both the clay and oxide fractions adsorb H^+ during acid titration of mixtures of clay minerals and oxides, pH titration curves do not provide an accurate measure of the positive charge on oxides below the ZPC. However, the extent of this positive charge can be determined by measuring the non-specifically adsorbed anions, such as NO_3^- and Cl^-, as a function of pH and electrolyte concentration.

The purpose of the study reported here was to determine the charge characteristics of two Andepts formed in volcanic ash in south-central Chile by the use of pH titration curves and NO_3^- adsorption. As will be pointed out, the surface charge on these soils is apparently the result of a complex mixture of constant and pH dependent charge materials, even though very little crystalline clay can be identified.

MATERIALS AND METHODS

Soil Characteristics

Composite soil samples of the Arrayan and Santa Barbara soil series were obtained from depths of 0–20 cm. Both soils are Andepts (Typic Dystrandepts) formed in volcanic materials in south-central Chile. The Arrayan soils are found at 300 m altitude where annual rainfall is about 1,150 mm and mean annual soil temperature is about 15C. The Santa Barbara soils are located at about 700 m where annual rainfall is about 1,350 mm and mean annual soil temperature is about 14C. The physical, chemical, and mineralogical properties of these soils have been discussed in detail elsewhere (Espinoza et al, 1974) so only the more pertinent information is given (Table 1). The soil characteristics most important for consideration in this study are: (i) the high organic matter content, (ii) the large pH dependence of the cation exchange capacity, and (iii) the high $< 2\mu m$ clay content.

Soil Fractionization Procedures

Surface soil samples of the Arrayan and Santa Barbara soils were subjected to progressive extraction of organic matter, free iron oxides and amorphous alumino-silicates resulting in the residues or "soil fractions" described in Table 2. Soil organic matter was removed using NaOCl adjusted to pH 9.5 as outlined by Lavkulich and Wiens (1970). Free iron oxides were removed from the organic-free samples using the sodium dithionite-citrate-bicarbonate method described by Jackson (1965). Amorphous alumino-silicates were removed from samples, free of organic matter and free iron oxides, using hot (90C) 0.5N NaOH as outlined by Wada and Greenland (1970).

Potentiometric Titration Curves

Potentiometric titration curves were made using the batch technique outlined by Van Raij and Peech (1972). Forty grams of washed soil and appropriate amounts of NaCl, 0.1N HCl or 0.1N NaOH were added to a 50-ml beaker along with sufficient water to give a final volume of 40 ml. The beakers were kept tightly covered to prevent evaporation and the contents were stirred occasionally. After 3 days, the pH of the suspension was measured. The final NaCl concentrations were either 1.0, 0.1, or 0.01N.

The amount of H^+ or OH^- adsorbed by the soil sample at any given pH value was taken as equal to the amount of HCl or NaOH added to the suspension minus the amount of acid or base required to bring the same volume and the same concentration of NaCl solution, without the soil sample, to the same pH. The ZPC was considered to be the common intersection point of the titration curves in the presence of the three NaCl concentrations. The net electric charge was calculated from the amount of H^+ and OH^- adsorbed with respect to the ZPC.

Nitrate Retention Curves

Nitrate retention was determined as a function of pH and electrolyte concentration (KNO_3) using the procedure outlined by Singh and Kanehiro (1969). Two-gram soil samples were placed in preweighed, 100-ml plastic centrifuge tubes along with 20 ml of KNO_3 solution having the appropriate concentration and which had been adjusted to the desired pH using HNO_3 or KOH. The tubes were agitated for 30 min and centrifuged for 15 min. This procedure, which was established to be adequate for attaining equilibrium, was repeated until there was no detectable change in either pH or KNO_3 concentration of the saturating solution. This involved at least four washings of each sample.

Following the last centrifugation, the supernatant was decanted, and the amount of solution in the soil pores calculated from the difference in the initial and final weight of the tubes. The concentration of KNO_3 in the entrapped solution was assumed to be the same as in the original solution. Nitrate was extracted by washing the sample four times with 20-ml aliquots of 1N KCl. The extracting solution was taken to 100 ml, and NO_3^- determined by reduction and steam distillation (Bremner, 1965). The NO_3^- retained by the soil was calculated as the difference between the total amount extracted and that entrapped after the last KNO_3 washing.

Table 1—Some properties of the 0–25 cm surface horizons of the Arrayan and Santa Barbara soils

	Arrayan Soil	Santa Barbara Soil
pH (H_2O)	5.8	6.0
pH (KCl)	5.3	5.4
O.M. (%)	14.8	16.2
$<2\mu$ Material (%)	36	44
SiO_2 (%)	48	47
Al_2O_3 (%)	24	25
Fe_2O_3 (%)	15	15
CEC, pH 3.5 (meq/100 g)*	22	18
CEC, pH 7.0 (meq/100 g)	45	48
CEC, pH 10.5 (meq/100 g)*	59	67
% Base Saturation	17	16
Surface Area (m^2/g)	225	225

* From Espinoza, W. (1969).

Table 2—Chemical composition of the soil fractions studied expressed as percent of the original soil

Soil Fraction		Organic matter	SiO_2	Al_2O_3	Fe_2O_3	Molar Ratio SiO_2/Al_2O_3
		Arrayan Soil				
A-I	Complete soil	14.8	48.3	23.9	14.7	3.7
A-II	O.M. removed	3.3	48.3	23.9	14.7	3.7
A-III	O.M. and Fe removed	0.7	48.3	18.9	7.1	4.8
A-IV	O.M., Fe, Al, and Si removed	0.0	34.6	3.6	6.8	16.1
		Santa Barbara Soil				
S-I	Complete soil	16.2	46.9	24.7	15.5	3.5
S-II	O.M. removed	6.0	46.9	24.7	15.5	3.5
S-III	O.M. and Fe removed	1.4	46.9	20.7	9.5	4.3
S-IV	O.M., Fe, Al, and Si removed	0.0	35.1	4.3	9.2	13.8

RESULTS AND DISCUSSION

Results (Fig. 1) show that the whole soils have a significant positive charge at low pH as evidenced by nitrate retention. Qualitatively at least, nitrate retention varied with both pH and electrolyte concentration as predicted by Eq. [2]. The amount of nitrate retained varied from nearly two at high pH and low KNO_3 concentration to about 14 and 19 meq $NO_3^-/100g$ for the Arrayan and Santa Barbara soils, respectively, at pH 2.0 and $0.3N$ KNO_3. The ZPC cannot be specifically identified from anion retention since it is the point where the quantities of positive and negative charge are equal, but not necessarily zero; i.e. the point at which the net surface charge is zero. However, the positive charge will be limited at this point, and the ZPC of pH dependent charge surfaces will be in the pH range where anion retention approaches zero. On this basis, curves in Fig. 1 suggest a ZPC in the pH range of 6 to 7.

Charge characteristics indicated by the pH titration curves differ in several ways from those shown by the nitrate retention curves (Fig. 2 and 3). First, the net positive charge predicted from the pH titration curves is much greater than the observed nitrate retention. This is probably the result of added H^+ being consumed for dissolution of Al during the titration with HNO_3 (Kinjo and Pratt, 1971a, 1971b). Secondly, the "apparent" ZPC from the titration curves is at a pH considerably lower than the ZPC indicated by the nitrate retention curves and the pH of the zero points of titration (ZPT): i.e. the pH of the washed soil in the respective KNO_3 solutions before any addition of acid or base.

The differences between the ZPT and "apparent" ZPC for the Arrayan and Santa Barbara soils is 34 and 20 meq/100g, respectively. As indicated previously, this difference is usually attributed to the presence of a permanent or structural negative charge (Van Raij and Peech, 1972). Assuming this is the case, and that this permanent charge is due to the presence of 2:1 type clay minerals with a CEC

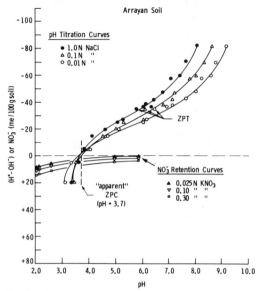

Fig. 2—pH titration curves and nitrate retention curves for Arrayan soil.

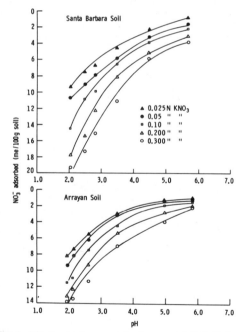

Fig. 1—Nitrate retention curves for Arrayan and Santa Barbara soils.

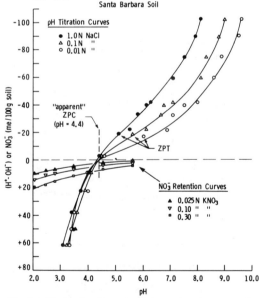

Fig. 3—pH titration curves and nitrate retention curves for Santa Barbara soil.

in the range of 1 meq/g, a clay content of at least 20 to 30% of the soil would be expected. A detailed mineralogical characterization of these soils, however, showed only barely detectable quantities of such clays (Espinoza et al. 1974). This is illustrated by the X-ray diffraction results (Fig. 4) for Mg- or K-saturated soil fractions of the Arrayan soil after: (i) removal of organic matter (A-III), (ii) removal of organic matter and Fe (A-III), and (iii) removal of organic matter, Fe, and soluble Al and Si (A-IV). Only in fraction A-IV was there any detectable 2:1 type clay minerals and even then, very little was present. Results were essentially the same for the Santa Barbara soil.

If the displacement of the ZPC to a lower pH than the ZPT is due to the presence of a permanent negative charge, and if there is little 2:1 type clays present, then the negative charge is apparently the result of ion substitutions or site vacancies within the X-ray amorphous material (Parks, 1967). Furthermore, the extent of this permanent negative charge is probably significantly greater than the 20 and 34 meq/100g difference between the ZPC and ZPT since (i) both soils have a relatively low percent base saturation, and (ii) the CEC at pH 7.0 is 45 and 48 meq/100g for the Arrayan and Santa Barbara soils, respectively (Table 1). If the soils had been 100% base saturated at the ZPT, considerably more H^+ would have been required to displace the saturating cations.

Since the ZPT in 0.01N NaCl and nitrate retention results indicate a ZPC of about pH 6–7 for the pH-dependent charge, the CEC values of 45 and 49 meq/100g at pH 7 are probably a fairly good measure of the permanent negative charge on the soil. Certainly the CEC of about 20 meq/100g at pH 3.5 (Table 1) must all be due to permanent negative charge for this pH is well below the ZPC for the pH dependent surfaces. The difference in CEC between pH 7.0 and 10.5 is 14 and 19 meq/100g of Arrayan and Santa Barbara soil, respectively. This is approximately the same as the maximum nitrate retention at low pH and high KNO_3 concentrations and is evidently due to the negative charge associated with the pH dependent charge surfaces.

It is difficult to establish whether a ZPC of pH 6–7 is reasonable for the pH dependent charge material in these soils since they are a complex mixture of organic matter and Fe, Al, and Si oxides. Organic matter and Si will tend to lower the ZPC while Fe and Al oxides will tend to raise it (Van Raij and Peech, 1972). However, all the available data tend to point to an average value of pH 6–7.

Nitrate retention by the various soil fractions at 0.1N KNO_3 and varying pH are shown in Fig. 5, expressed as meq NO_3^- adsorbed per 100g of residue. Curves for the whole soils, A-I and S-I, are in the same general range as those for the organic matter-free residues, A-II and S-II, indicating that NO_3^- retention by the organic matter is about the same, on a weight basis, as for the mineral fraction of the soil. Removal of both the free iron oxides and Al and Si resulted in a reduction in NO_3^- retention greater than the proportionate weight loss of the sample. These results indicate that the NO_3^- retention capacity or pH-dependent charge is associated with the organic matter, or extractable Fe, Al, and Si fractions of the soil with little retention by the final Si rich residues, A-IV and S-IV.

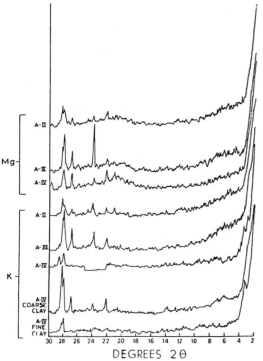

Fig. 4—X-ray diffraction patterns of Mg or K saturated soil fractions.

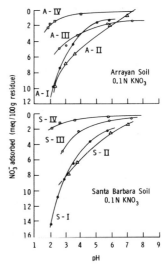

Fig. 5—Nitrate retention curves for various soil fractions of the Arrayan and Santa Barbara soils.

SUMMARY AND CONCLUSIONS

Charge characteristics of two Andepts, or volcanic-ash-derived soils from south-central Chile, were determined by a combination of nitrate retention, pH dependent CEC and pH titration curves. Results show the soils to consist of a complex mixture of pH dependent and permanent charged surfaces. The maximum pH dependent charge is in the range of 15 to 20 meq/100g at either very low or very high pH and high electrolyte concentration. The permanent negative charge is in the range of 45 meq/100 g and is apparently the result of ion substitutions and lattice vacancies in X-ray amorphous Fe, Al, or Si oxides and hydrous oxides rather than 2:1 type clay minerals.

LITERATURE CITED

1. Adamson, A. W. 1967. Physical chemistry of surfaces. 2nd ed. Interscience Publishers, N.Y.
2. Babcock, K.L. 1963. Theory of the chemical properties of soil colloidal systems at equilibrium. Hilgardia 34:417–542.
3. Blyholder, G., and E. A. Richardson. 1962. Infrared and volumetric data on the adsorption of ammonia, water and other gases on activated iron (III) oxide. J. Phys. Chem. 66:2597–2602.
4. Bremner, J. M. 1965. Inorganic forms of nitrogen. In C. A. Black (ed.). Methods of soil analysis, Part 2. Chemical and microbiological properties. Agronomy 9: 1179–1237. Amer. Soc. Agron., Madison, Wis.
5. de Bruyn, P. L. and G. E. Agar. 1962. Surface chemistry of flotation. In D. W. Fuerstenau (ed.) Froth flotation, 50th Anniversary volume. Am. Inst. Mining, Metallurgical and Petroleum Engineers, N.Y.
6. Espinoza, Waldo G. 1969. Determinacion de Alofan en suelos de Nuble mediante el valor Delta de la capacidad total de interambio cationico. Circ. Informativa No. 27, University De Concepcion, Chile.
7. Espinoza, W., R. H. Rust and R. S. Adams, Jr. 1974. Characterization of mineral forms in Andepts from Chile. Soil Sci. Soc. Am. Proc. 39:556–561.
8. Jackson, M. L. 1965. Free oxides, hydroxides and amorphous alumino silicates. In C. A. Black (ed.) Methods of soil analysis, Part I. Agronomy 9:578–601.
9. Jackson, M. L. 1969. Soil chemical anlysis—advanced course. Publ. by Author, Department of Soils, Univ. of Wisconsin, Madison, Wisconsin.
10. Kinjo, T., and P. F. Pratt. 1971a. Nitrate adsorption: I. In some acid soils of Mexico and South America. Soil Sci. Soc. Am. Proc. 35:722–725.
11. Kinjo, T., and P. F. Pratt. 1971b. Nitrate adsorption: II. In competition with chloride, sulfate, and phosphate. Soil Sci. Am. Proc. 35:725–728.
12. Lavkulich, L. M., and J. H. Wiens. 1970. Comparison of organic matter destruction by hydrogen peroxide and sodium hypochlorite and its effect on selected mineral characteristics. Soil Sci. Soc. Am. Proc. 34:755–758.
13. Marshall, C. E. 1949. The colloid chemistry of the silicate minerals. Academic Press, Inc. New York.
14. Parks, G. A. and P. L. de Bruyn. 1962. The zero point of charge of oxides. J. Phys. Chem. 66:967–973.
15. Parks, G. A. 1965. The isoelectric points of solid oxides, solid hydroxides, and aqueous hydroxo complex systems. Chem. Rev. 65:177–198.
16. Parks, G. A. 1967. Aqueous surface chemistry of oxides and complex oxide minerals. p. 121-160. In Robert F. Gould (ed.) Equilibrium concepts in natural water systems. Adv. Chemistry Series No. 67. American Chemical Soc., Washington D.C.
17. Pauling, L. 1930. The structure of micas and related minerals. Proc. Nat'l Acad. Sci. U.S. 16:123–129.
18. Rich, C. I. 1961. Calcium determination for cation-exchange capacity determinations. Soil Sci. 92:226–231.
19. Singh, B. R., and Y. Kanehiro. 1969. Adsorption of nitrate in amorphous and kaolinitic Hawaiian soils. Soil Sci. Soc. Am. Proc. 33:681–683.
20. Van Olphen, H. 1963. Introduction to clay colloid chemistry. Interscience Publishers, N.Y.
21. Van Raij, B., and Michael Peech. 1972. Electrochemical properties of some oxisols and alfisols of the tropics. Soil Sci. Soc. Am. Proc. 36:587–593.
22. Wada, K., and D. J. Greenland. 1970. Selective dissolution and differential infrared spectroscopy for characterization of amorphous constituents in soil clays. Clay Minerals 8:242–254.

14

Copyright © 1956 by Blackwell Scientific Publications
Reprinted from *Jour. Soil Sci.* **7**:130-147 (1956)

ION-EXCHANGE PHENOMENA IN SOME SOILS CONTAINING AMORPHOUS MINERAL CONSTITUENTS

K. S. BIRRELL AND M. GRADWELL

(*Soil Bureau, Department of Scientific and Industrial Research, New Zealand*)

(Soil Bureau Publication No. 69)

Summary

Soils containing allophane, 'palagonite', and amorphous oxides have been found to give cation-exchange capacity values as determined by conventional methods, which depend markedly upon concentration of leaching solution, the cation in the solution, the volume of the washing alcohol, and its water content. Experiments in which the soil is brought to equilibrium with solutions of barium acetate adjusted to pH 7, and containing an excess of barium with respect to the cation-exchange capacity value given by the leaching method, indicate that the uptake of cation conforms to the Brunauer, Emmett, and Teller theory of multi-layer physical adsorption. Similar results are also given by barium-chloride and lithium-acetate solutions at pH 7. Adsorption does not take place in the absence of water, and it is thought that it is the hydrated cation which is effective in covering the surface. Adsorption is strongly influenced by equilibrium pH values of the solutions. The adsorption of cations by these soils closely resembles the behaviour of hydrated alumina prepared by precipitation from sodium aluminate with CO_2. The acetate and chloride anions are taken up to a much smaller extent than the cations. Titration experiments indicate that soils containing allophane exhibit little if any tendency to hold exchangeable hydrogen ions in the same manner as montmorillonite.

Introduction

IN the course of an examination of volcanic ash soils containing the amorphous mineral allophane (Birrell and Fieldes, 1952), it was noted that a high exchange capacity for the cations NH_4^+, Na^+, K^+, and Ba^{++} appeared to be associated with the presence of this substance. It was also noted that these cations were fairly easily removed from the soils by leaching with water. Subsequently, it was also found that the substitution of 2 N solutions of ammonium acetate for N solutions in the standard laboratory determination of cation-exchange capacity (C.E.C.) used in the Soil Bureau (Metson, 1955) (in which the ammonia taken up by the soils is, after removing excess ammonium acetate with alcohol, estimated by distillation into standard acid) gave appreciably higher values. These facts seemed sufficiently different from those to be expected if the mechanism was the normal cation exchange shown by the crystalline clay minerals, to justify investigation. In addition to New Zealand volcanic ash soils known to contain much allophane, there were examined also some lateritic soils which showed high apparent cation-exchange capacity in the absence of crystalline clay minerals of known high exchange capacity.

Description of Soils

1. Volcanic ash soils having allophane as a major constituent of the clay fraction.
   (a) S447A and S503. Subsoils from Whenuapai Airfield, yielding about 60 per cent. of fraction below 2μ diameter.
   (b) S479. Subsoil of Tirau Ash Shower from Whakamaru Hydroelectric Station site, yielding about 20 per cent. below 2μ diameter.
   (c) S311. Subsoil of Egmont Ash, from the grounds of New Plymouth Hospital, yielding also about 20 per cent. below 2μ diameter.

In (a) and (b) only small amounts of kaolin and quartz in addition to allophane could be identified by X-ray and differential thermal analysis (D.T.A.) in the clay fraction. The clay fraction of (c) contained about equal parts of allophane and halloysite. Dispersion of soils for preparation of clay fractions was effected with sodium hydroxide at about pH 10, the organic-matter content being negligible. No deferration of the soil was attempted, in view of the risk of attacking the allophane (Birrell and Fieldes, 1952). It was unlikely that all the allophane was present in the clay fraction because the silt fraction of S479 was found to possess about 40 per cent. of the apparent cation-exchange capacity of the clay fraction, when this quantity was determined by the routine procedure.

2. Lateritic soils having hydrous oxides or 'palagonite' as the major constituent.
   (a) 1390B. Subsoil from Brown Loam soil group, Whangarei County. D.T.A. shows that mainly kaolin and limonite are present, with no detectable amounts of high exchange-capacity minerals.
   (b) 5810. Aitutaki Island, Cook Islands. This soil, containing mostly amorphous oxides, has been described by Fieldes, Swindale, and Richardson (1952).
   (c) 6168. Aitutaki Island. This soil was reported to contain the amorphous iron-aluminium silicate 'palagonite' (Dominion Laboratory, 1953). A light fraction (S.G. 2·2–2·3) separated from this soil had a cation-exchange capacity of 0·94 m.e. per gramme when neutral 2 N ammonium acetate was used in the standard method.

3. In addition a synthetic hydrated alumina, supplied by Mr. M. Fieldes, Soil Bureau, and prepared as described by Fieldes et al. (1952) was examined.

A characteristic of all these materials is a high surface area as shown in Table 5. This was determined by the adsorption of nitrogen at $-183°$ C. using the technique of Makower, Shaw, and Alexander (1937). The apparatus used enabled values of P/P_0 (relative pressure) up to 0·2 to be reached. For the Whakamaru soil clay fraction, check runs with argon at the same temperature (maximum P/P_0 value reached $= 0·7$) gave surface-area values within 5 per cent. of the nitrogen value. Application of the adjustment method of Joyner, Weinberger, and Montgomery

(1945) to the argon results indicated a value for the constant n in the general adsorption equation of Brunauer, Emmett, and Teller (B.E.T.) (1938) of at least 4·0, thus justifying the use of the simplified B.E.T. equation:

$$\frac{P}{V(P_0-P)} = \frac{1}{V_m C} + \frac{(C-1)P}{V_m C P_0} \quad \text{(which holds when } n = \infty\text{)}$$

for calculating surface areas from the nitrogen adsorption data.

Experimental

Cation-exchange capacity determinations by the leaching method were carried out with the apparatus described by Metson (1955), and using slight modifications of his procedure. In this, air-dry soil (10 g.) is leached first with neutral N ammonium acetate (350 ml.) and then with 80 per cent. alcohol (250 ml.) containing 0·5 ml. 2 N ammonium hydroxide per litre. All salt solutions were adjusted to pH 7·0 before use in this investigation. A suspension of magnesium oxide was used for liberating the ammonia taken up from ammonium-acetate solutions. When barium acetate was used in leaching-tube experiments the excess was removed by washing with plain 80 per cent. alcohol, the barium displaced with N acetic acid, and subsequently determined as sulphate.

For the equilibrium experiments, weighed amounts of air-dried material were added to accurately measured volumes of salt solutions. Tests at 20° C. were made in stoppered flasks, those at 95° C. in sealed tubes. Desorption tests were carried out by adding known volumes of water to the suspensions at equilibrium, and bringing to equilibrium again. Shaking intermittently over a period of 24 hours or continuously for 1 hour was found to be sufficient to attain equilibrium. After centrifuging for 10 min. the uptake of cation or anion was found by analysis of the supernatant liquid. Barium and lithium were determined gravimetrically as sulphate, acetate by distillation with sulphuric acid and titration of the distillate, chloride by titration with mercuric nitrate using diphenyl carbazone as indicator.

Solubilities of the salts used were determined on a weight/volume-of-solution basis, values found being:

Barium acetate   20° C.; 4·45 m.e. per ml.
„   „   95° C.; 4·61   „   „
„   chloride,   20° C.; 3·20   „   „
„   „   95° C.; 4·98   „   „
Lithium acetate, 20° C.; 5·8   „   „

Cation-exchange Capacity by Leaching Procedure

The results of these experiments are shown in Tables 1, 2, and 3. The departures from the standard procedure made in the course of these tests, which are apparent from the tables, are not considered to invalidate the results. For the Whenuapai soil, the washing alcohol was of 80 per cent. strength, adjusted to the neutral point of brom-thymol blue with ammonia as specified by Schollenberger and Simon (1945), but for all other materials alcohol of 95 per cent. strength was used without the addition of ammonia as recommended by Shaw (1940). Sections (*a*) and (*b*) of Tables 1 and 2 show that the cation-exchange capacity of these soils increases with increasing concentration of ammonium-acetate leaching solution, provided that the volume of washing alcohol relative to the amount of soil taken is constant. The change in C.E.C. is less marked with the lateritic soils than with the allophane soils. Illite and Wyoming bentonite showed no appreciable change in C.E.C. with changes in concentration of ammonium-acetate solution using the same analytical procedure, provided sufficient was present to saturate the exchange complex.

Increasing the relative volume of washing alcohol gives lower results, and if, as is shown by the footnote to Table 2, alcohol is replaced by water as washing medium, much lower results are obtained. It appears that the exchange complex in these soils must be extremely easily hydrolysed.

Table 3 shows that leaching the soil first with a strong solution of ammonium acetate, followed by a weaker solution, reduces the cation-exchange capacity to approximately the value which would have been obtained using the weaker solution originally, and if the second leaching is made with water, the C.E.C. is lowered still further.

When barium acetate was used in these experiments, the more concentrated the solution, the higher the C.E.C. value obtained, and the level of the values was greater than for ammonium acetate. Sodium acetate gave a lower level of C.E.C. values than ammonium acetate.

Equilibrium Experiments

The foregoing facts, namely, dependence of C.E.C. value on strength of leaching solution, ease of hydrolysis of the exchange complex, and variation in C.E.C. value with different cations, indicated some kind of adsorption mechanism rather than normal base exchange occurring with these soils, and it was accordingly determined to carry out equilibrium tests at constant temperature.

From these, the relationship of x/m, the amount of cation taken up per gramme to C, the equilibrium concentration was obtained. In Figs. 1 *a* and *b* and 2 *a*, x/m is shown plotted against C/C_0, where C_0 is the solubility of the salt used at the temperature of the experiment. For all soils and clay fractions tested, as well as for the synthetic hydrated alumina, the resulting curves resemble portions of gas adsorption isotherms.

A parallel experiment with a sample of calcium-saturated bentonite brought to equilibrium with barium-acetate solutions gave on the other hand an uptake of barium ion which was independent of equilibrium concentration (Fig. 1 *a*). The uptake of barium ion in this case agrees with the C.E.C. value obtained by Dr. R. B. Miller of Soil Bureau using the standard ammonium-acetate leaching procedure.

Neither the Freundlich nor the Langmuir adsorption equations were applicable to the results of the equilibrium experiments mentioned.

However, Ewing and Liu (1953), who recently studied the adsorption of the ionic dyes Crystal Violet and Orange II on certain pigments,

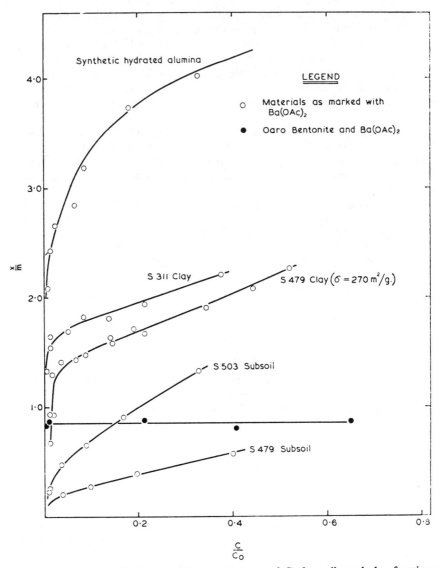

FIG. 1 (a) Adsorption isotherms of barium ion at 20° C. for soils and clay fractions, containing allophane, and for hydrated alumina.

found that, although these particular equations were likewise not applicable to their results, a modification of the multilayer B.E.T. gas-adsorption theory in which the reduced pressure P/P_0 was replaced by the reduced concentration C/C_0 and the volume adsorbed (v) was replaced by the amount adsorbed per gramme (x/m), for the case in which the

ION-EXCHANGE PHENOMENA IN SOME SOILS

constant $n = \infty$, was satisfactory for values of C/C_0 up to 0·25. That is to say, the B.E.T. equation quoted earlier, according to Ewing and Liu, can be written as:

$$\frac{C}{x/m(C_0-C)} = \frac{1}{(x/m)_m K} + \frac{K-1}{(x/m)_m K} \cdot \frac{C}{C_0}$$

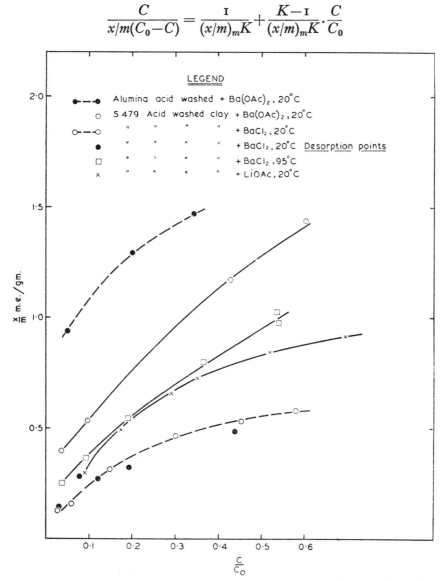

FIG. 1 (b) Adsorption isotherms of barium and lithium ions for acid-washed samples.

for adsorption from solution, where the constant K replaces constant C of the B.E.T. equation for the sake of clarity. (It should be noted that a misprint in their paper omits K from the denominator of the second term on the right-hand side.)

The linearity of the plot of $\dfrac{C}{(C_0-C)x/m}$ against C/C_0 for the allophane clay fractions shown in Fig. 1 c indicates that the above equation fits the uptake of cation by this substance. The synthetic alumina fits fairly well also. For conditions of lesser adsorption such as obtained with the

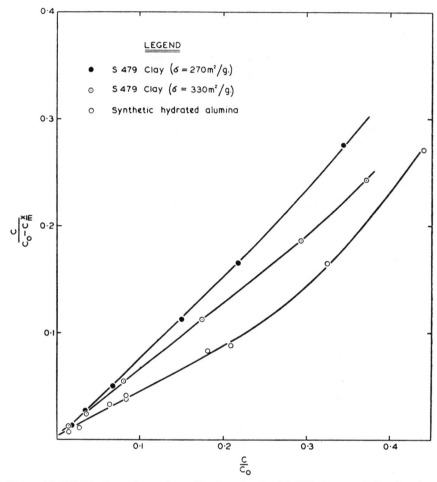

FIG. 1 (c) B.E.T. plot: adsorption of barium ion by NaOH-dispersed clay fractions and by hydrated alumina.

whole soils and the acid-washed clays the agreement is not quite as close, but can be considered as satisfactory (see Figs. 1 d and 2 b).

If the amounts of cation in a monolayer $(x/m)_m$ are calculated using the above equation, the results shown in Table 5 are obtained.

It can be seen that there is no simple relationship between total surface area and the amount of cation in a monolayer, such as Ewing and Liu found for their dyes. The pH range of the solutions at equilibrium appears to have a major effect on the monolayer values found. Even if this factor of variable pH is taken into account, the specific adsorbing power in relation to surface area is quite noticeably different throughout the materials examined.

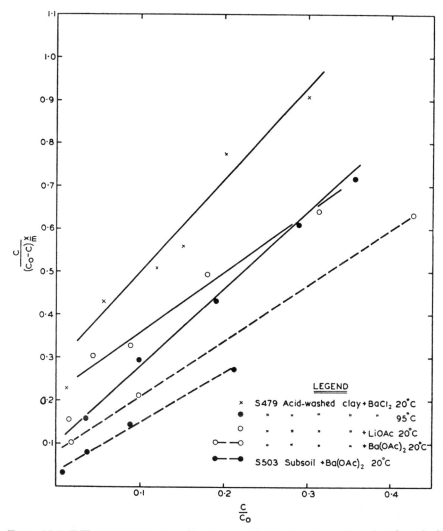

FIG. 1 (d) B.E.T. plot: adsorption of barium and lithium ions by soils and acid-washed clays.

The untreated soils and acid-washed clay fractions, and also the acid-washed alumina preparation, when brought to equilibrium with the acetates, gave solutions which were slightly acid. When barium chloride was used the equilibrium pH range was lower than with barium acetate, and the monolayer values were smaller. The relatively high equilibrium pH values obtained with the NaOH-dispersed clays and barium-acetate solutions were found to be due to sodium adsorbed by the clay. Sodium amounting to 1·29 m.e. per gramme could be extracted by N/10 acetic acid from the first Whakamaru clay fraction in Table 5, and 1·54 m.e.

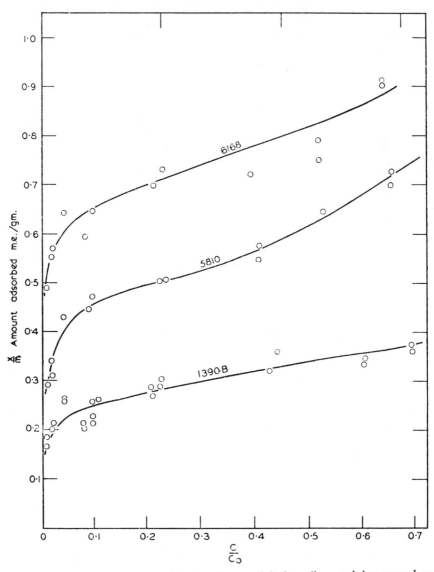

FIG. 2 (*a*) Adsorption isotherms of barium ion at 20° C. for soils containing amorphous oxides.

per gramme from the second. The extractable sodium in the alumina as prepared was 3·16 m.e. per gramme. As shown in the note under the heading in Table 6, less alkali was retained when lithium hydroxide was used for dispersing this soil.

ION-EXCHANGE PHENOMENA IN SOME SOILS 139

Table 6 shows the strong effect of original pH of the salt solution on the uptake of cation, although variations on the alkaline side have apparently less effect than those on the acid side of neutrality.

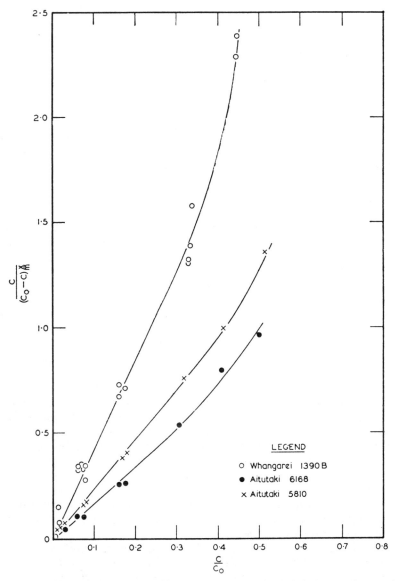

FIG. 2 (b) B.E.T. plot: adsorption of barium ion by soils containing amorphous oxides.

In spite of this complicating factor of pH, observations can be made from the results which lend support to the physical-adsorption theory.

1. Desorption points for the barium-chloride isotherm at 20° C. follow the adsorption curve fairly closely (Fig. 1 b).

2. The monolayer values for the acid-washed Whakamaru subsoil clay fraction with barium chloride at 20° C. and 95° C. are fairly close (Table 5). Owing to the rather acid reaction of the solutions at equilibrium and the presence of alumina in the solutions after heating to 95° C., it is considered that metallic salts of weak acids would have been more suitable for this comparison, but none appeared to be available which had a suitable solubility range.

3. The heat of adsorption of barium ions adsorbed from $BaCl_2$ solutions on acid-washed Whakamaru clay fraction, calculated from the B.E.T. plots in Fig. 1 d gave values of the order of 6,000–7,000 cals. per mole, which from published data on gas adsorption (Brunauer et al., 1938) is of suitable magnitude for a physical process. The calculation was based on the procedure used by Ewing and Liu except that the heat of solution of hydrated barium chloride was obtained from the International Critical Tables, instead of being calculated from solubility data. The heat of adsorption of nitrogen (and of argon) on this sample was approximately 2,100 cals. per mole.

4. The amount of lithium in a monolayer is less than that of barium by about 20 per cent. (Table 5), even although the equilibrium pH of the barium-acetate solution is rather less than that of the lithium acetate. This result is reasonable if it is the hydrated ion which is effective in covering the available surface. Using the values given by Nachod (1949) for hydrated ionic radii, the monolayer values for Li^+ and Ba^{++} should be in the ratio 1:2·5. Shortage of the prepared clay fraction has prevented repeating the experiment with adjustment of equilibrium pH values to be approximately the same in both cases. All that can be said at present is that there is a difference in the right direction for this hypothesis to hold.

5. The monolayer values of the clay fractions of the Whakamaru and New Plymouth soils appear to be closely related to their surface areas. These clays are predominantly allophane and would be expected to produce about the same pH range in the barium-acetate solutions at equilibrium, although shortage of material prevented this being checked with the New Plymouth sample.

Other points relating to the data in Table 5 which seem worth mentioning are:
1. The 'palagonite' sample gives a rather flat isotherm suggesting that it contains also some unidentified material exhibiting the normal type of cation-exchange capacity.
2. The high cation-adsorptive capacity of the precipitated alumina might suggest that 'allophane' is an intimate physical mixture of alumina and silica in which the alumina is in a form similar to that of the preparation used in this work.
3. That the adsorption is in fact ionic is shown by the absence of any appreciable uptake of cation by Whakamaru clay fraction which had been previously dried at 105° C., from anhydrous lithium chloride dissolved in absolute methanol.

Anion Uptake

In the equilibrium experiments with barium acetate, acetate ion was found to be taken up also, but in smaller amount than barium. Similarly, chloride ion was taken up from barium-chloride solutions, but in amount considerably less than acetate (Table 7). Owing to experimental difficulties in the precise determination of acetate the scatter of plotted

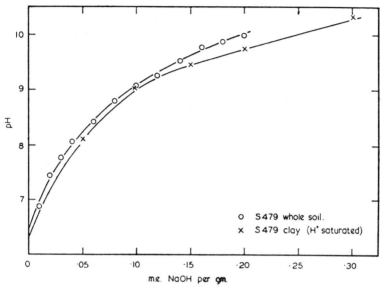

FIG. 3. Titration curves of allophane soil and clay fractions.

points is greater than for barium, but the relation between acetate taken up and residual-acetate concentration appears to be nearly linear. It can be seen that at the barium-uptake value corresponding to monolayer coverage for this clay sample (1·3 m.e. per gramme as given in Table 5) the contribution of acetate towards covering the surface will be very small, bearing in mind also that the hydrated acetate ion is generally regarded as being much smaller than the hydrated barium ion.

Titration of Allophane Clays

A clay fraction of the Whakamaru soil was prepared by dispersion of the soil with NaOH at pH 10, flocculated with NaCl after sedimentation, washed with 99 per cent. methyl alcohol, and air-dried. The clay so obtained was first washed with 100 c.c. of 0·004 N HCl, followed by 2,000 c.c. of 0·001 N HCl in 250-c.c. portions, and again washed with alcohol following essentially the procedure of Bower and Truog (1940), except for the addition of the washing with 0·004 N HCl which was found necessary with the Whakamaru soil clay fraction to ensure complete removal of sodium ion. Fig. 3 shows the titration curve for a 1 per cent. aqueous suspension of the clay so prepared. The outstanding differences in behaviour between the Whakamaru clay fraction and the materials investigated by Bower and Truog are the much higher initial pH of the clay suspension and the lack of any inflexion point on the titration curve in the pH range shown. The titration curve of a 1 per cent. aqueous suspension of the whole soil is also shown in Fig. 3. By the standard methods of soil analysis, i.e. determining total bases and C.E.C. by leaching with N ammonium acetate, this soil would be regarded as about 27 per cent. base-saturated. It can be seen that the titration curve of the whole soil is practically identical with that of the clay fraction. In addition the amount of NaOH needed to produce a suspension of pH 10 is for both soil and clay fraction far below the cation-exchange capacity found by the leaching method, whereas Bower and Truog (loc. cit.) found that the amount of NaOH added up to the inflexion point, which was close to pH 7, corresponded to the cation-exchange capacity as usually determined. It is apparent, therefore, that both the soil and the clay fraction have relatively weak acidoid properties in spite of the apparent high saturation of the exchange complex with hydrogen both in the natural soil and in the clay fraction as prepared.

Exchangeable Hydrogen and Bases in Whakamaru Soil

The ammonium-acetate leachates from the Whakamaru soil obtained in the experiment carried out to determine the C.E.C. values given in Table 1 *b* were divided into two equal portions, and total base and exchangeable hydrogen content estimated by the usual methods. These results are given in Table 4 along with the C.E.C. values from Table 1 *b*. It will be seen that there is little change in pH of the leachate irrespective of the strength of the leaching solution. The 'exchangeable hydrogen' is fairly constant while the total base content tends to fall with decreasing concentration. The sum of bases plus hydrogen falls short of the cation-exchange capacity, the deficiency being greater the stronger the leaching solution. This may be regarded as additional evidence that the normal base-exchange mechanism is not operating.

Role of Water in the Equilibrium Tests

Whether any adsorption of water as well as of cation takes place in the equilibrium experiments cannot, of course, be determined from the results. No obvious hydration of the soil occurs except perhaps in the tests with concentrated solutions of lithium acetate, where it was more difficult to separate soil from solution after bringing to equilibrium than with barium acetate. What is striking, however, is the difference in behaviour of the allophane soils compared with bentonite, as shown in Fig. 1 *a*, considering that both materials can hold large quantities of water in the natural state, and that the B.E.T. relation is followed for uptake of cations by the amorphous minerals, although any possible movement of water in or out of the soil system is ignored.

Conclusions

Physical adsorption of cations by soils containing allophane, 'palagonite', and certain amorphous oxides is responsible to a greater or less extent for the apparent high cation-exchange capacity shown by these soils in conventional determinations of this quantity by leaching with neutral salt solutions. Where the presence of such amorphous material is suspected in soils, the conventional C.E.C. methods should be supplemented by equilibrium experiments to test the significance of the C.E.C. values.

Acknowledgements

The authors offer their thanks to Mr. R. Q. Packard for valued assistance with the construction and operation of gas adsorption equipment, and to Dr. P. G. Harris, Dominion Laboratory, for making available a sample of 'palagonite'.

REFERENCES

BIRRELL, K. S., and FIELDES, M. 1952. Allophane in volcanic ash soils. J. Soil Sci. **3**, 156–66.

BOWER, C. A., and TRUOG, E. 1940. Base exchange capacity determination as influenced by nature of cation employed and formation of basic exchange salts. Soil Sci. Soc. Amer. Proc. **5**, 86–89.

BRUNAUER, S., EMMETT, P. M., and TELLER, E. 1938. Adsorption of gases in multimolecular layers. J. Amer. Chem. Soc. **60**, 309–19.

Dominion Laboratory, Wellington. 1953. Report from the Director, 4 Nov.

EWING, W. W., and LIU, F. W. J. 1953. Adsorption of dyes from aqueous solutions on pigments. J. Colloid Sci. **8**, 204–13.

FIELDES, M., SWINDALE, L. D., and RICHARDSON, J. P. 1952. Relation of colloidal hydrous oxides to the high cation exchange capacity of some tropical soils of the Cook Islands. Soil Sci. **74**, 197–205.

JOYNER, L. G., WEINBERGER, E. B., and MONTGOMERY, C. W. 1945. Surface area measurements of activated carbons, silica gel and other adsorbents. J. Amer. Chem. Soc. **67**, 2182–8.

MAKOWER, B., SHAW, T. M., and ALEXANDER, L. T. 1937. The specific surface and density of some soils and their colloids. Soil Sci. Soc. Amer. Proc. **2**, 101–8.

METSON, A. J. 1955. Methods of chemical analysis for soil survey samples. N.Z. Soil Bureau Bull. 12 (in press).

NACHOD, F. C. 1949. Ion Exchange, Theory and Application. Academic Press Inc., New York.

SCHOLLENBERGER, C. J., and SIMON, R. N. 1945. Determination of exchange capacity, and exchangeable bases in soil—NH_4OAc method. Soil Sci. **59**, 13–24.

SHAW, W. M. 1940. Determination of exchange bases and exchange capacity of soils. J. Off. Agric. Chem. **23**, 221–32.

(Received 4 May 1955)

TABLE I

Standard Cation-exchange Capacity Determinations on Allophane Soils

(a) *Whenuapai soil S447A, oven-dried*

Strength of ammonium-acetate solution	Volume used per 2-g. sample, ml.	C.E.C. m.e./g. (50 ml. washing alcohol)	C.E.C. m.e./g. (100 ml. washing alcohol)
1·74 N . . .	50	0·427	0·363
1·74 N . . .	100	0·425	..
0·87 N . . .	100	0·399, 0·409	..
	150	0·391	..
0·438 N . . .	100	0·347	0·331
	150	0·353	..
0·087 N . . .	60	0·221	..
	110	0·221	..
	210	0·244	..
	410	0·258	..

(b) *Whakamaru soil S479*

Strength of ammonium-acetate solution	Volume used per 10-g. sample, ml.	C.E.C. m.e./g. (100 ml. washing alcohol)	C.E.C. m.e./g. (275 ml. washing alcohol)
4·26 N . . .	60	0·347	..
2·16 N . . .	60	0·340	..
2·16 N . . .	375	..	0·253
1·145 N . . .	100	0·306	..
1·145 N . . .	375	..	0·248
0·558 N . . .	200	0·239	..
0·280 N . . .	400	0·179	..
0·1175 N . . .	500	0·146	..

TABLE 2

Standard Cation-exchange Capacity Determinations on Lateritic Soils

	Strength of ammonium-acetate solution	Volume used ml.	C.E.C. m.e./g.
(a)	2 N	50	0·465
	1 N	50	0·435
	0·5 N	80	0·399
	0·1 N	385	0·378
(b)	2 N	50	0·873
	1 N	100	0·850
	0·5 N	200	0·839
(c)	2 N	50	0·317
	1 N	50	0·296
	0·5 N	80	0·280
	0·1 N	385	0·267

(a) Aitutaki Island (5810) 1·19 g. (oven-dry basis), 8×4 c.c. portions 95 per cent. EtOH.
(b) Aitutaki Island (6168) 0·99 g. (oven-dry basis), 8×4 c.c. portions 95 per cent. EtOH.
(c) Whangarei County (1390B) 1·20 g. (oven-dry basis), 8×4 c.c. portions 95 per cent. EtOH.

Soil 5810 leached first with 50 ml. 2 N ammonium acetate, followed by 100 ml. water gave a C.E.C. of 0·268 m.e. per gramme and soil 1390B with the same treatment gave a C.E.C. of 0·186 m.e. per gramme.

TABLE 3

Effect on C.E.C. Values for Whenuapai Soil S447A of Subsequent Leaching with Weaker Solutions of NH_4OAc or Water

Strength of original NH_4OAc solution	Volume per 2-g. sample, ml.	Strength of second NH_4OAc solution	Volume ditto, ml.	C.E.C. m.e./g.
1·74 N	100	0·87 N	100	0·378
0·438 N	100	0·087 N	400	0·257
1·74 N	22	Distilled water only	22	0·080

TABLE 4

Exchangeable Hydrogen and Bases in Whakamaru Soil S479

Strength of neutral ammonium-acetate solution	pH of leachate	Standard C.E.C. from Table 1 (b), col. 3. m.e./g.	Exchangeable hydrogen m.e./g.	Total exchangeable bases m.e./g.	Sum of bases plus hydrogen m.e./g.
2·16 N	6·83	0·340	0·075	0·067	0·142
1·145 N	6·80	0·306	0·062	0·084	0·146
0·558 N	6·78	0·239	0·057	n.d.	..
0·280 N	6·80	0·179	0·070	0·049	0·119
0·1175 N	6·55	0·146	0·068	0·044	0·112

TABLE 5

$(x/m)_m$ *Values from Equilibrium Tests*

(20° C. unless otherwise stated)

Sample	Method of Preparation	Salt used in test	Equilibrium pH range	$(x/m)_m$ m.e./g.	Surface area m²/g.
S479 Whakamaru clay < 2 μ	Dispersion of soil with NaOH, flocculation with NaCl, washed EtOH, and acetone. Dried at < 35° C.	Ba(OAc)₂	7·7 for N.	1·3*	270
S311 New Plymouth clay < 2 μ		,,	7·6–8·4	1·6	330
		,,	..	1·5	300
S479 Whakamaru clay < 2 μ	As above followed by extraction with N/10 acetic acid till Na-free, then washed EtOH and acetone. Dried at < 35° C.	Ba(OAc)₂	6·0	0·72	330
		LiOAc	6·1–6·35	0·60	330
		BaCl₂	4·9–5·2	0·41	330
		BaCl₂, 95° C.	4·6–4·9	0·52	330
S479 Whakamaru subsoil < 2 mm.	Untreated, air-dried	Ba(OAc)₂	5·9–6·2	0·37	130
S503 Whenuapai subsoil < 2 mm.	,, ,,	,,	5·95–6·25	0·83	172
1390B Whangarei subsoil < 2 mm.	,, ,,	,,	6·3–6·55	0·23	75
5810 Aitutaki subsoil < 2 mm.	,, ,,	,,	6·6–6·7	0·425	94
6168 Aitutaki subsoil < 2 mm.	,, ,,	,,	6·3–6·6	0·59	102
Hydrated alumina	Ppn. by CO₂ from sodium aluminate³	Ba(OAc)₂	7·7	2·33	180
,, ,,	As above followed by extraction with N/10 acetic acid till Na-free, then washed EtOH and acetone. Dried at < 35° C.	,,	6·6	1·02	180
'Palagonite'	Separated from Aitutaki subsoil 6168 (ref. 4/7)	BaCl₂	4·2–4·8	0·435	86

* A monolayer value of 0·72 m.e. per gramme would be required to give uniform surface coverage of this clay fraction by the hydrated barium ion.

TABLE 6

Effect of Initial pH *on Barium Uptake for Whakamaru Clay Fraction*

(1 g. of clay separated by dispersion of soil with LiOH and containing 0·69 m.e. Li, brought to equilibrium with 10 ml. approx. N barium acetate in each case)

Original pH	Final pH	Amount of Ba adsorbed m.e./g.
4·3	4·4	0·606
6·0	6·1	0·988
7·0	7·3	1·372
11·1	7·45	1·395

TABLE 7

Anion Adsorptions at 20° C.

(a) *Acetate:* S479 *clay* (NaOH *dispersed*) *and* $Ba(OAc)_2$ *at* pH 7·0

Original strength	Barium uptake m.e./g.	Acetate uptake m.e./g.
3·166 N	2·30	1·31
2·477 N	2·07	0·92
2·018 N	1·91	0·78
1·072 N	1·70	0·43
0·485 N	1·47	0·29
0·096 N	1·28	0·16

(b) *Chloride:* S479 *clay* (NaOH *dispersed*) *and* $BaCl_2$ *at* pH 7·0

Original strength	Barium uptake m.e./g.	Chloride uptake m.e./g.
1·995	n.d.	0·068
1·00	1·202	0·02
0·50	1·103	0·002
0·253	1·085	nil
0·065	0·942	nil

15A

Copyright © 1974 by the Soil Science Society of America
Reprinted by permission from *Soil Sci. Soc. Am. Proc.* **38**:255-259 (1974)

Anion Adsorption by Allophanic Tropical Soils: I. Chloride Adsorption[1]

H. GEBHARDT AND N. T. COLEMAN[2]

ABSTRACT

The adsorption of Cl by Andepts from Mexico, Colombia, and Hawaii was measured in solutions of HCl, HCl + NaCl, and $AlCl_3$. Chloride adsorption varied from 0-8 meq/100 g at pH 6 to as much as 32 meq/100 g at pH 3.8. At given pH, adsorption was concentration-dependent in a manner consistent with the Langmuir adsorption equation. For a B-horizon sample from San Gregorio, Mexico, the Cl adsorption maxima, in meq/100 g, were 7.4 at pH 6; 10.7 at pH 4.8; 13.4 at pH 4.4; 17.3 at pH 4.2; and 31.6 at pH < 4. The average *a* from the Langmuir equation was 0.04 liter/meq. Adsorbed Cl was removed by leaching with water, and was exchanged by NO_3.

Chloride adsorption from HCl or HCl-NaCl was accompanied by the consumption of protons; adsorption from $AlCl_3$ resulted in the hydrolysis and precipitation of Al. Protons consumed or Al hydrolyzed exceeded Cl adsorbed by amounts corresponding closely to the effective CEC. The results suggest that protons are adsorbed to produce positively charged sites, which bind Cl nonspecifically.

Chloride adsorption capacity can be conveniently measured by shaking soil with $AlCl_3$ solution and measuring Cl uptake. Chloride capacities of soils and clay minerals, determined by the $AlCl_3$ procedure, were, in meq/100 g, 7-30 for Dystrandepts; 16-18 for Hydrandepts; 2-4 for acid soils containing crystalline clay and oxide minerals; 4 for kaolinite; zero for montmorillonite and illite.

Additional Index Words: Andept, Dystrandept, volcanic soil.

INORGANIC COLLOIDS of allophanic soils derived from volcanic ash carry electrical charges that vary with pH and may be positive or negative. Under acid conditions such soils may, depending on their weathering stage, exhibit appreciable anion adsorption capacity that becomes larger as pH is lowered. Large adsorption of Cl, as much as 13 meq/100 g of volcanic ash soil (4) and 50 meq/100 g of separated clay (12), has been reported.

Chloride adsorption has been used to determine the size and magnitude of electrical charges carried by clay and oxide mineral (9, 11). For allophanic soils, chloride adsorption, in contrast to that of sulfate, phosphate, or other anions that are specifically bound or can form insoluble compounds with Al or Fe^{3+}, may provide a convenient and reliable index of the positive charge that exists under given conditions and of the maximum positive charge that can develop at low pH.

---

[1] Contribution from the Dep. of Soil Science and Agr. Eng., Univ. of California, Riverside 92502. Support from the National Science Foundation, under Research Grant NSF-GB-11711-(Pratt), is gratefully acknowledged. Received 10 Sept. 1973. Approved 17 Sept. 1973.

[2] Postgraduate Research Soil Scientist and Professor, respectively, Dep. of Soil Science and Agr. Eng., Univ. of California 92502.

This report is part of a larger study on anion adsorption. The main points are:

1) Determination of the influence of pH and Cl-concentration on Cl adsorption by Dystrandepts and other soils containing allophane and/or hydrous sesquioxides.

2) Development of a procedure for determining the Cl adsorption capacity of soil.

3) Determination of the adsorption mechanisms involved.

MATERIALS AND METHODS

Soils

Most of the soils used were Andepts that occur in Mexico, Colombia, and Hawaii. The Mexican and Colombian soils represent a large population of volcanic ash-derived soils in Central and South America (8). The Hawaiian samples, from an elevation-rainfall transect, have been described by Flach et al. (K. W. Flach, R. E. Nelson, and S. Rieger. 1969. Agron. Abst. and personal communication from K. W. Flach, SCS-USDA, Riverside, California) and Lai and Swindale (7).

The Mexican soils are Dystrandepts (8) (personal communication from B. L. Allen, Texas Tech. Univ., Lubbock). They are derived from recent andesitic volcanic ash, and occur in the state of Michoacan at elevations between 2,250 and 2,730 m. Mean annual temperature is 16.5C; annual precipitation is 1,600 mm. The soils have bulk densities between 0.5–0.7. Clay contents of A and B horizons are between 10 and 30%. The clay is largely X-ray-amorphous, but appears to contain small amounts of gibbsite as indicated by X-ray diffraction. The organic matter content of A horizons is 8–17%; in B horizons it is 2–6%. Soil pH ranges from 5.6 to 7.

The Colombian soils have been described by Leon (L. A. Leon. 1959. Chemistry of some tropical acid soils of Colombia, S. A. Ph.D. Thesis, Univ. of California, Riverside). The La Selva soil from Rio Negro, Antioguia, is derived from volcanic ash. The clay is X-ray amorphous. The Tibaitata (Mosquera, Cundinamarco) and Portrerito (Jamundi, Valle de Cauca) soils are ash-affected but contain substantial amounts of kaolinite and vermiculite-intergrade minerals. The Portrerito also contains goethite.

Experimental Methods

Air-dried soils were used in all experiments. Limited work with the Mexican samples, which had been collected moist and stored moist in plastic bags, showed no appreciable effect of air drying on Cl adsorption. Five-gram samples (on an oven-dry basis, making allowance for occluded soil water) were shaken for 1 hour in 20 or 30 ml of solutions containing graded amounts of HCl, NaCl, HCl, and NaCl or $AlCl_3$. After shaking, the slurries were filtered and equilibrium solutions were analyzed as follows: pH with glass and calomel electrodes; Cl by titration with $AgNO_3$ using glass and Ag-AgCl electrodes to detect the end point; H_3O and Al by titration to pH 4 and 8 with NaOH; Ca + Mg by titration with EDTA using Eriochrome Black T and triethanolamine to mask interference from Al; Na and K by flame photometry. Amounts of Cl, H_3O, and Na adsorbed were calculated by difference between amounts added and those found in filtrates after reaction had taken place. Effective CEC's of the Mexican soils were estimated by leaching 5-g samples with 100 ml of $0.1N$ NH_4Cl and determining Ca + Mg, Na, and K (there was no exchangeable Al).

Chloride adsorption capacities of a number of Dystrandepts and other tropical soils were measured by shaking 5-g samples of air-dry soil for 1 hour in 20 ml of $0.17N$ $AlCl_3$, filtering, and determining Al and Cl in the filtrate as described above.

RESULTS AND DISCUSSION

Chloride Adsorption by a B-Horizon Sample of a Dystrandept from San Gregorio, Michoacan, Mexico

The quantities of H_3O and Cl adsorbed from HCl solutions are shown in Table 1. Chloride adsorption was concentration and pH dependent, with a maximum of 24 meq/100 g at low pH and high solution concentration. For HCl additions of 20 meq/100 g or less, H_3O was almost completely removed from solutions without appreciable solution of Al. Reactions with amounts of HCl above 20 meq/100 g resulted in the appearance of substantial amounts of Al in solution. For all systems, H_3O sorbed or H_3O + Al sorbed was greater than Cl sorbed by 6–8 meq/100 g. Consumption of around 5 meq/100 g of H_3O can be attributed to the displacement of exchangeable Ca, Mg, and K as shown by determinations of Ca + Mg and K in equilibrium solutions.

The data are consistent with the ideas that have been advanced by Wada and Ataka (12), Fieldes and Schofield (4), and Hingston et al. (5) that anion adsorption follows the protonation of allophane and oxide minerals to produce positive sites. This mechanism in soils with exchangeable cations would require that proton consumption equals the sum of Cl sorbed and cations displaced. The small apparent excess of H_3O consumed in these experiments, around 2 meq/100 g, may have been due to an unidentified side reaction, or to the imprecision in the procedure which involved determination by difference.

Table 2 and Fig. 1 summarize the outcome of a number of experiments in which NaCl, or mixtures of NaCl and HCl, were reacted with San Gregorio B horizon material. The data show that:

1) Capacity for Cl adsorption rises as pH is lowered to pH 4.0;

2) At a given pH, Cl adsorption increases with NaCl concentration to an apparent maximum;

3) When only NaCl is added, Na and Cl are adsorbed together, with Na sorption exceeding Cl sorption by an amount corresponding to the effective CEC at higher NaCl concentration.

4) Below pH 5, Na + H_3O sorbed exceeds Cl sorbed by an amount slightly greater than the effective CEC.

The sorption maximum measured at pH 3.9 was 27 meq/100 g, in good agreement with the value obtained with high concentrations of HCl alone reported in Table 1. Apparently Na was coadsorbed with Cl in the pH range 5.8–6.0 where only NaCl was added. At low pH's, however, the amounts of Na sorbed were small, averaging only 4 meq/100 g. Fieldes and Schofield (4) found constant amounts of Na adsorbed in the pH range 4–6 by allophanic soils.

Table 1—Reaction of HCl with 5-g samples of soil from the San Gregorio B-horizon

HCl added	pH	In solution			Adsorbed	
meq/100 g		H_2O	Al	Cl	H_2O	Cl
		— meq/100 g —		meq/liter	meq/100 g	
6	4.5	<2	0	5	6	4
10	4.3	<2	0	10	10	6
14	4.1	<2	0	15	14	8
20	3.8	<2	1	20	19	12
30	3.5	2	5	35	23	16
60	2.8	4	28	63	28	22
80	2.6	8	42	92	30	25
100	2.6	15	54	127	31	24

Table 2—Reaction of 5-g samples of soil from the San Gregorio B-horizon with NaCl-HCl

Added			In solution		Adsorbed		
HCl	NaCl	pH	Al	Cl	H₂O	Na	Cl
—meq/100 g—			meq/100 g	meq/liter	—meq/100 g—		
0	20	5.8	0	28	0	4	3
0	40	5.8	0	60	0	4	4
0	100	6.0	0	157	0	10	6
6	10	4.7	0	27	6	2	5
6	30	4.7	0	73	6	6	7
10	20	4.5	0	55	10	4	8
10	40	4.5	0	97	10	6	11
21	60	4.2	0	110	21	2	15
21	100	4.2	0	175	21	4	16
30	60	3.9	2	110	28	2	24
30	100	3.9	2	171	28	4	27

Samples of soil were also reacted with AlCl₃, on the assumption that hydrolysis of Al would provide protons to create positive charges on soil minerals, and high solution concentrations of Cl could be obtained at a pH near 4 without the extreme acidification that would be obtained from comparable addition of HCl. The data, given in Table 3, show a progressive increase in both Al and Cl adsorption with increase in concentration, with apparent maxima of 32 and 24 meq/100 g for Al and Cl, respectively. The difference, 8 meq/100 g, is similar to the value observed for excess cation adsorption in experiments with HCl and HCl-NaCl.

Adsorption Isotherms

All of the Cl sorption data obtained with the B-horizon sample from San Gregorio are summarized in Fig. 1, which shows the quantities of Cl adsorbed plotted against the final Cl concentrations in equilibrium solutions. Individual curves are drawn for five pH ranges: 5.8–6, 4.7–4.8, 4.4–4.5, 4.2–4.3, and 2.6–4.1. Sorption appears to approach a maximum at each pH. Lowering pH to 2.6 with HCl led to no greater Cl sorption than was obtained at pH 3.9 with HCl-NaCl or pH 3.7 with AlCl₃.

In each pH range, the data conform with the Langmuir equation

$$C/q = (1/ab) + (1/b)C$$

where b is the sorption capacity and a is proportional to sorption affinity. Langmuir equations and values for a and b given in Table 4 show

1) Calculated adsorption maxima are close to but slightly larger than the experimental values at C = 160 to 180 meq/liter; and
2) Sorption affinity for Cl⁻ was low in all cases and did not show a marked increase at low pH values.

Table 3—Reaction of 5-g samples of soil from the San Gregorio B-horizon with AlCl₃

AlCl₃ added	pH	In solution		Adsorbed	
meq/100 g		Al meq/100 g	Cl meq/liter	Al —meq/100 g—	Cl
20	4.1	4	15	16	11
40	3.9	16	37	24	18
60	3.8	31	65	29	21
80	3.7	48	93	32	24
100	3.7	68	127	32	24

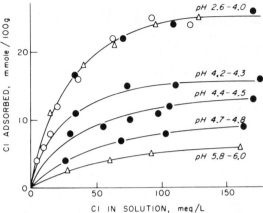

Fig. 1—Concentration-dependent Cl adsorption by a dystrandept (San Gregorio B-horizon). The three sets of data for the curve for pH 2.6-4.0 were obtained with three methods, i.e., adsorption of Cl at pH 2.6 with MCl, pH 3.9 with HCl-NaCl, and pH 3.7 with Al-Cl₃.

The data on the Mexican Dystrandept agrees with and expands on the findings of others who worked with volcanic ash soils. Fieldes and Schofield (4) found 13 meq/100 g Cl adsorbed at pH 4 from 0.2N NaCl by volcanic ash soils. At pH 7 and corresponding equilibrium concentrations, Imura (6) found the amount of Cl adsorbed by a volcanic ash subsoil was ~ 5 meq/100 g. Clay fractions from Japanese volcanic ash soils adsorb as much as 50 meq Cl/100 g clay at pH 5 from 0-0.2N NH₄Cl solution [Wada and Ataka (12)]. In our work, where maximum Cl concentrations were < 200 meq/liter, there was no evidence for the very large sorption of Cl observed by Imura (6) at pH 5 and below in > 0.5N NaCl and termed by him as "multimolecular adsorption."

Desorption of Cl from San Gregorio B Horizon

Five-gram samples of soil that had reacted with 0.17N NaCl or AlCl₃ were filtered and washed slowly with 200-ml increments of H₂O or 1N KNO₃. Making allowance for the occluded salt in the soil, KNO₃ displaced all of the adsorbed Cl from both NaCl- and AlCl₃-treated samples. The Al that had reacted along with Cl, presumably through hydrolysis, was not displaced by KNO₃. Washing with 100 ml of water removed all of the Cl adsorbed from NaCl.

Table 4—Adsorption of Cl by San Gregorio B-horizon: Langmuir equations and constants

pH	Equation	Correlation coefficient	a	b	Experimental maximum
			liter/meq	meq/100 g	
5.8–6.0	C/q = 6.26 + 0.136 C	0.98	0.027	7.4	6
4.7–4.8	C/q = 3.17 + 0.093 C	0.99	0.029	10.7	9
4.4–4.5	C/q = 1.57 + 0.075 C	0.98	0.047	13.4	13
4.2–4.3	C/q = 0.81 + 0.058 C	0.99	0.071	17.3	16
2.6–4.1	C/q = 1.07 + 0.032 C	0.98	0.030	31.6	27

C = Equilibrium concentration [meq Cl/liter]
q = meq Cl adsorbed/100 g soil.

Washing with three 200-ml increments of water removed 70% of the adsorbed Cl, along with a corresponding amount of Al, from AlCl$_3$-treated material. The remaining Cl was exchanged by $1N$ KNO$_3$. The different effects of H$_2$O and KNO$_3$ solution on Al desorption is probably related to pH effects, but more research is needed on this point.

Chloride Adsorption Capacity as Measured with AlCl$_3$

The data presented in the first section showed that for San Gregorio B-horizon, quantities of Cl adsorbed from 0.17N solutions below pH 4 were independent of pH between 2.6 and 4.1 and were the same for HCl, HCl-NaCl, and AlCl$_3$. Similar experiments with other Mexican Andosols gave essentially the same results. This suggested that the soil property responsible for Cl adsorption could be conveniently determined by reacting soils with AlCl$_3$ solutions and measuring the Al and Cl sorbed.

Chloride adsorption capacities of Mexican, Hawaiian, and Colombian Andosols, as measured in 0.17N AlCl$_3$, are presented in Table 5. Also given are amounts of Al sorbed and effective CEC's, which correspond closely to the sum of exchangeable Ca, Mg, and K. B-horizon samples from the Mexican Dystrandepts gave Cl adsorption capacities ranging from 9 to 30 meq/100 g, with an average of 20. A-horizon samples adsorbed smaller amounts of Cl, possibly because organic matter blocked potential sites. The La Selva B21 (Colombia) and the B-horizon samples from Hawaii have Cl adsorption capacities of the same magnitude observed for the Mexican soils. The two Colombian soils containing crystalline clay and oxide minerals (Tibiatata and Portrerito) adsorbed only small amounts of Cl.

A number of Ultisols and Oxisols from Colombia and Brazil adsorbed from 2–4 meq Cl/100 g, amounts in the range quoted by Sumner and Reeve (11) and Hingston et al. (5) for similar soils. Kaolinite (API 5) adsorbed 4 meq Cl/100 g, a quantity consistent with the observations of Quirk (9). Montmorillonite (API 26) and illite (API 36) adsorbed no Cl from 0.17N AlCl$_3$. Neither did several red California soils (Ultic Argixerolls and Haplic Durochrepts).

Among the Hawaiian soils, those in the intermediate weathering stages (Paauhau, Maile, Hanaipoe, and all Dystrandepts) had the largest Cl adsorption capacities. The less weather Apakui and the more highly weathered Akaka and Honakaa adsorbed considerably less Cl, although their capacities were large compared with most soils. Singh and Kanehiro (10) reported NO$_3$-adsorption capacities of Hawaiian soils at pH 3.5 to be only a few meq/100 g; however, their determinations were made in acetate buffers, with the probable result that essentially all anion adsorption sites were blocked by acetate ions. T. Kinjo in our laboratory finds NO$_3$ and Cl adsorption capacities of Dystrandepts to be nearly identical (personal communication from T. Kinjo, Univ. of California, Riverside).

With two exceptions, the excess of Al over Cl adsorption from 0.17N AlCl$_3$ by the soils listed in Table 5 is very nearly equal to the effective CEC. It appears that Cl, or H$_3$O from its hydrolysis, reacts with exchange sites to displace exchangeable ions, and in addition is coadsorbed with equivalent amounts of Cl. It is likely that the Al which leaves solution along with Cl is hydrolyzed, and that protons produced by its hydrolysis react with the soil to produce sites for anion adsorption.

Adsorption Mechanism

Our data are compatible with the reaction mechanism proposed by Hingston et al. (5), who refer to the reversible pH-dependent uptake of Cl by goethite and other oxide minerals as "nonspecific adsorption." Positive sites for nonspecific adsorption are thought to be created through the acceptance of protons by octahedrally coordinated Al or Fe^{3+}. The proton acceptors in the andepts from Mexico, Colombia, and Hawaii may be free aluminum or iron oxides or allophane. The Mexican soils contain gibbsite; all of the soils likely contain finely divided amorphous ferric and aluminum oxides. Chloride adsorption capacities of the soils, however, are large when compared with the capacities of goethite (5, 11), kaolinite (9), and nonallophanic soils (11). Egawa (2) and, more recently, Cloos et al. (1) make the interesting and plausible suggestion that allophane consists of a silica-alumina core coated with Al(OH)$_3$. Such a substance would exhibit the Cl-adsorption properties observed in the present work, particularly if the association of Al(OH)$_3$ in allophane prevents crystal growth as gibbsite.

Table 5—Chloride adsorption capacities as measured with AlCl$_3$ and some properties of soils used

Soil	Horizon	Depth	Soil group	pH (H$_2$O) 1:5	Effective CEC	Adsorbed from solution Al	Adsorbed from solution Cl
		cm			meq/100 g	meq/100 g	
San Gregorio (Mexico)	A	0–15	Dystrandept	4.9	3.6	11	7
	B	15–40		6.5	5.3	31	24
	B2b	50–65		6.6	3.9	37	30
La Guardia (Mexico)	Ap	0–15	Dystrandept	5.8	6.1	12	7
	B1	20–50		6.6	7.0	29	23
	B2	55–86		6.7	4.3	14	9
	A1b	86–100		6.6	10.8	13	3
Los Manzanillos (Mexico)	Ap	0–25	Dystrandept	5.7	5.8	12	7
	B1	25–50		6.5	9.0	23	15
	B2	50–80		6.6	8.7	23	16
La Palma (Mexico)	Ap	0–25	Dystrandept	5.5	3.8	13	13
	B1	25–75		6.1	6.2	24	18
	B2	75–108		6.4	5.9	32	25
Honakaa	B22	35–50	Hydrandept	5.4	1	16	16
Akaka	B22	38–58	Hydrandept	5.5	<1	18	18
Paauhau	B21	45–70	Dystrandept	6.4	10	42	32
Maile	II B2	35–73	Dystrandept	5.4	<1	30	30
Hanaipoe	B2	18–30	Dystrandept	6.7	3	29	29
Apakui	II Ab	20–30	Vitrandept	6.4	14	30	16
La Selva (Colombia)	Ap	0–29		5.1	9	7	0
	B21	42–90		6.2	0.5	14	15
Tibaitata	Ap	0–60	Aquic Eutrandept	5.0	16.7	12	2
	A3	60–90		5.1	14.2	16	2
Potrerito "Coco Rojo" (Colombia)	A2	25–40	Andic Typumbrult	5.1	1.7	4	4

LITERATURE CITED

1. Cloos, P., A. J. Leonard, J. P. Moreau, A. Herbillon, and J. J. Fripiat. 1969. Structural organization in amorphous silico-aluminas. Clays Clay Miner. 17:279–287.
2. Egawa, T. 1964. A study on coordination number of aluminum in allophane. Clay Sci. (Tokyo) 2:1–7.
3. Fieldes, M., and R. J. Furkert. 1966. The nature of allophane in soils. II. Differences in composition. N. Zeal. J. Sci. 9:608–622.
4. Fieldes, M., and R. K. Schofield. 1960. Mechanism of ion adsorption by inorganic soil colloids. N. Zeal. J. Sci. 3:563–579.

5. Hingston, F. J., R. J. Atkinson, A. M. Posner, and J. P. Quirk. 1968. Specific adsorption of anions on goethite. Int. Congr. Soil Sci., Trans. 9th (Adelaide, Aust.) 1:669–678.
6. Imura, J. 1961. Ion adsorption curves in allophane. Clay Sci. (Tokyo) 1:40–44.
7. Lai, Sung-Ho, and L. D. Swindale. 1969. Chemical properties of allophane from Hawaiian and Japanese soils. Soil Sci. Soc. Amer. Proc. 33:804–808.
8. Panel on Volcanic Ash Soils in Latin America. 1969. IAIAS, Turrialba, Costa Rica.
9. Quirk, J. P. 1960. Negative and positive adsorption of chloride by kaolinite. Nature 188:253.
10. Singh, B. R., and Y. Kanehiro. 1969. Adsorption of nitrate in amorphous and kaolinitic Hawaiian soils. Soil Sci. Soc. Amer. Proc. 33:681–683.
11. Sumner, M. E., and N. G. Reeve. 1966. The effect of iron oxide impurities on the positive and negative adsorption of chloride by kaolinites. J. Soil Sci. 17:274–279.
12. Wada, K., and H. Ataka. 1958. The ion uptake mechanism of allophane. Soil Plant Food (Tokyo) 4:12–18.

Anion Adsorption by Allophanic Tropical Soils: II. Sulfate Adsorption[1]

H. GEBHARDT AND N. T. COLEMAN[2]

ABSTRACT

Volcanic ash-derived soils (Andepts) from Mexico, Colombia, and Hawaii have large capacities for sulfate adsorption: 10-20 meq/100 g for surface soils and 15-60 meq/100 g for subsoils. The sulfate adsorption capacity is pH-dependent. For a typical case, B-horizon material from San Gregorio, Michoacan, Mexico, capacity in meq/100 g was 13 at pH 6.3, 22 at pH 5.1, 38 at pH 4.4, and 48 at pH 4.1. Sulfate adsorption was accompanied by and dependent upon the simultaneous adsorption, or consumption, of protons. Increasing solution concentration of sulfate beyond 5-10 meq/liter at a given pH resulted in relatively minor additional uptake by the soil. At a given pH, sulfate adsorption capacities, expressed as mmole/100 g soil, nearly equaled Cl adsorption maxima. This, together with the consumption of 1 meq H for each mmole of HCl or H_2SO_4 adsorbed, suggests that the two anions are adsorbed on the same sites and that site protonation is a prerequisite for adsorption.

Sulfate adsorbed by the San Gregorio soil at pH 4 was strongly bound against hydrolysis (removal by water leaching) and was only partly displaced by $1N$ KNO_3. It was completely displaced by $1N$ NH_4OAc, pH 7. Sulfate may be adsorbed by Andepts, specifically, by ligand exchange. Its affinity for soil is at least 10 times that of nonspecifically adsorbed anions such as Cl and NO_3.

Additional Index Words: Andept, volcanic soil.

A REVIEW of the pre-1964 literature on sulfate adsorption by soils was published by Harward and Reisenauer (11). They point out that: (i) many soils adsorb sulfate; (ii) capacity for sulfate adsorption varies widely with soil properties, being largest where there are substantial amounts of aluminum and iron oxides; (iii) with a given soil, sulfate adsorption increases with decreasing pH in the range 6.5-4; and (iv) sulfate is much less strongly adsorbed than phosphate.

Amounts of sulfate adsorbed by soils vary from a few parts per million to tens of mmoles/100 g. Barrow (2) reported a range of 0.2–1 mmole/100 g for a group of Australian soils, with capacity being taken as amount of adsorbed sulfate when solution concentration was 20 ppm. Chao, Harward, and Fang (7) found Oregon Latosols and one "Ando" soil to adsorb 0.6–0.8 meq sulfate/100 g from $0.03N$ K_2SO_4, while Bornemisza and Llanos (4) observed the adsorption of around 2 meq/100 g by Costa Rica Latosols and volcanic soils. The work of Chang and Thomas (6) showed that subsoils from the Virginia Piedmont, containing intergrade minerals as well as kaolinite and free iron oxides, adsorb several meq/100 g at pH 4, with the amount adsorbed decreasing rapidly at higher pH.

Hanson, Fox, and Boyd (10) found some Hawaiian Andepts to contain large amounts of adsorbed sulfate. For surface soils, isotopically exchangeable sulfate S varied from 10–1,000 ppm, with the largest amounts associated with highly weathered soils under high rainfall conditions. Sulfate S that could be extracted with Ca $(H_2PO_4)_2$ solution was as high as 7,200 ppm in an Akaka subsoil. In the pH range 5.2–7, the surface soils adsorbed additional sulfate from K_2SO_4 solution, with "first-stage" adsorption maxima, corresponding to final solution concentrations of 0.5–1 mmole/liter, being 0.5–8.5 meq/100 g (counting both isotopically exchangeable sulfate initially present and sulfate adsorbed from K_2SO_4). Corresponding adsorption maxima for Akaka subsoil samples were as large as 40 meq/100 g.

The range of sulfate adsorption capacities has not been established, nor are mechanisms known with certainty. It appears, however, that Dystrandepts have very large capacities and that in such soils, and to a lesser degree Oxisols and Ultisols, sulfate adsorption may have important direct and indirect effects on plant nutrition and fertilizer needs (2, 10). Aylmore, Karin, and Quirk (1) suggest that sulfate adsorption by aluminum and iron oxides and kaolinite follows proton acceptance by mineral surfaces, with consequent development of positive charge. Hingston et al. (12) have elaborated a theory for specific anion adsorption by sesquioxides that requires proton acceptance by the surface hydroxyl exchange to allow binding of the anion to surface Al or Fe, and that the adsorbed anion be a proton donor. Chang and Thomas (6) propose that in some soils containing partly hydrolyzed Al on cation-exchange sites, sulfate ions exchange hydroxyls that are consumed in further hy-

---

[1] Contribution from the Dep. of Soil Science and Agr. Eng., Univ. of California, Riverside 92502. Support from the National Science Foundation, under Research Grant NSF-GB-11711-(Pratt), is gratefully acknowledged. Received 10 Sept. 1973. Approved 17 Sept. 1973.

[2] Postgraduate Research Soil Scientist and Professor, respectively, Dep. of Soil Science and Agr. Eng., Univ. of California, Riverside.

drolyzing Al. They observed equivalent adsorption of sulfate and an accompanying metal cation, which would be required for such a reaction. Barrow (2) suggests that new adsorption sites are continually activated as the sulfate concentration in solution increases.

This report is the second in a series that describes anion adsorption by Dystrandepts and other allophanic soils. The first dealt with Cl, and established large anion adsorption capacities, 10–30 meq/100 g, at pH 4 (9). Experiments with sulfate, summarized in this paper, determined sulfate adsorption capacities of allophanic soils from Colombia, Mexico, and Hawaii, related sulfate adsorption to pH and solution concentrations, and gave some insight into adsorption mechanisms.

Table 1—Reactions of H_2SO_4 with 5-g San Gregorio B-horizon material, solution volume = 30 ml.

Added H_2SO_4	pH	In solution			Adsorbed		pKb*	pKg†
meq/100 g		H_3O	Al	SO_4	H_3O	SO_4		
		—meq/100 g—		meq/liter	—meq/100 g—			
6	5.1	<1	<1	0.67	6	6	--	--
10	5.0	<1	<1	1.0	10	9	--	--
14	4.9	<1	<1	2.0	14	12	--	--
20	4.7	<1	<1	3.3	20	18	--	--
30	4.4	<1	<1	5.0	30	27	--	--
40	4.1	<1	1	10.0	39	34	116.3	33.4
50	4.0	<1	3	16.7	47	40	116.0	33.4
60	3.8	<1	8	26.8	52	44	116.6	33.7
80	3.7	1	20	50.0	59	50	116.8	33.9
100	3.5	1	38	76.7	63	52	118.1	34.3
120	3.4	2	53	105.0	65	57	118.8	34.6
140	3.3	3	68	134.0	69	60	119.6	34.8

* pKb = 4 pAl + 10 pOH = pSO$_4$ (Basaluminite pKsp = 117.3).
† pKg = pAl + 3 pOH (Gibbsite pKsp = 34.03).

MATERIALS AND METHODS

The soils used were Mexican, Colombian, and Hawaiian Dystrandepts and other soils high in allophane, hydrous oxides, or both. Sample origins and general soil characteristics have been described in a companion paper (9). The Mexican and Colombian soils contained no sulfate extractable with $1N$ NH_4OAc at pH 7. Sulfate extracted from the Hawaiian samples was between 1 and 4 meq/100 g.

To measure sulfate adsorption, samples of air-dried soils to give 5 g of oven-dried material were shaken for 1 hour in 30 ml of solution (solution volume included water in the air-dried soils) containing known amounts of H_2SO_4, Na_2SO_4, or mixed H_2SO_4-Na_2SO_4. After shaking, the slurries were filtered and filtrates were analyzed for pH, Na by flame photometry, H_3O and Al by potentiometric titration with NaOH, and sulfate by precipitation with $BaCl_2$. In some cases sulfate desorption was measured by leaching samples of treated soil with H_2O, $1N$ KNO_3 or $1N$ NH_4OAc. When sufficient amounts of sulfate were present, aliquots of filtrate were passed through H-Dowex-50 columns to remove metal cations and then mixed with known amounts of $BaCl_2$. Barium remaining unprecipitated was determined by flame photometry and sulfate was taken as equivalent to the amount of Ba that had precipitated. For low sulfate concentrations, e.g., in NH_4OAc extracts or water extracts of soils, sulfate was determined by the turbidity method described by Jackson (13).

Ion-activity products for basaluminite, gibbsite, and H_2SO_4 were calculated from SO_4 and Al concentrations and equilibrium pH values. Activity coefficients were obtained from the nomogram given by Butler (5).

RESULTS AND DISCUSSION

Sulfate Adsorption by San Gregorio B-Horizon Material

San Gregorio B-horizon (Dystrandept from Mexico) adsorbed H_3O and sulfate from H_2SO_4 almost quantitatively for additions up to 40 meq/100 g (Table 1). For larger additions of H_2SO_4, there was proportionally less adsorption; amounts of bound H_3O and sulfate rose to apparent maxima near 69 and 60 meq/100 g, respectively, when 140 meq of H_2SO_4 per 100 g soil were added. For additions of $H_2SO_4 > 40$ meq/100 g, pH after 1 hour of contact between soil and solution was < 4 and appreciable amounts of Al were dissolved. Even at the largest addition of H_2SO_4 (140 meq/100 g), little H_2O remained in solution after 1 hour of contact with soil; the bulk of nonadsorbed H_3O was consumed in dissolving Al compounds. Very little Fe was dissolved.

The data in Table 1 are consistent with protonation of soil to create positively charged sites and adsorption of HSO_4 or SO_4 on those sites (1, 12). The excess of H_3O over sulfate adsorbed for the larger additions of H_2SO_4 corresponds reasonably well with the excess of H_3O over Cl adsorbed in parallel work, and is attributed to displacement of exchangeable Ca, Mg, and K by H_3O or Al.

It is not possible to unambiguously distinguish sulfate adsorption from the precipitation of basic aluminum sulfates. Singh and Brydon (15) found the mineral basaluminite to form when $Al_2(SO_4)_3$ solutions were neutralized in the presence of montmorillonite. They gave the solubility product of basaluminite as $(Al)^4(OH)^{10}SO_4 = 10^{-117.3}$. At an ion-activity product for sulfuric acid $(H)^2(SO_4) = 10^{-9.3}$ gibbsite and basaluminite co-exist. Ion-activity products (Table 1) suggest supersaturation with respect to basaluminite and saturation with respect to gibbsite for H_2SO_4 additions of 40–80 meq/100 g and pH values of 4.1–3.7. For larger H_2SO_4 additions, where pH was < 3.5, the solutions were undersaturated with respect to both basaluminite and gibbsite.

Amounts of sulfate adsorbed by the San Gregorio soil were strongly dependent upon pH and to a smaller degree upon solution concentration. The smooth curve to the left of Fig. 1 shows the adsorption of sulfate from sulfuric acid as a function of the final sulfate concentration in solution. The straight lines intersecting the H_2SO_4 curve show the effects of adding Na_2SO_4 to systems containing 0, 20, 40, and 60 meq H_2SO_4/100 g soil.

At pH 5.8–6.3, sodium and sulfate were adsorbed together from Na_2SO_4 in nearly equivalent amounts. Adsorption increased slowly with addition of Na_2SO_4. It was around 5 meq/100 g when final solution concentration was $0.02N$, and was 14 meq/100 g from $0.14N$ Na_2SO_4. Adsorption of sulfate from H_2SO_4-Na_2SO_4 mixtures was determined mainly by the amount of H_2SO_4 added. For example, 18 meq/100 g of sulfate was adsorbed from 20 meq/100 g of H_2SO_4, with a final solution concentration of 3.3 meq/liter. Adding Na_2SO_4 to the soil-acid mixture resulted in only slightly greater adsorption; 22 meq/100 g was adsorbed when solution concentration of sulfate was 130 meq/100 g. Similar results were obtained for the two other levels of H_2SO_4 addition, with 4–5 meq/100 g more sulfate adsorbed when sulfate ion concentration was increased to 120–130 meq/liter by addition of Na_2SO_4 at a given addition of H_2SO_4. Sodium adsorption (curve 7 in Fig. 1) paralleled additional sulfate uptake due to Na_2SO_4. Curve 7

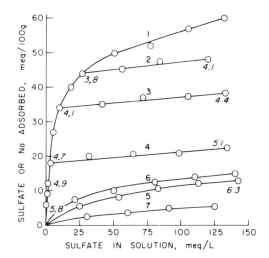

Fig. 1—Adsorption of sulfate and Na by San Gregorio soil: (1) SO$_4$ from H$_2$SO$_4$; (2) SO$_4$ from 60 meq H$_2$SO$_4$/100 g + Na$_2$SO$_4$; (3) SO$_4$ from 40 meq H$_2$SO$_4$/100 g + Na$_2$SO$_4$; (4) SO$_4$ from 20 meq H$_2$SO$_4$/100 g + Na$_2$SO$_4$; (5) SO$_4$ from Na$_2$SO$_4$; (6) Na from Na$_2$SO$_4$; and (7) Na from H$_2$SO$_4$ + Na$_2$SO$_4$, average for all H$_2$SO$_4$ additions. Decimal numbers are pH.

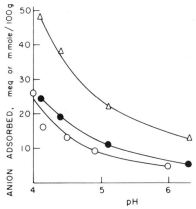

Fig. 2—Adsorption maxima for sulfate and chloride as related to pH. San Gregorio B-horizon. Open circles-mmole Cl; closed circles-mmole sulfate; and triangles-meq sulfate.

shows that average Na adsorption for the three acid additions was small compared with total sulfate adsorption and was smaller in the three lower pH ranges 3.8–4.1, 4.1–4.4, and 4.7–5.1 than in the range 5.8–6.3, which prevailed when only Na$_2$SO$_4$ was added to soil. In respect to coadsorption of Na with an adsorbed anion, the sulfate systems were like Cl systems (9), where there was little or no cation adsorption other than that attributable to cation-exchange. In contrast, the San Gregorio soil adsorbed as much as 20 meq Na/100 g when reacted with H$_3$PO$_4$-NaH$_2$PO$_4$ mixtures at pH 4.2–4.7.

As shown by the extreme pH values indicated in Fig. 1, pH in H$_2$SO$_4$-Na$_2$SO$_4$ was 0.3–0.5 units higher than when the same amount of acid was added alone. Hanson et al. (10) observed the same trend for Hawaiian Andepts.

The number of meq of sulfate adsorbed per 100 g of San Gregorio B-horizon at any pH was very nearly twice the meq of Cl adsorbed from 0.16N NaCl (Fig. 2). This is suggestive of adsorption of sulfate as HSO$_4$ ions, with the same sites reacting as for Cl. However, the adsorption affinity for sulfate added as H$_2$SO$_4$ was nearly ten times that for Cl added as HCl. This is deduced from the initial slopes of the adsorption isotherms, and is reflected by a much smaller tendency for adsorbed sulfate to be eluted by water or 1N KNO$_3$ than was the case for adsorbed Cl (9). Thirty-five percent of the sulfate adsorbed by San Gregorio B-horizon at pH 4.1 was removed by leaching 5 g of soil with 300 ml of H$_2$O. This contrasts with complete desorption of Cl under the same conditions. Similarly, 1N KNO$_3$, which rapidly and completely exchanged adsorbed Cl, removed sulfate only partially. All of the sulfate adsorbed at pH 4 was recovered when 5 g of soil were leached with 100 ml of 1N NH$_4$OAc.

Sulfate Adsorption Capacity as Measured with Mixtures of Na$_2$SO$_4$ and H$_2$SO$_4$ Solutions

The data obtained with the San Gregorio B-horizon and summarized in Fig. 1 suggest that the amounts of sulfate taken up by soil from mixed solutions of Na$_2$SO$_4$ and H$_2$SO$_4$ (approximately 0.1N) at pH 4 represent the sulfate adsorption capacity. If adsorption of HSO$_4$ is assumed, these values could represent the anion-adsorption capacity. Table 2 lists sulfate adsorption capacities of Mexican, Hawaiian, and Colombian soils as measured with mixtures of Na$_2$SO$_4$ solutions. The data were selected from results of a factorial experiment that combined rates of H$_2$SO$_4$ and Na$_2$SO$_4$, and reflect the combination (for the Andepts) that contained the minimum amount of acid to give a pH just below 4. Capacities were large, particularly for B-horizon samples of Mexican and Hawaiian Andepts. The sulfate adsorption capacities (in meq/100 g) of such samples were around twice the chloride adsorption capacities which are listed for comparison. The sulfate adsorption capacities of the Hawaiian subsoils are particularly large, as would be expected from the results of Hanson et al. (10) with similar materials. For all of the subsoil materials, mmoles S

Table 2—Sulfate adsorption capacities as measured with mixture of Na$_2$SO$_4$ and H$_2$SO$_4$ solutions; 5-g soil in 30 ml. Data expressed as meq/100 g air-dry soil

Soil	Horizon	Added			Adsorbed			Cl adsorption capacity (Ref. 9)
		Na$_2$SO$_4$	H$_2$SO$_4$	pH	H$_2$O	Na	SO$_4$	
		— meq/100 g —			— meq/100 g —			
San Gregorio (Mexico)	Ap	60	40	3.7	24	12	21	7
	B	40	60	4.0	50	4	48	24
	B2b	40	60	4.2	56	4	55	30
La Guardia (Mexico)	Ap	60	40	3.8	22	10	19	7
	B1	40	60	4.0	48	3	47	23
	B2	60	40	3.7	18	4	15	9
	A1b	60	40	3.5	10	3	9	3
Honakaa	B22	40	60	3.7	34	5	32	16
Akaka	B22	40	60	3.6	42	5	42	18
Paauhau	B21	40	80	3.8	68	7	62	32
Maile	II B2	40	80	3.8	60	5	59	30
Hanaipoe	B2	40	80	3.9	64	6	62	29
Apakui	II Ab	40	60	3.7	43	7	33	16
La Selva (Colombia)	Ap	80	20	3.8	10	5	10	0
	B21	40	60	3.7	34	12	32	15

adsorbed nearly equaled Cl adsorption capacity. This is consistent with HSO_4 and Cl adsorption on the same sites.

Surface samples (San Gregorio and La Guardia from Mexico and La Selva from Colombia) have small chloride adsorption capacities but adsorbed appreciable amounts of sulfate. Illite (API 36) and montmorillonite (API 26) exhibited no sulfate adsorption. Kaolinite (API 5) at low pH adsorbed sulfate in amounts up to 8 meq/100 g, around twice the amount found for Cl by Quirk (14) and the present authors (9).

Adsorption Mechanism

The large adsorption affinity of sulfate for allophanic soils as compared with Cl and the fact that bound sulfate was incompletely eluted by $1N$ KNO_3 suggests that sulfate may be specifically adsorbed in the sense proposed by Hingston et al. (12). They state: "specifically adsorbed anions must enter the inner coordination sphere of the metal ions on the oxide surface, through displacement of a coordinated hydroxyl ion. This displacement is a ligand exchange reaction." Hingston's theory requires that specifically adsorbed anions be able to donate or accept protons, and that maximum adsorption (or change in slope of the adsorption-pH curve) occurs when pH = pKa. In the case of sulfate, only the monovalent species, HSO_4, can be specifically adsorbed without creating additional negative charge at the surface. The pK_2 of H_2SO_4 is 1.92, and at pH > 4, HSO_4 makes up less than 1% of the ionic species. The adsorption of HSO_4 upon the addition of H_2SO_4 to soil would require the rapid protonation of a soil component and the simultaneous adsorption of HSO_4. The equivalence of proton and sulfate uptake, and the near-equality between sulfate and Cl adsorption capacities, assuming adsorption of HSO_4, lends credence to the suggestion that this is the adsorbed ion. That large concentrations of Na_2SO_4 hardly increase the amount of sulfate adsorbed in a given narrow range of pH shows the essentiality of H-ions to protonate either a potential adsorption site or a SO_4 ion. Perhaps the pH values in question, 4.1–5.1, are so far above pK_2 that only SO_4 can be adsorbed, with the extent of that reaction severely limited by the development of excess negative charge (12).

Sulfate adsorption and corresponding Na adsorption in the pH range between 5.8 and 6.3, when Na_2SO_4 was added to San Gregorio soil, may have proceeded according to the reaction proposed by Chang and Thomas (6), i.e., displacement of hydroxyls by sulfate, increase in pH, and consequent neutralization of weak-acid exchange sites. Both sodium and sulfate adsorbed in this pH range were completely removed by washing with water, consistent with the extremely small adsorption affinity deduced from Fig. 1.

Returning to the low-pH adsorption of sulfate, it is possible that contrary to Hingston's conclusions (12), HSO_4 is adsorbed nonspecifically as counterions opposing positively charged sites on oxide surfaces. However, the large affinity relative to nonspecifically adsorbed ions such as Cl on NO_3 and the lack of reversibility reported by Hanson et al. (10) for Hawaiian soils and Aylmore et al. (1) for aluminum oxides suggests that the adsorption reaction involves ligand exchange rather than simple ion-ion interaction. In that connection, sulfate adsorption by allophanic soil may proceed in the stages that have recently been elucidated for aluminum sulfate and other metal-sulfate complexes (3). Eigen (8) first gained information on these rapid reactions. Entering the outer coordination sphere requires a 10-fold longer time than required for simple diffusion; inner sphere complexes are formed at least 10^4 times more slowly, with displacement of the first water molecule of the hydration shell by an anion (sulfate or hydroxyl) rate determining. Further displacement of the second and subsequent water molecules proceeds much more rapidly. Therefore, partial hydrolysis of the metal cation, as likely will be the case in allophane, accelerates the formation of inner sphere complexes.

The capacities of Mexican and Hawaiian Andepts to adsorb sulfate are surprisingly large. Values of 50–60 meq/100 g may be compared on the one hand with maxima of a few meq/100 g that have been reported for Ultisols and Oxisols, and on the other with capacities measured for pseudoboehmite (84 meq/g) and goethite (13 meq/g) (1). For both of these synthetic minerals, an adsorbed sulfate ion occupied approximately 60 A^2 (1). Clearly, large surface areas of Fe-Al oxides are required for the sulfate adsorption that occurred in the Mexican and Hawaiian soils. The large specific surface of the amorphous clays in the Andepts, and the stabilization of hydrous oxide dispersions (an antigibbsite effect) likely are responsible.

LITERATURE CITED

1. Aylmore, L. A. G., M. Karin, and J. P. Quirk. 1967. Adsorption and desorption of sulfate ions by soil constituents. Soil Sci. 103:10–15.
2. Barrow, N. J. 1970. Comparison of the adsorption of molybdate, sulfate and phosphate by soils. Soil Sci. 109:282–288.
3. Behr, B., and H. Wendt. 1962. Schnelle Ionenreaktionen. I. Die Bildung des Alumininumsulfat Complexes. Z. Elektrochemie 66:223–228.
4. Bornemisza, E., and R. Llanos. 1967. Sulfate movement, adsorption, and desorption in three Costa Rican soils. Soil Sci. Soc. Amer. Proc. 31:356–360.
5. Butler, J. N. 1964. Ionic equilibrium. A mathematical approach. Addison-Wesley Publ. Co., Inc., Reading, Mass.
6. Chang, M. L., and G. W. Thomas. 1963. A suggested mechanism for sulfate adsorption by soils. Soil Sci. Soc. Amer. Proc. 27:281–283.
7. Chao, T. T., M. E. Harward, and S. C. Fang. 1962. Adsorption and desorption phenomena of sulfate ions in soils. Soil Sci. Soc. Amer. Proc. 26:234–237.
8. Eigen, M. 1960. Relaxations-spektren chemischer Umwandlungerz (Metallkomplexe and Protolytische Reaktionen in Wässriger Lösung). Z. Elektrochemie 64:115–123.
9. Gebhardt, H., and N. T. Coleman. 1974. Anion adsorption by allophanic tropical soils: I. Chloride adsorption. Soil Sci. Soc. Amer. Proc. 38:255–259 (this issue).
10. Hanson, S. M., R. L. Fox, and C. C. Boyd. 1970. Solubility and availability of sorbed sulfate in Hawaiian Latosols. Soil Sci. Soc. Amer. Proc. 34:897–901.
11. Harward, M. E., and H. M. Reiserauer. 1966. Reactions and movement of inorganic sulfur. Soil Sci. 101:326–335.
12. Hingston, F. J., R. J. Atkinson, A. M. Posner, and J. R. Quirk. 1967. Specific adsorption of anions. Nature 215:1459–1461.
13. Jackson, M. L. 1958. Soil chemical analysis. Prentice-Hall, Inc., Englewood Cliffs, N.J.
14. Quirk, J. P. 1960. Negative and positive adsorption of chloride by kaolinite. Nature 188:253.
15. Singh, S. S., and J. E. Brydon. 1969. Solubility of basic aluminum sulfates at equilibrium in solution and in the presence of montmorillonite. Soil Sci. 107:12–16.

15c

Copyright © 1974 by the Soil Science Society of America
Reprinted by permission from *Soil Sci. Am. Proc.* **38**:263-266 (1974)

Anion Adsorption by Allophanic Tropical Soils: III. Phosphate Adsorption[1]

H. GEBHARDT AND N. T. COLEMAN[2]

ABSTRACT

Andepts from Mexico and Hawaii bound 30-70 mmole P/100 g air-dry soil at pH 4.3-4.7 during a 1-hour reaction with 0.05M mixed H_3PO_4-NaH_2PO_4 solution. As much as 130 mmole P/100 g were taken up from H_3PO_4 at pH 2.4. A B-horizon sample, San Gregorio from Michoacan, Mexico, adsorbed P from H_3PO_4 and consumed H in nearly equimolar amount, with 38 mmole/100 g of each removed from solution in bringing soil from an initial pH of 5.4 to 4.3. Larger additions of H_3PO_4 dissolved appreciable Al, gave pH < 4, and yielded solutions that were supersaturated with respect to variscite. An apparent adsorption maximum of 38 mmole/100 g compares with maxima of 27 and 30 mmole/100 g for Cl and sulfate, respectively. High-affinity adsorption of P from H_3PO_4 appeared limited by the supply of H-ions to protonate sites or react with displaced OH. San Gregorio B horizon took up phosphate from NaH_2PO_4, but with low affinity. Phosphate uptake from NaH_2PO_4 was accompanied by coadsorption of Na. At a given pH and level of tightly bound P established with H_3PO_4, addition of NaH_2PO_4 resulted in low-affinity adsorption of P and coadsorption of Na. Around half of the P bound from NaH_2PO_4 was readily eluted with water. Phosphate bound from H_3PO_4 was virtually insoluble in water. Some P was eluted by 0.5M arsenate or selenite. The Andepts bound P through at least three mechanisms: high-affinity adsorption on protonated sites; low-affinity adsorption with coadsorption of Na; and formation of insoluble variscite-like substances.

Additional Index Words: Andept, Dystrandept, volcanic soil.

A NDEPTS and other soils containing large amounts of allophane and other amorphous minerals have high capacities for binding phosphate (2, 6, 13, 18). Such soils also adsorb significant amounts of Cl, NO_3, and sulfate (13). The two former ions are adsorbed nonspecifically in the sense that term is used by Hingston et al. (9, 10). Working with a collection of Dystrandepts from Mexico, Colombia, and Hawaii, we found nearly equal Cl and sulfate adsorption capacities (sulfate expressed as mmole/100 g, implying adsorption of HSO_4) with maxima at pH ≤ 4 (7, 8). It seemed that adsorption was on protonated sites and that each soil can develop a definite maximum anion adsorption capacity if enough protons are available.

The work reported in this paper was done to compare phosphate binding by Andepts with that of Cl and HSO_4, in terms of amounts and apparent affinities. Although precipitation of Al and Fe phosphates is recognized as an important reaction leading to phosphate retention by acid soils (14), there are situations in which P solubility appears to be controlled by adsorption rather than by the presence of slightly soluble compounds (1, 9, 12, 15). Muljadi, Posner, and Quirk (15) have given detailed descriptions of phosphate adsorption by aluminum hydroxides and kaolinite. They regarded H_2PO_4 to be the ion adsorbed. Hingston et al. (9) suggest that phosphate, in common with many anions of polybasic acids, is adsorbed specifically on oxide surfaces. Sites for phosphate adsorption, when protonated, bind ions such as Cl nonspecifically, i.e. through ion-ion interaction. Hingston et al. (9, 10) propose that metal cations may be coadsorbed with phosphate, although Muljadi et al. (15) observed coadsorption of K and phosphate by aluminum hydroxide only from concentrated solutions. Bar-Yosef, Kafkafi, and Lahav (1) and Kafkafi (12) interpret data obtained with kaolinite as showing specific adsorption of phosphate in the sense proposed by Hingston et al. (9).

The present paper gives data on phosphate binding by Andepts from H_3PO_4, NaH_2PO_4, or mixtures of acid and salt. In addition to comparing P adsorption with that of Cl and HSO_4, it examines the coadsorption of Na and the desorption of phosphate to water and solutions of arsenate or selenite.

MATERIALS AND METHODS

The soils used were Mexican, Colombian, and Hawaiian Dystrandepts and other soils high in allophane, hydrous oxides, or both. Sample origins and general soil characteristics have been described in a companion paper (7). The native bicarbonate extractable phosphate was less than 0.1 mmole P/100 g soil in all cases.

Two-gram samples of air-dried soils (on a 105C dry weight basis) were shaken in 20 ml of solution containing graded amounts of H_3PO_4, NaH_2PO_4, or mixtures of NaH_2PO_4 and H_3PO_4. Samples reacted with H_3PO_4 or NaH_2PO_4 were shaken with the phosphate solution for 1 hour. Contact times were limited deliberately to minimize decomposition-precipitation reactions and to correspond with time used in experiments with other anions. Samples reacted with mixtures of NaH_2PO_4 and H_3PO_4 were pH controlled by adding H_3PO_4 dropwise to stirred soil suspensions. The pH was not allowed to fall below 4.2. Total reaction times were about 70 min. After shaking or stirring, the slurries were filtered and filtrates were analyzed for pH, Na, Al, and P. Aluminum and phosphorus were determined using the aluminon and vanadomolybdo-phosphoric method, respectively (11). Sodium was determined flame photometrically. Phosphated samples were leached with water and 0.5M arsenate solution and in another series with water and 0.5M selenite solution. Phosphorus determination in arsenate leachates was carried out after removal of arsenic by bromide distillation (10). Ion activity products for variscite and gibbsite were calculated from P and Al concentrations and equilibrium pH values. Activity coefficients were obtained from the nomogram given by Butler (3).

---

[1] Contribution from the Dep. of Soil Science and Agr. Eng., Univ. of California, Riverside 92502. Support of the National Science Foundation through Grant no. NSF-GB-11711-(Pratt) is gratefully acknowledged. Received 10 Sept. 1973. Approved 17 Sept. 1973.

[2] Postgraduate Research Soil Scientist and Professor, respectively, Dep. of Soil Science and Agr. Eng., Univ. of California, Riverside 92502.

RESULTS AND DISCUSSION

Phosphate Sorption by a B-Horizon Sample of a Dystrandept from San Gregorio, Mexico

The quantities of phosphate retained by San Gregorio B horizon from H_3PO_4 solutions are shown in Table 1. For H_3PO_4 additions up to 50 mmole/100 g, the pH of the slurry was above 4 after 1 hour of shaking, and nearly all of the acid (both protons and phosphate) was removed from solution. Amounts of H_3PO_4 corresponding to 66 mmole/100 g soil or more resulted in pH < 4 and the solution of appreciable amounts of Al. Maximum P retention, approximately 125 mmole/100 g of soil occurred at pH 2.3–2.5, with > 50 mmole P and > 25 mmole Al/liter of solution. Below pH 4, the solution phase was supersaturated with regard to variscite, and Al phosphates may have precipitated. The strongly acid systems were undersaturated with respect to gibbsite.

Because of uncertainty in distinguishing adsorbed from precipitated phosphate, the data in Table 1 cannot be interpreted unambiguously in terms of an adsorption maximum. It is possible that the 38 mmole P/100 g soil retained at pH 4.3 without the solution of measurable Al, represents adsorbed rather than precipitated phosphate. If there is a phosphate adsorption maximum of this magnitude, then H_2PO_4, HSO_4 and Cl were adsorbed by this allophanic soil in near-equivalent amounts. Maxima for Cl and HSO_4 were 27 and 30 mmole/100 g, respectively (7, 8). Adsorption affinities for the three ions are markedly different, however, in the order $H_2PO_4 >> HSO_4 >> Cl$.

Table 2 shows the adsorption by San Gregorio B horizon of P and Na from NaH_2PO_4. Adsorption from NaH_2PO_4 was appreciably less than from an equal amount of H_3PO_4, and Na was adsorbed in almost equimolar amounts of P. The pH rose from 6.2 to 7.5 for small additions of NaH_2PO_4, presumably due to OH-exchange (9, 15). The progressive fall in pH with larger additions of NaH_2PO_4 is attributed to the buffering of phosphate salt remaining in solution.

Figure 1 shows the relationship between P adsorbed and P concentration of solution in the case of H_3PO_4 and NaH_2PO_4 additions to San Gregorio B horizon. The steep curve to the left (no. 5) depicts reaction with H_3PO_4 alone. The slope reflects large affinity for P up to a quantity adsorbed > 50 mmole/100 g. Curve 1 shows that reaction with NaH_2PO_4 resulted in a relatively small amount of

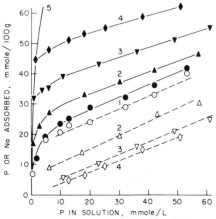

Fig. 1—Phosphate (closed symbols) and Na (open symbols) removed from solution by San Gregorio B horizon (2-g soil in 20 ml) as related to final solution concentrations of P. Phosphorus was added as follows: (1) NaH_2PO_4, pH = 6.2-7.15; (2) 17 mmole H_3PO_4/100 g + NaH_2PO_4, pH = 5.7-5.9; (3) 33 mmole H_3PO_4/100 g + NaH_2PO_4, pH = 4.8-5.1; (4) 47 mmole H_3PO_4/100 g + NaH_2PO_4, pH = 4.2-4.7; (5) H_3PO_4, pH = 3.8-5.4.

high-affinity adsorption (< 10 mmole/100 g) and a nearly linear adsorption isotherm of small slope between P solution concentrations of 10–60 meq/liter.

Data obtained with mixtures of H_3PO_4 and NaH_2PO_4 (amounts of H_3PO_4 were 47, 33, and 17 mmole/100 g) show that the adsorption process has two apparent parts: (i) high affinity uptake of H_3PO_4 and (ii) low-affinity uptake of P from NaH_2PO_4 accompanied by nearly equivalent adsorption of Na. Note in Fig. 1 that for solution concentrations larger than 5–10 mmole/liter, the P and Na adsorption lines were nearly parallel, regardless of the amount of H_3PO_4 in the reaction mixture.

Amounts of Na adsorbed at a given solution concentration of P were inversely proportional to the amounts of H_3PO_4 that had reacted. For an arbitrary solution concentration of 50 mmole/liter, adsorbed Na was 37, 28, 21, and 19 mmole/100 g for additions of 0, 17, 33, and 47 mmole H_3PO_4/100 g, respectively. The pH's were 6.3, 5.8, 5.0, and 4.6.

The data in Fig. 1 suggest two reactions:

1) High-affinity adsorption of H_2PO_4 on a sesquioxide surface (9, 15), as reflected by the steep initial portion of

Table 1—Reaction of H_3PO_4 with 2-g samples of soil from San Gregorio B-Horizon (Volume = 20 ml, data are expressed as mmole/100 g soil = mmole/liter)

Added H_3PO_4	pH	In solution		Adsorbed P	pKv*	pKg†
		Al	P			
6.7	5.4	<0.03	0.1	6.6	--	--
13.3	5.1	<0.03	0.2	13.1	--	--
20.0	4.8	<0.03	0.3	19.7	--	--
26.7	4.6	<0.03	0.5	26.2	--	--
33.3	4.5	<0.03	1.0	32.3	--	--
40.0	4.3	<0.03	1.7	38.3	--	--
50.0	4.1	0.03	2.2	47.8	26.8	33.9
66.6	3.8	0.10	5.7	60.9	26.9	34.9
83.3	3.4	0.27	7.0	76.3	27.4	36.6
133	2.7	4.3	23	110	27.4	36.9
167	2.5	12	44	123	27.4	37.3
233	2.3	41	107	126	27.4	37.7

* pKv = pAl + 2 pOH + pH_2PO_4 (variscite, pK_{sp}: 30.50).
† pKg = pAl + 3 pOH (gibbsite, pK_{sp}: 34.03).

Table 2—Reaction of NaH_2PO_4 with 2-g samples of soil from San Gregorio B-Horizon (Volume = 20 ml, data are expressed as mmole/100 g = mmole/liter)

Added NaH_2PO_4	pH	In solution		Adsorbed		P-Na
		Na	P	Na	P	
0	6.2	--	--	--	--	--
7	7.5	2	0.2	5	7	2
14	7.4	4	2	10	12	2
22	7.2	8	5	14	17	3
30	7.0	10	8	20	22	2
36	6.8	13	13	23	25	2
50	6.6	26	20	24	30	6
67	6.4	38	32	29	35	6
83	6.3	51	42	33	41	8
100	6.3	60	53	40	47	7

the adsorption isotherm. The amount of P that can be adsorbed in this way apparently depends upon the quantity of protons available for surface protonation and/or combination with displaced hydroxyl.

2) Low-affinity adsorption of H_2PO_4 and Na in nearly equivalent amounts, as shown by the gently sloping, nearly linear portions of the adsorption curves. Lack of protons for surface protonation or reaction with displaced OH does not prevent further adsorption of P as the solution concentration of H_2PO_4 increases, but does require coadsorption of Na to maintain electroneutrality (9).

The "nearly equi'pH" lines in Fig. 1 resemble smooth, continuous adsorption isotherms and, in fact, can be described reasonably well by Langmuir equations. The accord is fortuitous, since at least two disparate mechanisms are involved in phosphate retention from the acid-salt mixtures.

Phosphate Desorption

In an attempt to determine the amounts of bound phosphate that can be displaced by water or strongly adsorbed anions, 2-g samples of San Gregorio B horizon samples that had been reacted with H_3PO_4, H_3PO_4-NaH_2PO_4 mixtures, or NaH_2PO_4 were leached sequentially with 400 ml of deionized water and then with $0.5M$ selenite or arsenate solutions prepared by neutralizing the respective acids with NaOH to equal the pH of the soil-phosphate mixtures (2.4, 4.4, and 6.3, respectively). Leaching rate was 0.5–0.7 ml/min.

Selenite and arsenate were chosen as eluting anions because Hingston et al. (10) found $HSeO_3$ to be specifically adsorbed by goethite, and Dean and Rubins et al. (5) had shown arsenate to be effective in displacing phosphate from soil.

Results are graphed in Fig. 2 as mmole P remaining adsorbed against volume of leachate. Allowance was made for P occluded in solution. Water leaching removed half of the P adsorbed from NaH_2PO_4, and an equimolar amount of Na. In contrast, only 21% and 17% of the bound P was removed from the samples reacted with H_3PO_4-NaH_2PO_4 or H_3PO_4. Amounts of P remaining in the soil after the phosphate concentrations of water leachates fell to very low values were 24, 54, and 110 mmole/100 g for the samples reacted at pH 6.3, 4.4, and 2.4, respectively.

The amounts of water-insoluble phosphate removed by $0.5M$ selenite or arsenate solutions were substantial, suggesting that anion-exchangeable phosphate may have been present in each of the three samples. The amount of "exchangeable" phosphate ranged between 14 and 42 mmole/100 g, with arsenate being a more effective displacing agent than selenite. However, the results of the displacement experiment cannot be interpreted quantitatively in terms of a characteristic amount of anion-exchangeable P or a P adsorption capacity. Hingston et al. (9, 10) regard phosphate desorption induced by other specifically adsorbed anions as hydrolysis rather than ion-exchange. Half-molar sulfate and acetate solutions were ineffective in displacing P remaining after a water wash. Sodium bicarbonate (17) removed around half the soil P that remained after leaching with selenite or arsenate.

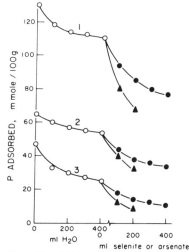

Fig. 2—Desorption of P from 2-g San Gregorio B horizon by water (open circles), $0.5M$ selenite (closed circles) and $0.5M$ arsenate (triangles). Phosphorus treatments as follows (2-g soil in 20 ml): (1) $0.25M$ H_3PO_4, pH = 2.4; (2) $0.05M$ NaH_2PO_4-$0.06M$ H_3PO_4, pH = 4.4; (3) $0.1M$ NaH_2PO_4, pH = 6.3.

Reaction Mechanisms

Although phosphate retention by allophanic soils is extremely complex, it is possible to suggest three separate reactions that are in accord with the present data and with earlier discussions of phosphate fixation by soils and minerals (9, 12, 14, 15, 16).

1) Precipitation of variscite- or strengite-like compounds. This reaction undoubtedly occurred in the low-pH (2.3–3.0) high-P solutions from which Dystrandepts removed up to 130 mmole P/100 g soil within 1 hour. The solubility product of variscite was exceeded in the reaction mixtures, and less than 1/3 of the bound P could be displaced by selenite or arsenate.

2) Specific, inner sphere, adsorption of H_2PO_4 on a protonated surface as suggested by Hingston et al. (9). Since one phosphate ion reacts with one site, capacity for specific adsorption should equal the nonspecific anion adsorption capacity measured with Cl (3). Phosphate adsorbed according to this reaction is tightly bound as compared with HSO_4 and Cl, but can be released by reaction of the soil with other specifically adsorbed anions (5, 16).

3) Adsorption of phosphate ions accompanied by Na, possibly through one of the following reactions: (a) Specific adsorption on a nonprotonated surface through donation of a proton of the adsorbed anion [Hingston, et al. (9)], or (b) Hydroxyl displacement by H_2PO_4 and subsequent reaction of Na with weakly acid cation-exchange material as proposed for sulfate adsorption by Chang and Thomas (4).

Reactions 3a and 3b can account for the mole-for-mole correspondence between P and Na adsorbed from NaH_2PO_4 by the San Gregorio soil, and for the ready desorption by water of large fractions of the coadsorbed P and Na.

Our results with allophanic soil differ from those of Muljadi, et al. (15) with synthetic aluminum hydroxide in a manner suggestive of a reaction similar to 3b. They found coadsorption of P and K from KH_2PO_4 solutions in the pH range 3–5 only at high solution concentrations. Gibbsite and pseudoboehmite adsorbed 40 and 140 mmole P/100 g, respectively, before any K was coadsorbed. In contrast, essentially all of the P adsorbed from NaH_2PO_4 by the allophanic San Gregorio soil was accompanied by a nearly equimolar amount of Na. On the other hand, it is difficult to imagine the allophanic San Gregorio soil as having a CEC of 19 + meq/100 g at pH 4.6, which it would need to have if all of the Na taken up was on exchange sites that existed before reaction with phosphate.

LITERATURE CITED

1. Bar-Yosef, B., U. Kafkafi, and N. Lahav. 1969. Relationships among adsorbed phosphate, silica, and hydroxyl during drying and rewetting of kaolinite suspensions. Soil Sci. Soc. Amer. Proc. 33:672–677.
2. Birell, K. S. 1961. Ion fixation by allophane. New Zeal. J. Sci. 4:393–414.
3. Butler, J. N. 1964. Ionic equilibrium, a mathematical approach. Addison-Wesley Publ. Co., Reading, Mass.
4. Chang, M. L., and G. W. Thomas. 1963. A suggested mechanism for sulfate adsorption by soils. Soil Sci. Soc. Amer. Proc. 27:281–283.
5. Dean, L. A., and E. J. Rubins. 1947. Anion exchange in soils. I. Exchangeable phosphorus and the anion exchange capacity. Soil Sci. 63:377–387.
6. Fassbender, H. W. 1968. Phosphate retention and its different chemical forms under laboratory conditions for 14 Costa Rica soils. Agrochimica 6:512–521.
7. Gebhardt, H., and N. T. Coleman. 1974. Anion adsorption by allophanic tropical soils: I. Chloride adsorption. Soil Sci. Soc. Amer. Proc. 38:255–259 (this issue).
8. Gebhardt, H., and N. T. Coleman. 1974. Anion adsorption by allophanic tropical soils: II. Sulfate adsorption. Soil Sci. Soc. Amer. Proc. 38:259–262 (this issue).
9. Hingston, F. J., R. J. Atkinson, A. M. Posner, and J. R. Quirk. 1967. Specific adsorption of anions. Nature 215:1459–1461.
10. Hingston, F. J., A. M. Posner, and J. P. Quirk. 1968. Adsorption of selenite on goethite. Advances in Chemistry Series 79, 87–90. (Amer. Chem. Soc. Publications).
11. Jackson, M. L. 1958. Soil chemical analysis. Prentice-Hall, Englewood Cliffs, N.J.
12. Kafkafi, U. 1968. Hydrogen consumption and silica release during initial stages of phosphate adsorption on kaolinite at constant pH. Israel J. Chem. 6:367–375.
13. Kingo, T., and P. F. Pratt. 1971. Nitrate adsorption. II. In competition with chloride, sulfate, and phosphate. Soil Sci. Soc. Amer. Proc. 35:725–728.
14. Kittrick, J. A., and M. L. Jackson. 1956. Electron-microscope observation of the reaction of phosphate with minerals, leading to a unified theory of phosphate fixation in soils. J. Soil Sci. 7:81–89.
15. Muljadi, D., A. M. Posner, and J. P. Quirk. 1966. The mechanism of phosphate adsorption by kaolinite, gibbsite, and pseudoboehmite, I–III. J. Soil Sci. 17:212–247.
16. Nagarajah, S., A. M. Posner, and J. P. Quirk. 1968. Desorption of phosphate from kaolinite by citrate and bicarbonate. Soil Sci. Soc. Amer. Proc. 32:507–510.
17. Olsen, S. R., C. V. Cole, S. R. Watanabe, and L. A. Dean. 1954. Estimation of available phosphorus in soils by extraction with sodium bicarbonate. USDA Circ. no. 939.
18. Wada, K. 1959. Reactions of phosphate with allophane and halloysite. Soil Sci. 87:325–330.

16

Copyright © 1981 by the Tropical Agriculture Research Center
Reprinted from *Japan Agric. Research Quart. (JARQ)* **15**(1):14–21 (1981)

Phosphorus Status of Some Andosols in Japan

By YOJI AMANO

Department of Soils and Fertilizers, National Institute of Agricultural Sciences

The belt around the Pacific Ocean is called the Circum-Pacific Volcanic Zone, or "Fire Ring," because of the existence of numerous active volcanoes. Japan is at the northwestern part of the ring. The Japan Islands Arc holds only one-thousandth of the earch surface, and sends out 7×10^{23} erg per year, or one-sixth of energy due to volcanic activity of the earth. Volcanic ejecta are spread throughout Japan. The ejecta alter to form Andosols, which occupy one-fourth of our arable land. Accordingly researches for Andosols are very important.

Distribution of Andosols in Japan

Andosols are mainly spread over Hokkaido, Tohoku, Kanto, Kyushu, and part of Chugoku District, as shown in Fig. 1. Their classification and distribution are set out in Table 1. According to humus content and thickness of humus horizons, they are divided into five subgroups, i.e., Thick High Humic, Thick Humic, High Humic, Humic, and Light-colored. The Light-colored Andosols include immatured ones, truncated ones, and exceedingly matured ones, all of which may be intergrades to other soil groups.

Distribution pattern of tephra* as parent materials of Andosols

Volcanoes eject pyroclastics usually upto the

---
* The term, "tephra," has been widely used since S. Thorarinsson proposed it as a collective term for all pyroclastics transported from the crater through the air, including both air-fall and flow pyroclastic materials. It originally means ashes in Greek.

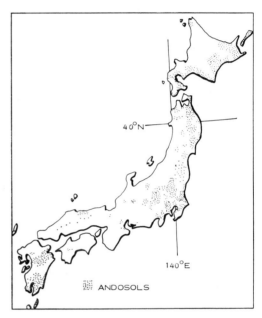

Fig. 1. Distribution of Andosols in Japan

Table 1. Distribution of Andosols in Japan

Soils	Total area	% of arable land
	ha	%
Andosols (Andepts)	954,279	18.8
Wet Andosols (Aquic Andepts)	348,820	6.9
Gleyed Andosols (Andaquepts)	52,610	1.0
Total arable land	5,084,065	100.0

stratosphere at the altitude between 12,000 and 18,000 m at the time of eruption. The pyroclastics extend by the westerly jet stream in the mid-latitude of the northern hemisphere. The isopachs, or thickness contour lines of tephra layers are commonly ellipses with major axes extending to the east of craters. Some examples in Hokkaido are shown in Fig. 2.

The tephra thin away from their eruption

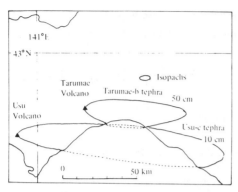

Fig. 2. Distributions of some tephra in Hokkaido

sources. Organic matter accumulates during the intervals between eruptions. Fig. 3 shows a schema of tephra stratigraphy. Andosols, extending wide to Kanto Plain in the central Japan, show that the stratigraphy abovementioned is not only a schema but an actual form (Fig. 4).

Plate 1. A profile of Andosol derived from tephra of Mt. Mashu

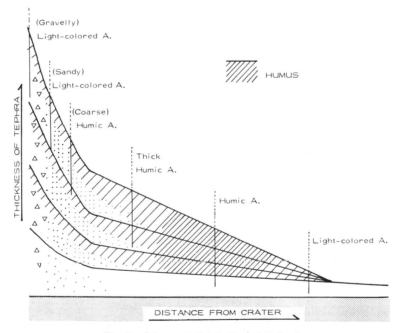

Fig. 3. Schema of distribution of Andosol

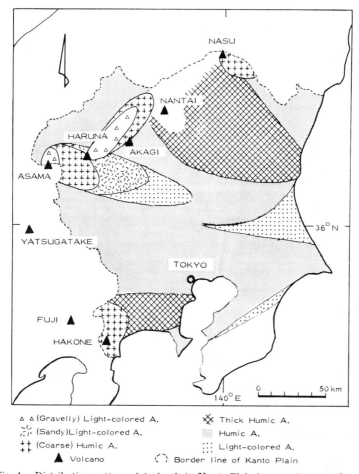

Fig. 4. Distribution pattern of Andosols in Kanto Plain in central part of Japan

Genesis of Andosols

Weathering of tephra is so rapid that differentiation of horizon occurs at an early stage, as —(A)/C—A/C—A/(B)/C—A/B/C. Accumulation of humus, allophanic clay, moisture retention, and phosphate fixation capacity increase simultaneously with weathering. At the latest stage of Andosol development decline the phosphate fixation capacity and humus accumulation.

However, rejuvenation of soil by tephra addition is usual and is a very important fact for soil formation in volacanic zone (Plate 1).

Andosols, including buried ones, show wide variations in the properties according to their ages after deposition. They cannot be defined only with such a diagnostic subsurface horizon as Argillic or Spodic horizon.

Fertility of Andosols

Farmers in Japan had regarded Gray and Brown Lowland Soils as the most productive and Andosol as one of the lowest-productive until the nineteenth century, although farmers in other countries might not always have regarded Andosols infertile.

Since commercial phosphate ferilizers applied

to Andosols at the beginning of the twentieth century, agricultural production from Andosols has markedly increased in Japan. The most important problem in fertility of Andosols has been considered to be phosphate supply. Many workers have studied phosphorus for higher crop production to find its available form, its determination, and physico-chemical equilibrium between soil and phosphorus, especially fixation of phosphorus by Andosols. However, few workers seem to have dealt with the fate of phosphorus in connection with soil genesis.

Consumption of large amount of phosphorus compounds is giving impact at present to the environment in some cases in Japan. For the solution of these problems investigations are necessary in relation to the behavior of phosphorus in ecosystem.

Phosphorus in Andosols

1) *Phosphorus in ecosystem*

In addition to native phosphorus derived from parent materials, soils receive phosphorus from aerosols, plant remains, animal wastes, manures and fertilizers. On the other hand, soils lose phosphorus in solution in ground water, erosion, crop production and so on.

Behavior of phosphorus under natural conditions may be investigated by choosing appropriate sites for study. Effect of manures and fertilizers may be removed in uncultivated soils, and the influence of erosion may be neglected at flat sites. Andosols may lose few amount of phosphorus for solution in water because of their high capacities of phosphorus fixation.

2) *Phosphate fixation*

Phosphate fixation is also called phosphate absorption, sorption or retention.

For the determination of phosphate sorption, a method by ammonium phosphate, 2.5% $(NH_4)_2HPO_4$ at pH=7.0, has long been used in Japan. Phosphate sorptions of Andosols by the method are usually higher than 1,500 mg P_2O_5/100 g soil.

Figure 5 shows a comparison of the traditional method with Blakemore's method which

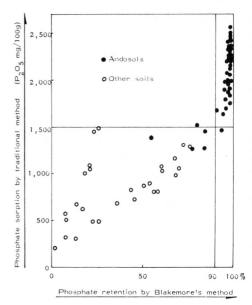

Fig. 5. Phosphate sorption of Andosols (A comparison of the traditional method of Japan with Blakemore's method)

was proposed of late. Phosphate retentions of Andosols by his method are mostly over 90%.

Status of phosphorus

1) *Well-drained Andosols*

Soil phosphorus is divided at first into three fractions; HCl-soluble inorganic phosphorus (IP), HCl-soluble organic phosphorus (OP), and hardly-soluble phosphorus (HP).**

The upper part of Fig. 6 shows an example of Light-colored Andosol corresponding to Typic Vitrandept of the U.S. System, well-drained and derived from basaltic tephra of Mt. Fuji. The surface layer was deposited in 1707 A.D., and the third layer in 1910 B.P.*** Total phosphorus (TP) contents of the soils are relatively high, some of which are more than 2,000 ppm P. TP contents are higher in humic horizons than in low-humic ones. HP is contrastive to TP. OP contents are dominantly high in A horizons and low in C horizons.

** HP=TP−(OP+IP)
*** B.P.=years before 1950 A.D., determined by radiocarbon dating.

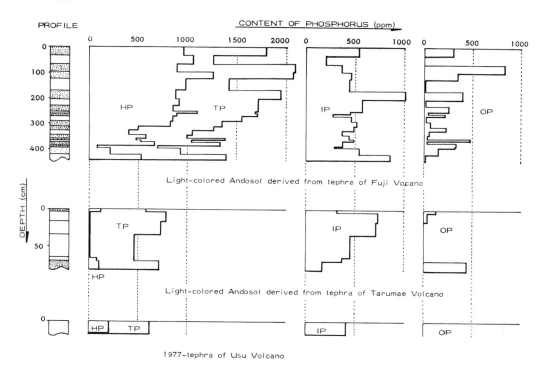

Fig. 6. Phosphorus distributions in well-drained Andosols

In the lower part of Fig. 6 are examples of Light-colored Andosol, well-drained and derived from andesitic tephra of Mt. Tarumae, and dacitic ones of Mt. Usu (Typic Vitrandept). The surface layer, so-called Tarumae-a tephra, was deposited in 1739 A.D. The second layer, so-called Tarumae-c tephra, is estimated to have deposited in 1,100—1,640 B.P. by radiocarbon dating. Usu tephra were sampled on the next day of eruption in 1977. Tarumae tephra have lower contents of TP than Fuji tephra and higher than Usu tephra. In general, the higher the content of silicon in tephra is, the lower is the content of TP. OP content of Usu tephra is zero. Tarumae tephra have markedly high OP in A hiorzons, especially in buried IIA horizon. The accumulation of OP seems to be pararell to the periods of weathering at the earth surface, i.e., appreciable amount of phosphorus seems to be retained in the organic fraction during the intervals between eruptions.

2) *Andosols under different water regime*

Poorly-drained Andosols have higher accumulation of organic matter than well-drained ones. Soils derived from the same tephra should have different distribution of phosphorus under different water regime. Phosphorus distributions were examined in a sequence of Andosols from the same tephra with different drainage, or a *hydrocatena*.

As shown in Fig. 7, TP contents are higher in poorly-drained soils than in well-drained ones. IP shows limited variation in each soil profile, and between profiles of different drainage. On the contrary, OP contents are extremely high in poorly-drained soils. The higher values of TP are related to the accumulation of OP.

In Table 2 is set out the result of subdivision of IP, comparing Light-colored Andosol (Typic Dystrandept) with Thick High Humic Wet Andosol (Aquic Dystrandept) sampled at Hokkaido. The specimen A represents the A horizon of Tarumae-b tephra deposited in 1667 A.D., and the IIA represents buried A horizon

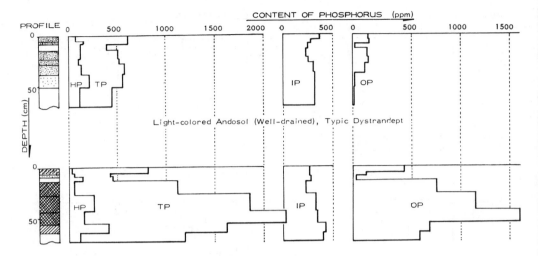

Fig. 7. Phosphorus distributions in Andosols with different drainage

Table 2. Subdivision of inorganic phosphorus in two Andosols

Sample	$M/2$ NH_4Cl	$M/2$ NH_4F	$N/10$ NaOH	(DCB)* N-NaOH	$N/2$ HCl	N HCl	Residual Org.	Residual Inorg.
				ppm P				
Light-colored Andosol (Well drained)								
A	31	76	33	11	110	5	34	58
C	1	1	2	5	128	4	3	22
IIA	1	56	13	14	20	4	36	68
Thick High Humic Wet Andosol (Somewhat poorly drained)								
A	22	90	45	8	108	3	55	91
C	1	6	2	6	141	4	6	21
IIA	1	53	49	15	31	3	68	70

* Dithionite-Citrate-Bicarbonate treatment

of Tarumae-c tephra. The period of soil development of the IIA is about 1,000 years longer than that of the A.

Small amount of easily-soluble P extracted by NH_4Cl can be found in surface A, but very few of them in buried IIA. The Ca-bound P extracted by HCl is high in the least weathered C, and low in IIA. The Al-bound P extracted by NH_4F and the Fe-bound P extracted by NaOH increase at an early stage of weathering and, then, gradually decline with time. The occluded P extracted by dithionite-citrate-bicarbonate and subsequent NaOH treatment gradually increase with increasing degree of soil development.

Organic phosphorus as a marker of Andosol age

Organic phosphorus in Andosols accumulates proportionally with increasing time upto about 8,000 B.P. Consequently, it is quite reasonable to postulate that in case when tephra are younger than 8,000 B.P., and have had no influence of erosion and heavy application of organic matter, their age can be estimated from the following equation, based on organic phos-

Table 3. Tephra ages estimated by the proposed equation

Tephra	Soil	Land use	Tephra age estimated by	
			Organic P	Radiocarbon
			y.B.P.	
Fuji	Light-colored A.	Deciduous forest	1,700	1,910±110
Tarumae-d	Light-colored A.	Deciduous forest	3,000	3,580±100
	High-Humic Wet A.	Deciduous forest	3,200	3,580±100
Daisen	Thick High Humic A.	Coniferous forest	2,760	2,490~3,200
	Thick High Humic A.	Vegetable field	3,060	2,490~3,200

phorus accumulation *in situ* of standard tephra of given age.

$$\frac{\sum_{i=1}^{c} b_i d_i p_i}{T} = \frac{\sum_{i=1}^{x} b_i d_i p_i}{y}$$

Therefore,

$$y = \frac{T}{\sum_{i=1}^{c} b_i d_i p_i} \sum_{i=1}^{x} b_i d_i p_i$$

where, y=unknown age of any tephra; T=given age of standard tephra; b=soil bulk density of each horizon; d=thickness of each horison; p=soil organic phosphorus content of each horizon; c=cth horizon (horizon No. of the standard tephra. The surface horizon counts 1st); x=xth horizon (horizon No. of tephra of unknown age); bdp is organic phosphorus accumulation per unit area of any horizon, and (the sum of bdp)/T is the accumulation rate of organic phosphorus per year *in situ*.

Some results obtained by above-mentioned procedure are set out in Table 3.

Phosphorus is an important element in soil fertility. Although large amount of consumption of phosphorus causes some environmental problems, its resources are poor in Japan. Phosphorus should be applied according to soil properties.

There is no epilogue in research on long-term behavior of phosphorus. The conversion rates of various forms of phosphorus might largely vary in other countries under the climate different from Japan.

References

1) Amano, Y.: Characteristics of clays in Andosols. Annual Report of the Third Division of Soils for 1977, National Institute of Agricultural Sciences, 71-81 (1978); ibid., for 1978, 59-63 (1979); ibid., for 1979, 73-80 (1980). [In Japanese].
2) Baker, R. T.: Changes in the chemical nature of soil organic phosphorus during pedogenesis. *J. Soil Sci.*, 27, 504-512 (1976).
3) Blakemore, L. C. et al.: Methods for chemical analysis. New Zealand Soil Bureau Scientific Report. (1977).
4) Egawa, T. & Sekiya, K.: Studies on the fractionations of soil phosphorus. Report of the Second Laboratory of Soil Chemistry, National Institute of Agricultural Sciences, 39-58 (1959) [In Japanese].
5) Hanada, S.: Studies on the sorption and desorption of phosphorus with special reference to organo-mineral complexes in volcanic ash soils. *Bull. Fac. Agr. Hirosaki Univ.*, 21, 102-184 (1973).
6) Hong, C. H. & Yamane, I.: A comparison of three methods used for determining total organic phosphorus in soil. *J. Sci. Soil and Manure, Japan*, 46, 185-191 (1975). [In Japanese with English Summary].
7) Legg, J. O. & Black, C. A.: Determination of organic phosphorus in soils: II. Ignition method. *Soil Sci. Soc. Am. Proc.*, 19, 139-143 (1955).
8) Matsuzaka, Y.: Major soil groups in Japan. Proc. of the International Seminar on Soil Environment and Fertility Management in Intensive Agriculture, 89-95 (1977).
9) Ministry of Agriculture, Forestry and Fisheries, Agricultural Production Bureau: General Features and management of soils in arable land of Japan, pp. 279 (1979). [In Japanese].
10) Miyazawa, K.: Response of crops to the levels of phosphorus accumulated in the surface layers derived from Andosols in Tokachi district. *Hokkaido Natl. Agr. Exp. Sta., Res. Bull.*, 126,

1-30 (1980). [In Japanese with English Summary].
11) Mizota, C.: Phosphate fixation by Ando soils different in their clay mineral composition. *Soil Sci. Plant Nutr.*, **23**, 311-318 (1977).
12) National Land Agency: Nature and land use of Japan. III. Kanto District, pp. 170 (1979). [In Japanese].
13) Onikura, Y.: An information on the behavior of phosphate in volcanic-ash soils. *Kyushu Natl. Agr. Exp. Sta., Bull.*, **9**, 51-68 (1963). [In Japanese with English summary].
14) Seki, T.: The chemical and mineralogical investigations on the infertile volcanogenous soils of the southern parts of the Prov. Shinano. *J. Agr. Chem. Soc., Japan*, **1**, 253-269 (1925). [In Japanese].
15) Seo, H. et al.: Soil survey report No. **18**, pp. 294. *Hokkaido Natl. Agr. Exp. Sta.*, (1968). [In Japanese].
16) The Third Division of Soils, National Institute of Agricultural Sciences: Soil series and their differentiae for arable land: The 2nd approximation, pp. 67 (1977). [In Japanese].
17) Wada, H.: Passage of phosphorus through ecosystems. *Chemistry and Chemical Industry*, **31**, 961-964 (1978). [In Japanese].
18) Williams, J. D. H., Syers, J. K. & Walker, T. W.: Fractionation of soil inorganic phosphate by a modification of Chang and Jackson's procedure. *Soil Sci. Soc. Am. Proc.*, **31**, 736-739 (1967).
19) Yoshida, M. & Miyauchi, N.: Method for determining phosphate fixation capacity of soils in acidic conditions. *J. Sci. Soil and Manure, Japan*, **46**, 89-93 (1975). [In Japanese with English summary].

(Received for publication, October 25, 1980)

17

Copyright © 1977 by Centro Internacional de Mejoramiento de Maíz y Trigo
Reprinted from pages 10-63 of *Soils Derived from Volcanic Ash in Japan,*
Y. Ishizuka and C. A. Black, eds., Centro Internacional de Mejoramiento de Maíz y
Trigo, Mexico City, 1977, 102p.

PROPERTIES OF SOILS DERIVED FROM VOLCANIC ASH

by

Tomoji Egawa

	Page
Introduction	11
Mineralogical Composition	12
Primary minerals	12
Weathering of volcanic materials and formation of clay minerals	14
Properties of allophane and allophanic clay	19
Organic Matter	28
Accumulation of organic matter	28
Nature and properties of organic matter	28
Physical Properties	33
Phase relationships	33
Structure	35
Water relations	37
Aeration	40
Chemical Properties	41
Reaction and base status	41
Cation-holding power and cation-exchange capacity	43
Phosphorus sorption	46
Biological Properties	51
Microflora	51
Influence of environmental factors on microbial population	54
Microbial activities	56
Literature Cited	59

INTRODUCTION

This chapter describes the mineralogical, physical, chemical, and biological properties of the dark colored soils derived from volcanic ash in Japan. Needless to say, the soil properties are a result of the nature and behavior of the soil components. Consequently, considerable space is devoted to these subjects, with due regard for recent progress made in Japan.

In this chapter, the term, "soils of the ando group," will be used as a collective designation of dark colored soils derived from volcanic ash. The nomenclature of these soils, however, is still debated among Japanese pedologists. Dark colored soils derived from volcanic ash have been designated in Japan by a variety of names, including the following: grassland brown earth (Seki, 1934), ando soils (Thorp and Smith, 1949), black soils (Soil Survey Staff, 1952) prairie-like brown forest soils (Kamoshita, 1958), Japanese volcanic ash soils (Kanno, 1956), humic allophane soils (Kanno, 1961), humus allophane soils (Miyazawa, 1962), and kuroboku soils (Matsui, Kurobe, and Kato, 1963). The term, "volcanic ash soils," is often used in Japan to designate soils in general that have been derived from volcanic ash.

Soils of the ando group occur under a wide range of climatic conditions from cold subalpine regions to humid equatorial tropics in various parts of the world (Wright, 1964). Environmental differences are often reflected in relatively minor variations in profile morphology, and different names are given to such soils in each country.

The common profile of soils of the ando group consists of a thick, loose A horizon with a high content of organic matter and a brownish structural B to BC horizon. In younger soils, the B horizon is often absent. Morphological features of podzolization and laterization do not appear. The "normal" profile is often disturbed by denudation, accumulation of surface soil materials, or addition of recent pyroclastic deposits, as described in chapter 1.

Most mature soils of the ando group are so highly weathered that the clay content often amounts to about one-half of the total soil mass. The mineral fraction of younger soils may consist mainly of sand and gravel. Chemical analysis shows that the mature soils have undergone severe weathering, resulting in marked loss of silica and bases and in the formation of clay rich in hydrous sesquioxides. The principal silicate mineral of the clay fraction is allophane, with some related minerals such as imogolite, halloysite, and some 14 angstrom minerals. Most of the special properties of soils of the ando group are a consequence of the properties of allophane, as will be explained in subsequent paragraphs.

The characteristic, thick A horizon contains large quantities of organic matter and has an appearance similar to that of the A horizon of a chernozem soil. The principal cause of the extraordinary accumulation of organic matter in soils of the ando group is thought to be the comparatively stable combination of organic matter with allophane. Soils with a high content of allophane characteristically contain more organic matter than do soils developed under similar conditions but without the allophane. The relatively high water-holding capacity of the soils after the ash has become sufficiently weathered to have a high content of clay is perhaps an additional factor of significance in the accumulation of organic matter.

The solid phase of the soils occupies only 20 to 30% of the soil volume. Consequently, the bulk density is low and the porosity and water holding capacity are high. These properties explain in part why soils of the ando group weather rapidly, why wind erosion and sheet erosion are important problems, and why plants grown on them are subject to frost damage.

The cation-exchange capacity of the soils at pH 7 is relatively high, but the effective cation-exchange capacity is considerably lower because the soils are generally acid and the cation-exchange capacity is highly pH-dependent. The cation-exchange sites show preferential sorption of calcium and magnesium over ammonium and potassium, presumably because of a predominance of organic over inorganic exchange sites.

Shiori (1934) used the term, "allitic," to describe soils of the ando group in Japan. An allitic soil, according to Shiori, is one in which aluminum in the clay fraction is readily activated when the soil becomes acid. Allophane in the clay fraction is now known to be the essential component that endows soils of the ando group with their allitic property. An important consequence of the allitic property is the strong binding of anions, notably phosphate.

From the viewpoint of soil productivity, most soils of the ando group in Japan are considered inferior. The allitic character of the soils is of primary concern. Amelioration of such soils involves a combination of suitable techniques, including liming and heavy applications of phosphate and compost. Although most cultivated soils have been improved, the low productivity due to the allitic character shows clearly in newly reclaimed fields. Even in cultivated soils that have been properly reclaimed, continual vigilance is needed to prevent reappearance of the allitic character. Because of the heavy rainfall in Japan, the exchangeable bases in the soils are readily lost by leaching, with the consequence that the soils become acid in a few years and the aluminum is reactivated.

This introduction provides a preview of the subject matter of the remainder of the chapter. The sections that follow will treat the various properties in more detail.

MINERALOGICAL COMPOSITION
Primary minerals

Mineralogical composition. Volcanic activity in Cenozoic Japan was fairly strong, with the consequence that soils occupying more than half of the land area have at least some admixture of volcanic ash. The ash occurs not only in soils under forest but also in cultivated soils of the uplands and rice fields. The volcanic ash content is one of the distinctive features of soils of Japan.

In the Japanese Islands, seven volcanic zones have been distinguished. The main rock types in each of these zones are as follows: (a) Chishima volcanic zone: pyroxene andesite. (b) Nasu volcanic zone: pyroxene andesite, pyroxene dacite, and, in part, hornblende andesite. (c) Chokai volcanic zone: hornblende andesite, dacite. (d) Fuji volcanic zone: hornblende andesite, dacite, and, in part, basalt. (e) Norikura volcanic zone: biotite dacite, biotite or hornblende andesite. (f) Daisen volcanic zone: alkali rocks. (g) Ryukyu volcanic zone: pyroxene andesite, dacite. Most of the recent volcanic rocks are andesitic. Basaltic or rhyolitic rocks are next in order of abundance.

The available data on the mineralogical composition of soils derived from volcanic ash in Japan seem to reflect the abundance of andesite (Kidachi, 1964). Some data are presented in Table 1. Generally speaking, the fine sand fraction of soils of the ando group includes light minerals such as quartz, plagioclase, biotite, pumice, volcanic glass fragments, and plant opal as well as heavy minerals such as olivine, pyroxene, hornblende, magnetite, and some opaque minerals. These minerals occur in certain associations, the nature of which is affected by the composition of the original volcanic ash and by the degree of weathering. Based on the microscopic examination of the fine sand fraction (0.2-0.02 mm effective diameter) separated from more the 20 soil profiles, Kanno (1961) identified the following four mineral associations in soils of the ando group in Japan: (a) Olivine association, in which a relatively high content of olivine is observed. (b) Volcanic glass and plant-opal association, in which acid glasses (n=1.50) are markedly abundant in rhyolitic ash sublayers and are accompanied by abundant plant opal (n=1.45) in organic-matter-rich surface layers derived from pyroxene-andesitic ash. (c) Hornblende association, in which hornblende predominates among heavy minerals. (d) Pyroxene association, in which pyroxene predominates among heavy minerals.

Because these four mineral associations were connected with the color of the B horizon, Kanno divided soils of the ando group into three genera, namely, (a) the brown genus (soils derived from basaltic ash), (b) the onji genus (glassy or rhyolitic ash layers below the surface horizon), and (c) the light yellowish brown genus (soils derived from biotite-hornblende andesitic ash).

Kobo and Ohba (1963) divided Japanese soils of the ando group into the following four subgroups on the basis of the relative contents of olivine, hornblende, and pyroxene in heavy minerals found in the fine sand fraction (0.02-0.05 mm effective diameter): (a) Olivine subgroup, in which the ratio of olivine to pyroxene is more than 1 to 10. (b) Pyroxene subgroup, in which both the ratio of olivine to pyroxene and the ratio of hornblende to pyroxene are less than 1 to 10. (c) Pyroxene-hornblende subgroup, in which the ratio of hornblende to pyroxene is between 1 to 10 and 1 to 1. (d) Hornblende subgroup, in which the hornblende content exceeds the pyroxene content.

The distribution of each subgroup is locally restricted because the dominant minerals are inherited from the parent volcanic ash. The olivine subgroup is restricted to the southern part of the Kanto district, the area near Mt. Iwate in the Tohoku district, and the area near Mt. Aso and Mt. Kaimon in the Kyushu district. The pyroxene subgroup is the one that is most widely distributed. The pyroxene-hornblende subgroup is common in the area along the Japanese Sea in the Tohoku district and in the northern part of the Kanto district. The hornblende subgroup is distributed in the San-in district and in the area along the Japanese Sea in the Tohoku district.

Degree of weathering. The degree of weathering of soils derived from volcanic ash in Japan varies with the time since deposition and with the weathering conditions, and it has a great influence upon the properties of the soils. Kobo and Ohba (1964) divided each of the four mineralogical subgroups described in the preceding section into four stages of weathering. The criteria used for the degree of weathering were the etching of pyroxene and the clay content. Relationships among parent material, degree of weathering, and soil properties will be described in later paragraphs.

Volcanic glass. Volcanic glass plays an important role as the origin of allophane and imogolite, the nature and properties of which will be described in the section on weathering of volcanic materials and formation of clay minerals. The fact that the content of volcanic glass decreases with increasing weathering indicates that the glass is unstable.

TABLE 1. Mineralogical Composition of the Fine Sand Fraction of Some Soils Derived from Volcanic Ash in Japan (Kanno, 1961).

Percentage of mineral grains of 0.2 to 0.02 mm diameter identified in each of the indicated mineral groups, soils, and horizons

Mineral group	Soil derived from basaltic volcanic ash from Koganei (Tokyo)				Soil derived from pyroxene-andesitic volcanic ash from Utsonomiya (Tochigi)				Soil derived from amphibole-andesitic volcanic ash from Daisen (Tottori)				Soil derived from pyroxene-andesitic volcanic ash with rhyolite sublayer from Kodonbaru (Kumamoto)		
	0-25 cm (A_{11})	25-50 cm (A_{12})	50-90 cm (B_1)	90-150 cm (B_2C)	0-23 cm (A_{11})	23-53 cm (A_{12})	53-68 cm (B_1)	68-100 cm (B_2)	0-10 cm (A_{11})	10-35 cm (A_{12})	35-60 cm (B_1)	60-90 cm (B_2C)	0-30 cm (A)	30-55 cm (D)	55-65 cm (E)
Olivine	3	2	2	1											2
Pyroxene	16	11	17	12	15	16	14	16	1	3	1	3	1	2	2
Amphibole	3	1	2	1	6	3	1	1	5	7	14	14			
Magnetite	8	6	8	11	8	5	6	3	2	3	1	3			
Plagioclase	22	33	28	30	39	20	39	35	25	29	26	26	7	5	53
Weathered micas									7	7	7	4			
Quartz					3	4	1	3	4	1	2	3			
Volcanic glass[1]	5	11	15	17	13	20	19	25	13	7	12	18	4	83	28
Plant opal[2]	30	20	6	10	8	29	13	11	36	31	31	19	88	4	6
Other minerals	13	16	22	18	8	3	7	6	7	12	6	10		6	11
Heavy minerals[3]	34	24	36	28	34	25	24	25	8	14	17	21	1	2	2

[1] Index of refraction = 1.50 to 1.55.
[2] Index of refraction = 1.45.
[3] Specific gravity >2.8. These values are per cent by weight.

Plant opal. Kanno and Arimura (1954, 1955, 1957a,b, 1958) were the first to find amorphous silica particles in soils of the ando group, especially those of the "onji" genus, which are distributed widely in the southern part of the Kyushu district. They thought originally that these particles originated from volcanic activity, but eventually the particles were proved to be plant opal (Kato, 1960; Kanno, 1961). The source of plant opal ($n=1.45$) found in the soil is considered to be the leaves, stems, and rootlets of certain grass species, such as **Miscanthus, Imperata, Zoysia,** and **Sasa.** These plant species are also a source of the soil organic matter. The stabilization of silica as plant opal may have an important influence on the silica cycle in the soils.

Weathering of volcanic materials and formation of clay minerals

The clay minerals of soils of the ando group have attracted great interest on the part of Japanese soil scientists. From the results of chemical analysis, Seki (1913) inferred the presence of allophane, the nature of which was somewhat ambiguous at that time. Shioiri (1934) also inferred the presence of allophane in soils of the ando group from optical and chemical studies of clay fractions. Kawamura and Funabiki (1936) were the first to apply the X-ray diffraction technique to the study of soil clay minerals in Japan.

Since 1950, rapid progress in clay mineralogy of soils of the ando group has been achieved with the aid of X-ray diffraction, thermal analysis, electron microscopy, infrared absorption spectrometry, and other modern techniques. From the geological and mineralogical viewpoint, researches by Sudo (1953, 1954, 1956, 1959) shed light on the process of formation of allophane and its alteration to halloysite by the weathering of volcanic ash, and this concept was developed by geologists in the clay mineralogical study of "kanto" soils derived from volcanic ash (Tsuchiya and Kurabayashi, 1959; Kurabayashi and Tsuchiya, 1960, 1961, 1965). From the standpoint of soil science, the mineralogical composition of clays of volcanic ash deposits and of soils of the ando group has been investigated intensively by a number of soil scientists, including Aomine and Yoshinaga (1955), Egawa, Watanabe, and Sato (1955), Kuwano and Matsui (1957), Kanno (1959), Kanno, Kuwano, and Honjo (1960), Matsui (1959, 1960), Ishii and Sasaki (1959), Ishii and Kondo (1962a, b), and Ishii (1963).

Much has now been learned of the occurrence and distribution of allophane and related minerals in soils of the ando group. According to Aomine (1958) and Kanno (1961), the location of the various minerals in the weathering sequence seems to be:
Volcanic glass and Feldspar→Allophane→Halloysite and Gibbsite→Metahalloysite.

Within soil profiles, the weathering of individual strata tends to decrease with depth, but the weathering of successive strata increases with depth because of the greater weathering time since deposition. The rate of weathering, however, depends also on local factors. For example, in examination of volcanic materials on the western slope of the central cone of Mt. Aso, in the Kyushu district, Aomine and Wada (1962) observed that weathering had progressed to the halloysite stage in one area but only to the allophane stage in another. As a result of the mode of occurrence in the field and the findings from chemical analyses, the authors concluded that a combination of local differences in leaching and biotic effects may well have been responsible. Chemical analysis of the corresponding portions showed that the weathering process involved loss of silica and bases and that there was essentially no loss of aluminum and iron. The chemical transition from allophane to halloysite was regarded as a continuous process of dehydroxylation through condensation and replacement with silica tetrahedra.

Another significant feature of the weathering of volcanic ash is the formation of gibbsite. Wada and Aomine (1966) studied the occurrences of gibbsite in the weathering of volcanic materials at Kuroishibaru, in the Kyushu district, by means of field observation, chemical analysis, X-ray diffraction, and microscopic techniques. Gibbsite was found as white concretions and coatings in lower sand and gravel layers, where allophane dominated as a weathering product, and also as unsegregated particles of clay size in one of the upper volcanic ash layers. Nearly equal amounts of gibbsite and amorphous iron oxides and a relatively small amount of allophane were produced from this acid andesitic (dacitic) ash. Layer silicates consisted mainly of illite, chlorite, and vermiculite. This investigation showed that gibbsite aggregates will form under temperate and fairly humid climatic conditions, as they do under tropical and subtropical conditions, provided that the texture and petrological nature of the parent volcanic material are suitable.

The mineralogical composition of the clay fraction in soils of the ando group shows a zonality or regionality as affected by climatic conditions under which weathering and clay formation proceed. Miyazawa (1966) found that, in the warm, temperate zone of Japan, a considerable amount of allophane is formed in soils of the ando group by severe chemical weathering. Stable aggregates are produced, probably due to dehydration under seasonally dry conditions. In this case, gibbsite may be derived from allophane and appears often in the surface soil layer. In soils of the ando group in the cooler zone, as in Hokkaido or in the highlands of Honshu, allophane may be produced from volcanic glass. Under these circumstances, however, allophane has high dispersibility and is easily moved downward, to be altered to gibbsite by desilication.

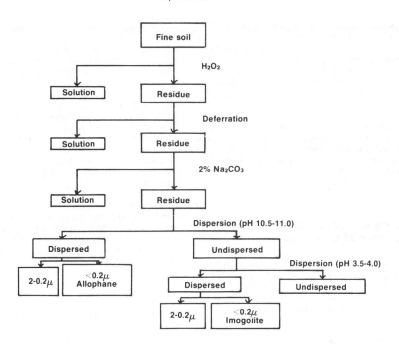

Fig. 1—Flow diagram for separation of allophane and imogolite (Yoshinaga and Aomine, 1962a)

Imogolite. In the course of investigations on soil allophane, a new clay mineral, "imogolite," was discovered in some soils of the ando group in Japan. The name, imogolite, was first used by Yoshinaga and Aomine (1962a, b). "Imogo" means soils derived from glassy volcanic ash. "Imogolite" was approved as a name for the new mineral by the Committee on Nomenclature of the Association Internationale Pour l'Etude des Argiles on the occasion of the International Clay Conference held in Tokyo, Japan, in September, 1969.

Allophane disperses in both acid and alkaline media, whereas imogolite disperses only in acid media. This difference is used in the separation of the two mineral colloids according to the technique shown in Figure 1.

The technique just described facilitates investigation of the properties of imogolite. Imogolite is characterized by several broad but intense X-ray diffraction bands, an endothermic peak at 410 to 430°C in differential thermal analysis, and its appearance under the electron microscope as a thread 100 to 200 angstroms in diameter (Wada, 1967; Yoshinaga, Yotsumoto, and Ibe, 1968; Wada and Yoshinaga, 1969). The results of investigations thus far suggest that imogolite is structurally distinct from allophane and from other known layer and chain minerals.

Although more work is needed to elucidate in detail the mechanism of formation of imogolite, some insight is provided by the results of investigations cited here. Imogolite seems to be a common constituent of weathered pumice and of soils derived from volcanic ash (Miyauchi and Aomine, 1966; Kawasaki and Aomine, 1966; Kanno, Onikura, and Higashi, 1968). Wada and Matsubara (1968) found that, within the coarse pumice grains in the Kitakami pumice bed in Iwate Prefecture, fine granules of allophane with the Si/Al ratio 1/1.17 appeared in the crust concentric to the unweathered core of micron-size "subgrains" of the pumice and that imogolite appeared exclusively as macroscopic gel films covering the surface of the pumic grains or filling their interstices. A few white concretions appeared on the rock fragments, together with some black surface concentrations rich in manganese oxides. The concretions were found to consist of peculiarly wrapped, micronsize rods of gibbsite. These findings suggest that parallel formation of various minerals may occur on a microscopic or submicroscopic scale in the weathering of finely comminuted volcanic ash.

On the basis of field observations and laboratory work on the formation of imogolite and other clay minerals, Yoshinaga (1968, 1970) suggested that imogolite is formed under conditions of severe leaching (desilication) and that it co-exists with allophane or gibbsite in pumice beds, scoria layers, and soils derived from volcanic ash. He suggested also that there may be two weathering sequences, one a parallel formation of allophane, imogolite, and gibbsite from volcanic glasses or feldspars, and the other a sequence, allophane→ imogolite→ gibbsite.

Amorphous silica. It is generally accepted that the early weathering stage of soils derived from volcanic ash is characterized by a relatively abundant supply of bases and silica because of the rapidity of weathering of the ash. Therefore, it is supposed that amorphous clay minerals formed in this stage may be different in chemical composition and properties from those found in highly weathered soils.

On the basis of evidence from differential thermal analysis and infrared spectrophotometry concerning allophane and related minerals in New Zealand soils, Fieldes (1955) suggested the following weathering sequence for clays formed in a single weathering cycle from volcanic ash:

Allophane B→ Allophane AB→ Allophane A→ Metahalloysite.

In allophane B, amorphous silica is considered a discrete component, and clay particles are ultra-fine.

Some researchers (Kanno, 1959; Matsui, 1959; Egawa and Sato, 1960; Ishii and Kondo, 1962a, b) obtained similar evidence on clay minerals in Japanese soils derived from volcanic ash, but Miyauchi and Aomine (1964) raised some questions about the existence of allophane B. They found that the fine clay fraction ($< 0.2\mu$) of six young Japanese soils from volcanic ash had differential thermal analysis patterns with strong exothermic peaks near 900°C and infrared spectrophotometric patterns without absorption maxima near 800 cm^{-1}, thus corresponding to allophane A and geologic allophane. The coarse clay fraction (0.2 to 2μ) of these soils had differential thermal analysis patterns with the 900°C exothermic peak greatly reduced or absent and infrared spectrophotometric patterns showing absorption near 800 cm^{-1}, thus corresponding to the description of allophane B. They found X-ray diffraction evidence of feldspar and cristobalite (SiO$_2$) in the coarse clay fractions and concluded that the exothermic reaction of the whole clay less than 2μ depended on the fine clay content, whereas the infrared absorption at 800 cm^{-1} was due to cristobalite in the coarse fractions. This evidence led them to question the presence of allophane B in soils derived from volcanic ash in Japan.

Fieldes and Furkert (1966) reconsidered this viewpoint for a range of New Zealand soils and confirmed the presence of allophane A in the fine clay fractions (less than 0.2μ) of some of these soils but found insufficient cristobalite in the coarse clay fractions (0.2 to 2μ) to account for the magnitude of the infrared absorption of the whole clay at 800 cm^{-1}. They confirmed the presence of discrete amorphous silica in the coarse clays and in some of the fine clays. They suggested that some effects of amorphous silica could be derived from domain zones of vitreous silica in volcanic glass or from related hydrous silica zones in hydrous glass. According to Furkert and Fieldes (1968), the evidence of the papers by Fieldes and Furkert (1966) and by Miyauchi and Aomine (1964) indicates "that coarse fractions of the clays are likely to include amorphous hydrous derivatives of volcanic glass or feldspars, which fit the definition of allophane as amorphous hydrous aluminosilicate, and which have surface properties like allophane A." From these results, Furkert and Fieldes (1968) proposed "that the term allophane B be retained generically for this kind of material, thus recognizing the possibility of association with crystalline minerals and also the possibility of differences between hydrous derivatives of feldspars, hydrous glasses and vitreous glasses and allowing for the possible presence of discrete amorphous hydrous oxides of aluminum and silicon."

Uchiyama, Masui, and Shoji (1968) found that the amorphous materials of the clay fractions of young soils derived from volcanic ash were highly siliceous. The average molecular ratio of SiO_2 to Al_2O_3 was 10.3 in the clay of soils younger than 600 years and 5.0 in the clay of those 600 to 1,200 years old.

Shoji and Masui (1969a,b,c,d) investigated samples of four soils derived from volcanic ash about 220 years old in Hokkaido, the northernmost island of Japan, and found that the fractions finer than 2μ were rich in amorphous siliceous materials. The amorphous materials were more siliceous in the specimens of the A horizons than in those of the C horizons and became more siliceous with an increase in particle size. Discrete amorphous silica was present in all the size-fractions finer than 2μ, and its amount was relatively greater in 0.4 to 2μ material than in finer fractions. They studied the nature of the discrete amorphous silica by means of microscopy, electron microscopy, infrared absorption spectroscopy, differential dissolution techniques, and so on, and found that there were at least three kinds of amorphous silica, namely, plant opal, opaline silica with unique shapes like discs or ellipsoids, and aggregated amorphous silica. Plant opal was small in amount, and its existence was restricted to the coarse silt and fine sand fractions of the A horizons. Opaline silica was present in the 0.2 to 5μ fractions and was most abundant in the 0.4 to 2μ fractions. There was little aggregated amorphous silica in the clay fractions. Most of the discrete amorphous silica in the whole clay fractions was opaline silica. The size of opaline silica particles calculated from electron micrographs ranged from approximately 0.2μ to 15μ.

Discrete amorphous silica was not detected in any of the volcanic glass contained in the soil samples, fresh volcanic ash and pumice being used for comparison. Opaline silica particles, however, were relatively abundant in many young soils derived from volcanic ash in which the 0.5 **N** NaOH-soluble materials were highly siliceous, but such particles were rarely found in the strongly weathered soils, as seen in Table 2.

On the basis of this information, Shoji and Masui concluded that the disc-shaped and ellipsoid-shaped particles of opaline silica they found were not primary mineral particles but were formed during the initial weathering of volcanic ash, when the environment was rich in bases and highly siliceous. In this stage, the predominant crystalline clay mineral seemed to be montmorillonite, and the clay was highly siliceous. The soils then entered an "aluminous" weathering stage, in which there was intensive formation of allophane and accumulation of aluminum in interlayer positions in montmorillonite.

Montmorillonite and related 14 angstrom minerals. Considerable amounts of montmorillonite and related 14 angstrom minerals have been found in soils derived from volcanic ash. A number of scientists (Masui, 1960, 1966; Masui et al., 1963, 1966; Masui and Shoji, 1967, 1969; Uchiyama et al., 1962) have reported that montmorillonite predominates in the clay fraction of a soil derived from dacitic ash at Kawatabi, Miyagi Prefecture. They found that the major crystalline clay minerals in soils derived from volcanic ash in the Tohoku district are montmorillonite, vermiculite, montmorillonite-vermiculite integrade, and chlorite. The weathering sequence they proposed for volcanic materials in northeastern Japan is as follows:

Volcanic ash→amorphous materials→ montmorillonite→montmorillonite-vermiculite integrade→Al chlorite.

Generally speaking, the conditions favoring the formation of montmorillonite are different from those under which kaolin minerals are formed. Montmorillonite is usually formed in a medium characterized by a high Si/Al ratio, a relative abundance of bases with correspondingly low concentration of H+ ions, and moist or reducing conditions. The environment in the early weathering stage of volcanic materials in Japan should be favorable for the formation of montmorillonite because soluble silica and bases are relatively abundant and the soils are nearly neutral or only slightly acid. Uchiyama, Masui, and Shoji (1968a,b,c,d,e,f) used X-ray diffraction to identify the clay minerals in very young soils derived from volcanic ash in Hokkaido in northern Japan and found that expanding layer silicates made up a large proportion of the crystalline clay minerals irrespective of the soil horizon, drainage, and texture. Chlorite, illite, and possibly kaolin minerals, on the other hand, were found only in small amounts. Montmorillonite was dominant in the soil groups younger than 500 years and had little interlayering with aluminum. In soils older than 500 years, vermiculite increased, and both montmorillonite and vermiculite had more interlayering with aluminum. These expanding layer silicates were considered by the researchers to be formed pedochemically because the parent materials of the soils did not contain micas and hornblende.

Kondo (1969) carried out clay-mineralogical studies on the weathering of Quaternary pyroclastic deposits in the "Usu" volcanic district in southwestern Hokkaido Island, where a number of pyroclastic deposits derived from late Pleistocene to Holocene times are widely distributed. The petrographic character of these pyroclastic deposits ranges from basalt to rhyolite; hence, differences in mode of weathering might be expected. On the basis of geological and tephrochronological studies, about 20 clay samples less than 2μ in effective diameter were separated from the weathered pyroclastic materials and examined by means of chemical analysis, X-ray diffraction, differential thermal analysis, infrared absorption spectroscopy, and electron microscopy. He recognized two different weathering processes in the pyroclastic deposits composed mostly of mafic materials. One is the formation of iron-rich montmorillonite, and the other is the formation of halloysite, as shown in the following diagram.

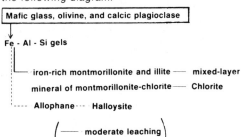

In this case, 2:1 type minerals are predominant. They are considered to have been formed under alkaline and wet conditions below the surface.

Similarly, two processes in the weathering of the older felsic pyroclastic deposits in this district were recognized as follows:

TABLE 2. Composition of NaOH-Soluble Materials and Relative Contents of Opaline Silica Particles in the <2-Micron Fractions of Soils Derived from Volcanic Ash in Different Districts of Japan (Shoji and Masui, 1969b).

District[1]	Sample No.	Ash layer	Age of ash, years	Depth, cm	Horizon	Texture[2]	NaOH soluble materials[3]		SiO_2/Al_2O_3[4]	Relative content of opaline silica[5]
							SiO_2 %	Al_2O_3 %		
Hokkaido I	145	Tarumae-a	218	0- 9	A	SL	29.4	4.8	10.4	++
	146			9- 13	C	SL	20.6	5.7	6.1	++
Hokkaido II	175	Meakan-a	220	2- 8	A	L	37.0	2.9	21.7	++
	176			8- 10	C	L	31.0	3.2	16.5	++
	177	Kamui-nupri-2a	<500	10- 20	C	L	30.4	4.8	10.8	++
	178			26- 38	C	L	13.1	9.1	2.4	+
	179	Kamui-nupri-4a		39- 42	C	SL	13.4	6.2	3.7	+
	181	Kamui-nupri-c		50- 58	C	L	21.0	9.8	3.6	+
	182			65- 73	C	L	12.5	10.6	2.0	+
	183	Kamui-nupri-d		73- 80	A	L	16.7	11.5	2.5	+
	184			80- 93	C	SL	12.2	10.8	1.9	+
	185	Kamui-nupri-e		93- 99	A	SL	17.4	13.1	2.3	+
	186	Kamui-nupri-2f	1150	99-111	C	SCL	19.9	15.8	2.1	+
	187			111-127	A	SCL	16.0	17.6	1.5	+
Kanto I	701			0- 45	A	SiCL	22.4	19.9	1.9	+
	702			45- 60	A	LiC	22.2	22.8	1.7	+
	703			60- 80	A	LiC	21.9	23.6	1.7	+
	704			80-130	A	LiC	19.3	24.6	1.3	—
	705	Schichihon-zakura pumice		130-160	C	—	23.5	28.3	1.4	—
	706	Imaichi pumice	<6000	160-220	C	—	26.9	26.8	1.7	—
	707			220-280	C	HC	19.6	26.8	1.2	—
	708			280-300	A	HC	22.3	20.5	1.8	—
	709			300-370	B	LiC	28.9	27.3	1.8	—
	710			370-	C		28.0	28.3	1.7	—
Kanto II	720			0- 10	A	SL	22.4	16.9	2.3	++
	721			10- 30	A	SL	25.3	14.8	2.9	++
	722	Futatsudake pumice	1400	30- 60	C					
	723			60- 75	C	LiC	20.6	19.1	1.8	++
	724			75- 83	A	LiC	20.5	23.0	1.5	+
	725			83-110	B	LiC	20.2	22.0	1.5	—
	726			110-133	A	LiC	20.0	24.1	1.4	—
	727	Itahana pumice	13130±230	133-	C	LiC	21.9	21.8	1.7	—
Kyushu I[6]	813	Bora	50	0- 10	C(A)	LS	14.7	3.3	7.6	++
	814			10- 27	A	SL	17.5	4.3	6.9	++
Kyushu II[6]	817	Kuroboku		0- 40	A	SL	22.0	18.7	2.0	++
	818	Kuroniga	4640±80	40- 90	C	L	18.0	22.7	1.4	—
	819	Akahoya		100-150	A	SL	22.0	24.8	1.5	—
	820		6360±90	150-195	C	L	23.8	23.4	1.7	—
	821			195-235	C	SL	22.5	25.8	1.5	—
	822			235-	A	CL	22.1	19.6	1.9	—

[1] Locality of soil samples: Hokkaido I-Tsukisam, Sapporo City, Hokkaido; Hokkaido II-Hagino, Teshikaga Town, Kawakami Country, Hokkaido; Kanto I-Mamiana, Utsunomiya City, Tochigi Pref.; Kanto II-Mizorogi, Shibukawa City, Gunma Pref.; Kyushu I-Kurogami, Sakurajima, Kagoshima Pref.; and Kyushu II-Karoya City, Kagoshima Pref.
[2] Texture designations: SL = Sandy loam, L = Loam, SCL = Sandy clay loam, SiCL = Silty clay loam, LiC = Light clay, HC = Heavy clay, LS = Loamy sand, and Cl = Clay loam. The significance of these texture designations may be seen in the appendix figure at the end of this chapter.
[3] NaOH-soluble materials were extracted by boiling the specimens for 2.5 minutes with 0.5 N NaOH solution after the removal of free oxides.
[4] Molecular ratio of SiO_2 to Al_2O_3.
[5] The relative contents of opaline silica particles were determined from electron micrographs of the <2-micron fractions and are indicated by the following symbols: ++ = relatively large amount, + = small amount, and — = none.
[6] Analytical data for Kyushu I and II were published by Kanno (1961).

From the results, Kondo concluded that the weathering in this district was characterized by the development of 2:1-type clay minerals from the overlying deposits under conditions of high pH and abundant supplies of silica, magnesium, and alkalies.

Properties of allophane and allophanic clay

As already mentioned, the distinguishing chemical properties of soils of the ando group are traceable to the characteristics of the inorganic mineral colloid, allophane. Recent progress by researchers in several parts of the world has thrown much new light on the nature and behavior of the amorphous clay in soil. Researches on chemical composition and chemical properties of allophanes, with supplementary evidence from use of modern instrumental methods, such as X-ray diffraction, differential thermal analysis, and infrared absorption spectroscopy, have shown that allophanes are amorphous hydrous aluminosilicate gels. Allophanes have much in common with synthetic hydrous aluminosilicate gels that may be prepared in the laboratory. This section contains a brief account of important information on the chemical or physicochemical properties of allophane or allophanic clay recently obtained by Japanese researchers.

Chemical composition. Many analytical data on the chemical composition of allophane were accumulated in earlier years. There is no guarantee, however, that these previously obtained data really represent the chemical composition of allophane because of the probability that most of the samples contained some impurities such as free hydrous oxides of iron and aluminum and crystalline minerals.

Yoshinaga (1966) studied the chemical composition of allophane in detail using 18 clay samples separated from soils of the ando group and from weathered pumice. These clay samples were considered to contain only "pure" allophane because the coexisting crystalline clay mineral, imogolite, was separated first by differential dispersion, as described earlier in this chapter. Contamination by primary minerals was controlled by using only the fraction from 0 to 0.2μ in effective diameter. The free hydrous oxides were removed by repeated treatments of deferration and washing.

The chemical composition of these presumably pure allophanes is shown in Table 3, and the calculated molecular ratios among the three major components, viz., SiO_2, Al_2O_3, and H_2O, are listed in Table 4.

As may be seen from Table 3, all samples consisted principally of silica, alumina, and water, with minor amounts of all other constituents except sodium. (The values for sodium are for total sodium, which the author considered almost entirely exchangeable. The clays were exchange-saturated with sodium and had previously been treated with alkali, which increases the cation-exchange capacity.) Molecular ratios of silica to alumina ranged from about 1.3 to 2.0. Molecular ratios of ignition loss $H_2O(+)$ and total water $H_2O(\pm)$ to alumina were nearly constant, with average values of 2.5 and 5.7, respectively. The average chemical composition of these allophanes may be expressed approximately as follows:

$$(SiO_2)_{1.5} \cdot Al_2O_3 \cdot (H_2O)_{2.5}$$

A ternary plot of the weight proportions of SiO_2, Al_2O_3, and H_2O in Fig. 2 shows that the composition of the various samples was restricted to a relatively small area. In other words, the chemical composition of these allophanes did not vary widely but fell in a relatively narrow range. Figure 3 shows another way of representing the same data.

TABLE 3. Chemical Composition of Allophanes Separated from Soils and Weathering Pumice in Japan (Yoshinaga, 1966).

Lab. No.	Chemical composition on oven-dry basis, %												
	SiO_2	Al_2O_3	Fe_2O_3	TiO_2	P_2O_5	MnO	CaO	MgO	K_2O	Na_2O	$H_2O(+)$[1]	Total	$H_2O(-)$[1]
1006	35.31	39.18	0.66	0.79	0.12	tr.	0.16	0.44	0.13	5.73	16.71	99.23	23.08
1008	31.72	40.89	0.59	0.79	0.14	tr.	0.11	0.29	0.13	5.35	19.18	99.19	25.21
1010	34.32	40.94	0.70	0.94	0.16	tr.	0.07	0.14	0.13	5.08	17.65	100.13	22.16
1014	35.58	39.82	0.35	0.45	0.22	tr.	0.10	0.17	0.16	5.90	16.67	99.42	22.10
1015	34.01	39.73	0.83	0.86	0.28	tr.	0.54	0.32	0.10	4.77	18.38	99.82	24.02
1016	34.25	39.59	0.77	0.89	0.19	tr.	0.16	0.15	0.13	4.77	18.34	99.24	23.20
1019	35.07	39.89	0.63	0.88	0.09	tr.	0.05	0.21	0.14	4.56	16.92	98.44	22.19
1033	37.16	38.41	0.97	1.82	0.23	tr.	tr.	tr.	0.15	5.69	15.64	100.07	23.03
1040	36.14	39.32	0.48	0.88	0.08	tr.	0.09	0.16	0.19	6.11	16.14	99.59	22.85
1041	34.24	40.12	0.37	0.34	0.38	tr.	0.36	0.06	0.14	4.76	17.56	98.33	23.10
1049	32.25	41.55	0.62	0.53	0.65	tr.	tr.	0.09	0.11	5.48	18.38	99.66	26.21
1051	31.38	39.15	0.49	0.17	0.50	tr.	0.85	0.60	0.13	4.96	19.41	98.64	24.26
1052	34.07	38.69	0.84	1.00	0.31	tr.	0.70	0.15	0.13	6.11	18.08	100.08	22.31
1056	36.38	37.51	1.12	0.41	0.19	0.01	0.58	0.20	0.13	6.12	16.62	99.27	21.00
1057	37.90	36.18	1.21	0.35	0.12	tr.	0.56	0.21	0.13	6.09	16.54	99.29	20.12
1058	38.53	36.73	1.21	0.75	0.14	0.01	0.10	0.26	0.12	6.19	16.25	100.29	18.31
1059	39.92	34.78	0.96	1.06	0.11	tr.	0.16	0.23	0.13	6.56	15.62	99.53	19.08
1066	34.64	39.37	0.88	0.43	0.04	tr.	0.24	0.24	0.23	5.30	18.65	100.02	23.00

[1] $H_2O(-)$ is the loss in weight when the air-dry clay is oven-dried at 110°C, and $H_2O(+)$ is the loss in weight when the oven-dry clay is ignited.

Acidic properties. Iimura (1966) characterized the nature of allophane as a clay acid on the basis of ultimate pH, exchange acidity, and titration curves. As shown in Table 5, two different samples of allophane showed no exchange acidity after they were electrodialyzed to remove exchangeable bases. The ultimate pH value of both samples was considerably above the ultimate pH of montmorillonite. One of the electrodialyzed allophanes was almost neutral.

The titration curves in Fig. 4 show that allophane in water had little buffering capacity when small amounts of acid or alkali were added. This behavior is a consequence of the weak dissociation of hydrogen and hydroxyl ions from allophane at pH values near the ultimate pH. When potassium chloride was added, the buffering capacity near the ultimate pH was much increased, presumably because potassium and chloride ions were adsorbed at pH values near the ultimate pH, replacing weakly dissociating hydrogen and hydroxyl ions, respectively. The hydrogen and hydroxyl ions replaced by potassium and chloride neutralized each other. The hydrogen and hydroxyl ions added in the acid and alkali then replaced come of the adsorbed potassium and chloride ions and were retained in weakly dissociated form by the clay. Iimura represented in the following manner the reactions involved in the titrations:

With addition of alkali to a suspension of clay in water,

$\equiv$SiOH + NaOH ——> $\equiv$SiONa + H$_2$O
($\equiv$SiO)$_3$Al + NaOH ——> [($\equiv$SiO)$_3$AlOH] Na

With addition of alkali to a suspension of clay in a solution of potassium chloride,

K(Clay)Cl + KOH + K(Clay)OH + KCl

With addition of acid to a suspension of clay in water,

=AlOH + HCl ——> =AlOH$_2$Cl

With addition of acid to a suspension of clay in a solution of potassium chloride,

K(Clay)Cl + HCl ——> H(Clay)Cl + KCl

In addition, Iimura used Schofield's method to determine the electrical charges on allophane as a function of the pH of equilibrium solutions. The results obtained with one sample of allophane are shown in Fig. 5. The amount of ammonium adsorbed is taken as a measure of the number of negative charges, and the amount of chloride adsorbed is taken as a measure of the number of positive charges. The central curve is the net charge and is obtained by algebraic addition of the equivalents of positive and negative charges. In other experiments, Iimura found that results similar to the curves for ammonium adsorption and chloride adsorption in Fig. 5 were obtained experimentally when the methods used on allophane were applied to synthetic gels of silica and hydrous aluminum oxide, respectively. The silica gel showed slight negative ad-

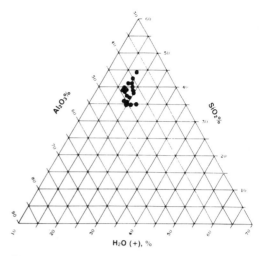

Fig. 2—Ternary diagram showing partial chemical composition of allophanes separated from soils and weathering pumice in Japan. The sum of the contents of SiO$_2$, Al$_2$O$_3$, and H$_2$O(+) (loss of water from oven-dry clay on ignition) by weight has been recalculated to represent 100%.
(Yoshinaga, 1966)

sorption of chloride, and the hydrous aluminum oxide gel showed slight negative adsorption of ammonium. A synthetic, coprecipitated aluminosilicate gel yielded three curves similar to those in Fig. 5.

From these results, Iimura inferred that the negative charge or cation-exchange capacity results mainly from $\equiv$SiOH groups. He attempted to measure the total number of such groups by titrating allophane with barium hydroxide. He called the values obtained the "absolute" cation-exchange capacity. Numerical values he obtained for the absolute cation-exchange capacity were 6 to 23 times as great as those obtained for the conventional cation-exchange capacity at pH 7. He pointed out that the ratio of these two exchange-capacity values in soils derived from volcanic ash is in the range of 6:1 to 23:1. In contrast, the ratio of the exchange capacities for minerals other than allophane in Table 6 is 1.7 for montmorillonite and 2.6 for kaolinite. He pointed out that the ratio is about 4 in red and yellow soils in which kaolinitic minerals are dominant. He suggested, therefore, that the ratio of the two exchange capacities might provide a useful test to determine whether the parent material of a given soil contains volcanic ash.

Cation-exchange capacity. The conventional Schollenberger and Simon (1945) method for determination of cation-exchange capacity of soils and clays involves three steps: the first is saturation of the sample with ammonium from a neutral, 1-normal solution of ammonium acetate, the second is removal of the excess ammonium acetate by washing

TABLE 4. Molecular Ratios of Major Constituents of Allophanes Separated from Soils and Weathering Pumice in Japan (Yoshinaga, 1966).

Lab. No.	$\dfrac{SiO_2}{Al_2O_3}$	$\dfrac{H_2O(+)}{Al_2O_3}$	$\dfrac{H_2O(-)}{Al_2O_3}$	$\dfrac{H_2O(\pm)}{Al_2O_3}$	$\dfrac{H_2O(+)}{SiO_2}$	$\dfrac{H_2O(-)}{SiO_2}$	$\dfrac{H_2O(\pm)}{SiO_2}$
1006	1.53	2.42	3.34	5.76	1.58	2.18	3.76
1008	1.32	2.66	3.49	6.15	2.02	2.65	4.67
1010	1.42	2.45	3.06	5.51	1.71	2.15	3.86
1014	1.52	2.37	3.15	5.52	1.56	2.07	3.63
1015	1.46	2.63	3.42	6.05	1.80	2.36	4.16
1016	1.47	2.63	3.31	5.94	1.78	2.26	4.04
1019	1.50	2.40	3.15	5.55	1.61	2.11	3.72
1033	1.64	2.31	3.40	5.71	1.40	2.07	3.47
1040	1.56	2.33	3.29	5.62	1.48	2.11	3.59
1041	1.45	2.48	3.26	5.74	1.71	2.25	3.96
1049	1.32	2.51	3.57	6.08	1.90	2.71	4.61
1051	1.36	2.81	3.51	6.32	2.06	2.58	4.64
1052	1.50	2.65	3.27	5.92	1.77	2.18	3.95
1056	1.64	2.51	3.17	5.68	1.52	1.92	3.44
1057	1.78	2.59	3.15	5.74	1.45	1.77	3.22
1058	1.79	2.51	2.82	5.33	1.41	1.58	2.99
1059	1.95	2.55	3.10	5.65	1.30	1.59	2.89
1066	1.49	2.68	3.32	6.00	1.79	2.21	4.00

TABLE 5. Ultimate pH and Exchange Acidity of Various Clays (Iimura, 1966).

Clay mineral	Treatment for removal of exchangeable bases	Ultimate pH[1]	Exchange acidity (extracted by 1N KCl)		
			Acidity per 100 g of air-dry clay, m.e.	H+ %	Al³⁺ %
Kaolinite	H-resin[2]	4.5	5.8	58	42
	Electrodialysis	5.5	0.7	–	–
Montmorillonite	H-resin[2]	3.4	42.3	20	80
	Electrodialysis	3.3	43.4	36	64
Allophane (from weathered pumice)	Electrodialysis	6.3	0	–	–
Allophane (from subsoil derived from volcanic ash)	Electrodialysis	4.3	0	–	–

[1] Concentration of suspensions 1% for kaolinite and montmorillonite and 2% for allophane.
[2] Hydrogen-saturated Amberlite IR-120.

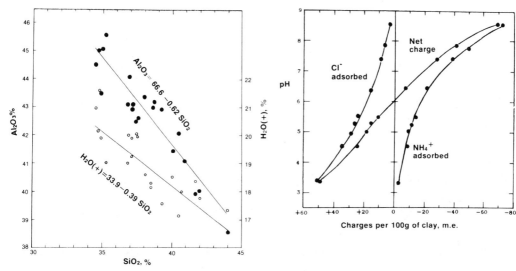

Fig. 3—Relationships among the three major chemical components of 21 allophanes separated from soils and weathering pumice in Japan. The sum of the contents of SiO_2, Al_2O_3, and $H_2O(+)$ loss of water from oven-dry clay on ignition) by weight has been recalculated to represent 100%.
(Yoshinaga, 1966; Yoshinaga and Aomine, 1962a)

Fig. 5—Electrical charges on allophane in 0.2 N ammonium chloride. The allophane used was separated from the Misotsuchi subsoil.

(Iimura, 1966)

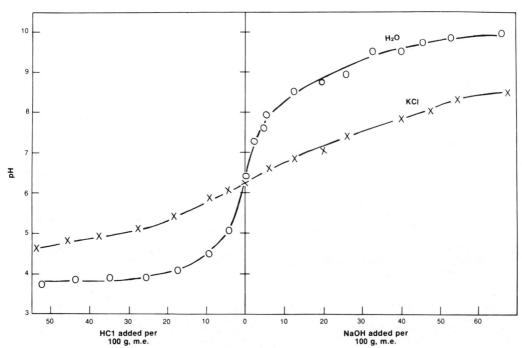

Fig. 4—Titration curves obtained with allophane from a sample of Misotsuchi subsoil in the presence of water and 1N potassium chloride.

(Iimura, 1966)

TABLE 6. Cation-Exchange Capacity and Surface Area of Clays Less than 2 Microns in Effective Diameter (Iimura, 1966).

Clay sample no.	Dominant clay mineral[1]	Cation-exchange capacity per 100 g. of clay, m.e.		Surface[4] per gram, m^2	Number of ≡ SiOH groups per $m\mu^2$ of surface area
		Conventional[2]	Absolute[3]		
1	Allophane	29.2	667	429	9.36
2	Allophane	42.0	597	496	7.25
3	Allophane	31.3	478	409	7.04
4	Allophane	25.2	162	136	7.17
5	Kaolinite	5.8	15	23.7	3.81
6	Montmorillonite	87.3	147	298	1.45
7	Silica gel	31.0	831	369	13.6

[1] Sample 1 was from Misotsuchi weathered pumice, sample 2 was from Kanumatsuchi weathered pumice, sample 3 was from Shakujii subsoil, sample 4 was from Kuroishibara subsoil, sample 5 was from Ibusuri kaolin, and sample 6 was from Gunma bentonite.
[2] Determined by $1N$ ammonium acetate at pH 7.
[3] Determined by consumption of hydroxyl ions from barium hydroxide.
[4] Determined by the glycerol sorption method of Kinter and Diamond (1958a,b).

TABLE 7. Cation-Exchange Capacity of Allophanes and Other Materials with Neutral, 1-Normal Calcium Acetate After Different Pretreatments, and Dissolved Matter Associated with the Pretreatments (Aomine and Jackson, 1959).

Minerals	Cation-exchange capacity, m.e./100 g*				Dissolved matter, %*			
	Pretreatment with 2% Na_2CO_3 (B)	Pretreatment with NaOAc at pH 3.5 (A)	B/A	Δ — Value (B—A)	By Na_2CO_3		By NaOAc	
					SiO_2	Al_2O_3	SiO_2	Al_2O_3
Allophane (Choyo)	157.2	55.6	2.83	101.6	0.31	3.40	0.73	4.65
Allophane (White)	143.3	60.6	2.36	82.7	1.19	4.29	0.23	0.16
Allophane (Brown)	154.9	44.1	3.51	110.8	0.54	2.14	0.23	0.21
Halloysite (Indianite)	30.2	12.3	2.46	17.9	0.50	0.15	0.22	0.11
Montmorillonite (Wyoming)	88.0	77.9	1.13	10.1	0.91	0.12	0.22	0.04
Kaolinite (South Carolina)	4.6	4.6	1.00	0	—	—	—	—
Gibbsite	3.1	2.6	1.19	0.5	—	—	—	—
Quartz	0.0	0.0	—	0	—	—	—	—

* Oven-dry basis.

with alcohol, and the third is displacement and determination of the ammonium retained. Cation-exchange capacities for allophane ranging from about 20 to 100 m.e. per 100 g of clay have been reported by investigators using methods of this general type.

The marked fluctuation of values for cation-exchange capacity obtained for allophane by methods using the general approach described in the preceding paragraph was first pointed out by Birrell and Fieldes (1952) and Birrell and Gradwell (1956). They found that the cation-exchange capacity of allophane was greatly influenced by the species of cations and anions in the salt solution used for saturation, the concentration and pH of this solution, and the volume and water content of the alcohol used to remove the salt solution before determination of the cation retained. On the basis of the dependence of the cation-exchange capacity of allophane on the experimental conditions, Birrell and Gradwell reasoned that physical adsorption of cations is responsible for a part of the high apparent cation-exchange capacity of allophane. The concept that two different adsorption mechanisms (electrostatic and physical) are operating in ion uptake by allophane was introduced by them. Japanese researchers (Aomine, 1958; Wada and Ataka, 1958; Egawa, Watanabe, and Sato, 1959) similarly noticed a fluctuation in the cation-exchange capacity values for allophane.

Aomine and Jackson (1959) made use of the fluctuation of cation-exchange capacity values obtained for allophane as the basis for a quantitative estimate of the allophane content of soils and clays. As shown in Table 7, the cation-exchange capacity of allophane measured with neutral, 1 N calcium acetate after preliminary treatment of the allophane with hot 2% sodium carbonate solution was higher, by about 100 m.e. per 100 g of oven-dry material, than was the cation-exchange capacity measured in the same way after treatment of the allophane with 1 N sodium acetate solution adjusted to pH 3.5. The increase in cation-exchange capacity due to the pretreatment with the alkaline solution was designated the "cation-exchange capacity delta value." Table 7 shows the delta values obtained with several materials used as standards. A preliminary treatment with

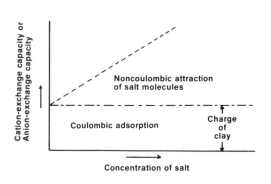

Fig. 6—Schematic representation of the relationship between coulombic and noncoulombic adsorption of ions by clays.
(Wada and Ataka, 1958)

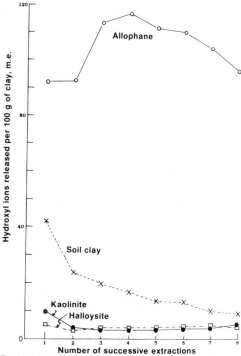

Fig. 7—Release of hydroxyl ions from various clays when 1-g samples were treated repeatedly overnight at 30°C with 50-ml portions of neutral, 1 N sodium fluoride solution. The hydroxyl ions released were determined by titration with standard sulfuric acid. (Egawa et al., 1960)

hydrogen peroxide to remove organic matter was recommended for use in application of the method to estimation of the allophane content of soils.

Inaccuracies are inevitable in measurements of cation-exchange capacity by the conventional method. Several researchers have stressed the importance of avoiding removal of the saturating salt with alcohol as a means of avoiding errors due to hydrolysis and salt retention.

Wada and Ataka (1958) determined the cation-exchange capacity and anion-exchange capacity of allophane in equilibrium with ammonium chloride solutions of various concentrations and pH values (Schofield's method). Their results confirmed that two different mechanisms of ion retention by allophane were discernible, namely, retention by coulombic and noncoulombic forces. A schematic representation of the relationship thought to exist between these two forms of attraction in ion uptake by allophane is shown in Fig. 6.

Wada and Harada (1969) recently investigated further the effects of salt concentration and cation species on the cation-exchange capacity of various clays. They found that the increase in cation-exchange capacity with increasing equilibrium concentration of saturating salt solution was most marked with allophane and imogolite having a SiO_2/Al_2O_3 ratio of about 1.0, less marked with halloysite and with allophane having a $SiO_2Al_2O_3$ ratio of about 2.0, and least marked with montmorillonite. They considered that the relatively rapid increase of the cation-exchange capacity with salt concentration at concentrations above 0.1 **N** was probably due to noncoulombic adsorption of cations with an equivalent amount of anions. They found that when they equilibrated clays with a 0.05 **N** salt solution they were able to avoid what seemed to be errors due to retention of cations by noncoulombic adsorption, and at the same time they did not lose exchangeable cations by hydrolysis. They observed also that the species of cations in the saturating salt solution had an influence on the cation-exchange capacity of allophane and imogolite. The explanation for this effect remains to be determined.

Reaction with anions. The reaction of clays with anions may involve not only a true ion exchange but also the additive formation of new solid compounds. The reaction between F—ions and OH—ions of clay minerals has been studied by a number of researchers. Egawa and colleagues (1960) investigated the release of hydroxyl ions from clays by titrating with standard acid the alkalinity produced when the clays were treated with a neutral solution of ammonium or sodium fluoride. As seen in Fig. 7, hydroxyl ions continued to be released in relatively large quantities when allophane was treated repeatedly with neutral, 1-normal sodium fluoride. Fig. 8 shows the relationship between the concentration of neutral ammonium fluoride solution and the release of hydroxyl ions. The results show that allophane and synthetic aluminosilicate gel released far more hydroxyl than did kaolinite and montmorillonite. The high pH produced in the reaction was suggested by Fieldes and Perrott (1966) as the basis of a rapid field and laboratory test to determine the presence of allophane.

The reaction between fluoride solutions and allophane involves more than an exchange between fluoride and hydroxyl ions. New reaction products such as $(NH_4)_3AlF_6$ or Na_3AlF_6 may be identified by means of X-ray diffraction and differential thermal analysis. Fig. 9 and Fig. 10 show the X-ray diffraction and differential thermal analysis results in the case of allophane.

The reaction of allophane with phosphate ions is the most important of the anion reactions from the viewpoint of agricultural practice. Wada (1959) found that ammonium phosphate reacts rapidly with allophane at pH 4 to produce an ammonium-substituted taranakite (a hydrous aluminum ammonium phosphate). Crystals of the newly formed phosphate could be observed with the aid of an ordinary light microscope after only a few days of contact of these minerals with the phosphate solution at room temperature. Allophane may be completely converted to the taranakite-like phosphate within 3 weeks. The reaction was retarded at pH 7. This observation helps to explain the importance of pH in phosphate fixation by soils derived from volcanic ash.

Allophane reacts also with "humic acid" (Kobo and Fujisawa, 1963, 1964). This reaction is probably the principal explanation for the accumulation of organic matter in abundant amounts in soils containing allophane. The so-called allitic properties of allophanic clays are depressed by reaction with organic matter because the organic matter reacts with the aluminum that is the source of these properties. The positive charges on allophanic clays are neutralized by humic acid (Watanabe, 1966; Kobo and Fujisawa, 1966). Fig. 11 gives the results of Watanabe's (1966) measurements of the electrophoretic behavior of allophane as affected by reaction with soil organic matter. In this experiment, the allophane was separated from "Misotsuchi" (weathered pumice), and the humic acid fraction of the organic matter was separated from the Koganei soil, a soil of the ando group. As shown in the figure, allophane was amphoteric, but humic acid and the complex of allophane with humic acid were electronegative. The isoelectric pH of allophane was lowered by drying of the clay. Similar results were obtained by Kobo and Fujisawa (1966).

Laboratory experiments simulating the early reactions that occur when organic substances derived from decomposing clover leaves are added to soils indicate that the reactions are affected by the nature of clay minerals (Wada and Inoue, 1967). As may be seen in Table 8, the adsorption capacity of soil for organic carbon and the stability of the adsorbed organic substances against leaching and microbial degradation were greater in allophanic soil than in montmorillonitic soil. In their succeeding study (Inoue and Wada, 1968), the adsorption maximum, expressed in grams of carbon per 100 grams of clay, was 14.5 to 16.7 for the allophanes (including imogolite) with a Si/Al ratio less than 1 to 1.5, 6.1 for allophane with a Si/Al ratio of 1 to 1, 0.7 for halloysite, and 1.4 to 2.1 for montmorillonites. These differences between amorphous allophane and crystalline clay minerals are considered to explain the difference in the tendency for accumulation of organic matter between soils of the ando group and other soils that are not derived from volcanic ash.

Coordination Number of Aluminum in Allophane. The coordination number of aluminum in allophane has been a matter of some interest because of its possible connection with the properties and process of formation of allophane. Egawa (1964) used the X-ray fluorescence method to determine the coordination number of aluminum in an allophanic clay sample separated from weathered pumice and obtained the results shown in Table 9 and Fig. 12. The angle of the Al Kα radiation from allophane was intermediate between that from clays with known four-fold and six-fold coordination and corresponded to five-fold coordination (Fig. 12). Egawa interpreted these results to mean that, in the allophane in question, the

TABLE 8. Characteristics of Reactions Between Organic Matter in Extracts of Decomposing Clover Leaves and Two Soils Differing in Principal Clay Mineral Species (Inoue and Wada, 1968).

	Nakajyo soil (montmorillonitic)	Choyo soil (allophanic)
Rate of attainment of adsorption equilibrium	Rapid	Slow
Adsorption capacity	0.65 g carbon per 100 g soil	2.8 to 4.4 g carbon per 100 g soil
Relative stability of adsorbed organic substances	Low	High
Relative degree of humification of adsorbed organic substances	High	Low

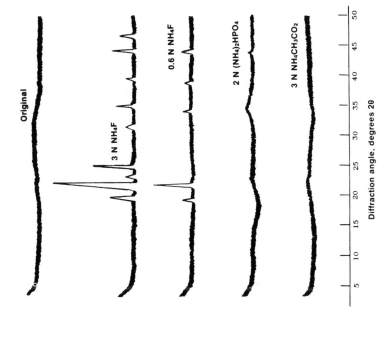

Fig. 9.—X-ray diffraction patterns of reaction products obtained upon treatment of allophane with solutions of various ammonium salts. Fe Kα radiation was employed.

(Egawa et al., 1960)

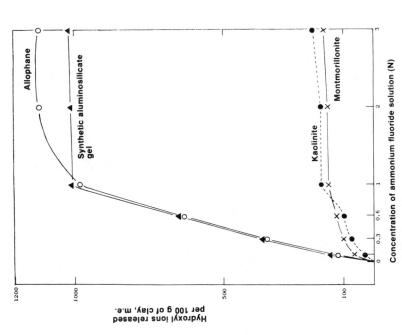

Fig. 8.—Release of hydroxyl ions from clays and a synthetic aluminosilicate gel when 1-g samples were treated overnight at 30°C with 50-ml portions of ammonium fluoride solutions differing in concentration. The hydroxyl ions released were determined by titration with standard sulfuric acid.

(Egawa et al., 1960)

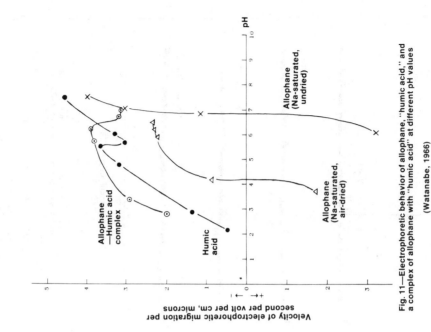

Fig. 11—Electrophoretic behavior of allophane, "humic acid," and a complex of allophane with "humic acid" at different pH values

(Watanabe, 1966)

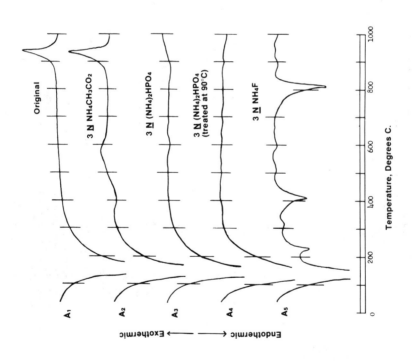

Fig. 10—Differential thermal analysis curves of reaction products after treatment of allophane with solutions of various ammonium salts.

(Egawa et al., 1960)

aluminum was partly in four-fold coordination and partly in six-fold coordination. The sample he used contained 26.6% SiO_2 and 19.1% Al_2O_3. Samples with different chemical composition may have a different angle of Al $K\alpha$ radiation, which would indicate different proportions of four- and six-fold coordination in Egawa's view. Further investigations along this line are to be expected.

TABLE 9. Mean Angles of Al $K\alpha$ Radiation Measured on Different Aluminum-Bearing Substances (Egawa, 1964).

Substance	Al coordination number	Angle of Al $K\alpha$ radiation $(2\theta°)$
Metallic aluminum	0	142.4960 ± 0.0066*
Anorthite	4	142.4540 ± 0.0066
Kaolinite	6	142.4138 ± 0.0057
Halloysite	6	142.4117 ± 0.0066
Allophane	?	142.4357 ± 0.0066

* 95% confidence limits

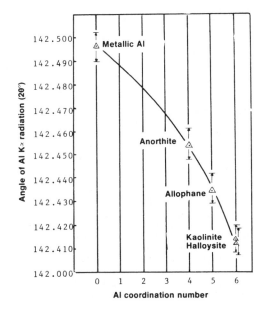

Fig. 12—Relation between the angle of Al $K\alpha$ radiation and the Al coordination number of several substances. Confidence limits (95%) are shown with arrows. (Egawa, 1964).

ORGANIC MATTER

Accumulation of organic matter

Soils of the ando group are characterized by an abundant accumulation of organic matter in the surface horizon. According to various soil surveys of arable lands in Japan, an abundant accumulation of organic matter occurs in soils of the uplands of the Japan-Sea side of Honshu, northern Kanto, and southern Kyushu, as well as in soils of the hilly lands of northern Kanto, Sanin, and central Honshu (Adachi, 1963a). The thickness of the layer of organic matter accumulation seldom exceeds 100 cm.

Kanno (1961) analyzed 23 samples of A and A_1 horizons of soils of the ando group and found that the organic matter content ranged from 7.6% to 40.3% with an average of 18.1%. Kosaka (1963) found that the average organic matter content in the surface horizons was 20.9% and that soils derived from volcanic ash generally contained more than 5% organic carbon, whereas soils derived from other geologic sources generally contained less than 5%. The main source of organic matter was considered to be grasses, especially **Miscanthus sinensis.**

Kobo and Ohba (1964) noticed a close relationship between the content of organic matter in the surface layer and the weathering degree of the soil. They attributed their observation mainly to the clay minerals associated with the different degrees of weathering. Shinagawa (1962) investigated soil samples selected in the district covered by recent volcanic ash deposits of Volcano Sakurajima in Kagoshima Prefecture. He found that accumulation of organic matter in the soil was associated with soil development, as evidenced by (a) the alteration of soil texture to finer status, (b) the luxuriance of the vegetation, (c) the disintegration of labradorite in the parent rocks, (d) the increase in content of free sesquioxides, and (e) intensification of the allophanic character of the clay.

The general consensus in Japan with regard to the accumulation of relatively large amounts of organic matter in soils derived from volcanic ash is that the humid subtropical conditions in Japan and the favorable nutrient status prevailing in the early stages of weathering result in a luxuriant growth of grasses, which are the primary source of the organic matter (Kawamura, 1950; Uchiyama et al., 1954; Kumada, 1956b; Tokudome and Kanno, 1964a, b, 1968). As weathering proceeds, the decomposition products of the grasses interact with allophane and sesquioxides and are thereby stabilized against further decomposition (Aomine and Kodama, 1956; Kosaka and Iseki, 1957; Aomine and Kobayashi, 1964, 1966). Kyuma and Kawaguchi (1964) suggested that the allophane and sesquioxides may also have a catalytic action that promotes the development of stable forms of organic matter.

Nature and properties of organic matter

Since 1950, many papers on the nature and properties of the organic matter of soils of the ando group in Japan have been published (Hayashi, 1951, 1953a,b; Hayashi and Nagai, 1953, 1955, 1956a,b, 1957, 1959, 1960a,b, 1961, 1962a,b; Kumada, 1955 a,b,c,d,e,f,g, 1956a,b, 1958a,b, 1959, 1963, 1965a,b; Hosoda, 1938; Hosoda and Takada, 1953a,b,c, 1957a,b,c, 1967a,b,c,d,e,f; Kosaka, 1953, 1963; Kosaka and Iseki, 1957; Kosaka and Honda, 1957, Kosaka, Honda, and Iseki, 1961a,b; Okuda and Hori, 1952a,b,c, 1955; Shinagawa, 1953, 1954, 1958; Hanai and Shinagawa, 1952; Kobo and Tatsukawa, 1959; Miyoshi, 1964; Ohba, 1965).

The papers just cited deal mostly with the proportions of different empirically defined chemical fractions in the organic matter. Many of the investigations were influenced by the views of German researchers, especially those of Springer and Simon. Recently, methods used by Russian researchers, viz., Kononova, Tyurin, and Ponomareva, were introduced into Japan by Kawaguchi and Kyuma (1959). Results obtained by the different methods are not comparable, and the significance of the separations is not as clear as desired because of the limited understanding of the chemical nature of the fractions. A further limitation of the available knowledge of the organic matter in soils derived from volcanic ash is that the research has dealt mostly with the fractionation as such and not with the properties or functions in relation to soil productivity. In view of these limitations and the objective of this publication, only a few recent papers will be considered here. Those wishing further information should refer to the original papers cited and to reviews by Kobo (1951/52), Kumada (1958b), Kosaka (1964), and Kobo et al. (1968).

Kawaguchi and Kyuma (1959) were the first to apply Tyurin's method to soils in Japan, and they found that in most soils the organic matter is combined with sesquioxides. Kosaka and Iseki (1957) investigated the combination of organic matter with inorganic matter in soils of the ando group based on the solubility of organic matter at different pH values. They found similarly that the organic matter was combined with aluminum and iron.

Hayashi and Nagai (1953, 1955) separated soil organic matter into four fractions by means of column chromatography using active alumina as adsorbent. Fractions 1 and 2 possessed the character of so-called "Nähr Humus" (of Scheffer), and fractions 3 and 4 possessed the character of so-called "Dauer Humus" (of Scheffer). The contents of carbohydrates and nitrogen, especially of easily hydrolyzable nitrogen, differed between the two groups. Nagai (1969) developed further the Hayashi and Nagai fractionation system. The four groups, namely, group **a**, including a_1, a_2 and a_3, group **b**, including b_1 and b_2, group **c**, and an insoluble group, were separated by the modified procedure. Among these fractions, group **a** was considered to be combined with amorphous mineral colloids, mainly allophane, and group **b** was considered to be combined with crystalline clay minerals. Group **c** was considered to be combined with vermiculite-like minerals. The strength of combination with the inorganic part increased in the order, $a_1 < a_2 < a_3$ and $b_2 < b_1$. His analytical data on samples of 26 soils derived from Daisen volcanic ash showed that the contents of groups **a, b, c,** and the insoluble fraction were 4.7 to 10.0%, 2.8 to 7.0%, 2.2 to 5.3%, and 7.4 to 12.8%, respectively. On the average, about 38% of the total organic matter could not be extracted and occurred in the insoluble fraction.

Kumada (1955a,b,c,d,e,f,g, 1956a,b) investigated the relationships between chemical properties of various soil types and the character of the fraction of soil organic matter that dissolves in alkali and precipitates on acidification — the so-called "humic acid" fraction. The degree of "humification" (the degree of alteration from fresh organic matter to the relatively stable type) was characterized by various criteria. From analyses made on a number of samples of humic acids, he found that values of cation-exchange capacity, stability to oxidation by permanganate, and sensitivity of suspensions to precipitation by electrolytes were correlated, and all increased with the progress of decomposition and presumably of humification. The ratio of the amount of hydrolyzable nitrogen to the total amount of nitrogen in humic acid was negatively correlated with the values just mentioned and hence was negatively correlated with humification. The degree of humification was much higher in the case of humic acids separated from soils of the ando group than from soils of the alluvial or diluvial groups.

The content of methoxyl was used by Kumada (1955e) and Kosaka (1963) as an index of humification. According to Kosaka (1963), the methoxyl content, the ratio of carbon to nitrogen, the ratio of carbon in humic acid to total carbon, and the proportion of A-type humic acid in total humic acid are closely related. On the basis of these properties, he identified the five degrees of humification shown in Table 10. Certain other properties, including the cation-exchange capacity, the content of carboxyl, and the ratio of hydroxyl to carboxyl, were similarly considered to be related to the degree of humification in the manner shown in Table 11.

Kosaka determined in the following way the various organic fractions named in the tables. "Total humic acid" was determined by treating 1 g of soil overnight at 30°C with 100 ml of 0.5% sodium hydroxide, centrifuging the suspension, adding concen-

TABLE 10. Degree of Humification and Properties of Soil Organic Matter (Kosaka, 1963).

Degree of humification	Content of methoxyl, %	C/N	True humic acid-C × 100/total organic C
A	>1.4	>20	<5
B	1.0	<20	10-20
C	0.6	<20	20-40
D	0.3	>20	40-60
E	0.1	30-40	60-80

TABLE 11. Schematic Diagram Showing Relationships Between Properties of Soil Organic Matter and Degree of Humification (Kosaka, 1963).

Property	Trend of property with indicated degree of humification			
	B ———>	C ———>	D ———>	E
True humic acid/Total humic acid		———> increasing		———>
A-type humic acid/Total humic acid	0%	70%	80-90%	90%
Carboxyl		———> increasing		———>
Hydroxyl/Carboxyl	<———	decreasing	<———	
Cation-exchange capacity		———> increasing		———>
Light absorbance index		———> increasing		———>

trated sulfuric acid to the solution to precipitate the humic acid, and washing the precipitate several times with water. The total humic acid was separated into "A-type humic acid" and "B-type humic acid" in the following way. The total humic acid was dissolved in 1% sodium acetate solution and then treated with 1 **N** magnesium sulfate. The resulting precipitate was washed with dilute sulfuric acid (1 ml of acid + 99 ml of water) and water to remove the magnesium. The precipitate was then dissolved in 0.5% sodium hydroxide. A-type humic acid was precipitated from this solution by addition of sulfuric acid. The filtrate contained the B-type humic acid. To obtain "true humic acid", a 4-g sample of soil was treated at 40 to 45°C for 24 hours with 40 ml of a mixture of 200 ml of glacial acetic acid, 200 ml of acetic anhydride, and 40 ml of concentrated sulfuric acid. The suspension was then centrifuged, and the soil was washed consecutively with glacial acetic acid, 5% hydrochloric acid, and distilled water. The soil was then extracted overnight at 30°C with 400 ml of 0.5% sodium hydroxide. The solids were removed by centrifuging, and the humic acid precipitated from the solution by addition of sulfuric acid was designated as true humic acid.

The light absorbance index mentioned in Table 11 is a measure of the intensity of color of a sodium hydroxide solution containing 0.1 g of soil organic carbon per 1,000 ml, as determined with a filter photometer fitted with a 610 mμ filter.

Kosaka determined the degree of humification of organic matter in a number of soils and related the results to the classification of the soils. Some of his analytical data are shown in Table 12.

Tokudome and Kanno (1965a,b, 1968) used the methods of the Russian school to fractionate the organic matter in various soils of the ando group in Japan. The scheme for obtaining fractional components of soil organic matter by the method of Tyurin is presented in Fig. 13. Ponomareva's fractionation

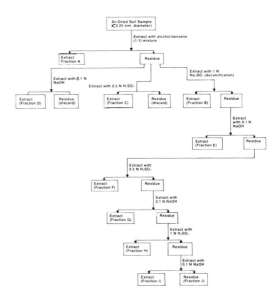

Fig. 13—Fractionation of soil organic matter by the method of Tyurin (1951). The organic carbon in each extract and in the residue (Fraction J) is determined and expressed as a percentage of the total organic carbon. The correspondence between the fractions in the diagram and those indicated in Table 13 is as follows: A = bitumen. C − B = fraction 1a. D − C = fraction 1. E − (D − B) = fraction 2. G + I = fraction 3. F + H = N H_2SO_4 hydrolyzable organic matter. J = residue unextracted. The organic matter in each of the sodium hydroxide extracts is separated into two fractions by acidifying the extracts to pH 2 or below. The resulting precipitate is analyzed for organic carbon and is represented as "humic acid." The organic carbon remaining in solution is determined and is represented as "fulvic acid." The several fractions of humic and fulvic acid are thus closely related.

TABLE 12. Content of Methoxyl and Degree of Humification in Soils of Different Groups (Kosaka, 1963).

Soil group	Locality	Degree of humification	Total organic C %	C/N[a]	Content of metroxyl[a] %
Ando	Aonohara	C	6.20	13.8	0.77
	Okunakayama	C	8.16	13.2	0.67
	Kawatabi	C	15.1	19.6	0.67
	Asamizo	C	8.78	16.0	0.63
	Sanperizan	D	15.7	22.1	0.41
	Kawaski	D	9.05	16.8	0.34
	Yonaizawa	D	11.5	22.0	0.20
	Daisen	E	12.5	24.1	0.16
	Kataji	E	15.5	44.3	0.05
Diluvial	Nihondaira	B	5.30	16.1	1.38
	Kano	B	4.49	19.5	1.07
Mountain podzol	Kira	A	29.9	23.8	1.41

[a] In total soil organic matter

scheme is simpler than the one proposed by Tyurin. The principal differences between the methods are as follows:

1. In the Tyurin method, 1 N Na_2SO_4 solution and 0.5 N H_2SO_4 solution are used for decalcification and extraction of fraction 1a, whereas in the Ponomareva method the decalcification and extraction of fraction 1a are carried out at the same time by extraction with 0.1 N H_2SO_4.

2. In the Tyurin method, each fraction is obtained by repeated extraction of the soil by centrifuging with NaOH solution until the extracts give a light yellowish coloration, whereas in the Ponomareva method the extractions with NaOH solution are not carried out repeatedly, but the ratio of extractant to solid is varied with the organic matter content.

3. In the expression of analytical results, the 1 N Na_2SO_4-soluble fraction and the 1 N H_2SO_4 hydrolyzable fraction are shown separately in the Tyurin method, whereas in the Ponomareva method the 1 N H_2SO_4-hydrolyzable fraction is added to the fulvic acid fraction as fraction 4.

In both methods, the organic carbon in the extracts is determined by oxidation with a known excess of potassium dichromate in sulfuric acid and back-titration of the residual dichromate with standard ferrous ammonium sulfate in the presence of phenylanthranilic acid as indicator.

Values for the various fractions of organic matter obtained by the two schemes are shown in Tables 13 and 14. Tokudome and Kanno (1965a) pointed out that the amount of organic matter in humic acid fraction 2 by the Ponomareva fractionation scheme is inconsistent with the exchangeable calcium status of the soils. They felt this behavior indicated that, for the soils they were investigating, the Ponomareva fractionation scheme was less suitable than the Tyurin fractionation scheme. Values for the fulvic acid fraction are higher in the Ponomareva scheme than in the Tyurin scheme, but the difference in this respect is mostly a consequence of incorporation of the N sulfuric acid-hydrolyzable organic matter as fraction 4 of fulvic acid in the Ponomareva scheme but not in the Tyurin scheme.

The data in Table 14 suggest that much of the organic matter extracted was associated with aluminum in the soils and that the proportion of humic acid carbon in the total organic carbon increased somewhat with depth. The total organic carbon and the proportion of humic acid carbon in the total organic carbon were less in young soils than in old soils.

The ratios of humic acid carbon to fulvic acid carbon in organic matter separated by Tyurin's method from the surface horizons of ten well developed soils of the ando group were found by Tokudome and Kanno (1965a) to be 0.9 to 1.2 in the subboreal zone of northern Japan (southern Hokkaido and northern Honshu), 0.8 to 1.0 in the humid subtropical zone of the Kanto region of central Japan and 1.4 to 1.9 in the humid subtropical zone of southwestern Japan (southern Kyushu). On the basis of these findings and a comparison of their

TABLE 13. Fractionation of Organic Matter by Tyurin's Method in the Particle-Size Fraction of Diameter Less than 0.25 Millimeter in Soils of the Ando Group in Japan (Tokudome and Kanno, 1965a).

Item		Numerical values for indicated soil sample number*						
		13	23	8	9	29	1	7
Total organic carbon (%)		5.68	8.29	19.27	23.57	15.94	1.67	11.88
Proportion of total organic carbon (%)								
Bitumen		2.64	3.62	2.49	2.72	3.26	2.99	1.85
NNa$_2$SO$_4$-soluble		1.94	1.45	1.21	0.96	1.10	2.39	0.98
Humic acid	Fraction 1	28.34	22.79	20.48	25.21	39.60	8.98	37.44
	Fraction 2	0	6.51	0.14	0.57	0.31	3.59	0.33
	Fraction 3	4.57	6.51	1.80	2.75	1.88	7.19	3.10
	Total	32.91	35.81	22.42	28.53	41.79	19.76	40.87
Fulvic acid	Fraction 1a	8.62	13.02	20.12	14.32	16.97	26.35	16.66
	Fraction 1	14.43	9.05	9.96	7.82	6.05	9.58	6.41
	Fraction 2	2.46	0	0.11	1.37	0.69	0	0.15
	Fraction 3	5.98	8.32	8.16	5.92	4.68	11.98	5.52
	Total	31.49	30.39	38.35	29.43	28.39	47.91	28.74
N H$_2$SO-hydrolyzable (α)		4.05	5.43	9.02	8.62	5.57	10.78	6.69
Residue unextracted		26.97	23.30	26.51	29.74	19.89	16.17	20.87
Humic acid C/fulvic acid C		1.05	1.18	0.58	0.97	1.47	0.41	1.42
Humic acid C/ (fulvic acid C+α)		0.93	1.00	0.47	0.75	1.23	0.34	1.15

* Sample descriptions as follows: No. 13 = 2 to 13 cm. depth from a soil derived from andesitic ash, Memuro, Kawanishigun, Hokkaido. No. 23 = 4 to 15 cm. depth from a soil derived from andesitic ash, Aisaka, Towadashi, Aomori Pref., Northern Japan. No. 8 = 0 to 10 cm. depth from a young soil derived from basaltic ash, Obuchi, Yoshiwarashi, Shizuoka Pref., Central Japan. No. 9 = 0 to 10 cm. depth from a moderately old soil developed from basaltic ash, Fumoto, Abegun, Shizuoka Pref., Central Japan. No. 29 = 0 to 15 cm. depth from an old soil developed from Onji andesitic ash, Kodonbaru, Uemura, Kumagun, Kumamoto Pref., Southern Kyushu. No. 1 = 0 to 30 cm. depth from a very young soil developed from andesitic ash, Hikinohira, Nishi-sakurajima, Kagoshima Pref., Southern Kyushu. No. 7 = 35 to 70 cm. depth from an old soil developed from andesitic ash, Kasanbaru, Kanoyashi, Kagoshima Pref., Southern Kyushu.

TABLE 14. Fractionation of Organic Matter by Ponomareva's Method in the Particle-Size Fraction of Diameter Less than 0.25 Millimeter in Soils of the Ando Group in Japan (Tokudome and Kanno, 1965a).

Item		Numerical values for indicated soil sample number*								
		13	23	8	9	29	1	7	G13	G14	G15
Total organic carbon (%)		5.68	8.29	19.27	23.57	15.94	1.67	11.88	14.05	13.54	7.55
Proportion of total organic carbon (%)											
Bitumen		2.64	3.62	2.49	2.72	3.26	2.99	1.85	5.2	1.6	1.0
Humic acid	Fraction 1	25.70	25.21	14.63	14.30	33.37	13.77	18.18	21.4	27.5	27.0
	Fraction 2	2.99	4.22	0.73	5.22	6.27	1.80	14.39	2.7	2.7	3.1
	Fraction 3	3.87	4.10	2.49	2.97	4.45	3.59	4.29	5.2	6.5	9.5
	Total	32.56	33.53	17.85	22.49	44.09	19.16	36.86	29.3	36.7	39.6
Fulvic acid	Fraction 1a	6.69	6.75	11.36	9.55	7.46	11.38	11.95	11.2	8.8	7.5
	Fraction 1	17.78	15.44	26.88	21.13	14.80	22.16	17.93	11.8	11.0	13.8
	Fraction 2	1.41	8.32	0.05	1.49	0.94	8.98	2.53	3.3	4.8	3.6
	Fraction 3	3.87	3.86	2.91	2.97	3.95	4.38	2.78	2.0	3.0	2.1
	Fraction 4	7.39	6.39	8.20	7.26	5.58	10.18	5.39	14.7	11.6	12.4
	Total	37.14	40.76	49.40	42.40	32.73	61.68	40.58	43.0	39.2	39.4
Residue unextracted		27.66	22.09	30.26	32.39	19.92	16.17	20.71	22.5	22.5	20.0
Humic acid C/fulvic acid C		0.88	0.82	0.36	0.53	1.35	0.31	0.91	0.68	0.94	1.01
Inorganic constituents extracted by 0.1 NH$_2$SO$_4$ (% of soil)											
Al$_2$O$_3$		2.19	4.02	11.20	10.68	10.75	1.52	11.60	7.83	8.80	8.85
Fe$_2$O$_3$		0.09	0.07	0.89	1.12	0.08	0.06	0.48	0.37	0.36	0.27
CaO		0.14	0.40	0.38	0.65	0.12	0.14	0.39	0.15	0.15	0.13
MgO		0.06	0.16	0.22	0.17	0.09	0.06	0.18	0.06	0.06	0.06
Inorganic constituents extracted by 1 N H$_2$SO$_4$ (% of soil)											
Al$_2$O$_3$		3.20	1.96	4.41	2.61	6.21	2.58	2.54	4.45	5.34	7.18
Fe$_2$O$_3$		0.55	0.89	4.80	4.33	2.54	0.60	1.81	4.19	5.22	5.97

* See Table 13 for description of samples Nos. 1, 7, 8, 9, 13, 23, and 29. Samples Nos. G13, G14, and G15 are from the 0 to 21, 21 to 54, and 54 to 68+ cm depth from a soil derived from andesitic ash, Tateyama, Goshimura, Kikuchigun, Kumamoto Pref., Central Kyushu.

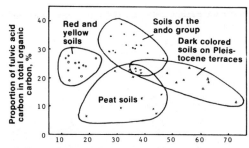

Fig. 14—Proportions of fulvic acid carbon and humic acid carbon in the total organic carbon of soils of different groups. The dark colored soils on Pleistocene terraces are thought to be developed on a mixture of volcanic ash and Pleistocene sediments. One relatively young soil was omitted from the group of soils of the ando group. (Tokudome and Kanno, 1968)

results with data published by investigators in other countries, they came to the tentative conclusions that the ratio of humic acid carbon to fulvic acid carbon of the organic matter in soils derived from volcanic ash (a) is higher in the humid subtropical zone than in the boreal, subboreal, and humid tropical zones and (b) is different from the corresponding ratio in organic matter of red and yellow soils and of brown forest soils. Fig. 14 shows the ratios in various groups of soils.

Other investigators (Kobo and Ohba, 1964; Ohba, 1965; Adachi, 1963a,b, 1966a,b,c, 1968) also have observed trends in composition of soil organic matter among geographic regions in Japan. Although the numerical values reported by the various researchers are not entirely consistent, probably because of differences in the methods adopted, the evidence indicates that, in the soils developed on volcanic ash in the humid, subtropical parts of Japan, the conditions are most favorable for formation of allophane, the amounts of organic matter accumulated are the greatest, and the ratios of humic acid carbon to fulvic acid carbon are highest. The suggestion has been made that allophane catalyzes the alteration of fulvic acids to humic acids.

Relatively little work has been done to investigate the significance of the various soil organic matter fractions in soil fertility. According to Nagai (1969), the amount of "group a" organic carbon in his fractionation scheme (previously described) is correlated with the mineralization of nitrogen, the allitic character of the soil due to active alumina, and the development of stable aggregates. He considered the organic carbon in group a less stable than that in group b or group c. He found that the amount of group a organic carbon is influenced by liming or phosphate application in the process of reclamation. The transference of organic carbon from fraction a_1 to fraction a_2 was found to take place with increasing accumulation of exchangeable Ca. Transference of organic carbon from fraction a_2 to a_1 was found to occur with increasing accumulation of soil phosphate. This change was thought to accelerate the mineralization of organic nitrogen.

Honda and Nishimune (1969) investigated the effect of treatment with silicate or phosphate on the proportions of certain fractions of the organic matter in soils derived from volcanic ash in the Tokachi district of Hokkaido. They measured the amounts of organic carbon extracted by phosphate buffers at pH 5.0, 9.0, and 12.5 when the soils were heated with the buffers for 2 hours at 50°C. Addition of sodium silicate, superphosphate, or diammonium phosphate 3 years before sampling decreased the amount of organic carbon extracted at pH 5, and addition of sodium silicate decreased the amount of organic carbon extracted at pH 9.0. The amounts of organic carbon extracted at pH 5.0 were closely correlated with the capacity of the various soils to react with phosphate. The authors considered that the changes in extractable organic carbon associated with the silicate and phosphate treatments resulted from increased microbial activity brought about by the improved phosphate supply in the soils.

PHYSICAL PROPERTIES

As compared with ordinary mineral soils, soils derived from volcanic ash have a relatively high porosity. Consequently, the bulk density is low and the water-holding capacity is high, especially if calculated on a weight basis. These and related subjects will be developed in this section.

Phase relationships

The phase relationships of soils derived from volcanic ash differ from those of other mineral soils. The difference is illustrated in Fig. 15, which shows the vertical distribution of solids, water, and air in two soil profiles under field conditions. The soil derived from volcanic ash, a soil of the ando group, has only from one-third to one-half as great a volume of solids but has about twice the volume of water per 100 cm³ as does the soil of the red and yellow group derived from diluvial materials.

Fig. 16 represents the phase relationships in another way. The sides of the equilateral triangle represent the proportions of solids, water, and air as a percentage of the total volume, and the proportions of the three phases in a given soil at a given time are represented by a single point. The figure shows points for a number of soils of different groups that were sampled when the proportions of water and air were those that occur under ordinary field conditions in Japan. The soils of the ando group may be seen to be in a class by themselves on account of their low proportion of solids. Soils developed on alluvium have a higher proportion of solids and soils of the red and yellow group a still higher proportion. Soils developed on sand dunes are notable for their low proportion of water, and those used for paddy are notable for their high

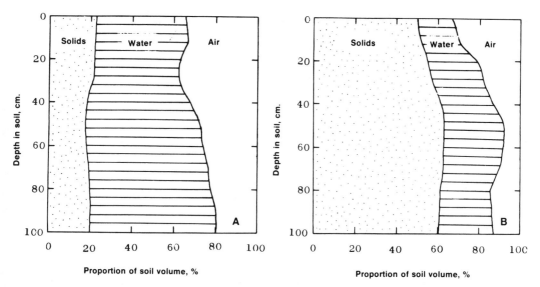

Fig. 15—Proportion of the volume of two soils consisting of solids, water, and air under field conditions in Japan. A Tanashi soil derived from volcanic ash. B. Shinjogahara soil of the red and yellow group.

(Misono, 1964)

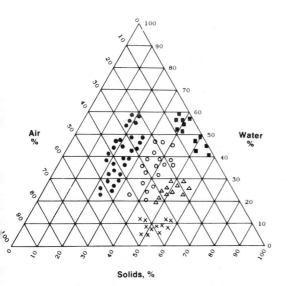

Fig. 16—Relative volumes of solids, water, and air in soils under ordinary field conditions in Japan. The legend for the different groups of soils is as follows: • = soils of the ando group, ° = soils derived from alluvium, ■ = paddy soils derived from alluvium, △ = red and yellow soils, and × = soils derived from dune sand.

(Misono, 1964)

proportion of water. Similar results were obtained by Masujima (1968), whose data in Table 15 represent average values derived from many determinations of solid, water, and air components of soils developed on volcanic ash and of other soils not developed on volcanic ash.

Misono and Kawajiri (1967, 1968) reported additional values for various groups of soils, as shown in Table 16. As in preceding illustrations, one may note that the volume occupied by solids is relatively low in soils derived from volcanic ash. The ratio of volume occupied by water to volume occupied by solids in

TABLE 15. Average Proportions of Solid, Water, and Air Components of Soils Developed on Volcanic Ash and of Other Mineral Soils * (Masujima, 1968).

Parent mineral material	Proportion of total soil volume, %					
	Plowed layer			Subsoil		
	Solids	Water	Air	Solids	Water	Air
Volcanic ash	25.6	35.9	38.5	23.0	43.3	33.8
Other than volcanic ash	38.8	29.7	31.5	42.1	32.1	25.8

* Numbers of samples were as follows: 106 and 102 for the plowed and subsoil layers of soils developed on volcanic ash and 145 and 134 for the plowed and subsoil layers of soils developed on mineral materials other than volcanic ash.

Properties of Soils Derived from Volcanic Ash   35

TABLE 16. Values of Weight of Solids plus Water, Volume of Solids plus Water, Volume of Solids, and Volume of Water in Soils of Different Groups Under Ordinary Field Conditions in Japan (Misono and Kawajiri, 1967).

Soil parent material	Solids + water per 100 cm³ of soil, g		Solids + water per 100 cm³ of soil, cm³		Solids per 100 cm³ of soil, cm³		Water per 100 cm³ of soil, cm³	
	Surface soil	Subsoil	Surface soil	Subsoil	Surface soil	Subsoil	Surface soil	Subsoil
Volcanic ash	80-120	80-130	45-75	55-90	18-30	16-22	24-50	42-70
Diluvium, nonvolcanic	120-175	157-200	56-85	73-98	36-58	47-67	15-36	18-45
Alluvium	108-180	141-197	50-90	67-100	40-54	45-60	20-43	20-60
Residuum from tertiary deposits	120-170	156-200	75-90	70-100	36-51	44-66	13-45	20-47
Dune sand	140-200	——	55-100	——	——	——	——	——

soils derived from volcanic ash is greater in the subsoil than in the surface soil, and it is greater in soils derived from volcanic ash than in soils derived from other parent materials.

Structure

Microstructure. Kawai (1969) made an extensive investigation of the microstructure of soils of the ando group on the basis of microscopic examination of soil thin sections together with supplementary analyses of soil composition. He classified the microstructure, or so-called fabric, of soils of the ando group into seven divisions, as shown in Table 17.

Compacted fabric is characterized by skeleton grains and plasma (colloidal and relatively soluble materials not contained in the skeleton grains) that are not associated with each other but are randomly distributed and compacted. There are few pores. The clay fraction of such soils is composed mainly of amorphous material with only an endothermic peak at 130 to 140°C. The soils contain less than 10% organic carbon that is easily decomposed by hydrogen peroxide.

Fine-grained porous fabric is characterized by high porosity and micro-aggregates or micropeds that occur as discrete entities. The soil color is 7.5 YR to 10 YR in hue, with a chroma of 1, the organic carbon content is 13 to 21%, and the bulk density is 0.36 to 0.54. The clay in most instances is amorphous, having only an endothermic peak at 140 to 150°C. In some instances, however, the clay has an exothermic peak at 900°C.

Blocky loose fabric is the most characteristic microstructure of soils of the ando group. The distinguishing features are loosely aggregated primary peds, small pores in the plasma, and numerous channel-type pores. The organic carbon content of the soil ranges from 1.4 to 13.2%. The clay fraction is dominated by allophane or allophane and gibbsite. The soil color usually has a hue of 7.5YR or 10YR and a chroma of 3 or 1, but the chroma of soils containing gibbsite is 2 to 6.

Coarse-grained porous fabric is characterized by a dominance of primary peds 0.15 to 0.3 mm in diameter. These peds are 3 to 5 times larger than those in the fine-grained porous fabric. An additional distinguishing feature is the occurrence of pedotubules. The organic carbon content is 3.8 to 8.4%, and the clay is dominantly gibbsite with small quantities of allophane. The soil is 5YR to 7.5YR in hue and 2 to 4 in chroma.

Foam-like compound fabric is characterized by the occurrence of micropores in the plasma that were inherited from volcanic glass. The pore walls are still unweathered glass.

The smoothed dense fabric is characterized by the dominance of pores of intermediate and large size in which the pore walls are usually smooth. This kind of microstructure occurs in B horizons of soils in which the clay is composed mainly of halloysite with some gibbsite or allophane. It does not occur in A horizons. The organic carbon content ranges from 0.6 to 5.5%.

The cemented porous fabric is characterized by the presence of large pores and mineral grains that are coated with clay and are linked together to form a porous but cemented fabric. This kind of microstructure is generally found in hard pans, but it occurs also in weathered pumice that is relatively soft.

36 T. Egawa

TABLE 17. Classification of Microstructure or "Fabric" of Soils as Proposed by Three Investigators.

K. Kawai (1969)	W. L. Kubiena (1938)	R. Brewer (1969)
Compacted fabric	Agglomeratic fabric	Agglomeroplasmic fabric
Fine grained porous fabric		
Blocky loose fabric		
Coarse-grained porous fabric		
Foam-like compound fabric		
Smoothed dense fabric	Porphyropectic fabric	Porphyroskelic fabric
Cemented porous fabric	Intertextic fabric	Intertextic fabric

Macrostructure. The surface layer of soils of the ando group commonly has granular structure. Table 18 shows that a substantial proportion of the weight of three soils of the ando group occurred in water-stable aggregates of diameter greater than 0.25 mm. These values for water-stable aggregates imply fairly good aggregation and permeability to water when compared with results obtained by other investigators with different types of soil. The Koganei soil in Table 18 showed the usual effect of cultivation, namely, a decrease in the proportion of the soil in large aggregates and an increase in the proportion in small aggregates. Cultivation seemed to have no major effect in this direction with the Godai soil. The low percentage of material exceeding a diameter of 3.5 mm indicates that the large clods collapse easily in water.

Table 19 shows that the percentage content of water-stable aggregates of different sizes found in soils of the ando group by the method of wet-sieving is markedly affected by the treatment of the sample before analysis. Soil samples in field-moist condition at the time of analysis had a greater proportion of large aggregates and a smaller proportion of small aggregates than did soil that had been air-dried before analysis. The usual procedure is to use air-dried soil. The way in which the dried sample is remoistened has a considerable effect on the results. The data in Table 19 were obtained by immersing the air-dried soil in water. This procedure causes maximum disruption of aggregates. Remoistening the soil under vacuum or under tension results in less disruption of aggregates than occurs with direct immersion at atmospheric pressure.

Although the practice of air-drying samples of soil before determining the content of water-stable aggregates may be defended on the basis of convenience, the validity of the results as an indication of the condition of soil in the field may be questioned, particularly in a humid region such as Japan. Table 19 provides some evidence on this point.

Several papers have been published on the mechanisms of formation of water-stable aggregates in soils (Egawa and Sekiya, 1956; Iimura and Egawa, 1956; Misono and Kishita, 1957b). All these authors emphasize the importance of organic matter. Two stages are mentioned. In the first stage, single particles are bound to each other by the action of colloidal clay and organic matter into small aggregates (primary aggregates or microaggregates), and in the second stage these small aggregates are bound to each other by the action of living organisms, including plant roots and microbial tissue, and also by organic substances such as polyuronides or colloidal proteins that are derived from microbial activities. The improvement of soil structure resulting from growth of perennial grasses is thought to be effected by these two steps (Kitagishi and Okita, 1956; Egawa et al., 1957).

TABLE 19. Water-Stable Aggregates in Air-Dry and Field-Moist Samples of Two Soils of the Ando Group (Misono, Kishita, et al., 1953).

		Percentage of soil by weight in water-stable aggregates in indicated size class	
Soil	Pretreatment	>2.5 mm	<0.25 mm
Godai	Field-moist	78.2	11.8
	Air-dried	20.6	27.1
Kurihara	Field-moist	51.4	29.3
	Air-dried	6.6	53.0

TABLE 18. Size-Distribution of Water-Stable Aggregates as Found by Wet-Sieving in Some Cultivated and Uncultivated Soils of the Ando Group (Mishono and Kishita, 1957a).

	Percentage of soil by weight in indicated size classes					
Soil	3.5—2.5 mm	2.5—1.0 mm	1.0—0.5 mm	0.5—0.25 mm	0.25—0.1 mm	<0.1 mm
Godai (cultivated)	4.1	12.1	17.8	20.9	31.0	14.1
Godai (uncultivated)	2.3	11.0	18.4	20.7	25.9	21.7
Kurihara (cultivated)	0.9	10.5	11.8	17.8	37.7	21.3
Koganei (cultivated)	1.9	8.8	11.9	11.2	16.7	49.5
Koganei (uncultivated)	7.5	18.8	12.1	9.5	11.0	41.0

Properties of Soils Derived from Volcanic Ash 37

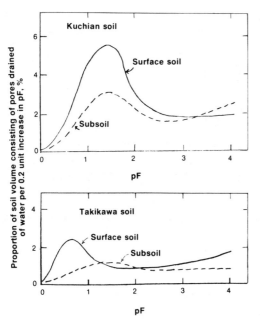

Fig. 17—Proportion of soil volume consisting of pores drained of water per 0.2 unit increase in pF in surface and subsoil samples of two soils. The Kuchian soil is from the ando group, and the Takikawa soil is derived from alluvium.
(Kishita and Tanoue, 1969)

Pore-size distribution. Soils of the ando group usually have relatively high total porosity, as already mentioned. Kishita and Tanoue (1969) investigated the distribution of pore space among pores of different sizes in some soils derived from different kinds of parent material in Hokkaido. The pore-size distribution curve was obtained by determining the amount of water removed from the soil at successively greater suctions, expressed as pF. The value of pore size was calculated from Jiren's formula, $D = 0.3/h$, where D is the pore diameter in centimeters and h is the height of the center of the soil sample above the level of water in a manometer attached to the tension table on which the soil sample is equilibrated. If $h = 10, 100$, and 1000 cm (pF = 1, 2, and 3), for example, the corresponding pore diameters are 0.3, 0.03, and 0.003 mm. Some results on two soils are given in Figure 17. As may be inferred from the figure, the Kuchian soil derived from volcanic ash contained more pore space than the Takikawa soil derived from alluvium. The most abundant pore sizes were smaller in the Kuchian soil than in the Takikawa soil.

The same researchers related the pore size distribution to the nature of the soil based on field observations. The classification system they proposed is given in Table 20. Pore distributions of types F and G are commonly found in mature soils of the ando group. The structural development is less pronounced in soils with pore-size distribution characterized by type F than type G. The soils derived from volcanic ash and having pore-size distributions characterized as types F and G have more capillary (fine) pores than do the other soils, and they generally have more total pore space as well. The volume of noncapillary (large) pores is intermediate. The characteristic distribution of pore size provides some explanation for the high water permeability and relatively low air permeability of soils of the ando group.

Water relations

During the past two decades, research on the water relations of soils of the ando group has progressed rapidly in Japan. The behavior of water in these soils has been clarified by measurements of water content versus pF and by field tests on water movement (Misono, Terasawa, et al., 1953; Misono and Terasawa, 1957; Terasawa, 1963).

Fig. 18 gives a comparison of the water content versus pF curves for a soil of the ando group and a soil of the red and yellow group. In each case the water content is calculated as a percentage of the weight of dry soil and as a percentage of the soil volume. The two curves lie close together for the soil derived from nonvolcanic materials but far apart for the soil derived from volcanic ash. Irrespective of the basis for calculating the water content, however, the soil derived from volcanic ash contained considerably more water at a given pF value than did the soil derived from nonvolcanic materials.

The significance of the data in Fig. 18 may be clarified by examining the relative water content of the two soils in different pF ranges. The content of water in soil in the range from pF 0 to pF 1.8 correlates mainly with the volume of large pores and hence with the permeability or drainage of the soil. In the case of the two soils represented in Fig. 18, the amount of water in this range was greater in the soil of the ando group than in the soil of the red and yellow group as a consequence of the relatively large total volume of pore space corresponding to this pF range in the former soil. (As will be indicated later, the high water content of the soil of the ando group in the

276

38 T. Egawa

Table 20. Pore Space, Bulk Density, and Structure of Soils Derived from Different Parent Materials in Hokkaido, Japan

(Kishita and Tanoue, 1969)

Pore distribution type	Soil No.	Parent material	Structure	Bulk density	Total porosity %	Non-capillary porosity (pF 0-1.6) %	Capillary porosity (pF 1.6-3.0) %
Type A	1 [a]	Dune sand	Single grain	1.25	54.0	15.0	13.9
Type A	2	Pumice	Single grain	1.33	53.3	30.1	11.6
Type B	3	Diluvium	Blocky	1.20	53.6	6.0	6.9
Type B	4	Diluvium	Blocky	1.15	56.0	9.4	11.6
Type C	5	Alluvium	Blocky	0.97	61.9	13.7	7.7
Type C	6	Alluvium	Blocky	1.14	55.8	13.4	12.6
Type D	7	Diluvium	Nutty	1.06	58.0	21.0	5.8
Type D	8	Diluvium	Nutty	0.97	61.2	18.4	5.8
Type E	9	Alluvium	Nutty	0.91	65.2	25.4	10.8
Type E	10	Diluvium	Nutty	1.08	58.2	15.2	6.1
Type F	11	Volcanic ash	Granular	0.71	69.6	10.0	24.5
Type F	12	Volcanic ash	Granular	0.81	68.3	15.0	14.3
Type G	13	Volcanic ash	Granular	0.72	68.7	11.9	22.6
Type G	14	Volcanic ash	Granular	0.97	62.1	8.4	22.5

Y-axis: Proportion of soil volume consisting of pores drained of water per 0.2 unit increase in pF, %
X-axis: pF

[a] This soil contained about 6% organic matter, probably as a consequence of a combination of admixture of young volcanic ash and the conditions of low temperature and poor drainage under which it developed.

277

Properties of Soils Derived from Volcanic Ash 39

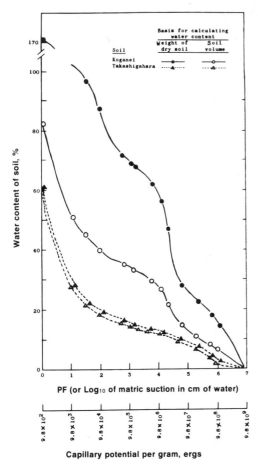

Fig. 18—Water content of two soils on the dry-weight basis and the volume basis versus pF and capillary potential of soil water. The Koganei soil is a member of the ando group, and the Takashigahara soil is a member of the red and yellow group developed on diluvial parent material.
(Terasawa, 1963)

low pF range does not seem typical.) The porosity in this pF range is generally less in the subsoil but the decrease with depth is usually smaller in soils of the ando group than in other types of soil. Thus the permeability to water tends to be relatively high in soils of the ando group.

The value of 1.8 as the boundary between the portion of the soil water just considered and the portion next to be considered is somewhat arbitrary. It is intended to represent the field capacity of soil for water, which has been defined as the water content of the moistened portion of the soil after the excess water has drained away and the rate of downward movement has decreased materially. Table 21 shows the pF values corresponding to the water content of two soils of the ando group at different times after wetting and at different depths under field conditions. The values for the surface layer of these two soils are somewhat higher than the values the same author determined for the surface layer of other soils not derived from volcanic ash (1.6 and 1.5 for the Taketoyo and Takashigahara soils derived from diluvial parent material and 1.7 for the Kashima soil, a sandy soil derived from parent material of Tertiary origin) but lower than the value 2.7, which Russell (1939) found applied to the water content of several soils at the moisture equivalent (a measurement used at one time to estimate the field capacity of soil), and lower than the value of 2.5, which corresponds to the one-third bar percentage now commonly used in the U.S. to estimate the field capacity.

The second range of significance lies between pF 1.8 and 4.2. A pF value of 4.2 is considered to correspond approximately to the permanent wilting percentage. The difference between the water content of soil at pF 1.8 and at pF 4.2 may be considered the capacity of soil to hold water in a form available to plants under conditions of free drainage in the field, a soil characteristic that is of great importance for agricultural purposes. Although the water in this range does not drain freely from the soil, it does move readily through short distances into adjacent soil that has been dried by evaporation or by removal of water by plants. Reference to Fig. 18 will show that the soil of the ando group contained considerably more

TABLE 21. pF Values Corresponding to the Field Capacity at Different Depths and at Different Times After Moistening Two Soils of the Ando Group (Iwata, 1968).

	Kumamoto soil				Saitama soil	
	pF value after indicated time in hours				pF value after indicated time in hours	
Depth, cm.	24	48	92	Depth, cm	24	48
10	1.98	2.02	2.11	10	1.81	1.89
20	1.89	1.95	2.06	20	1.76	1.84
30	1.83	1.91	2.00	40	1.62	1.72
50	1.86	1.93	2.03	70	1.40	..
100	..	..	1.84	90	1.34	1.45

TABLE 22. Water Removed in Two pF Ranges from Surface and Subsoil Samples of Nishigahara Soil of the Ando Group After Various Pretreatments (Misono, Terasawa, et al., 1953).

Sample	Treatment	Percentage content of water removed on the oven-dry basis by increasing the pF through the ranges indicated	
		4.2 to 5.5	5.7 to 7.0
Surface	Original soil	14.2	16.5
	Air-dried	9.2	15.9
	Air-dried and crushed	10.0	15.7
	Heated to 600° C	7.3	4.4
Subsoil	Original soil	47.5	25.2
	Air-dried	10.3	21.8
	Air-dried and crushed	10.4	23.3
	Heated to 600° C	8.3	10.7

water in this range than did the soil derived from diluvium. Because of the relatively large volume of water held in this range by soils of the ando group, the water removed by roots from the immediately adjacent soil should be replaced more readily by movement from more distant soil than is the case in soils holding a low volume of water. Terasawa (1963) demonstrated experimentally that upward movement of water from moist soil into dry soil occurred in a soil of the ando group when the upper soil layers were dried. Kira, Ambo, et al. (1963) and Kira, Soma, et al. (1963) made similar observations.

The third range of significance is above pF 4.2. Water in this range is retained primarily by the colloidal fraction of soils, and the amount retained varies with the nature and quantities of colloidal materials. As may be seen in Fig. 18, the water retained at pF values above 4.2 by the soil of the ando group exceeded that retained by the soil derived from diluvium. Misono, Terasawa, et al. (1953) named the water in the range of pF 4.2 to 5.7 "swelling water" and stressed the importance of this category of water in soils of the ando group. They found that more swelling water is retained by subsoils than by surface soils and that the amount retained is greatly decreased after the soil has been dried or heated, especially in subsoils, as seen in Table 22. A similar tendency was recognized also by Takenaka (1963). The removal of amorphous inorganic colloids by washing with Tamm's solution greatly decreased the content of swelling water (Masujima, 1962). Moreover, the amount of swelling water retained by soils of the ando group much exceeded the amount retained by young soils derived from volcanic ash.

These facts suggest that the swelling water is associated with the characteristic inorganic mineral colloid, allophane, and that the water-retaining properties of allophane are modified by the treatments mentioned. Iwata (1968) suggested that the swelling water is physically adsorbed on the surface of allophane particles. Further studies are needed, however, to verify the physical condition of the water in the pF range from 5.2 to 5.7.

The content of water held at pF values above 6 (hygroscopic water) similarly is higher in soils of the ando group than in other soils. As suggested by Table 22, the effect of a preliminary heat treatment on the amount of water that may be removed from the soil above pF 6 is substantial, but a preliminary air-drying treatment has little effect.

Soils of the ando group clearly have relatively favorable water relations for most agricultural plants. Their available water capacity is exceptionally high, their permeability to water is fairly adequate, and there is also substantial movement of water from moist soil to adjacent dry soil. There are circumstances, of course, in which these generalizations for mature soils of the ando group may not apply. For example, where a soil is developed in a relatively young layer of volcanic ash that overlies older deposits, hard and compact sublayers may interrupt the movement of water. A condition of this kind occurs extensively in Hokkaido and locally in Honshu. Special tillage practices are recommended to overcome this defect, as will be explained in Chapter 3.

Aeration

The exceptionally high porosity of soils of the ando group suggests immediately that these soils have adequate aeration. On the other hand, the high capacity of the soils to hold water, together with the large populations of anaerobic bacteria and the poor development of plant roots sometimes observed, suggest that aeration might after all be a problem.

Properties of Soils Derived from Volcanic Ash 41

Kishita (1955) investigated the relationship between air permeability and porosity of a number of soils derived from different parent materials. His findings in Fig. 19 indicate that, despite the high porosity, the soils of the ando group derived from volcanic ash had lower permeability to air than soils of any of the other groups.

Kishita (1959) measured also the concentrations of CO_2 and O_2 in samples of air withdrawn from two soils on which upland rice and soybeans were grown. Fig. 20 shows that the CO_2 concentration was higher in the Nishigahara soil of the ando group than in the Takashigahara soil of the red and yellow group, indicating that the aeration was poorer in the former soil than in the latter. Both of the soils used in this experiment were prepared by sieving to obtain coarse and fine aggregates. As expected, the CO_2 content of the air was generally somewhat higher in the soils artificially constituted of fine aggregates than in the soils artificially constituted of coarse aggregates. In another experiment, Kishita determined the composition of air in the same two soils to which pulverized rice straw had been added. Again the CO_2 concentration was greater in the Nishigahara soil of the ando group than in the Takashigaara soil of the red and yellow group.

The results of these experiments on the aeration of soils of the ando group are generally consistent with the data in Table 20, which indicate that the proportion of the total soil volume occupied by large pores in soils of the ando group is modest and is generally smaller than it is in other soils. The relatively large volume of large pore space indicated for the soil of the ando group in Fig. 18 would seem exceptional.

CHEMICAL PROPERTIES

Reaction and base status

Reaction and base status are influenced by the parent materials, the environmental conditions, and the time of weathering, as reflected in the weathering stage. Some data on the soil reaction and base status of soils of the ando group in Japan are presented in Table 23.

In general, soils of the ando group show an acid reaction in the surface horizons, and the pH tends to increase in lower horizons. The pH (H_2O) values of the A horizons are generally in the range from 4.0 to 6.5. The acidity originates mainly from the organic exchange sites and not from the inorganic sites because allophane is a very weak acid, as mentioned previously. Young soils derived from volcanic ash are commonly neutral because little organic matter has accumulated and leaching of bases has not been extensive. Occasional young soils derived from volcanic ash may be strongly acid because of the presence of sulfur compounds and the oxidation of these compounds to produce sulfuric acid. The acidity of buried soils derived from volcanic ash is generally not strong, with pH (H_2O) values of about 5.0 to 6.5. Soils of the brown genus and onji (glassy) genus show a rather high pH value of about 6.0, due in the

Table 23. pH and Cation-Exchange Properties of Three Soils of the Ando Group.

Investigator	Soil Name and location	Soil Parent material	Horizon	Depth cm	pH In H_2O	pH In 1N KCl	Exchange acidity y_1[a]	Cation-exchange capacity per 100g. m.e.	Exchangeable cations per 100g. m.e. Ca	Mg	K	Na	Degree of base saturation %
Kanno (1961)	Koganei (Tokyo)	Basaltic ash	A_{11}	0-25	5.4	4.7	0.9	49.4	6.1	1.7	0.1	0.9	18
			A_{12}	25-50	6.3	5.0	0.7	51.5	14.4	3.1	0.3	2.1	39
			B_1	50-90	6.4	5.2	0.7	53.6	10.8	4.0	0.3	1.6	31
			B_2C	90-150	6.5	5.6	0.7	39.9	4.9	2.5	0.2	1.4	22
Kanno (1961)	Rokuhara (Iwate)	Pyroxene-andesitic-ash	A_{11}	0-10	5.3	4.0	11.0	37.2	1.0	0.8	0.4	1.3	10
			A_{12}	10-20	4.9	4.1	11.1	33.1	0.5	0.5	0.3	1.2	7
			B_1	20-30	4.7	4.1	11.4	27.9	0.4	0.4	0.2	1.0	7
			B_2	30-	4.9	4.0	22.1	19.1	0.9	0.5	0.2	1.1	14
Miyazawa (1966)	Hirusen-bara (Okayama)	Hornblende-quartz-andesitic ash	A_{11}	0-13	5.3	4.4	13.0	51.0	0.5	1.3	0.7	0.4	6
			A_{12}	13-46	5.4	4.5	8.5	48.0	0.5	0.5	0.5	0.2	3
			B_1	46-64	5.5	5.0	1.6	15.0	0.4	0.2	0.3	0.2	8
			B_2	64-90	5.5	5.5	0.6	10.6	0.4	0.2	0.3	0.1	10
			C	90-	5.5	6.0	0.9	4.6	0.5	0.3	0.4	0.1	27

[a] See the text for a definition of y_1

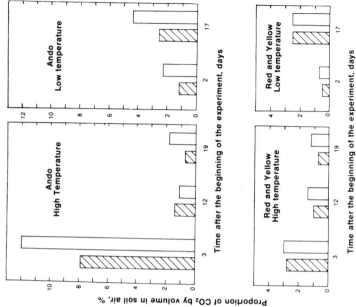

Fig. 20—Carbon dioxide concentration in samples of air withdrawn from cultures of Nishigahara soil of the ando group and Takashigahara soil of the red and yellow group at different times after the start of an experiment. The high temperature was 30°C in the daytime and 20°C at night. The low temperature was the out-of-door temperature in February (−5 to +10°C). The cross-hatched and open bars refer to measurements on soil artificially constituted of coarse granules (∼1 mm diameter) and fine granules (∼1 mm diameter), respectively. All cultures received 80 g of pulverized rice straw and were maintained at a water content corresponding to pF 2.7.

(Kishita, 1959)

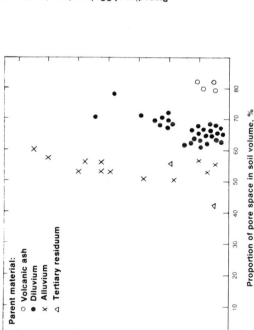

Fig. 19—Permeability of various soils to air versus total pore space. The soils derived from volcanic ash were all members of the ando group.   (Kishita, 1955)

former case to a relatively high content of exchangeable calcium and magnesium and in the latter case to a low content of organic matter and to the high isoelectric point of allophane.

Generally speaking, the exchange acidity increases with the content of organic matter. The exchange acidity in soils of the ando group, however, is not always closely related to the pH or the degree of base saturation.

Values for exchange acidity in Table 23 are milliliters of 0.1 N sodium hydroxide solution used to titrate 125 ml of an extract obtained by equilibrating 250 ml of 1 N potassium chloride solution for 5 days with 100 g of soil. This method was proposed by Daikuhara (1911), and in Japan the values are conventionally represented by the symbol, "y_1", the reason for the subscript being that the method as described was only the first in a series of equilibrations and titrations leading to a value for the total exchange acidity. After removing and titrating the 125-ml aliquot, Daikuhara added another 125 ml of 1 N potassium chloride, equilibrated the solution with the soil, and then removed 125 ml of the solution for a second titration, which was designated "y_2." This process was repeated a number of times to obtain the total exchange acidity.

When the cation-exchange capacity and degree of base saturation of the soils in Table 23 are taken into account, the values shown for exchange acidity are relatively low. In soils of the ando group, the total acidity obtained by equilibration of a sample of soil repeatedly with 1 N potassium chloride solution and titration of the extract in the manner described is sometimes more than ten times as great as the value of y_1, whereas in soils derived from parent materials other than volcanic ash the total acidity is considered to be 3 to 3.5 times the value of y_1. The relatively low ratio of exchange acidity to total acidity in soils of the ando group is consistent with the fact that most of the cation-exchange capacity is due to organic matter and allophane, in both of which the number of cation-exchange sites increases rapidly with an increase in pH value.

A second explanation for the relatively low ratio of exchange acidity to total acidity in soils of the ando group is provided by the behavior of allophane, which occurs in these soils. Allophane does not have the strong-acid character that is imparted by isomorphous substitution to montmorillonite and certain other minerals. Rather, allophane is an amphoteric substance with an isoelectric pH near 6. It holds anions as well as cations in exchangeable form. Presumably, therefore, when allophane or soils containing allophane are treated with potassium chloride to release the exchange acidity, the hydrogen ions released by exchange with potassium are neutralized in part by hydroxyl ions released concurrently by exchange with chloride.

TABLE 24. pH, Exchange Acidity, and Exchangeable Calcium and Magnesium in Soils of the Ando Group Differing in Degree of Weathering (Kobo and Ohba, 1964).

Degree of weathering	pH in 1 N KCl	Exchange acidity y_1[a]	Exchangeable calcium per 100 g, m.e.	Exchangeable magnesium per 100 g, m.e.
I	4.70	2.8	4.2	0.6
II	4.43	3.5	2.4	0.6
III	4.47	8.8	0.9	0.4
IV	4.13	16.0	1.1	0.3

[a] See the text for a definition of y_1.

Among the exchangeable cations, calcium is generally most abundant, and magnesium is next. The contents of potassium and sodium are usually very low. The sodium content is usually equal to, or a little higher than, that of potassium, perhaps because alkali plagioclase is fairly abundant in volcanic ash. Some soils are rich in magnesium, which is attributed mainly to the presence of olivine. Exchangeable manganese is present in concentration lower than that of potassium.

Kobo and Ohba (1964) related the reaction and base status of 36 soils of the ando group to the degree of weathering, with results as shown in Table 24. Four weathering classes were used, according to the degree of etching of augite (determined microscopically), the ratio of coarse sand (2 to 0.2 mm diameter) to clay (<.002 mm diameter), and the clay content. The findings show that the pH in 1 N potassium chloride solution decreased, the exchange acidity increased, and the content of exchangeable calcium and magnesium decreased with increasing weathering of the soils according to the criteria employed. Table 25 shows the same trends of pH, exchange acidity, and exchangeable calcium and magnesium where soils were classified according to the mineralogical nature of the parent volcanic ash. A relatively high content of exchangeable magnesium was found in the soils derived from olivine-type volcanic ash.

Cation-holding power and cation-exchange capacity

The two terms, cation-holding power and cation-exchange capacity, would ordinarily be considered synonymous. In the case of soils in which an important part of the cation-exchange capacity is derived from allophane or from organic matter, however, the cation-exchange capacity varies markedly with the manner in which it is determined. In soils of the ando group, the cation-exchange capacity resides mostly in allophane and organic matter. Consequently, the

TABLE 25. pH, Exchange Acidity, and Exchangeable Calcium and Magnesium in Soils of the Ando Group Derived from Different Types of Volcanic Ash (Kobo and Ohba, 1964).

Type of parent volcanic ash	pH in 1 N KCl	Exchange acidity y_1 [a]	Exchangeable calcium per 100 g, m.e.	Exchangeable magnesium per 100 g, m.e.
Olivine	4.78	1.7	3.1	0.8
Pyroxene	4.30	7.2	2.3	0.4
Pyroxene-hornblende	4.35	9.2	1.0	0.4
Hornblende	4.14	18.8	0.6	0.2

[a] See the text for a definition of y_1.

variation of the cation-exchange capacity with experimental conditions is especially pronounced in these soils, and Japanese investigators have given considerable attention to the matter because of its great importance in soils of Japan. The term, cation-holding power, has been used to describe the variations that occur.

The basic phenomenon responsible for the characteristic behavior of organic matter and especially of allophane, where cation-retention is concerned, is the weak-acid character of the cation-exchange positions. The proportion of the potential sites that is active in holding cations in exchangeable form depends rather strongly on the concentrations of basic cations and hydrogen ions in solution.

Harada and Kutsuna (1955) and Yoshida (1953, 1956) used two different methods to characterize the cation-holding power of soils. In the first method a soil was leached with a 1-normal or a 0.1-normal solution of ammonium acetate or calcium acetate until equilibrium was attained. The amount of the added cation retained by the soil was then determined. In the case of the 1-normal solutions, the excess ammonium acetate or calcium acetate was removed by washing with alcohol before determination of the adsorbed ammonium. In the case of the 0.1-normal solutions the excess of ammonium acetate or calcium acetate remaining after exchange saturation was determined by weighing the leaching tube, and the exchangeable and soluble forms of ammonium were then displaced and determined together. This procedure eliminated loss of ammonium from exchangeable form by hydrolysis during leaching. The ratio of the amount of the cation adsorbed from the 0.1-normal solution to the amount adsorbed from the 1-normal solution was taken as an index of the capability of the soil to hold the cation against hydrolysis. The second method was based on the effect of pH on cation adsorption. A soil was equilibrated by leaching with several salt solutions, the pH values of which were adjusted to 7, 6, 5, or 4. The amount of the cation retained by the soil decreased with decreasing pH, and the amount retained at a given pH value below 7 relative to the amount of the cation retained at pH 7 was adopted as an index of the capability of the soil to hold cations against acidification.

They used their method with various cation species and soils and found that soils of the ando group had lower index values for cation retention than did other soils. Table 26 gives some of the data obtained by Harada and Kutsuna (1955) on cation retention.

Kutsuna and Nomoto (1961b) determined the cation-exchange capacity of various soils at pH 5 and pH 7 and made an associated investigation of the principal sources of cation-exchange properties. Their results, summarized in Table 27, illustrate the substantial distinction that exists between montmorillonite and kaolinite on the one hand and allophane and organic matter on the other.

The results of the laboratory studies of the cation-retention properties of various soils shown in Tables

TABLE 26. Cation-Exchange Capacity of Two Soils as Determined at pH 5 and pH 7 with Normal and Tenth-Normal Solutions of Calcium and Ammonium Acetates (Harada and Kutsuna, 1955).

		Cation-exchange capacity per 100 g of soil as found with indicated solution, m.e.							
		1N NH_4OAc		0.1 N NH_4OAc		1 N $Ca(OAc)_2$		0.1 N $Ca(OAc)_2$	
Soil parent material	Total C, %	pH 7	pH 5	pH 7	pH 5	pH 7	pH 5	pH 7	pH 5
Volcanic ash	6.4	27.4	20.6	23.0	11.3	32.4	17.9	26.5	17.5
Diluvium	1.1	18.5	16.7	18.5	14.9	19.3	15.1	18.5	16.0

26 and 27 indicate that soils of the ando group tend to lose cations from the exchangeable form comparatively readily by hydrolysis when the concentration of cations in solution is low and, moreover, that hydrogen ions in solution have a strong replacing power for basic cations. These observations aid in explaining the rapid loss of cations by leaching from soils derived from volcanic ash and the low content of exchangeable bases that develops in these soils in a humid region such as Japan.

Investigation has shown also that soils of the ando group adsorb calcium preferentially over ammonium and potassium, whereas other soils and clays show the reverse. Some of Yoshida's (1961) data on this subject are given in Tables 28 and 29. To obtain the values in Table 28, the various samples were first exchange-saturated by leaching them with an excess of a solution 0.5 N to calcium chloride and 0.5 N to ammonium chloride. The pH of this solution was near 7.0. Then the samples were extracted with 1 N sodium nitrate to remove the excess of soluble calcium and ammonium and the exchangeable cations. Determination of the calcium, ammonium, and chloride in the extract and allocation of the chloride equally to calcium and ammonium made it possible to calculate the calcium and ammonium retained in exchangeable form without loss of part of either cation by the hydrolysis that would have occurred if the soluble cations had been removed before extraction of the exchangeable cations. The data in Table 29 were obtained in the same way. The 0.1 N solution was obtained by diluting one part of the mixture of 0.5 N calcium chloride and 0.5 N ammonium chloride with nine parts of water. The pH of the diluted solution was 6.5.

Yoshida (1957, 1960a,b) classified the negative charges into two categories, namely **i** (inside) and **o** (outside). He identified **i** charges with preferential adsorption of ammonium and potassium over calcium and attributed these charges to locations inside the crystal structure of clay minerals, i.e., to isomorphous substitution. He considered that the preferential adsorption of ammonium and potassium was a consequence of their having the appropriate ion-radius to fit the pores formed in the hexagonal plane of oxygen ions on the surface of the individual molecular layers of clay. He attributed the **o** charges to organic matter and amorphous inorganic colloids. The method he used to determine the two classes of charges was to leach a 2- to 5-g sample of soil with 100 ml of a neutral (pH 7) solution that was 0.5 N to sodium and 0.5 N to ammonium. The anions were acetate and chloride. Then the soil was leached with methanol until the leachate was free of chloride, after which the soil was leached with 50 ml of 1 N magnesium sulfate solution to extract the exchangeable sodium and ammonium. The **i** charges were identified with the exchangeable ammonium, and the **o** charges identified with the exchangeable sodium. Table 30 gives Yoshida's data on the different kinds of exchange positions in several substances.

TABLE 27. pH-Dependence of Cation-Exchange Capacity of Various Soils Classified According to the Main Sources of Negative Charges (Kutsuna and Nomoto, 1961b).

Principal source of negative charges	Cation-exchange capacity at pH 5 as percentage of cation-exchange capacity at pH 7
2:1 clays, dominantly montmorillonite	90
1:1 clays, dominantly kaolinite	85 to 90
Allophane	<70
Organic matter, allophane	<75
Organic matter, 1:1 clays	<75

TABLE 28. Preferential Adsorption of Ammonium and Calcium by Various Soils and Bentonite from a Solution 0.5 Normal to Ammonium Chloride and 0.5 Normal to Calcium Chloride (Yoshida, 1961).

	Soil		$\dfrac{NH_4 \times 100[1]}{NH_4 + Ca}$	$\dfrac{Ca \times 100[1]}{NH_4 + Ca}$
Location	Parent material	Dominant clay mineral		
Morioka	Volcanic ash	Allophane	28	72
Tokyo	Volcanic ash	Allophane	23	77
Tottori	Volcanic ash	Allophane	17	83
Miyazaki	Volcanic ash	Allophane	23	77
Morioka	Alluvium	—	60	40
Ehime	Schist	Vermiculite	76	24
(Gunma bentonite)		Montmorillonite	80	20

[1] Values for NH_4 and Ca used in the calculations were m.e. of the cations retained in exchangeable form per 100 g of soil.

TABLE 29. Proportions of Potassium and Calcium in Exchangeable Form in Different Substances After Equilibration With an Equinormal Solution of Calcium Chloride and Potassium Chloride at Two Concentrations at pH 6.5 (Yoshida, 1960).

Concentration of solution[a]	Sample[b]	Exchangeable bases per 100g of material, m.e.[c]			$\frac{K \times 100^d}{K + Ca}$	$\frac{Ca \times 100^d}{K + Ca}$
		Potassium	Calcium	Total		
1.0 N	133	27.6	34.8	62.4	44	56
	134	14.3	8.0	22.3	64	36
	Bentonite	49.9	10.8	60.3	83	17
	Exchange resin	0.68	0.73	1.41	48	52
0.1 N	133	7.5	17.1	24.6	30	70
	134	10.1	9.7	19.8	51	49
	Bentonite	31.5	25.1	56.6	56	44
	Exchange resin	0.26	1.06	1.32	20	80

a The 1.0 N solution was 0.5 N to potassium chloride and 0.5 N to calcium chloride, and the 0.1 N solution was 0.05 N to potassium chloride and 0.05 N to calcium chloride, as explained in the text.
b No. 133 was a surface sample of a soil from Morioka derived from volcanic ash. No. 134 was a surface sample of a soil from Morioka derived from alluvium. The bentonite was Gunma bentonite, a commercial product. The exchange resin was Amberlite IR-120.
c Values for the Amberlite IR-120 exchange resin are milligram equivalents per 0.3 ml of resin.
d Values for potassium and calcium used in the calculations were milligram equivalents of the cations retained in exchangeable form per 100 g of soils and bentonite and per 0.3 ml of Amberlite IR-120 exchange resin.

Perhaps the most widely accepted method in Japan for determining the cation-exchange capacity of soils is Yoshida's (1953) modification of the Schollenberger and Simon (1945) method. The Schollenberger and Simon method involves saturating the exchange positions with ammonium by leaching the soil with an excess of neutral, 1 N ammonium acetate, removing the excess ammonium acetate by leaching the soil with 80% methyl or ethyl alcohol treated with enough ammonium hydroxide to produce the pH 7 tint of bromthymol blue, transferring the residual soil to a flask, and distilling off the ammonium as ammonia in the presence of 1 N sodium hydroxide. The ammonia is absorbed in acid and determined titrimetrically. A blank distillation is made on soil that has been leached to remove exchangeable ammonium to provide a correction for the release of ammonium by decomposition of the soil organic matter in the presence of hot sodium hydroxide. Yoshida's (1953) modification of this method is similar in principle but uses smaller equipment and permits leaching in the absence of atmospheric carbon dioxide.

Generally speaking, cation-exchange capacity values for soils of the ando group are greater than those of other soils, they are greater in the surface soils than in the subsoils, and they usually decrease a great deal as a result of treatment with hydrogen peroxide to decompose the organic matter. The cation-exchange capacity per gram of organic carbon is about 2.6 m.e. This value is smaller than the cation-exchange capacity of organic matter that has been separated from soil. It is assumed that the cation-exchange capacity of the surface layer of soils of the ando group is carried mostly by organic matter but that the organic matter is combined with clay to form a complex and loses some of its exchange capacity as a result of the combination (Kutsuna and Nomoto, 1961a).

TABLE 30. Proportions of i and o Charges in Soils, Kanto Bentonite, and IR-120 Cation-Exchange Resin (Yoshida, 1960a).

	M.E. per 100 g		
	i Charges (A)	o Charges (B)	$\frac{100A}{A + B}$
Soil 133[a]	4	44	8
Soil 134[a]	11	11	50
Kanto bentonite	44	17	72
IR-120 cation-exchange resin	0	100	0

a See footnote to Table 29.

Phosphorus sorption

Soils of the ando group are characterized by their capability to react rapidly with large amounts of phosphorus, particularly under acid conditions. Consequently the availability to plants of soluble phosphates applied as fertilizer is quickly depressed, and only about 10% of the applied phosphorus is utilized by most upland crops.

The low availability of native and applied phosphorus for plants in soils of the ando group is undoubtedly one of the most important limiting factors in crop production on these soils. How to reduce the loss of availability of applied phosphorus and how to increase the availability of native and previously applied phosphorus are very important problems in agricultural practice on soils of the ando group. In a later chapter, recent progress in the forms of chemical fertilizers and in application methods designed to alleviate these problems will be described.

In Japan, the "phosphorus sorption coefficient" of the soil is commonly used as an index of the ability of the soil to react with applied phosphorus. It is measured in the following way. A 50-g sample of soil is treated with 100 ml of a neutral, 2.5% $(NH_4)_2HPO_4$ solution for 24 hours at room temperature with occasional shaking. The suspension is then filtered, and the phosphorus content of the filtrate is determined. The amount of phosphorus sorbed, calculated as mg of P_2O_5 per 100 g of soil, is called the phosphorus sorption coefficient. Kosaka (1963) proposed a modified rapid method in which the weight of soil is reduced to 10 g and the residual phosphorus in solution is determined colorimetrically by the vanadomolybdate method after decomposition of the dissolved organic matter by nitric acid and hydrogen peroxide.

Some data are shown in Table 31. With the exception of one sample of coarse, recently erupted volcanic material, the phosphorus sorption coefficient of the samples from soils derived from volcanic ash exceeded 1300 mg of P_2O_5. All the samples from soils derived from other parent materials had phosphorus sorption coefficients less than 700 mg of P_2O_5.

Kobo and Ohba (1964) investigated also the relationships between the kind of parent material, the degree of weathering, and the phosphorus sorption capacity. In this investigation, the amount of phosphorus sorbed was determined when 10 g of soil were equilibrated with 20 ml of a 4% $(NH_4)_2HPO_4$ solution for 24 hours at 30° C. The results in Table 32 suggest that the mineralogical character of the ash may not be of great significance but that the phosphorus sorption increases with increasing degree of weathering (and increasing amount of clay) through the first three stages of weathering but that it decreases at weathering stage IV. Because the clay content of soils at weathering stage IV commonly exceeds that at stage III, the indications are that the clay minerals present at stage IV are different from those at stage III.

The strong affinity of soils of the ando group for phosphorus is a consequence of the allophane they contain. Wada (1959) found that ammonium phosphate reacts with allophane to form an insoluble phosphate (an ammonium-substituted taranakite). The reaction takes place rapidly at pH 4 but more slowly at pH 7.

The reaction between soil and phosphate solution continues over a period of time. The first stage is a surface reaction, and it takes place rapidly. The

TABLE 31. Coefficient of Phosphorus Sorption of Soils Classified According to Parent Material (Kosaka, 1963).

Soil parent material	Soil series	Depth, cm	Soil texture	Phosphorus sorbed per 100 g of soil, mg P_2O_5
Volcanic ash	Taisho	0-7	Loamy fine sand	423
		7-10	Fine sandy loam	1604
		10-28		1378
	Kamisato	0-15	Fine sandy loam	2270
		15-35	Loam	2670
		35-80	Fine sandy loam	2530
	Yuzuhara	0-45	Light clay	2134
		45-100	Sand	2793
	Wakahara	0-12	Clay loam	2399
	Miyagasaki	0-26	Clay loam	1887
		26-40	Clay loam	2422
		40-100	Loam	2312
Other than volcanic ash	Komukai	0-10	Light clay	289
		10-25	Silty clay	289
		25-40	Light clay	418
	Nishikai	0-15	Clay loam	263
		15-75	Clay loam	330
		75-90	Clay loam	655

T. Egawa

TABLE 32. Phosphorus Sorption by Soils Derived from Different Mineralogical Types of Volcanic Ash at Different Degrees of Weathering (Kobo and Ohba, 1964).

Degree of weathering	Phosphorus sorbed per 100 g of soil derived from indicated type of volcanic ash, mg of P_2O_5			
	Olivine	Pyroxene	Pyroxene-hornblende	Hornblende
I	2940	1620	—	3020
II	3970	3590	3350	—
III	4410	4620	4660	4100
IV	—	2870	4110	—

aluminum phosphate thus formed separates from the surface and changes gradually to a stable crystalline phase. The so-called phosphorus sorption coefficient may be considered to represent the phosphate sorbed on the clay surface in the first stage. It is not intended to represent the maximum amount of phosphate that could be sorbed by the soil.

The relationship between pH and phosphate sorption has been studied by several researchers (Shioiri, 1934; Kobo and Akatsuka, 1951; Sekiya, 1970), and their results show that lowering the pH of soils of the ando group causes an increase of phosphate sorption. The increased reactivity of the soils toward phosphate as the pH is decreased is thought to be a result of activation of aluminum on the surface of the inorganic mineral colloids.

The forms of phosphorus present in soils of the ando group have been investigated in some detail. Sekiya and Egawa (1961) found a high content of total and organic phosphorus in such soils, as shown in Table 33. These were all uncultivated virgin soils. The soil from the red and yellow group included for comparison was developed on diluvium and was from a control plot that had not received phosphate fertilizer.

Sekiya and Egawa (1959, 1960, 1961) examined the applicability of the method of Chang and Jackson (1957) for determining the forms of phosphorus in soils of the ando group in Japan. They synthesized artificial aluminum phosphate ($AlPO_4 \cdot 2H_2O$) and ferric phosphate ($FePO_4 \cdot 2H_2O$) and determined their solubility in the 0.5 **N** NH_4F (pH = 7.0) and 0.1 **N** NaOH extractants proposed by Chang and Jackson. Although these phosphate compounds were completely dissolved by the respective reagents, the recoveries of phosphorus in the extracts were not satisfactory when the compounds were added to a sample of soil of the ando group. In this study, the phosphorus was labeled with the radioactive isotope, ^{32}P, and 2.5 mg of the labeled phosphate compounds were added to 2 g of soil. Where the phosphates were added to a soil of the ando group, only about 20% of the P added as $AlPO_4 \cdot 2H_2O$ was extracted by the 0.5 **N** NH_4F solution recommended by Chang and Jackson. About 25% of the P added as $AlPO_4 \cdot 2H_2O$ was removed by a second extraction with 0.5 **N** NH_4F solution, and 40 to 50% was removed by a subsequent 0.1 **N** NaOH extraction. Finally, 7 to 9.% was extracted by 0.5 **N** H_2SO_4. On the other hand, the phosphorus of $FePO_4 \cdot 2H_2O$ added to a soil of the ando group was extracted only by 0.1 **N** NaOH solution, and the extraction was complete, as shown in the experiment by Chang and Jackson.

On the basis of these data, Sekiya and Egawa considered that the method of Chang and Jackson is not directly applicable for soils of the ando group, in which the main clay mineral is allophane. They conducted additional experiments on the recovery of aluminum and iron phosphates and finally adopted a procedure in which 100 ml of **N** NH_4F (pH 7) were added to 1 g of soil to extract the Al-type phosphorus. One more important modification was concerned with the concept and method of extraction of the so-called Ca-type phosphorus. According to Chang and Jackson, Ca-type phosphorus is considered to be the stable apatite type and is soluble in 0.5 **N** H_2SO_4 after extraction of the Al- and Fe-type phosphorus. The modified method is based on the concept that, in cultivated soils of the ando group that have received large amounts of soluble phosphate fertilizers, the calcium phosphates present are probably not apatites but rather are intermediate forms of greater solubility that may be dissolved by 2.5% acetic acid, as recommended by Mattson, Williams and Barkoff (1950). Sekiya and Egawa adopted 2.5% acetic acid as the reagent to dissolve Ca-type phosphorus prior to extraction of Al- and Fe-type phosphorus. The complete flow sheet of the fractionation of soil phosphorus proposed by Sekiya and Egawa is shown in Fig. 21. Some data on the amounts of various forms of phosphorus found by the modified method in some Japanese soils are

TABLE 33. Organic and Inorganic Phosphorus Content of Various Soils (Sekiya and Egawa, 1961).

Soil group	Soil name and location	Phosphorus per 100 g of soil, mg of P_2O_5		
		Total	Organic	Inorganic
Ando	Ibaragi (Kanto District)	256	82	174
	Kariya (Tohoku District)	140	57	83
	Kikyogahara (Kanto District)	220	44	176
Red and yellow	Toyohashi (Tokai District)	20	0	20

Properties of Soils Derived from Volcanic Ash 49

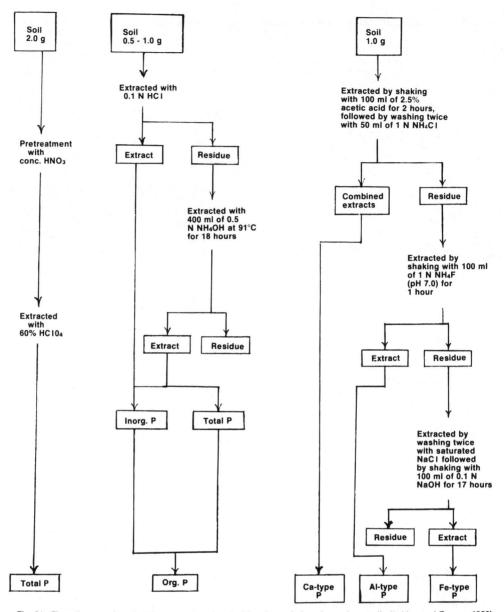

Fig. 21—Flow diagram of method for extracting chemical fractions of phosphorus from soils (Sekiya and Egawa, 1959)

TABLE 34. Chemical Fractions of Phosphorus in Some Cultivated Upland Soils of Japan (Sekiya and Egawa, 1961).

		Phosphorus per 100 g of dry soil, mg of P_2O_5						
				Inorganic fractions				
Soil name and location	Soil group	Total	Organic	Ca-type	Al-type	Fe-type	Insoluble	Inorganic extracted by 0.002 N H_2SO_4[1]
Kashima (Ibaragi pref.)	Ando	430	92	9	221	50	58	0.56
Shibusawa (Iwate pref.)	Ando	208	57	9	91	22	99	0.06
Nonoshima (Kumamoto pref.)	Ando	498	128	9	201	52	108	0.69
Kikyogahara (Nagano pref.)	Ando	488	68	8	231	65	116	0.22
Toyooka (Kumamoto pref.)	Ando	505	152	7	214	48	144	0.56
Toyohashi (Aichi pref.)	Red and yellow[2]	164	8	25	33	37	61	—
Higashiura (Aishi pref.)	Red and yellow[2]	140	26	36	27	28	23	10.1

[1] Extracted by method of Truog (1930).
[2] Diluvial parent material.

TABLE 35. Average Contents of Chemical Fractions of Phosphorus in Upland Soils of Japan (Sekiya and Egawa, 1961).

			Phosphorus per 100 g of dry soil, mg of P_2O_5						
					Inorganic fractions				
Parent material	Cultivation	Number of samples	Total	Organic	Ca-type	Al-type	Fe-type	Reductant soluble	Inorganic extracted by 0.002 NH_2SO_4[1]
Volcanic ash	Virgin	5	183	52	6	48	18	58	0.08
Volcanic ash	Cultivated	10	342	94	8	167	43	86	0.49
Other than volcanic ash	Virgin	2	20	0	7	6	7	1	0.50
Other than volcanic ash	Cultivated	5	140	13	37	31	28	32	17.1

[1] Extracted by method of Truog (1930).

TABLE 36. Distribution of Phosphorus Added as KH_2PO_4 Among Chemical Forms in Two Soils (Sekiya and Egawa, 1960).

		Proportion of added phosphorus found in indicated fraction in 2-g soil sample					
		Ca-type		Al-type		Fe-type	
Soil group[1]	Phosphorus added per 2 g of soil, mg of P_2O_5	mg of P_2O_5	%	mg of P_2O_5	%	mg of P_2O_5	%
Ando	7.5	0.32	4.3	5.10	68.0	1.40	18.7
	15.0	0.97	6.4	10.7	71.4	1.96	13.0
	22.5	1.99	8.8	16.5	73.3	2.68	11.8
	30.0	3.76	12.6	21.2	70.6	3.32	11.0
	60.0	12.29	20.5	39.7	66.2	5.20	8.7
Red and yellow	7.5	5.63	75.0	0.86	10.6	0.22	2.9
	15.0	11.83	79.0	1.53	10.4	0.43	2.9
	22.5	16.23	72.1	2.08	9.2	0.49	2.2
	30.0	23.53	72.8	2.80	9.4	0.69	2.3
	60.0	48.23	80.0	5.50	9.2	0.85	1.4

[1] The sample representing the ando group is the Kikyogahara soil from Nagano Prefecture, and the sample representing the red and yellow group is the Miyoshi soil from Aichi Prefecture. The latter soil is derived from diluvium.

presented in Table 34, and average contents of various forms of phosphorus in upland soils of Japan are presented in Table 35.

Sekiya and Egawa studied also the distribution of phosphorus added as KH_2PO_4 among the various chemical fractions of phosphorus in the soil. The effect of the magnitude of the phosphorus addition on the distribution of added phosphorus among the various chemical fractions of phosphorus in the soil is shown in Table 36. In the case of the soil of the ando group, 66 to 73% of the added phosphorus was extracted with the Al-type fraction and 9 to 19% was extracted with the Fe-type fraction. Only a little was extracted with the Ca-type fraction except where the largest additions of phosphorus were made. In the case of the soil of the red and yellow group, most of the added phosphorus was extracted with the Ca-type phosphorus fraction irrespective of the addition of phosphorus. This soil was derived from diluvial parent material.

Various chemical reagents have been used to obtain an index of phosphorus availability in soils. The author considers, however, that the most suitable reagent depends on the crop. For instance, with wheat and onion, the concentration of phosphorus in the soil solution, especially in the early stage of growth, has a great influence on the growth and yield. For such crops, easily soluble forms of phosphorus such as the Ca-type may be especially important. On the other hand, for some summer crops such as soybean, maize, or upland rice, and for perennial grasses, the solubility of Al- and Fe-type phosphorus is increased by the higher microbial activity and root activity at the higher soil temperature. Al- and Fe-type phosphorus, therefore, may be used to some degree by these crops. Eventually, the concept of "availability" of soil phosphorus must be related to the physicochemical behavior, e.g., the rate of dissolution of phosphorus in the soil solution and the rate of demand of the crops.

Sekiya and Egawa (1961) carried out a pot experiment using 17 soils of different origins and studied the relationships between the chemical forms of soil phosphorus and the amount of phosphorus absorbed by wheat. A high coefficient of correlation between the amounts of phosphorus absorbed by the wheat and the amounts of Ca-type phosphorus extracted from the soils was obtained with both mineral soils and soils of the ando group. Thus they adopted the Ca-type phosphorus as an index of phosphorus availability for wheat.

Yamamoto et al. (1966) and Takahashi and Yamamoto (1969) made a similar investigation on a group of samples of the Kuriyagawa soil of the ando group in the Tohoku district and used corn as the test plant. In their work, the highest correlation between phosphorus absorbed by the test plants and the chemical fractions of phosphorus in the soils was obtained with the Al-type phosphorus and not the Ca-type. As shown in Table 37, the correlation coefficient involving the Al-type phosphorus increased with the length of time the plants were allowed to grow before harvest. The reason was considered to be that the important period for nutrient absorption by corn plants was rather late in the growth of the crop.

The investigators noted also that the amounts of phosphorus absorbed by the corn plants were correlated negatively with the amounts of aluminum extracted from the soils by 1 **N** sodium acetate at pH 4 (see Table 37). This observation verifies the significance attached to reactive aluminum in controlling the availability of phosphorus and in reacting with soluble forms of phosphorus added as fertilizer.

BIOLOGICAL PROPERTIES

Microflora

Characteristics. The existence of inherent differences in microflora among soil types was reported by Mishustin (1956). Because soils derived from volcanic ash have certain chemical and physical characteristics that set them apart from other soils, the existence of certain distinguishing characteristics of the microflora might be expected.

To investigate the microflora of soils derived from volcanic ash on a comparative basis with soils not derived from volcanic ash, Ishizawa and Toyoda (1964) collected soil samples from both classes of soil from the same area and at the same season. Soil samples were taken from both cultivated and uncultivated land. The areas from which the soil samples were collected covered most of Japan. including Hokkaido and Kyushu. A summary of the results in Table 38 shows that soils derived from volcanic ash had significantly fewer total bacteria, significantly more anaerobic bacteria, and significantly more actinomycetes than soils derived from other parent materials. The trends with regard to these three groups were the same in cultivated soils as in virgin soils. Evidently there were some clear differences in microbial counts between soils derived from volcanic ash and those from other parent materials.

Yoshida and Sakai (1962); studied the microflora of four kinds of typical soils in Hokkaido, namely, Taisho soil derived from volcanic ash, Bibai peat soil, Komukai clay soil, and Kotoni alluvial soil. The results obtained showed that each of the soils had characteristic microflora, and a relatively high number of actinomycetes was observed in the soil derived from volcanic ash.

The relationship between the microflora and the particle-size distribution of a series of soils derived from volcanic ash was studied by Tsuru, Nishio, and Furusaka (1966). The samples used in this investigation were collected from northern Hokkaido, where

Table 37. Correlation Between Phosphorus Absorbed by Corn Plants From Soils of the Ando Group and (a) the Aluminum Extracted from the Soils and (b) the Phosphorus Extracted From the Soils in the Al-P Fraction. (Takahashi and Yamamoto, 1969).

Age of corn plants at harvest days	Correlation coefficient (r)	
	Aluminum extracted	Phosphorus extracted in Al-P fraction
39	−0.671**	0.901***
59	−0.540**	0.925***
90	−0.556**	0.947***
121	−0.581***	0.938***
149	−0.587***	0.968***

*** Significant at 0.1 % level.
** Significant at 1% level.

TABLE 38. Counts of Four Groups of Microorganisms in Soils Derived From Volcanic Ash and in Other Soils in Japan (Ishizawa and Toyoda, 1964).

Microbial group	Numbers of microorganisms per gram of dry soil[a]										
	Cultivated soils						Virgin soils				
	From volcanic ash n = 31		Not from volcanic ash n = 27		Difference	From volcanic ash n = 23		Not from volcanic ash n = 17		Difference
	Average	S.D[b]	Average	S.D		Average	S.D	Average	S.D	
Bacteria	181	116	218	114	37[c]	84	45	121	79	37[c]
Actinomycetes	80	33	48	25	32[cc]	47	25	23	12	24[cc]
Anaerobic bacteria	234	129	147	95	87[cc]	148	80	83	64	65[cc]
Fungi	241	93	231	96	10	118	90	135	61	17

[a] Numbers in the table should be multiplied by 10^5 in the case of bacteria and actinomycetes, by 10^4 in the case of anaerobic bacteria, and by 10^3 in the case of fungi.
[b] Standard deviation.
[c] Significant at 5% level.
[cc] Significant at 1% level.

TABLE 39. Counts of Certain Groups of Microorganisms in Soils Derived from Volcanic Ash and Differing in Particle-Size Distribution (Tsuru et al., 1966).

Soil No.	MWD[a] mm	pH (KCl)	H$_2$O[b] %	O$_2$ absorbed[c] μl/4g dry soil	Total C %	Total N %	C/N	Total bacteria[d]		Gram negative bacteria[d]		Fungi[d]	
								Per g of dry soil	Per mg of nitrogen	Per g of dry soil	Per mg of nitrogen	Per g of dry soil	Per mg of nitrogen
1	0.929	5.1	19.0	7.7	2.56	0.150	17.1	6.2	41	2.5	16.7	37.0	25.3
2	0.849	6.2	25.0	13.1	2.96	0.147	20.1	8.0	55	13.3	90.5	53.3	36.2
3	0.682	5.1	18.0	9.3	3.10	0.207	15.0	9.8	47	12.2	59.0	61.0	29.4
4	0.631	5.0	29.0	9.6	2.86	0.183	15.6	8.5	46	14.1	77.0	70.5	38.6
5	0.589	5.2	34.0	11.7	2.78	0.197	14.1	46.0	280	15.1	76.7	60.7	30.8
6	0.547	4.7	30.0	14.5	3.03	0.274	11.1	29.0	106	23.6	86.6	14.3	5.26
7	0.475	5.4	32.0	15.7	3.18	0.309	10.3	29.0	94	58.8	190	7.4	3.39
8	0.500	5.9	31.0	12.3	3.40	0.276	12.3	220	800	58.0	210	14.5	5.26

[a] Particle-size distribution expressed as mean weight-diameter.
[b] The water content of the natural sample of soil as collected.
[c] Quantity of oxygen absorbed by the soil per hour as found by Warburg manometry.
[d] Values in the table for counts of bacteria per gram of dry soil should be multiplied by 10^6, those for bacteria per milligram of nitrogen should be multiplied by 10^5, and those for fungi should be multiplied by 10^3.

the soils are derived from young volcanic ash very recently erupted from Mt. Tarumae. The soil samples differed in particle-size distribution because of their different distances from the crater. The results are presented in Table 39. The numbers of total and of Gram-negative bacteria evidently tended to increase with decreasing particle size.

The distribution of **Azotobacter** in Japanese soils was studied originally by Yamagata (1925), who stated that the frequency of detecting this organism was lower in soils of northern Japan than in those of southern Japan. Ishizawa and Toyoda (1964) re-examined the question of seeming north-south distribution and concluded that the observed trend was a consequence of the kinds of soils. They found that **Azotobacter** was not detected in upland soils derived from volcanic ash, which are more common in northern Japan than in southern Japan. The differences in distribution of **Azotobacter** between soils derived from volcanic ash and those derived from other materials seems to be related in some way to the characteristics of upland soils derived from volcanic ash because **Azotobacter** is easily detected in soils of rice fields derived from volcanic ash (Ishizawa and Toyoda, 1964).

The abundance of actinomycetes in soils derived from volcanic ash (Table 38) does not necessarily mean that these microorganisms are exceptionally active. The reason is that the numbers mentioned here were determined by plate counts, which do not distinguish between actinomycete spores and mycelium. When organic matter is added to soil to induce microbial change, the density of actinomycete mycelium is higher in soils derived from parent materials other than volcanic ash than in soils derived from volcanic ash. Therefore, the abundance of actinomycetes in soils derived from volcanic ash may be considered to be caused by the stimulation of spore formation in this type of soil. The abundance of anaerobic bacteria in soils derived from volcanic ash (234×10^5 per gram as indicated in Table 38) may seem unrealistic, but the observation may suggest the existence of anaerobic locations in otherwise aerated soils derived from volcanic ash.

Soil aggregates as a habitat for microorganisms. As a microhabitat for soil microorganisms, soil crumbs or aggregates may be divided into inner and outer parts. The inner part would present the anaerobic environment and the outer part the aerobic environment in wet soil, and the inner part would present the more moist environment and the outer part the less moist environment when a soil is becoming dry.

Hattori (1967, 1968) fractionated the bacterial cells in soil aggregates by a method involving shaking the soil gently with sterilized water to remove the outer portion of the aggregates followed by a sonic treatment to disrupt the inner portion. He verified that Gram-negative bacteria, which are relatively sensitive to desiccation, were more abundant in the inner portion of the aggregates than in the outer portion. Furusaka's (1968) conception of the relationship between aggregates and bacteria is shown in Figure 22.

Furusaka (1968) percolated a solution of glycine through two soils and then used the method described in the preceding paragraph to determine the distribution of bacteria in the inner and outer portions of the aggregates, with results as shown in Fig. 23. As may be seen in the figure, the bacterial numbers were higher in the inner part than in the outer part in the soil derived from volcanic ash, whereas the opposite result was obtained with the soil derived from diluvium. The difference in behavior between the two soils in this regard is presumably a consequence of a greater abundance of small pores in the aggregates of the soil derived from volcanic ash than in the soil derived from diluvium. This observation is in accord with the findings of Ishizawa and Toyoda (1964) described at the beginning of the section on microflora.

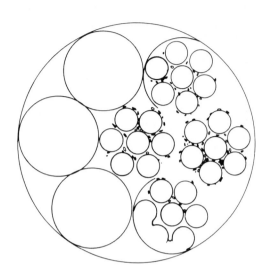

○ Gram-negative bacteria
• Gram-positive bacteria

Fig. 22—Schematic cross-section of soil showing the relation of bacteria to soil particles and the distribution of Gram-positive and Gram-negative bacteria in soil aggregates.

(Furusaka, 1968)

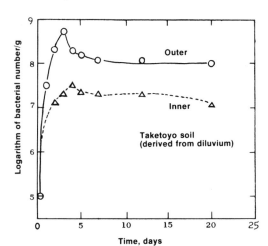

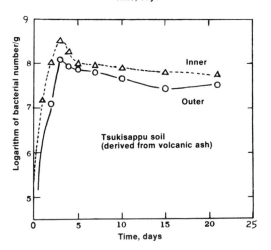

Fig. 23—Bacterial numbers in outer and inner portions of aggregates of two soils as found different lengths of time after percolation of a solution of glycine through the soils. The Taketoyo soil is a member of the red and yellow group.

(Furusaka, 1968)

Influence of environmental factors on microbial population

Water content of soil. It is well known that the composition of the soil population depends on soil conditions, including the content of water and air. Bacterial numbers commonly increase with an increase in water supply.

Ishizawa and colleagues (1958) studied the influence of water supply on the microflora and on the mineralization of organic nitrogen in some soils of the ando group. In the case of Nishigahara uncultivated soil, the water content of separate samples was adjusted to pF 1.7, 2.4, 3.4, or 3.5, and each soil sample was treated with 1% rice straw together with 0.1% superphosphate and 0.03% NH_4NO_3. The abundance of microorganisms was estimated by the direct method and the plate method, and the amounts of mineralized nitrogen were determined at weekly intervals. At the same time, the effect of acidity on the microflora was examined in the series at pF 2.4 and 3.4. The results indicated that the number of bacteria generally increased and that the numbers of fungi and actinomycetes decreased with an increase in water content of the soil.

The effect of acidity on the microflora was more evident in the plate count than in the mycelial density or the total count. The ratios of fungi to bacteria and of actinomycetes to bacteria were greater at pH 6.4 than at pH 7.4. The amounts of nitrogen mineralized in the first 5 weeks of incubation increased with the water content and were greater in the series at pF 2.7 and pH 7.4. The results clearly showed that the change from the mycelia-dominant type of microflora to the bacterial-dominant type occurred in the pF range of 2.4 to 3.5.

Yoshida and Sakai (1963a,b) determined the effect of soil water content on the microflora of various types of soil in Hokkaido. As may be seen from their results in Fig. 24, the bacterial numbers in the various soils were generally greater under wet conditions than under dry conditions irrespective of the nature of the soil. The reverse tendency was observed with fungi and actinomycetes. The terms, "wet" and "dry", in Fig. 24 refer to the natural condition of the soils and not to conditions induced by treatment in given samples. The "wet" and "dry" samples were thus from different locations.

Organic substances. Needless to say, the addition of organic substances exerts a great influence on the soil microflora. Ishizawa and colleagues (1958) added various types of organic substances to certain soils of the ando group and found that the effects on the microflora varied with the nature of the organic matter added. Counts of microorganisms made after incubation of samples of Kikyogahara soil from Nagano Prefecture at 15 and 28°C with 1% alfalfa, 1.2% glucose, or 2.7% manure showed, for example, that glucose was the most effective of all the organic materials in increasing the numbers of microorganisms except for actinomycetes at 15°C. Addition of alfalfa produced a smaller increase in numbers of microorganisms than did addition of glucose, where plate counts were concerned, but addition of alfalfa greatly increased the growth of fungi as indicated by the contact-slide method. At 28°C the difference in effects of alfalfa and glucose on the growth of the

Properties of Soils Derived from Volcanic Ash 55

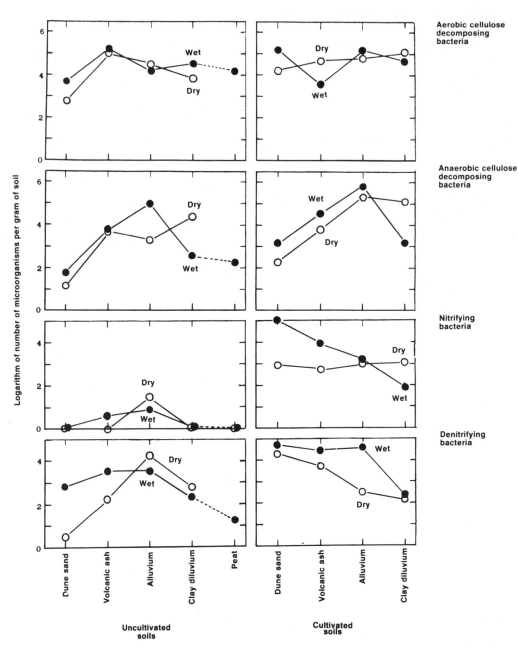

Fig. 24—Numbers of different groups of bacteria in wet and dry soils derived from various kinds of parent material

(Yoshida and Sakai, 1963a).

microorganisms was not so distinct; however, **Penicillium** was dominant in the glucose series and the white type of fungi was dominant in the alfalfa series. The addition of manure had almost no effect on the growth of microorganisms regardless of temperature. CO_2 evolution was greatest in the glucose series followed by the alfalfa series at both temperatures.

An injurious effect on the growth of young plants caused by the decomposition of recently added green manure was investigated by Sawada, Nitta, and Igarashi (1964, 1965). In both field and laboratory experiments, an injurious effect on the growth of oat seedlings appeared in the early stage of decomposition of green manure. From field observations and the results of an experiment in which the soil was aerated or treated with CO_2, it seemed unlikely that the injury was due to the composition of the soil air. Partial sterilization of the soil by heating or by formalin, however, prevented the injury to the seedlings. A fungus belonging to the **Pythium** species associated with the decomposition of the green manure was isolated from the injured tissues as a causal organism. This fungus was believed by the researchers to be primarily a saprophytic organism that occurred as one of the pioneers in the ecological succession of organisms colonizing decaying organic materials. They confirmed that the injurious effect of green manure on plant seedlings was due to this type of fungus from an experiment with ten kinds of cultivated soils collected from various regions of Japan. The harmful effect of green manure on plant seedlings was found to be dependent on the rate of its decomposition. The decomposition was considerably accelerated by increasing the population of the **Pythium** species in the soil, and a high incidence of the species on decaying clover leaves was found in the soils in which the added clover was rapidly decomposed. It was recognized further that the progressive ammonification of the green manure was due mainly to the saprophytic activity of the **Pythium** species. From this information, the researchers concluded that the **Pythium** species (the causal fungus of the injurious effect of green manure) was a dominant saprophyte in the decomposition process.

Sawada and Koyanagawa (1969) confirmed the existence of severe infection of seedling plants by **Pythium** on nonsterile soil in which recently added crop residues were undergoing rapid decomposition. In addition, however, they obtained evidence for a highly toxic effect on the growth of wheat seedlings when green residues such as alfalfa, sugar beet leaves, red clover, or timothy were added to soil under sterile conditions. No correlation was found between phytotoxic substances in the residues and the infection of the seedlings by **Pythium.**

Generally speaking, intensive competition among soil microorganisms occurs at all stages of decomposition of crop residues. The species involved and the rate of the activity, however, depend on the stage of decomposition of crop residues.

Change of fungal flora during reclamation. Suzuki and Ishizawa (1965) studied the change of fungal microflora associated with reclamation of uncultivated soils of low fertility. Some of their data in Table 40 show that numbers of fungi were greater in virgin soils than in cultivated soils. The decrease in numbers caused by addition of 0.3% boric acid was much greater in uncultivated soils than in cultivated soils, indicating the existence of a difference in kinds of fungi present. Fungi in virgin soils, without addition of boric acid, consisted mainly of **Penicillium** and a few **Fungi Imperfecti.** In cultivated soils, various types of fungi were found, including **Penicillium, Aspergillus, Phycomycetes,** and **Fungi Imperfecti.** The greater resistance of the fungi to boric acid in cultivated soils than in uncultivated soils may be a consequence of loss of the less stable soil organic matter associated with cultivation. The same researchers found also that, when the virgin and cultivated soils were incubated under similar conditions, fungal mycelia were prevalent in the former, whereas bacteria, particularly the gelatin-liquefying group, were dominant in the latter.

Cropping system. The effect of cropping system on soil microflora was investigated by Suzuki and Ishizawa (1965). They found that in field plots growing Sudan grass the ratio of dye-tolerant bacteria to total bacteria was greater in the root zone than in other parts of the soil. The better was the growth of Sudan grass, the higher was the ratio of dye-tolerant bacteria to total bacteria in the root zone.

The number of microorganisms in the soil was less under perennial grass than under tilled crops. The decrease was particularly marked in the case of nitrifying and denitrifying bacteria and **Azotobacter.** In the case of obligate anaerobes, however, the reverse trend was observed.

Microbial Activities

Mineralization of soil nitrogen. Akatsuka (1966) found that, when soils of coarse texture derived from volcanic ash in Hokkaido were incubated under uniform conditions in the laboratory, the amount of nitrogen mineralized increased with a decrease in mean weight diameter of the soils. The results are shown in Fig. 25. This observation is probably attributable primarily to an increase in content of organic matter and total nitrogen in the soils with a decrease in mean weight diameter. Suzuki and Ishizawa (1965) similarly found that nitrate production in young soils of coarse texture derived from volcanic ash was low in comparison with that of other soils derived from volcanic ash.

Suzuki and Ishizawa (1965) incubated samples of soils in the laboratory under uniform conditions to determine the changes in nitrate production in soils derived from volcanic ash after reclamation from the virgin, uncultivated state. The data obtained are shown in Table 41. As may be seen from the table, nitrate production increased as a result of both an increase in the number of years of reclamation and the addition of manure and minerals. They observed also that nitrate production was correlated positively with soil fertility and with the degree of calcium saturation of the soil. The latter relationship is shown in Fig. 26. The rate of increase of nitrate production with the degree of calcium saturation was distinctly greater in soils derived from volcanic ash than in other soils.

Sakai (1959, 1960a,b,c,d,e,) found that the amount of nitrate produced in soils derived from volcanic ash in the Tokachi District of Hokkaido was greater in soils of the "wet" class than in those of the "dry" class. A lower population of nitrifying organisms under the dry conditions was suggested as a cause of the observed behavior.

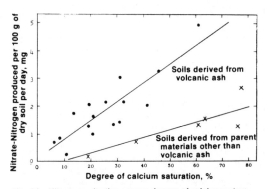

Fig. 26—Nitrate production versus degree of calcium saturation of two classes of soils. The soils were incubated in the laboratory under uniform conditions.

(Suzuki and Ishizawa, 1965)

Soil respiration. The Warburg manometric method was used by Tanabe and Ishizawa (1969) to measure soil respiration as an index of total soil microbial activity. Measurements were made on 28 soils derived from volcanic ash and on 14 other soils. Half of the samples in each group were from cultivated soils, and the other half were from uncultivated soils. When the water content was adjusted to 55% of the water-holding capacity and the temperature was kept at 30°C, the average rate of oxygen uptake was 4.4 + or − 1.9 and 2.7 + or − 1.1 μl/g/hr for the virgin

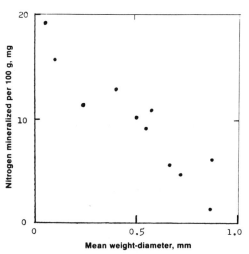

Fig. 25—Mineralization of nitrogen in soils of coarse texture derived from volcanic ash in Hokkaido. The soils were incubated under uniform conditions in the laboratory.

(Akatsuka, 1966)

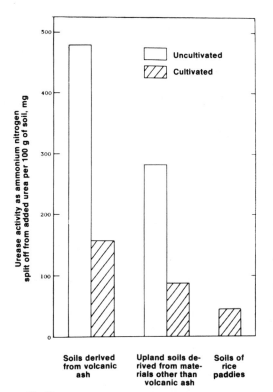

Fig. 27—Urease activity in different groups of soils.

(Tanabe and Ishizawa, 1969)

TABLE 40. Numbers of Fungi and Resistance to Boric Acid in Seven Pairs of Uncultivated and Cultivated Soils (Suzuki and Ishizawa, 1965).

Soil[1]	Numbers of fungal mycelia per gram of soils by direct observation, thousands	Numbers of fungi per gram of soil in plate counts, thousands		$(A - B)(100)$
		Control (A)	Boric acid (B)	A
Fukushima				
Uncultivated	331	360	86	76
Cultivated	68	122	89	27
Fukushima				
Uncultivated	346	380	6.7	99
Cultivated	119	283	232	18
Saitama				
Uncultivated	153	252	129	94
Cultivated	177	360	196	45
Saitama				
Uncultivated	156	203	1.9	99
Cultivated	138	245	186	24
Saitama				
Uncultivated	220	249	8.7	97
Cultivated	86	124	142	−15
Aichi				
Uncultivated	131	207	9.3	96
Cultivated	161	317	199	37
Aichi				
Uncultivated	279	313	3.9	99
Cultivated	92	118	80	40

[1] The first five pairs of soils were derived from volcanic ash. The last two pairs of soils were derived from diluvial deposits.

TABLE 41. Production of Nitrate by Soils Different Lengths of Time After Reclamation During Incubation Under Uniform Conditions in the Laboratory With and Without Addition of Manure and Minerals (Suzuki and Ishizawa, 1965).

Soil	Reclamation		Nitrate-nitrogen formed in mg per 100 g of dry soil during incubation for indicated time in days				
	Years	Treatment[1]	3	6	9	12	15
Nasu	5	None	tr.	2	2	12	26
	5	Yes	1	8	46	64	73
	10	None	tr.	5	9	18	31
	10	Yes	4	17	59	67	74
Daisen	5	None	tr.	3	6	15	44
	5	Yes	2	11	38	61	71
	20	None	7	25	46	62	69
	20	Yes	9	36	56	72	75
Kawanami	3	None	1	4	4	18	19
	3	Yes	2	8	42	61	70
	10	None	tr.	8	12	42	47
	10	Yes	6	35	65	—	70

[1] Where treatment is indicated, the additions prior to incubation in the laboratory were 2% manure, 0.1%K_2HPO_4, 0.2%$MgSO_4 \cdot 7H_2O$, micronutrients, and enough $CaCO_3$ to produce a pH of 6.5.

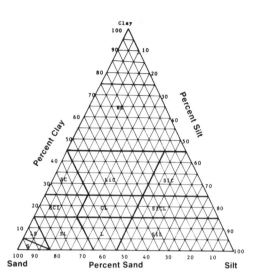

Appendix Figure. Ternary diagram for soil textures used in Japan

and cultivated soils derived from volcanic ash, and 4.5 + or − 2.7 and 2.0 + or − 0.3 μl/g/hr for the virgin and cultivated soils derived from other parent materials. The difference due to cultivation was significant at the 1% level, but the over-all difference due to parent material was not significant. The difference due to parent material in the cultivated soils was significant at the 20% level.

Urease activity. The urease activity of soils as affected by parent material, reclamation, and soil management was investigated by Tanabe and Ishizawa (1969). They determined the amounts of ammonium nitrogen split off from urea incubated with soil under standard conditions in the laboratory and used the number of milligrams of ammonium nitrogen released per 100 g of dry soil as an index of urease activity. The average values they obtained in this way for urease activity (see Fig. 27) were 482 + or − 267 and 161 + or − 71 for 21 virgin and 21 cultivated soils derived from volcanic ash, 284 + or − 219 and 88 + or − 58 for 16 virgin and 16 cultivated upland soils from parent materials other than volcanic ash, and 44 + or − 15 for 30 soils of rice paddies. In both the soils derived from volcanic ash and the upland soils derived from other parent materials, the urease activity of the virgin soils was significantly higher than that of the cultivated soils. The difference between soils derived from volcanic ash and upland soils derived from other parent materials was significant at the 5% level in the case of virgin soils and at the 1% level in the case of cultivated soils.

LITERATURE CITED [1]

Adachi, T. 1963a. Pedologist 7:1-14 (J-E).
Adachi, T. 1963b. Jour. Sci. Soil and Manure, Japan 34:287-290 (J).
Adachi, T. 1966a. Pedologist 10:49-55 (J).
Adachi, T. 1966b. Jour. Sci. Soil and Manure, Japan 37:207-212 (J).
Adachi, T. 1966c. Jour. Sci. Soil and Manure, Japan 37:505-510 (J).
Adachi, T. 1968. Jour. Sci. Soil and Manure, Japan 39:343-348 (J).
Akatsuka, K. 1966. Unpublished. See Furusaka's article, "Soil structure and soil microorganisms." In Furusaka, C. Soils and Microorganisms, No. 8, 1-9) (J).
Aomine, S. 1958. Jour. Sci. Soil and Manure, Japan 28:508-516 (J).
Aomine, S., and Jackson, M. L. 1959. Soil Sci. Soc. Amer. Proc. 23:210-214 (E).
Aomine, S., and Kobayashi, Y. 1964. Soil Sci. Plant Nutr. 10:28-32 (E).
Aomine, S., and Kobayashi, Y. 1966. Soil Sci. Plant Nutr. 12:7-12 (E).
Aomine S., and Kodama, I. 1956. Kyushu Univ. Fac. Agr. Jour. 10:325-344 (J-E).
Aomine, S, and Wada, K. 1962. Amer. Miner. 47:1024-1048 (E).
Aomine, S., and Yoshinaga, N. 1955. Soil Sci. 79:349-358 (E).
Birrell, K. S., and Fieldes, M. 1952. Jour. Soil Sci. 3:156-166(E).
Birrell, K.S., and Gradwell, M. 1956. Jour. Soil Sci. 7:130-147 (E).
Brewer, R. 1960. Jour. Soil Sci. 11:280-292 (E).
Chang, S. C., and Jackson, M. L. 1957. Soil Sci. 84:133-144 (E).
Daikuhara, G. 1911. Nat. Agric. Exp. Sta., Japan, Rept. 37:66-72 (J).
Egawa, T. 1964. Clay Sci. 2:1-7 (E).
Egawa, T., and Sato, A. 1960. Adv. Clay Sci. Japan 2:385-393 (J-E).
Egawa, T., Sato, A., and Nishimura, T. 1960. Adv. Clay Sci., Japan 2:252-262 (J-E).
Egawa, T., and Sekiya, K. 1956. Soil and Plant Food 2:75-82 (E).
Egawa, T., Sekiya, K., and Iimura, K. 1957. Nat. Inst. Agric. Sci., Japan, Bull. B7:53-76.
Egawa, T., Watanabe, Y., and Sato, A. 1955. Nat. Inst. Agric. Sci., Japan, Bull. B5:39-106 (J-E).
Egawa, T., Watanabe, Y., and Sato, A. 1959. Adv. Clay Sci., Japan 1:260-272 (J-E).
Fieldes, M. 1955. New Zealand Jour. Sci. Tech. B37:336-350 (E).
Fieldes, M., and Furkert, R. J. 1966. New Zealand Jour. Sci. 9:608-622 (E).
Fieldes, M., and Perrott, K. W. 1966. New Zealand Jour. Sci. 9:623-629 (E).

J in parentheses at the end of a citation means that the publication is written in Japanese, E means that it is in English. J-E means that it is in Japanese with a summary in English, G means that it is in German.

Furkert, R. J., and Fieldes, M. 1968. Trans. 9th Intern. Cong. Soil Sci. 3:133-141 (E).
Furusaka, C. 1968. Amino Acid and Nucleic Acid 17:15-26 (J).
Hanai, H., and Shinagawa, A. 1952. Kagoshima Univ. Fac. Agric. Bull. 1:63-71 (J-E).
Harada, T., and Kutsuna, K. 1955. Nat. Inst. Agric. Sci., Japan, Bull. B5:1-26 (J-E).
Hattori, T. 1967. Tohoku Univ. Inst. Agric. Res. Bull. 18:159-193 (J-E).
Hattori, T., 1968. Chemistry and Biology 6:235-242 (J).
Hayashi, T. 1951. Tottori Univ. Fac. Agric. Jour. 1:1-52 (E).
Hayashi, T. 1953a. Tottori Univ. Fac. Agric. Jour. 1:125-134 (E).
Hayashi, T. 1953b. Tottori Univ. Fac. Agric. Jour. 1:135-147 (E).
Hayashi, T., and Nagai, T. 1953. Jour. Sci. Soil and Manure, Japan 24:212-216 (J-E).
Hayashi, T., and Nagai, T. 1955. Jour. Sci. Soil and Manure, Japan 25:285-290 (J-E).
Hayashi, T., and Nagai, T. 1956a. Jour. Sci. Soil and Manure, Japan 26:371-375 (J-E).
Hayashi, T., and Nagai, T. 1956b. Jour. Sci. Soil and Manure, Japan 27:305-308 (J).
Hayashi, T., and Nagai, T. 1957. Jour. Sci. Soil and Manure, Japan 28:369-373 (J).
Hayashi, T., and Nagai, T. 1959. Jour. Sci. Soil and Manure, Japan 29:567-572 (J).
Hayashi, T., and Nagai, T. 1960a. Jour. Sci. Soil and Manure, Japan 31:197-200 (J).
Hayashi, T., and Nagai, T. 1960b. Tottori Soc. Agron. Trans. 12:69-76 (J-E).
Hayashi, T., and Nagai, T. 1961. Jour. Sci. Soil and Manure, Japan 32:280-283 (J).
Hayashi, T., and Nagai, T. 1962a. Jour. Sci. Soil and Manure, Japan 33:250-254 (J).
Hayashi, T., and Nagai, T. 1962b. Jour. Sci. Soil and Manure, Japan 33:255-258 (J).
Honda, C., and Nishimune, A. 1969. Hokkaido Nat. Agric. Exp. Stat. Res. Bull. 95:1-10 (J-E).
Hosoda, K. 1938. Tottori Agric. Coll. Memoire 6:1-238 (J-E).
Hosoda, K., and Takada, H. 1953a. Jour. Sci. Soil and Manure, Japan 24:65-69 (J-E).
Hosoda, K., and Takada, H. 1953b. Jour. Sci. Soil and Manure, Japan 24:70-73 (J-E).
Hosoda, K., and Takada, H. 1953c. Jour. Sci. Soil and Manure, Japan 24:153-156 (J-E).
Hosoda, K., and Takada, H. 1957a. Jour. Sci. Soil and Manure, Japan 27:498-502 (J).
Hosoda, K., and Takada, H. 1957b. Jour. Sci. Soil and Manure, Japan 28:23-26 (J).
Hosoda, K., and Takada, H. 1957c. Jour. Sci. Soil and Manure, Japan 28:64-68 (J).
Hosoda, K., and Takada, H. 1967a. Jour. Sci. Soil and Manure, Japan 38:149-155 (J).
Hosoda, K., and Takada, H. 1967b. Jour. Sci. Soil and Manure, Japan 38:156-160 (J).
Hosoda, K., and Takada, H. 1967c. Jour. Sci. Soil and Manure, Japan 38:179-182 (J).
Hosoda, K., and Takada, H. 1967d. Jour. Sci. Soil and Manure, Japan 38:183-186 (J).
Hosoda, K., and Takada, H. 1967e. Jour. Sci. Soil and Manure, Japan 38:223-227 (J).
Hosoda, K., and Takada, H. 1967f. Jour. Sci. Soil and Manure, Japan 38:228-231 (J).
Iimura, K. 1966. Nat. Inst. Agric. Sci., Japan, Bull. 17B:101-157 (J-E).
Iimura, K., and Egawa, T. 1956. Soil and Plant Food, 2:83-88 (E).
Inoue, T., and Wada, K. 1968. Trans. 9th Intern. Congr. Soil Sci. 3:289-298 (E).
Ishii, J. 1963. Hokkaido Univ. Fac. Sci. Jour., Ser. 4, 11:545-569 (J-E).
Ishii, J., and Kondo, Y. 1962a. Earth Sci. 62:29-45 (J).
Ishii, J., and Kondo, Y. 1962b. Adv. Clay Sci., Japan, 4:193-212 (J).
Ishii, J., and Sasaki, T. 1959. Geol. Committee Hokkaido Bull. 38:13-21 (J-E).
Ishizawa, S., Suzuki, T., Koda, T., and Sato, O. 1958. Nat. Inst. Agric. Sci., Japan, Bull, B8:1-211 (J-E).
Ishizawa, S., and Toyoda, K. 1964. Nat. Inst. Agric. Sci., Japan, Bull. B14:203-284 (J-E).
Iwata, S. 1968. Soil Physical Conditions and Plant Growth, Japan 18:18-26 (J).
Kamoshita, Y. 1958. Nat. Inst. Agric. Sci., Japan, Misc. Pub. B5 (E).
Kanno, I. 1956. Rapp. VIe Cong. Intern. Sci. Sol. E:105-109 (E).
Kanno, I. 1959. Adv. Clay Sci., Japan 1:213-233 (J-E).
Kanno, I. 1961. Kyushu Agric. Exp. Sta. Bull. 7:1-185 (J-E).
Kanno, I., and Arimura, G. 1954. Kyushu Agric. Exp. Sta. Bull. 2:307-320 (J-E).
Kanno, I., and Arimura, G. 1955. Jour. Sci. Soil and Manure, Japan 26:41-45 (J).
Kanno, I., and Arimura, G. 1957a. Jour. Sci. Soil and Manure, Japan 27:492-494 (J).
Kanno, I., and Arimura, G. 1957b. Jour. Sci. Soil and Manure, Japan, 28:138-140 (J).
Kanno, I., and Arimura, G. 1958. Soil and Plant Food 4:62-67 (E).
Kanno, I., Kuwano, Y., and Honjo, Y. 1960. Adv. Clay Sci., Japan 2:355-356 (J-E).
Kanno, I., Onikura, Y., and Higashi, T. 1968. Trans. 9th Intern. Cong. Soil Sci. 3:111-122 (E).
Kato, Y. 1960. Jour. Sci. Soil and Manure, Japan 31:25-28 (J).
Kawaguchi, K., and Kyuma, K. 1959. Jour. Sci., Soil and Manure, Japan 29:527-530 (J).
Kawai, K. 1969. Nat. Inst. Agric. Sci. Japan, Bull. B20:77-154 (J-E).
Kawamura, K. 1950. Agric. and Hort. 25:11-14 (J).
Kawamura, K., and Funabiki, S. 1936. Jour. Sci. Soil and Manure, Japan 10:281-294 (J).
Kawasaki, H., and Aomine, S. 1966. Soil Sci. Plant. Nutr. 12:144-150 (E).
Kidachi, M. 1964. **In** Volcanic Ash Soils in Japan, pp. 27-28. Ministry of Agric. and For., Japan (E).

Kinter, E. B., and Diamond, S. 1958a. Clays and Clay Miner., 5th Conf. Nat. Acad. Sci. Nat. Res. Council (U.S.A.) Pub. 556:318-333 (E).

Kinter, E. B., and Diamond, S., 1958b. Clays and Clay Miner., 5th Conf. Nat. Acad. Sci. Nat. Res. Council (U.S.A.) Pub. 556:334-347 (E).

Kira, Y., Ambo, F., Soma, K., and Ito, K. 1963. Trans. Agric. Engin. Soc. Japan 7:76-78 (J-E).

Kira, Y., Soma, K., and Takenaka, H. 1963. Res. Agric. Eng., Special Publ., Soc. Agric. Eng., Japan 7:81-85 (J-E).

Kishita, A. 1955. Nat. Inst. Agric. Sci., Japan, Ann. Rept., Sec. 1, Soil, pp. 134-136 (J).

Kishita, A. 1959. Jour. Sci. Soil and Manure, Japan 29:545-548 (J).

Kishita, A., and Tanoue, M. 1969. Hokkaido Nat. Agric. Exp. Sta., Ann. Rept., Dept. Agr. Chem., Part 4, 1-11 (J).

Kitagishi, K., and Okita, T. 1956. Tohoku Univ. Agric. Exp. Sta. Res. Bull. 8:62-71 (J-E).

Kobo, K. 1951/52. Jour. Sci. Soil and Manure, Japan 21:125-128, 306-309; 22:145-147 (J).

Kobo, K., and Akatsuka, K. 1951. Jour. Sci. Soil and Manure, Japan 22:151 (J).

Kobo, K., and Fujisawa, T. 1963. Jour. Sci. Soil and Manure, Japan 34:13-17 (J).

Kobo, K., and Fujisawa, T. 1964. Jour. Sci. Soil and Manure, Japan 35:40-46 (J).

Kobo, K., and Fujisawa, T. 1966. Jour. Sci. Soil and Manure, Japan 37:284-288 (J).

Kobo, K., Kumada, K., Shiga, H., Matsuo, Y., and Mori, N. 1968. Jour. Sci. Soil and Manure, Japan 39:10-16 (J).

Kobo, K., and Ohba, Y. 1963. Trans. Cong. Sci. Soil and Manure, Japan 9:30 (J).

Kobo, K., and Ohba, Y. 1964. In Volcanic Ash Soils in Japan, pp. 29-33. Ministry of Agric. and For., Japan (E).

Kobo, K., and Tatsukawa, R. 1959. Zeitschr. Pflanzenernähr. Düng. Bodenkunde 84:137 (G).

Kondo, Y. 1969. Obihiro Zootech. Univ. Res. Bull. Ser. 2, 6:74-111 (E).

Kosaka, J. 1953. Nat. Inst. Agric. Sci., Japan, Bull. B2:49-67 (J-E).

Kosaka, J. 1963. Nat. Inst. Agric. Sci., Japan, Bull. B13:253-352 (J-E).

Kosaka, J. 1964. In Volcanic Ash Soils in Japan, pp. 99-101. Ministry of Agric. and For., Japan (E).

Kosaka, J., and Honda, C. 1957. Jour. Sci. Soil and Manure, Japan 27:435-438 (J).

Kosaka, J., Honda, C., and Iseki, A. 1961a. Jour. Sci. Soil and Manure, Japan 32:333-337 (J).

Kosaka, J., Honda, C., and Iseki, A. 1961b. Jour. Sci. Soil and Manure, Japan 32:447-450 (J).

Kosaka, J., and Iseki, A. 1957. Nat. Inst. Agric., Sci., Japan, Bull. B7:161-184 (J-E).

Kubiena, W. L. 1938. Micropedology. Collegiate Press, Ames, Iowa, U.S.A. (E).

Kumada, K. 1955a. Jour. Sci. Soil and Manure, Japan 25:217-221 (J-E).

Kumada, K. 1955b. Jour. Sci. Soil and Manure, Japan 25:263-267 (J-E).

Kumada, K., 1955c. Jour. Sci. Soil and Manure, Japan 26:5-10 (J-E).

Kumada, K. 1955d. Jour. Sci. Soil and Manure, Japan 26:97-100 (J-E).

Kumada, K. 1955e. Jour. Sci. Soil and Manure, Japan 26:179-182 (J-E).

Kumada, K. 1955f. Jour. Sci. Soil and Manure, Japan 26:231-234 (J-E).

Kumada, K. 1955g. Jour. Sci. Soil and Manure, Japan 26:287-290 (J-E).

Kumada, K. 1955h. Soil and Plant Food 1:29-30 (E).

Kumada, K. 1956a. Jour. Sci. Soil and Manure, Japan 27:79-82 (J).

Kumada, K. 1956b. Jour. Sci. Soil and Manure, Japan 27:119-122 (J-E).

Kumada, K. 1958a. Soil and Plant Food 3:152-159 (E).

Kumada, K. 1958b. Agr. and Hort. 33:873-876, 1019-1022, 1169-1172, 1333-1336, 1487-1490 (J).

Kumada, K. 1959. Soil and Plant Food 4:181-188 (E).

Kumada, K. 1963. Jour. Sci. Soil and Manure, Japan 34:417-422 (J).

Kumada, K. 1965a. Soil Sci. Plant Nutr. 11:151-156 (E).

Kumada, K. 1965b. Jour. Sci. Soil and Manure, Japan 11:367-372 (J).

Kurabayashi, S., and Tsuchiya, T. 1960. Adv. Clay Sci., Japan 2:178-196 (J).

Kurabayashi, S., and Tsuchiya, T. 1961. Adv. Clay Sci., Japan 3:204-213 (J).

Kurabayashi, S., and Tsuchiya, T. 1965. Adv. Clay Sci., Japan 5:19-31 (J).

Kutsuna, K., and Nomoto, K. 1961a. Jour. Sci. Soil and Manure, Japan 32:183-186 (J).

Kutsuna, K., and Nomoto, K. 1961b. Jour. Sci. Soil and Manure, Japan 32:243-246 (J).

Kuwano, K., and Matsui, T. 1957. Inst. Sci. Natural Resources, Japan. Bull. 45:33-42 (J-E).

Kyuma, K., and Kawaguchi, K. 1964. Soil Sci. Soc. Amer. Proc. 28:371-374 (E).

Masui J. 1960. Jour. Japan. Assoc. Miner. Petrol. Econ. Geol. 44:236-271 (J).

Masui, J. 1966. Jour. Japan. Assoc. Miner. Petrol. Econ. Geol. 55:221-241 (J).

Masui, J., Onikura, Y., and Uchiyama, N. 1963. Adv. Clay Sci., Japan 4:281-290 (J).

Masui, J., and Shoji, S. 1967. Pedologist 11:33-45 (J).

Masui, J., and Shoji, S. 1969. Proc. Intern. Clay Conf., Tokyo 1:383-392 (E).

Masui, J., Shoji, S., and Uchiyama, N. 1966. Tohoku Univ. Jour. Agric. Res. 17:17-36 (E).

Masujima, T. 1962. Hokkaido Nat. Agric. Exp. Sta. Res. Bull. 77:40-47 (J-E).

Masujima, T. 1968. Soil Physical Conditions and Plant Growth, Japan 18:10-17 (J).

Matsui, T. 1959. Adv. Clay Sci., Japan 1:244-259 (J).

Matsui, T. 1960. Adv. Clay Sci., Japan 2:229-241 (J).

Matsui, T., Kurobe, T., and Kato, Y. 1963. The Quaternary Res. 3:40-58 (J-E).

Mattson, S., Williams, G., and Barkoff, E. 1950. Kgl. Lantbrukshögskol. Ann. 17:107-120 (E).

Mishustin, E. M. 1956. Soils and Fert. 19:358-392 (E).

Misono, S. 1964. In Volcanic Ash Soils in Japan, pp.

75-85. Ministry of Agric. and For., Japan (E).
Misono, S., and Kawajiri, M. 1967. Nat. Inst. Agric. Sci., Japan, Bull. B18:49-128 (J-E).
Misono, S., and Kawajiri, M. 1968. Nat. Inst. Agric. Sci., Japan, Bull. B19: 1-68 (J-E).
Misono, S., and Kishita, A. 1957a. Nat. Inst. Agric. Sci., Japan, Bull. B7:105-122 (J-E).
Misono, S., and Kishita, A. 1957b. Nat. Inst. Agric. Sci., Japan, Bull. B7: 123-160 (J-E).
Misono, S., Kishita, A., Sudo, S., and Terasawa, S. 1953. Nat. Inst. Agric. Sci., Japan, Bull. B2:125-145 (J-E).
Misono, S., and Terasawa, S. 1957. Nat. Inst. Agric. Sci., Japan, Bull. B7:77-103 (J-E).
Misono, S., Terasawa, S., Kishita, A., and Sudo, S. 1953. Nat. Inst. Agric. Sci., Japan, Bull. B2:95-124 (J-E).
Miyauchi, N., and Aomine, S. 1964. Soil Sci. Plant Nutr. 10:199-203 (E).
Miyauchi, N., and Aomine, S. 1966. Soil Sci. Plant Nutr. 12:187-190 (E).
Miyazawa, K. 1962. Jour. Sci. Soil and Manure, Japan 33:317-321 (J).
Miyazawa, K. 1966. Nat. Inst. Agric. Sci., Japan, Bull. B17:1-100 (J-E).
Miyoshi, Y. 1964. Jour. Sci. Soil and Manure, Japan 35:377-380 (J).
Nagai, T. 1969. Tottori Univ. Fac. Agric. Jour. 5:53-90 (E).
Ohba, Y. 1965. Pedologist 9:26-30 (J).
Okuda, A., and Hori, S. 1952a. Jour. Sci. Soil and Manure, Japan 22:339 (J).
Okuda, A., and Hori, S. 1952b. Jour. Sci. Soil and Manure, Japan 22:357 (J).
Okuda, A., and Hori, S. 1952c. Jour. Sci. Soil and Manure, Japan 23:72 (J).
Okuda, A., and Hori, S. 1955. Jour. Sci. Soil and Manure, Japan 26:201-203 (J-E).
Russell, M. B. 1939. Soil Sci. Soc. Amer. Proc. 4:51-54 (E).
Sakai H. 1959. Jour. Sci. Soil and Manure, Japan 30:149-153 (J).
Sakai, H. 1960a. Jour. Sci. Soil and Manure, Japan 31:149-151 (J).
Sakai, H. 1960b. Jour. Sci. Soil and Manure, Japan 31:207-210 (J).
Sakai, H. 1960c. Jour. Sci. Soil and Manure, Japan 31:253-255 (J).
Sakai, H. 1960d. Jour. Sci. Soil and Manure, Japan 31:281-284 (J).
Sakai, H. 1960e. Jour. Sci. Soil and Manure, Japan 31:331-334 (J).
Sawada, Y., and Koyanagawa, T. 1969. Hokkaido Nat. Agric. Exp. Sta. Res. Bull. 94:1-6 (J-E).
Sawada, Y., Nitta, K., and Igarashi, T. 1964. Soil Sci. Plant Nutr. 10:163-169 (E).
Sawada, Y., Nitta, K., and Igarashi, T. 1965. Soil Sci. Plant Nutr. 10:241-245 (E).
Schollenberger, C. J., and Simon, R. H. 1945. Soil Sci. 59:13-24 (E).
Seki, T. 1913. Landw. Versuch. Stat. 79/80:871-890 (G).
Seki, T. 1934. Jour. Sci. Soil and Manure, Japan 8:3-18 (J).
Sekiya, K. 1970. Methods for Determination of Nutrient Elements in Soils. Yokendo Pub. Co., Tokyo, pp. 225-257 (J).
Sekiya, K., and Egawa, T. 1959. Nat. Inst. Agric. Sci., Japan, Ann. Rept., Sec. 1, Soil, pp. 39-58 (J).
Sekiya, K., and Egawa, T. 1960. Nat. Inst. Agric. Sci., Japan, Ann. Rept., Sec. 1, Soil, pp. 33-38 (J).
Sekiya, K., and Egawa, T. 1961. Nat. Inst. Agric. Sci., Japan, Ann. Rept., 2nd Lab., Soil Chem., pp. 66-81 (J).
Shinagawa, A. 1953. Kagoshima Univ. Fac. Agric. Bull. 2:141-149 (J-E).
Shinagawa, A. 1954. Kagoshima Univ. Fac. Agric. Bull. 3:80-87 (J-E).
Shinagawa, A. 1958. Kagoshima University Fac. Agric. Bull. 7:172-176 (J-E).
Shinagawa, A. 1962. Kagoshima University Fac. Agric. Bull. 11:155-205 (J-E).
Shioiri, M. 1934. Assoc. Sci. Acad., Japan, Rept. 10:694-699 (J).
Shoji, S., and Masui, J. 1969a. Soil Sci. Plant Nutr. 15:161-168 (E).
Shoji, S., and Masui, J. 1969b. Soil Sci. Plant Nutr. 15:191-201 (E).
Shoji, S., and Masui, J. 1969c. Jour. Sci. Soil and Manure, Japan 40:441-447 (J).
Shoji, S., and Masui, J. 1969d. Jour. Sci. Soil and Manure, Japan 40:448-456 (J).
Soil Survey Staff. 1952. Collection of Profile Illustrations of Forest Soils in Japan. Part. 1. Nat. For. Exp. Sta., Tokyo (J).
Sudo, T. 1953. Clay Mineralogy, pp. 174-178. Iwanami Pub. Co., Tokyo (J).
Sudo, T. 1954. Clay Miner. Bull. 2:96-106 (E).
Sudo, T. 1956. Tokyo Univ. of Education, Sci. Rept., Sec. C, 5:39-55 (E).
Sudo, T. 1959. Mineralogical Study on Clays in Japan. Maruzen Pub. Co., Tokyo (E).
Suzuki, T., and Ishizawa, S. 1965. Nat. Inst. Agr. Sci., Japan, Bull. B15:91-196 (J-E).
Takahashi, T., and Yamamoto, T. 1969. Tohoku Agric. Exp. Sta. Res. Bull. 37:139-156 (J).
Takenaka, H. 1963. Res. Agr. Eng., Spec. Pub. (Soc. Agr. Eng.) 7:68-75 (J-E).
Tanabe, I., and Ishizawa, S. 1969. Nat. Inst. Agr. Sci., Japan, Bull. B21:115-253 (J-E).
Terasawa, S. 1963. Nat. Inst. Agric. Sci., Japan, Bull. B13:1-116 (J-E).
Thorp, J., and Smith, G. D. 1949. Ann. Assoc. Amer. Geographers 30:163-194 (E).
Tokudome, S., and Kanno, I. 1964a. Kyushu Agric. Exp. Sta. Bull. 10:185-193 (E).
Tokudome, S., and Kanno, I. 1964b. Kyushu Agric. Exp. Sta. Bull. 10:195-204 (E).
Tokudome, S., and Kanno, I. 1965a. Soil Sci. Plant Nutr. 11:185-192 (E).

Tokudome, S., and Kanno, I. 1965b. Soil Sci. Plant Nutr. 11:193-199 (E).
Tokudome, S., and Kanno, I. 1968. Trans. 9th Int. Cong. Soil Sci. 3:163-173 (E).
Truog, E. 1930. Jour. Amer. Soc. Agron. 28:874-888 (E).
Tsuchiya, T., and Kurabayashi, S. 1959. Adv. Clay Sci., Japan 1:113-120 (J-E).
Tsuru, N., Nishio, D., and Furusaka, C. 1966. Unpublished. See Furusaka's article, "Soil structure and soil organisms." In Furusaka, C. Soils and Microorganisms, No. 8, pp. 1-9 (J).
Tyurin, I. V. 1951. Trudy Pochv. Inst. Dokuchaeva 38:5-25.
Uchiyama, N., Abe, K., and Tsuchiya, T. 1954. Nat. Inst. Agr. Sci., Japan, Bull. B3:43-138 (J-E).
Uchiyama, N., Masui, J., and Onikura, Y. 1962. Soil Sci. Plant Nutr. 8:13-19 (E).
Uchiyama, N., Masui, J., and Shoji, S. 1968. Jour. Sci. Soil and Manure, Japan, 39:149-153 (J).
Uchiyama, N., Masui, J., and Shoji, S. 1968a. Soil Sci. Plant Nutr. 14:125-132 (E).
Uchiyama, N., Masui, J., and Shoji, S. 1968b. Soil Sci. Plant Nutr. 14:133-140 (E).
Uchiyama, N., Masui, J., and Shoji, S. 1968c. Jour. Sci. Soil and Manure, Japan 39:101-109 (J).
Uchiyama, N., Masui, J., and Shoji, S. 1968d. Jour. Sci. Soil and Manure, Japan 39:110-115 (J).
Uchiyama, N., Masui, J., and Shoji, S. 1968e. Jour. Sci. Soil and Manure, Japan 39:149-153 (J).
Uchiyama, N., Masui, J., and Shoji, S. 1968f. Jour. Sci. Soil and Manure, Japan 39:154-160 (J).
Wada, K. 1959. Soil Sci. 87:325-330 (E).
Wada, K. 1967. Amer. Miner. 52:690-708 (E).
Wada, K., and Ataka, H. 1958. Soil and Plant Food 4:12-18 (E).
Wada, K., and Aomine, S. 1966. Soil Sci. Plant Nutr, 12:151-157 (E).
Wada, K., and Harada, Y. 1969. Proc. Intern. Clay Conf., Tokyo, 1:561-571 (E).
Wada, K., and Inoue, T. 1967. Soil Sci. Plant Nutr. 13:9-16 (E).
Wada, K., and Matsubara, I. 1968. Trans. 9th Intern. Cong. Soil Sci. 3:123-131 (E).
Wada, K., and Yoshinaga, N. 1969. Amer. Miner. 54:50-71 (E).
Watanabe. Y. 1966. Nat. Inst. Agr. Sci., Japan, Bull. B16:91-148 (J-E).
Wright, A. C. S. 1964. FAO World Soil Resources Rept. 14:9-22 (E).
Yamagata, U. 1925. Inst. Bacterial Fixation of Atmospheric Nitrogen Res. Bull. 1:3-42 (J).
Yamamoto, T., Shimada, A., Miyasato, S., Hakoishi, T., Ogasawara, K., and Saski, K. 1966. Tohoku Agr. Exp. Sta. Res. Bull. 33:311-423 (J).
Yoshida, M. 1953. Jour. Sci. Soil and Manure, Japan 23:213-215 (J-E).
Yoshida, M. 1956. Jour. Sci. Soil and Manure, Japan 27:241-244 (J-E).
Yoshida, M. 1957. Jour. Sci. Soil and Manure, Japan 28:195-198 (J).
Yoshida, M. 1960a. Jour. Sci. Soil and Manure, Japan 31:415-418 (J).
Yoshida, M. 1960b. Jour. Sci. Soil and Manure, Japan 31:447-450 (J).
Yoshida, M. 1961. Science (Japan) 31:310-313 (J).
Yoshida, T., and Sakai, H. 1962. Hokkaido Nat. Agric. Exp. Sta. Res. Bull. 79:36-44 (J-E).
Yoshida, T., and Sakai, H. 1963a. Jour. Sci. Soil and Manure, Japan 34:155-160 (J).
Yoshida, T., and Sakai, H. 1963b. Jour. Sci. Soil and Manure, Japan 34:197-202 (J).
Yoshinaga, N. 1966. Soil Sci. Plant Nutr. 12:47-54 (E).
Yoshinaga, N. 1968. Soil Sci. Plant Nutr. 14:238-246 (E).
Yoshinaga, N. 1970. Jour. Clay Sci., Japan 9:1-11 (J).
Yoshinaga, N., and Aomine, S. 1962a. Soil Sci. Plant Nutr. 8, No. 2, pp. 6-13 (E).
Yoshinaga, N., and Aomine, S. 1962b. Soil Sci. Plant Nutr. 8:114-121 (E).
Yoshinaga, N., Yotsumoto, H., and Ibe, K. 1968. Amer. Miner. 53:319-323 (E).

Part VI
MINERALOGICAL CHARACTERISTICS

Editor's Comments
on Papers 18 Through 22

18 BIRRELL and FIELDES
Allophane in Volcanic Ash Soils

19 YOSHINAGA and AOMINE
Allophane in Some Ando Soils

20 YOSHINAGA AND AOMINE
Imogolite in Some Ando Soils

21 BESOAIN
Imogolite in Volcanic Soils of Chile

22 ESWARAN
Morphology of Allophane, Imogolite and Halloysite

The mineralogy of Andosols covers two important aspects: mineralogy of the sand fraction and mineralogy of the clay fraction. Most of the information in the literature is about the clay mineralogy; not much has been published on the mineralogy of the sand fraction, which is somewhat surprising since the latter is essential in determining the origin of Andosols. Ash ejected by different volcanoes, or by the same volcano at different times, may be quite different in mineral composition and consequently in chemical composition. Differences in the chemical composition may lead to the formation of Andosols with different chemical characteristics and different fertility status. Some volcanoes in Indonesia—the Kelut volcano, for example—have produced ash with little or no pumice or volcanic glass at one eruption, whereas others—the Krakatao volcano, for example— have yielded ejecta containing 70 percent pumice and 21 percent volcanic glass (Mohr and Van Baren, 1960). In a subsequent eruption, however, the Kelut volcano yielded ash containing 80 percent volcanic glass (Baak, 1948, 1949).

The type of ash also appears to make a great difference in mineralogical composition. Acidic ash—rhyolitic and dacitic volcanic ash, for example—contains the following major minerals: volcanic glass, feldspars, hornblende, pyroxenes, quartz, and biotite. On the other hand, intermediate ash—andesitic volcanic ash, for example—is characterized by smaller amounts of volcanic glass, little or no quartz, smaller amounts of hornblende, and little or no biotite (Mohr and Van Baren, 1960; Tan and Van Schuylenborgh, 1961; Birrell, 1964).

Of the minerals listed, volcanic glass is generally considered the main source for formation of allophane, although other primary minerals can also contribute toward formation of this clay mineral (Birrell, 1964). Volcanic glass is reported to be formed from the remainder of the magma after crystallization of the other minerals. Therefore it is composed of surplus elements that have not been used in the formation of the other minerals (Mohr and Van Baren, 1960). Consequently different types of volcanic glasses can be found. Acid glass is usually light in color and may contain high amounts of silica (SiO_2) or may be composed of pure silica. Intermediate and basic volcanic glasses are darker in color due to higher Fe content. They also contain appreciable amounts of P, K, Ca, and Mg. Therefore these types of glasses are excellent sources of the plant nutrients listed. For formation of allophane, acid and intermediate volcanic glass, containing sufficient amounts of Si and Al, are required. On weathering, basaltic glass produces high amounts of Fe, but relatively little Al. The resulting condition is not favorable for the formation of allophane since the mineral is an aluminum silicate compound (Wright, 1964).

The clay mineralogy of Andosols has been the subject of numerous investigations. Most scientists believe that allophane is the principal mineral in the clay fraction of Andosols. Although imogolite has also been detected as an important mineral in these kinds of soils, it appears that imogolite occurs only in certain types of Andosols; evidence for its universal occurrence in all Andosols has yet to be established. Paper 18 discusses the occurrence of allophane as the principal clay mineral in Andosols derived from andesitic and rhyolitic volcanic ash in New Zealand. Papers 19 and 20 discuss the identification of allophane and imogolite, respectively, in Andosols in Japan. Paper 21 shows the occurrence of imogolite in Andosols in Chile, and Paper 22 shows the morphology of allophane and imogolite as studied by electron microscopy.

Both allophane and imogolite were thought to be specific clay minerals of Andosols only, but recently these minerals have also been detected in Spodosols (Brydon and Shimoda, 1972; Tait, Yoshinaga, and Mitchell, 1978; Ross and Kodama, 1979; Ross, 1980). Imogolite or imogolite-like material has also been found in a podsol-braunerde of nonvolcanic parent material in Finland (Alberto and Jaritz, 1966).

Editor's Comments on Papers 18 Through 22

Allophane is classified as a noncrystalline (amorphous) alumino silicate with a molar SiO_2/Al_2O_3 ratio ranging from 1 to 2 (Yoshinaga, 1966; Henmi and Wada, 1976), but recent investigations by high-resolution electron microscopy show allophane to be a hollow spherule of the size of 35 to 50 Å (3.5 to 5.0 nm) in diameter (Kitagawa, 1971; Wada, 1980). According to Fieldes (1955) and Fieldes and Furkert (1966), three types of allophane can be distinguished: allophane B, AB, and A. Allophane B is the first form of allophane that develops from the weathering of volcanic ash minerals, and after further weathering, this mineral can be transformed into the other types of allophane and eventually into halloysite. The weathering sequence as anticipated by Fieldes and coworkers is as follows: allophane B → allophane AB → allophane A → metahalloysite (see also Paper 2). Fieldes and Perrott (1966) have also devised the so-called NaF test for the presence of allophane in soils. Although this test has been adopted by the U.S. *Soil Taxonomy* for the identification of amorphous materials, examples have been reported several times in the literature for the inaccuracy of the test, in which soils containing no allophane gave a positive NaF reaction (Mizota and Wada, 1980; Paper 12). According to Wada (1980), the NaF test is not specific for allophane and imogolite but can be used to indicate the presence of active Al-OH groups in soils.

Many Japanese scientists also questioned the presence of allophane B (Wada and Aomine, 1973; Miyauchi and Aomine, 1964). They noted that the mineral called allophane B had similar DTA (differential thermal analysis) and IR (infrared) characteristics as cristobalite, quartz, or feldspar, while Shoji and Masui (1969a, 1969b) claimed it to be opaline silica.

Imogolite, the second most important mineral in Andosols, has been included in the amorphous clay group, although X-ray diffraction analysis and electron microscopy reveal the mineral to exhibit some crystallinity. As observed with the electron microscope, imogolite has a smooth thread-like structure (Papers 20 and 21), varying in size from 100 to 300 Å (10 to 30 nm) in diameter. Wada (1980) indicated that the threads consisted of fine hollow tubes with an inner and outer diameter of 10 and 20 Å (1 and 2 nm), respectively. The external surface of the tube is composed of a gibbsite-like material; the internal surface consists of silica tetrahedra. Unlike allophane, the Al in the gibbsite layer of imogolite is present in octahedral coordination (Henmi and Wada, 1976), and each silica tetrahedron at the internal surface occurs as an isolated unit. Siloxane bonds, Si-O-Si bonds, connecting the silica tetrahedra together, are absent (Brown et al., 1978; Parfitt, 1980).

Imogolite is considered an intermediate phase in the weathering sequence of allophane to crystalline layer silicates. In this respect, Wada and Aomine (1973) suggested the following hypothesis:

$$\text{allophane} \begin{array}{c} \nearrow \text{halloysite} \searrow \\ \\ \searrow \text{imogolite} \nearrow \end{array} \text{gibbsite}$$

The authors indicated that these were the most plausible reactions due to the fact that a desilication process seemed to prevail in most Andosols.

Other amorphous minerals, such as amorphous silica, have also been detected in Andosols. Shoji and Matsui (1969a, 1969b) were able to distinguish three kinds of amorphous silica: plant opal, opaline silica, and aggregates of amorphous silica. Of the three, opaline silica was considered the dominant amorphous mineral.

In addition to the amorphous materials, Andosols recently have been found to contain varying amounts of crystalline silicate clays. Halloysite and kaolinite have frequently been detected in the soils by Kanno (Paper 2), Fieldes (1955), and others. More recently montmorillonite (Uchiyama, Masui, and Chikura, 1962; Uchiyama, Masui, and Shoji, 1968a, 1968b) and 14 Å minerals (Kawasaki and Aomine, 1966) have been added as mineral constituents of the crystalline clay fraction of Andosols. These crystalline minerals are especially important in Andosols of older pedogenic age. Wada and Aomine (1973) noted that allophane and imogolite had a relatively short life in an open leaching environment under warm humid conditions. Many of the Japanese scientists postulated that in the process of crystallization, the globular allophane particle grew into an onion-like mass, which finally burst into tubular halloysite particles. Such a hypothesis is in contrast with that of most U.S. scientists (Jackson and Sherman, 1953) who claim that kaolin minerals decomposed into allophane.

REFERENCES

Alberto, G. F., and G. Jaritz, 1966, Amorphe und kristalline Bestandteile einiger typischen Bodenbildungen Skandinaviens, *Z. Pflanzenernahr., Düng. Bodenk.* **114**:27-46.

Baak, J. A., 1948, De Mineralogische Samenstelling van enkele recente vulkanische assen op Java, *Landbouw* (Buitenzorg, Indonesia) **20**:269-274.

Baak, J. A., 1949, A Comparative Study on Recent Ashes of the Java Volcanoes Smeru, Kelut and Merapi, *Meded. Alg. Proefsta. Landbouw, Bogor* (Indonesia), No. 83, 37p.

Birrell, K. S., 1964, Some Properties of Volcanic Ash Soils, Meeting on the Classification and Correlation of Soils from Volcanic Ash, Tokyo, Japan, June 11-27, 1964. *Food and Agric. Org., United Nations, World Soil Resources Rept.* **14:**74-81.

Brown, G., A. C. D. Newman, J. H. Rayner, and A. H. Weir, 1978, The Structures and Chemistry of Soil Clay Minerals, in *The Chemistry of Soil Constituents,* D. J. Greenland and M. H. B. Hayes, eds., John Wiley & Sons, New York, pp. 29-178.

Brydon, J. E., and S. Shimoda, 1972, Allophane and Other Amorphous Constituents in a Podzol from Nova Scotia, *Canadian Jour. Soil Sci.* **52:**465-475.

Fieldes, M., 1955, Clay Mineralogy of New Zealand Soils, Part 2, *New Zealand Jour. Sci. Tech.* **B37:**336-350.

Fieldes, M., and R. J. Furkert, 1966, The Nature of Allophane in Soils, Part 2, *New Zealand Jour. Sci.* **9:**608-622.

Fieldes, M., and K. W. Perrott, 1966, The Nature of Allophane in Soils, Part 3, Rapid Field and Laboratory Test for Allophane, *New Zealand Jour. Sci.* **9:**623-629.

Henmi, T., and K. Wada, 1976, Morphology and Composition of Allophane, *Am. Mineralogist* **61:**379-390.

Jackson, M. L., and G. D. Sherman, 1953, Chemical Weathering of Mineral Soils, *Adv. Agronomy* **5:**219-318.

Kawasaki, H., and S. Aomine, 1966, So-called 14 Å Minerals in Some Ando Soils, *Soil Sci. Plant. Nutri.* (Japan) **12:**144-150.

Kitagawa, Y., 1971, The "Unit Particle" of Allophane, *Am. Mineralogist* **56:**465-475.

Miyauchi, N., and S. Aomine, 1964, Does "Allophane B" Exist in Japanese Volcanic Ash Soils? *Soil Sci. Plant Nutr.* (Japan) **10:**199-203.

Mizota, C., and K. Wada, 1980, Implications of Clay Mineralogy to the Weathering and Chemistry of Ap Horizons of Ando Soils, *Geoderma* **23:**49-63.

Mohr, E. C. J., and F. A. Van Baren, 1960, *Tropical Soils: A Critical Study of Soil Genesis as Related to Climate, Rock and Vegetation,* Les Editions A. Manteau S. A., Brussels, 498p.

Parfitt, R. L., 1980, Chemical Properties of Variable Charge Soils, in *Soils with Variable Charge,* B. K. G. Theng, ed., New Zealand Soc. Soil Sci., Lower Hutt, New Zealand, pp. 167-194.

Ross, G. J., 1980, The Mineralogy of Spodosols, in *Soils with Variable Charge,* B. K. G. Theng, ed., New Zealand Soc. Soil Sci., Lower Hutt, New Zealand, pp. 127-143.

Ross, G. J., and H. Kodama, 1979, Evidence for Imogolite in Canadian Soils, *Clays and Clay Minerals* **27:**297-300.

Shoji, S., and J. Masui, 1969a, Amorphous Clay Minerals of Recent Volcanic Ash Soils in Hokkaido. I, *Soil Sci. Plant Nutr.* (Japan) **15:**161-168.

Shoji, S., and J. Masui, 1969b, Amorphous Clay Minerals of Recent Volcanis Ash Soils in Hokkaido. II, *Soil Sci. Plant Nutr.* (Japan) **15:**191-201.

Tait, J. M., N. Yoshinaga, and B. D. Mitchell, 1978, The Occurrence of Imogolite in Some Scottish Soils, *Soil Sci. Plant Nutr.* (Japan) **24:**145-151.

Tan, K. H., and J. Van Schuylenborgh, 1961, On the Classification and Genesis of Soils Developed over Acid Volcanic Materials under Humid Tropical Conditions, II, *Netherlands Jour. Agric. Sci.* **9:**41-54.

Uchiyama, N., J. Masui, and Y. Onikura, 1962, Montmorillonite in a Volcanic Ash Soil, *Soil Sci. Plant Nutr.* (Japan) **8:**13-19.

Uchiyama, N., J. Masui, and S. Shoji, 1968a, Crystalline Clay Minerals of the Soils Derived from Recent Volcanic Ashes in Hokkaido, Japan, I. *Soil Sci. Plant Nutr.* (Japan) **14:**125-132.

Uchiyama, N., J. Masui, and S. Shoji, 1968b, Crystalline Clay Minerals of the Soils Derived from Recent Volcanic Ashes in Hokkaido, Japan, II, *Soil Sci. Plant Nutr.* (Japan) **14:**133-140.

Wada, K., 1980, Mineralogical Characteristics of Andisols, in *Soils with Variable Charge,* B. K. G. Theng, ed., New Zealand Soc. Soil Sci., Lower Hutt, New Zealand, pp. 81-107.

Wada, K., and S. Aomine, 1973, Soil Development on Volcanic Materials during the Quaternary, *Soil Sci.* **116:**170-177.

Wright, A. C. S., 1964, The "Andosols" or "Humic Allophane" Soils of South America, Meeting on the Classification and Correlation of Soils from Volcanic Ash, Tokyo, Japan, June 11-27, 1964, *Food and Agric. Org., United Nations, World Soil Resources Rept.* **14:**9-21.

Yoshinaga, N., 1966, Chemical Composition and Some Thermal Data of Eighteen Allophanes from Andosols and Weathered Pumices, *Soil Sci. Plant Nutr.* (Japan) **12:**47-54.

ALLOPHANE IN VOLCANIC ASH SOILS
K. S. BIRRELL AND M. FIELDES
(*Soil Bureau, Department of Scientific and Industrial Research, New Zealand*)

Introduction
WITH ONE PLATE

IN the course of examination of volcanic ash soils from a hospital site at New Plymouth as a foundation for proposed new buildings, and from the sites of hydro-electric stations at Whakamaru and Atiamuri on the Waikato River for earthwork construction, the high water-holding capacity, high shrinkage, and irreversible drying of these soils were very striking. A moderately high compressibility under load was also shown by some of the New Plymouth soils. It is intended to present data in a subsequent paper on the physical properties of these ash soils, but the object of the present investigation is to examine the finer fractions of these soils in order to identify the minerals responsible for the above-mentioned physical properties. The parent material of the soils from the New Plymouth site is largely andesitic, and from the Whakamaru and Atiamuri sites is both andesitic and rhyolitic (Grange, 1931).

Clay fractions were separated from these soils after using the deferration procedure of Dion (1944), and on differential thermal analysis gave curves resembling those published for allophane (Berkelhamer, 1944).

A subsoil sample of heavy clay texture obtained from the site of a proposed hard-standing area at Whenuapai airfield, 13 miles north-west of Auckland City, showed similar physical properties to the andesitic ash soils but had much higher water-content and shrinkage. The parent material of this soil is thought to be water-sorted volcanic ash, and although the rainfall at Whenuapai is lower than at the other sites mentioned, the original swampy nature of the portion of the airfield where the sample was found could account for the ash being in a similar state of weathering. The clay fraction was identified as a fairly pure allophane, practically free of sesquioxides.

Allophane was also found by differential thermal analysis in admixture with kaolin in a heavy clay weathered from the volcanic ash beds at another location on the hospital site in New Plymouth, and in the clay fraction of a soil formed from the rhyolitic Mamaku ash shower on the eastern side of Lake Rotorua, although the clay content of this latter soil was not high enough to have much influence on physical soil properties.

Mr. N. H. Taylor (1933), who had found that the clay fractions prepared by dispersion with water or very dilute NaOH from certain 'waxy pans' occurring in old ash beds found on the Mairoa Plateau, west of Te Kuiti, gave silica-alumina ratios close to 1·0 and were accordingly presumed to be allophane, suggested examination of this material by differential thermal analysis. The 'waxy pan' had similar physical properties to the Whenuapai soil, but the clay fraction separated from it was found to contain appreciable gibbsite as well as allophane. The subsoil of the top ash shower on the Mairoa Plateau resembled the New Plymouth soil and contained mostly allophane and some limonite.

The predominance of allophane in the clay fractions in all these soils and also its probable presence in some silt fractions was later inferred from the largely amorphous nature of the material when examined with a Philips Geiger X-ray spectrometer.

The deferration procedure of Jeffries (1946) was used in preparing clay fractions from the Whenuapai, Mamaku ash, and Mairoa Plateau soils.

Allophane was also identified in clay fractions prepared from some of these soils by methods which did not involve deferration.

Preparation of Clay and Silt Fractions

The parent materials and locations of the soil samples used in this work are set out in Table 1, below.

TABLE 1
Origin and Location of Volcanic Ash Soils

Clay or silt sample Nos.	Ash shower	Location	Depth from surface
1 a–c	Egmont (andesitic)	New Plymouth Hospital	16–17 ft.
2	Tirau (andesitic + rhyolitic)	Whakamaru Hydro, G18	..
3 a	,, ,, ,,	Atiamuri Hydro, Shaft 29	5 ft.
3 b	,, ,, ,,	,, ,, Shafts 25 and 26	
4 a–c	Mairoa (andesitic + rhyolitic)	Ngapenga Rd., Mairoa	9–15 in.
5	pre-Hamilton	,, ,, ,,	6 ft.
6 a–c	Water-sorted ash	Whenuapai airfield	6 ft.
7	Mamaku (rhyolitic)	East side, Lake Rotorua	16 in.

The procedures used in obtaining the clay and silt fractions from the above soils are set out in Table 2.

TABLE 2
Soil Pretreatments and Yields of Clay

Sample No. (see Table 1)	Treatment	Clay %
1 a	Dion deferration, Na_2CO_3 dispersion	6·8
1 b	,, ,, ,, ,,	13·8
1 c	,, ,, N/500 HCl dispersion	14·0
2	,, ,, Na_2CO_3 dispersion	10·9
3 a	,, ,, ,, ,,	7·9
3 b	No deferration, N/100 NH_4OH dispersion	11·3
4 a	Dion deferration, Na_2CO_3 dispersion	27·0
4 b	,, ,, N/500 HCl dispersion	21·2
4 c	No deferration, N/500 HCl dispersion	30
5	N/5 HCl leaching, Jeffries deferration, Na_2CO_3 dispersion	c. 0·5
6 a	Jeffries deferration, Na_2CO_3 dispersion	22·6
6 b	,, ,, thrice repeated, Na_2CO_3 dispersion	< 5
6 c	No deferration, N/100 NH_4OH dispersion	64
7	,, ,, N/500 HCl dispersion	5

Notes: 1. Undried soil was used as starting material, except in the case of sample 1a. Yields are calculated on oven-dry soil basis.

2. The Na_2CO_3 solution had a pH of approximately 8·0.

Discussion on Table 2

The reduced yields with sample 1*a* as compared with 1*b* as a result of oven-drying are due to the irreversible drying effects shown by these soils. Even air-drying renders the soil much less dispersible than in its natural state.

N/500 HCl as a dispersing medium was tried in view of the reports by Davies (1933) and the Imperial Bureau of Soil Science (1933) that a dilute acid caused dispersion of some New Zealand volcanic ash soils when the alkaline media used for mechanical analysis produced flocculation. It can be seen that such treatment did not improve the yields of clay, either with or without deferration, in comparison with the results obtained by deferration followed by sodium carbonate dispersion.

The New Plymouth and Mairoa ash soils would not disperse satisfactorily in an alkaline medium without deferration treatment. The yield of clay in the case of the Whenuapai soil agreed closely with the result of mechanical analysis by the Bouyoucos method using sodium silicate and sodium oxalate as dispersing agents. The decreased yield of clay with this last-mentioned soil as a result of the normal Jeffries deferration and

TABLE 3

Differential Thermal Analysis and X-Ray Diffraction Data

Sample	Fraction	D.T.A. conclusions	X-ray conclusions
1 *b*	Clay	Allophane with 3% kaolin	Mostly amorphous with a little felspar.
1 *b*	Silt	..	Mostly amorphous with some quartz, felspar, and orthoclase.
2	Clay	Allophane with 3% kaolin	Mostly amorphous.
3 *a*	,,	,, ,, 5% quartz	,, ,,
3 *b*	,,	,, ,, 2% quartz	,, ,,
4 *a*	,,	,, ,, 5% kaolin and some limonite	,, ,, with small amounts of kaolin and felspar.
4 *a*	Silt	..	More than 50% quartz with some albite.
4 *c*	Clay	Allophane with a little limonite	..
5	,,	,, ,, much gibbsite and 5% quartz	Some amorphous with much gibbsite.
6 *a*	,,	Allophane with 5% quartz.	Mostly amorphous.
6 *b*	,,	Small amount of quartz and unidentified material with endothermics at 135° and 210° but no exothermics. Allophane absent	,, ,, with small amount of unidentified crystalline material.
6 *c*	,,	Allophane with 3% kaolin and 2% quartz	Mostly amorphous.

Notes: 1. Figures for percentages of minor constituents in the D.T.A. column are estimates only.

2. Quartz was identified in the D.T.A. curves by the normal α–β inversion, although it is possible that it has been missed in some samples, since it has been found in the course of work in this laboratory that some forms of quartz sometimes fail to show this inversion. Small amounts of quartz were less easily detected in the X-ray records.

the still smaller yield following the triple deferration treatment are very probably due to attack on the allophane by the reagents.

The soil, of heavy silt loam texture, which was derived from the pre-Hamilton ash shower (a 'waxy pan') required leaching with N/5 HCl

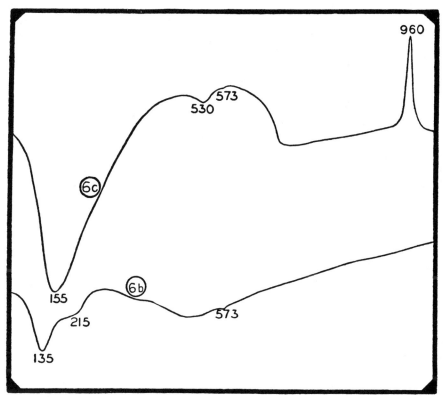

FIG. 1. Differential Thermal Analysis curves for clays containing Allophane
6c. (undeferrated)   6b. (triply deferrated)

prior to deferration to remove the greater part of the gibbsite present before any dispersion could be effected. The yield of clay was about 5 per cent., and a yield of a similar order was obtained from another sample by heating with N/10 NaOH, followed by dispersion with water only, no deferration treatment being used. This procedure also dissolved some alumina. The low yields of clay from this soil are indicative of only partial dispersion.

The results of differential thermal analysis (D.T.A.) and X-ray diffraction measurements on various clay and silt preparations as detailed in Table 2, are given in Table 3, page 158.

The D.T.A. curves for deferrated and non-deferrated samples were similar, with the important exception that the Whenuapai clay fraction sample 6b which had been subjected to the triple deferration showed the absence of the characteristic exothermic peak at about 960° and the initial endothermic low-temperature water-loss effect was much reduced (see Fig. 1). Normal deferration treatment as prescribed by Dion or Jeffries may apparently not prevent the identification of allophane from the D.T.A. curves, but this mineral may fail to be detected if the treatment should be unduly prolonged. In this connexion it is to be noted that a sample of the 'waxy pan' which was deferrated once by Jeffries's method, but from which the clay fraction was not subsequently separated, behaved on thermal analysis in a similar way to sample 6b clay fraction. Reference is made to this 'waxy pan' sample in the later section on dye-adsorption. It seems desirable, therefore, that when seeking to identify allophane in soils, clay fractions should be prepared where possible by methods which do not involve deferration.

The fact that allophane appears amorphous to X-rays suggests that its particle size is below 100 Å. A specimen of what was considered to represent true allophane from Whenuapai, sample 6c, was sent to H. L. Nixon of Rothamsted Experimental Station, who has kindly supplied the electron micrograph reproduced in Fig. 2. The smallest particles show no external regularity and are 50 Å or less in equivalent diameter. The larger particles appear to be aggregates.

Chemical Analyses of Clay and Silt Fractions

The results of chemical analyses made of the clay prepared in the course of this investigation are shown in Table 4, the sample numbers being the same as in the previous tables.

It will be seen that the clay fraction samples 3b, 4c, and 6c which have not been deferrated have lower silica-alumina ratios than the samples from the same soil which have been so treated. There is therefore a relative loss of alumina by deferration which becomes almost complete in the triply deferrated sample 6b. The excess silica in the deferrated samples is apparently not detectable by thermal analysis.

The low silica-alumina ratio of sample 5 is due to the high gibbsite content. Another clay fraction from the same soil which had more leaching with N/5 HCl before dispersion gave a silica alumina ratio of 0·99.

Because both the Dion and Jeffries deferration treatments involve heating the samples with slightly acid solutions, some attack on the allophane is to be expected, inasmuch as the decomposition of this mineral by fairly concentrated acid has been observed by several investigators. With the majority of clay minerals the effect of deferration on composition should be slight, although Williams (1950) has reported that the Jeffries method gives appreciable losses of alumina and considerable loss of iron from biotite.

The silica-alumina ratios for allophane specimens reported by other workers cover a considerable range. Ross and Kerr (1934) refer to twelve analyses where this ratio varied from 0·74 to 1·98, about one-half being not far from 1·0. Salmang is reported (Pask and Davies, 1943) as recording compositions for the mineral covering the range Al_2O_3, 0·3–1·3 SiO_2, 1·8–8·5 H_2O. Some still more recent analyses (1936, 1948) give silica-alumina ratios between 1·0 and 1·29, with water-contents somewhat over 30 per cent. These specimens, with the possible exception of those described by Salmang, whose monograph was not available, are from sources other than soil clays. That the silica-alumina ratio for allophane derived from N.Z. volcanic ash soils may show a fairly wide range even without deferration is shown by the analysis of samples 3b, 4c, and 6c (Table 4), bearing in mind that any other constituents are present in only a very small amount.

TABLE 4

Analyses of Clay Fractions of Soils Extracted According to Treatments in Table 2

Sample No.	1b	2	3a	3b	4a	4b	4c	5	6a	6b	6c
SiO_2	47.17	52.81	50.31	43.5	40.91	43.04	29.33	22.64	44.21	90.29	41.24
Al_2O_3	43.24	37.04	39.32	41.7	43.41	42.33	49.36	70.26	45.73	5.01	57.71
Fe_2O_3	5.27	5.45	4.82	9.5	8.29	8.36	18.40	2.36	0.63	0.62	0.72
TiO_2	0.74	0.47	0.49	0.9	1.55	1.45	1.72	0.93	0.45	0.91	0.49
CaO	tr.	tr.	tr.	tr.	tr.	tr.	tr.	n.d.	nil	nil	nil
MgO	0.28	0.66	0.48	nil	0.62	0.64	tr.	0.65	n.d.	nil	nil
Na_2O	2.38	3.04	n.d.	n.d.	4.57	n.d.	0.69	n.d.	n.d.	1.71	0.15
K_2O	0.46	n.d.	n.d.	n.d.	0.62	n.d.	0.29	n.d.	n.d.	0.05	nil
P_2O_5	0.35	n.d.	n.d.	n.d.	0.21	n.d.	n.d.	n.d.	n.d.	n.d.	n.d.
Total constituents determined	99.89	99.56	95.42	95.6	100.02	95.82	99.79	96.84	91.03	98.59	100.31
SiO_2/R_2O_3	1.73	2.22	2.02	1.55	1.44	1.53	0.82	0.54	1.64	28.4	1.20
SiO_2/Al_2O_3	1.85	2.42	2.17	1.78	1.61	1.73	1.01	0.55	1.64	30.6	1.21
Ignition loss*	21.2	15.02	19.7	24.8	24.7	22.05	30.8	27.0	24.0	9.42	31.5
S. G.	2.11	2.30	2.21	n.d.	2.10	n.d.	n.d.	n.d.	n.d.	2.37	n.d.	1.98

* Expressed as percentage of weight of sample dried at 105° C.

Refractive index determination on the clay fraction of sample 6c by Mr. A. Steiner of the N.Z. Geological Survey gave 1·483, which is close to the average value of 1·480 reported by Ross and Kerr (1934) for specimens of allophane from seven localities.

Base-Exchange Capacity of Allophane

Table 4 shows in several instances deferrated clay fractions containing an appreciable amount of sodium and smaller amounts of potash. Sodium chloride having been added as saturated solution to flocculate the clay suspension, the removal of chloride ion from the clay by washing first with 80 per cent. alcohol followed by 90 per cent. alcohol was taken as freeing it from all alkali salts. Although allophane has been stated to have a low base-exchange capacity, the relatively high alkali content of these clays resulting from the treatment with alkali tartrates or oxalates would be expected to arise by some reaction which is at least analogous to the base-exchange reactions of the crystalline clay minerals. Confirmation of the high capacity of allophane to take up alkali metals was obtained by leaching some of the deferrated clay and silt fractions with neutral potassium acetate solution, removing excess of this salt with alcohol, and determining by flame photometer the potassium which could be subsequently leached out by ammonium acetate solution.

Some soda plus potash contents of the soil fractions have been calculated as milli-equivalents per cent. on the oven-dry basis and tabulated in Table 5 together with the figures for potassium taken up from neutral potassium acetate solution by the same soil fraction.

TABLE 5

Alkali Content of Deferrated Clay Fractions and Potassium taken up from Neutral N Potassium Acetate

Sample	Fraction	Soda plus potash from fusion analysis (m.e. %)	Adsorbed potassium (m.e. %)
1 b	Clay	61	75
2	,,	Na_2O only = 82	74
3 a	,,	n.d.	96
4 a	,,	121	78

Sufficient of preparation 1b clay fraction was available for a single determination of base-exchange capacity and total bases by the standard soil method using neutral N ammonium acetate. With the other preparations only sufficient material was left over for determining adsorbed potassium by the flame photometer method. Results by the ammonium acetate method were base-exchange capacity 54 m.e. per cent., total bases 36·5 m.e. per cent., of which 34·7 m.e. per cent. was sodium.

The alkalies in the allophane clays which had been deferrated appear to be rather more loosely held than is the case with exchangeable bases held by other clay minerals. Leaching with distilled water (0·5 g. sample to 250 c.c.) removed about one-half of the total sodium, but a smaller proportion of potassium from the clay fractions. The remainder of the sodium and most of the potash could be leached out with 250 c.c. neutral N ammonium acetate.

These indications of high base-exchange capacity for allophane are comparable with the values of base-exchange capacity towards 0·5 neutral N barium acetate, found by Mattson for iso-electrically precipitated aluminium 'silicates' of similar chemical composition to the clay fractions described above.

Mattson (1931) states that a precipitate of the composition $Al_2O_3(SiO_2)$ 1·63 took up 47 m.e. per cent. Ba while another corresponding to the composition $Al_2O_3(SiO_2)$ 1·09 had a base-exchange capacity measured in the same way of 30 m.e. per cent.

Further work on the whole soil from Whenuapai has shown that the base-exchange capacity as determined with neutral N/2 barium acetate is somewhat higher than the values found using N potassium or ammonium acetate, but is still within the order of the results given in Table 5.

Adsorption by Allophane of Alizarin Red S

Hardy and Rodrigues (1939) have described a method for the estimation of gibbsite alumina in tropical soils by using the adsorptive power of the ignited mineral for the acidic dye Alizarin Red S. When this method was used for quantitative determination of the gibbsite present in some of the soil fractions where it had been clearly detected by D.T.A. and X-ray analysis, it was found that amounts of this mineral were indicated far in excess of that estimated to be present by the two last-mentioned methods. Tests on clay fractions of the Whenuapai soil which appears from the evidence given earlier to be practically pure allophane showed that this mineral was also capable of adsorbing the dye after ignition under the conditions specified by Hardy in his method. The adsorptive power of ignited allophane is about 70 per cent. of that of ignited gibbsite. Table 6 gives the results of some tests made on volcanic ash soils with this method.

TABLE 6

Adsorption Tests with Alizarin Red S

Sample	Gibbsite content by dye adsorption		Gibbsite content estimated from D.T.A.	Minerals present	SiO_2/Al_2O_3
	Oven-dry material	Ignited material			
Whenuapai clay 6c	25	69	nil	Allophane about 95%	1·21
Whenuapai whole soil	35	72	n.d.	Mostly allophane	1·52
Ngapenga Rd., Mairoa, 6 ft. whole soil (cf. No. 5, Table 1)	51	81	10–20	Allophane + 10% gibbsite +1% kaolin, +10% silica,	1·13
Ditto but deferrated	n.d.	1·8	2–5	2% gibbsite, 3% kaolin, 25% silica, no allophane.	13·3

It is to be noted that the difference between the gibbsite contents of the Whenuapai clay and the Ngapenga Road soil as measured by dye adsorption is approximately equal to the actual gibbsite content of the latter soil as estimated from the D.T.A. curves.

The deferrated Ngapenga Road soil is comparable with the triply deferrated Whenuapai clay fraction 6a in its D.T.A. curve (see Fig. 1), and the silica-alumina ratio in Table 6 confirms that the allophane has been largely decomposed. Decomposition of the allophane has largely destroyed the dye-adsorptive power.

The conclusion to be drawn is that this dye adsorption test for gibbsite must be used with caution when allophane is known to be present in soils.

Dispersion of Soils containing Allophane and Free Sesquioxides

The difficulties in dispersing some volcanic ash soils have been discussed by Davies (1933) and he suggested N/500 HCl as a suitable dispersing agent.

The following observations were made in the course of the present investigation with respect to the andesitic ash soils.

1. The Whenuapai soil, which contains at the most only a trace of gibbsite, dispersed satisfactorily in dilute ammonia solution, the pH of the suspension being 10·4. The soil was first treated with hydrogen peroxide as in the International method of mechanical analysis, but the treatment with N/5 acid was omitted. The clay content found was 64·6

per cent. The same figure was obtained when analysed by the hydrometer method of Bouyoucos using sodium oxalate and sodium silicate as dispersing agents. N/500 HCl was much less effective as dispersing agent, the clay content found being only 34·6 per cent.

2. The stable ammonia dispersion of the Whenuapai soil had added to it a suspension made from undried soil equivalent to about 10 g. dry weight of the Ngapenga Road soil containing, as estimated from the thermal analysis curve, about 10–20 per cent. gibbsite. After daily shaking for 1 week, very definite flocculation had occurred. Examination of the filtrate showed, however, a negligible amount of alumina in solution.

3. A dispersion of the clay fraction of the soil from the Mairoa ash shower in N/200 HCl (pH of dispersion = 4·1) was strongly flocculated by the addition of N/10 ammonium hydroxide when the pH reached 6·6, signs of flocculation beginning at pH = 5·9. Further addition of ammonia to bring the pH to 11·4 did not produce dispersion. On boiling with N/10 NaOH, centrifuging off, and shaking with water, dispersion was obtained. Addition of N/5 HCl to this NaOH dispersion produced flocculation at pH 9·6, but redispersion did not occur at any stage on adding acid sufficient to give a final pH of 2·5. On centrifuging off and shaking with N/200 HCl, dispersion was again obtained.

A probable explanation of the observations made by Davies and those described above is that we are dealing with two colloidal systems of different iso-electric points capable of mutual coagulation under the appropriate pH conditions, these being allophane on the one hand and aluminium or ferric hydroxide on the other. The third experiment above suggested that the iso-electric point of allophane is slightly on the acid side of neutrality, and this receives support from the work of Mattson (1930) on iso-electrically precipitated aluminium silicates. In fact he states that the iso-electric point of a floc with a composition $Al_2O_3(SiO_2)$ 1·09 is pH 6·6, which is the required hydrogen-ion concentration for strong flocculation found in experiment 3. It appears to be accepted that the iso-electric points of $Al(OH)_3$ and $Fe(OH)_3$ are between pH 7 and pH 8. Consequently dispersion of a soil containing both allophane and free sesquioxides should be possible in a weakly acid medium (provided the anion concentration is sufficiently low) or in a strongly alkaline medium (provided the cation concentration is low enough). There seems to be also some critical concentration of iron or aluminium hydroxide in the soil, above which neither acid nor alkaline dispersing media are effective.

The observation of Davies that 'a soil which has been motor dispersed flocculates more rapidly and completely on the addition of any of the usual deflocculants than one that has not' has a likely explanation in that the sesquioxides which help cement the soil aggregates are set free thereby to exert their flocculating action.

There is no practical method in sight for completely dispersing soils containing all but traces of oxides of iron and aluminium as well as allophane. If the soil is known to be almost entirely free of sesquioxides, pretreatment with hydrogen peroxide, washing with water, and dispersion in a rotary shaker using an alkaline medium of pH about 10 appears to be the most effective procedure. No acid pretreatment should be given with the intention of removing free sesquioxides as this is likely to cause partial decomposition of the allophane.

Summary

Allophane has been found to be the principal mineral in the clay fractions of soils derived from andesitic and rhyolitic volcanic ash showers in New Zealand. It is considered to be the common factor responsible for the characteristic physical properties of the andesitic ash soils. The chemical composition of the purest allophane separated from these soils varies similarly to that of geological specimens described by other workers. When deferration treatments are applied to the soil before separation of the clay fractions there can be considerable loss of alumina, although the characteristic differential thermal analysis curve of the mineral may still be obtained. If the deferration treatment is unduly prolonged the characteristic high temperature exothermic peak of allophane in the differential thermal curve may disappear.

The deferrated allophane clays take up appreciable amounts of alkali metals and could accordingly be considered to have high base-exchange capacity, although allophane has been considered previously to be in the low base-exchange capacity group of clay minerals. Allophane after ignition to 750° C. has a high capacity for adsorption of the dye Alizarin Red S, taking up about 70 per cent. of the quantity of dye adsorbed by gibbsite ignited under the same conditions. The difficulties found by previous workers in dispersing certain New Zealand volcanic ash soils are considered to be due to the association of allophane and free sesquioxides. Experiments by the writers lead to the conclusion that the isoelectric point of allophane is much closer to neutrality than is the case with the other clay minerals. Dispersion of soils containing allophane appears to be practicable only if the free sesquioxide content is low and the pH of the medium should at least be 10. Acid pretreatment should not be given owing to possible attack on the allophane, and a dilute acid medium does not appear to be a satisfactory solution to the problem of dispersing these soils.

Acknowledgements

The authors express their thanks to the following members of the Soil Bureau staff: Mr. J. P. Richardson for differential thermal analyses, Messrs. L. B. Robinson, R. F. Thomas, and M. Gradwell for preparation of clay and silt fractions of the soils examined.

REFERENCES

BERKELHAMER, L. H. 1944. An apparatus for differential thermal analysis. U.S. Bur. Mines Res. Investigation 3762.
Chem. Abstr. 1936, **30**, 993[4].
Chem. Abstr. 1948, **42**, 4877[e].
DAVIES, E. B. 1933. Studies in the dispersion and deflocculation of certain soils. N.Z. J. Sci. Tech. **14**, 228–32.
DION, H. G. 1944. Iron oxide removal from clays and its influence on base-exchange properties and X-ray diffraction patterns of the clays. Soil Sci. **58**, 411–24.

GRANGE, L. I. 1931. Volcanic ash showers. N.Z. J. Sci. Tech. **12,** 228–40.
HARDY, F., and RODRIGUES, G. 1939. Soil genesis from andesite in Grenada (B.W.I.). Soil Sci. **48,** 361–84.
Imp. Bur. Soil Sci. 1933. The dispersion of soils in mechanical analysis. Tech. Comm. No. 26, 31 pp.
JEFFRIES, C. D. 1946. A rapid method for the removal of free iron oxides in soils prior to petrographic analysis. Proc. Soil Sci. Soc. Amer. **11,** 211–12.
MATTSON, S. 1930. The laws of soil colloidal behaviour. III. Isoelectric precipitates. Soil Sci. **30,** 459–95.
—— 1931. The laws of soil colloidal behaviour. V. Ion adsorption and exchange. Soil Sci. **31,** 311–31.
PASK, J. A. and DAVIES, B. 1943. Thermal analysis of clay minerals and acid extraction of alumina from clays. U.S. Bur. Mines Res. Investigation 3737.
ROSS, C. S., and KERR, P. F. 1934. Halloysite and Allophane. U.S. Geol. Surv. Prof. Paper 185–G: 135–48.
TAYLOR, N. H. 1933. Soil processes in volcanic ash-beds. N.Z. J. Sci. Tech. **14,** 338–52.
WILLIAMS, C. H. 1950. Aust. J. Agr. Res. **1,** 156–64.

(Received 14 November 1951)

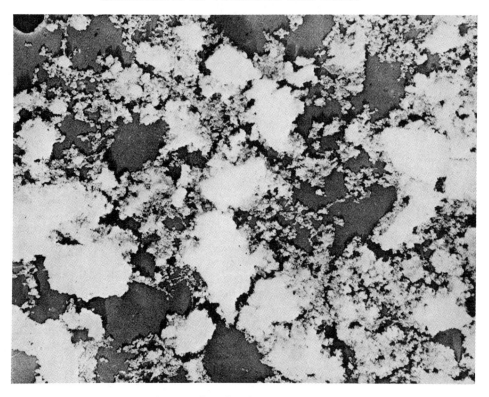

Fig. 2. Clay fraction 6c × 40,000

ALLOPHANE IN SOME ANDO SOILS*

Naganori YOSHINAGA AND Shigenori AOMINE**

Faculty of Agriculture, Kyushu University, Fukuoka, Kyushu.

Allophane has been known to occur widely in volcanic ash soils in Japan and New Zealand. However, exact knowledge of its nature has not been well established, mainly because of extreme difficulty to separate it in pure state and of its x-amorphous nature. In the course of the studies on soil allophane, it was noticed that certain Ando soils contained two different mineral colloids together, in addition to crystalline clay minerals and free sesquioxides. X-ray examination revealed that one was x-amorphous colloid which would be called allophane, and the other an unknown colloid of low crystallinity. Imogolite*** was proposed as the name of the latter colloid by the present authors after imogo in which imogolite was first found. Imogo is a brownish yellow, volcanic ash soil in the Kuma basin in the Kumamoto Prefecture[3]. When deferration treatment is applied to the soils, allophane disperses both in an acid and alkaline media, whereas imogolite disperses in an acid medium and flocculates in an alkaline one.

In the study reported here, allophane was obtained free from crystalline minerals and imogolite from three Ando soils, and these specimens of allophane were examined by modern techniques for clay mineralogy. It would be worthwhile to record information for these specimens, since earlier data on soil allophane are mostly those derived from allophanic soil clays in which a greater or lesser amount of crystalline minerals, and in some cases possibly imogolite, are intermixed with allophane.

Materials and Methods

Soil samples Four Ando soil samples collected at Uemura, Choyo, Kawanishi, and Okamoto were used for separation of allophane (Table 1).

The former three soils containing imogolite were in-

Table 1 Soil Samples Used.

No.	Source	Depth (cm)	Characteristics
905	Uemura, Kumamoto	25–55	"Imogo", vitreous yellowish brown loam
1041	Choyo, Kumamoto	30–75	Light brown sandy clay loam
77	Okamoto, Tokyo	50–80	"Kanto loam", brown silt loam
1024	Kawanishi, Hokkaido	25–30	Yellowish brown silt loam

vestigated previously by AOMINE and YOSHINAGA[3], who reported that the clay fraction (-2μ) of all these soils consisted principally of allophane with a small amount of some 2:1 layer type minerals. The Okamoto soil was investigated early by TSUCHIYA and KURABAYASHI[14], who reported the predominance of allophane in its clay fraction from the data of x-ray and differential thermal analyses.

Procedure for separation of allophane The procedure for separation of allophane, as well as of imogolite, is shown in a flow sheet (Figure 1).

The air-dried fine soil sample (less than 1 mm) was deferrated by the citrate-dithionite procedure (7, 47 pp.) after removal of organic matter with H_2O_2. The deferrated soil was boiled in 2% Na_2CO_3 solution for 5 minutes to make the following dispersion effective (7, 72 pp.). Soil treated in this way was kneaded thoroughly using a rubber ball mounted on the tip of a glass rod, and dispersed in a dilute NaOH solution of pH 10.5 to 11.0. The suspension was centrifuged for the separation of -2μ particles. Repeating the treatment of dispersion and centrifugation, the -2μ particles dispersed in this alkaline medium were removed exhaustively from the soil in a centrifugal tube, and clay suspensions obtained were collected in a flask and flocculated by the addition of saturated NaCl solution.

The residue in the centrifugal tube was washed several times with dilute HCl solution of pH 3.5 to 4.0 and clay particles, which remained undispersed in the preceding alkaline medium, were brought to dispersion. A the

---
* The results were used by the senior author in partial requirement for the degree D. Agr., The Kyushu University, 1961.
** Senior author's present address is Faculty of Agriculture, Ehime University, Matsuyama, Japan.
*** Imogolite will be described in a subsequent paper by the present authors.

Results and Discussion

Separation of Allopane

For convenience, the clay separated in the alkaline medium was given the prefix AK, and that separated subsequently in the acid medium, AC, respectively.

Uemura soil (Figure 2) The AK -2μ size fraction gave no distinct diffraction peak suggesting the predominance of allophane. On the contrary, the AC -2μ size fraction produced two broad but pronounced diffraction peaks at 13.5 A and 8.04 A, suggesting the presence of considerable amounts of some crystalline minerals.

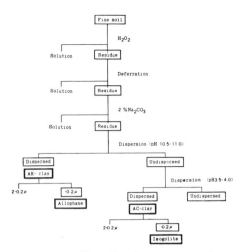

Fig. 1. *Flow sheet for separation of allophane and imogolite.*

beginning of the washing, clay particles were rather flocculated, presumably because NaCl being produced by neutralization of NaOH. But they were soon dispersed by additional two or three washings. The content of the tube was again subjected to kneading, dispersion in the acid medium, and centrifuge separation of -2μ particles in the same way as described above. The clay suspensions obtained were collected in another flask and flocculated with NaCl.

Two -2μ clay fractions separated above were fractionated separately into two or three subsize fractions at 0.2μ or at 1μ and 0.2μ by centrifugation after redispersion in the same medium as used for their separation. Each subfraction obtained was examined by a Geiger counter x-ray diffractometer "Geigerflex". Parallel orientation specimens were prepared by the procedure described by JACKSON (7, 184 pp.).

Clay mineralogical analysis of separated allophane. The clay preserved in a normal NaCl solution after separation was used for electron microscope photographing (8, 420 pp.), and the air-dried Na-clay for other analyses. Random orientation specimens for x-ray analysis were prepared on the clays which were air-dried and heated at 1000°C and at about 1200°C for two hours, by the procedure of McGREERY[9]. Differential thermal curves were obtained on an 0.2 g sample with or without piperidine treatment[2], at the heating rate of 10°C per minute. Integral dehydration curves were obtained by heating an 0.5 g sample at various temperatures up to 1000°C for 10 to 15 hours, cooling in a desiccator, and weighing. Temperature in the electric furnace was elevated at 25°C intervals between 250°C and 500°C, and 50°C intervals in other temperature regions. Infrared absorption measurement was made with KBr technique.

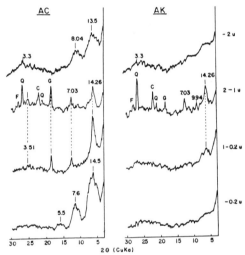

Fig. 2. *X-ray diffraction patterns of Mg-saturated specimens of clay separates from the Uemura soil. G: Gibbsite; Q: Quartz; C: Cristobalite; F: Feldspar.*

The -2μ clay fraction usually contains a more or less amount of the primary minerals. The 2-1 μ size fraction of both AK- and AC-clays exhibited several distinct diffraction peaks. Some of them indicated quartz (3.34 A and 4.21 A), cristobalite (4.04 A), feldspar (3.20 A), and gibbsite (4.83 A). The 14.26 A, 9.94 A, and 7.03 A diffraction maxima suggested the presence of some layer lattice type minerals. These diffractions, however, disappeared almost entirely from the AK 1–0.2μ size fraction, leaving the greatly weakened 14.26 A diffraction maximum. And the AK -0.2μ fraction did not show any diffraction peak excepting a very diffused halo near 3.3 A. Thus, it would be considered that the AK -0.2μ fraction is practically x-amorphous, that is, nearly pure allophane.

On the other hand in the AC 1–0.2μ size fraction, there were seen 4.83 A (gibbsite), 7.03 A, and 14.26 A diffraction maxima; the increase in the intensity of the 14.26 A peak as compared with that of the AC 2–1μ fraction would indicate some layer lattice mineral being concentrated into the 1–0.2μ fraction. The AC -0.2μ size fraction gave broad, relatively intense 14.5 A and 7.6 A, and weak 5.5 A diffractions. As will be shown in a subsequent paper by the present authors*, there is positive proof that the mineral generating these broad diffraction peaks in the -0.2μ fraction and that generating the sharp diffraction peaks in the 1–0.2μ fraction (AC 14.26 A and 7.03 A) are distinctly different kinds. The contribution of the diffractions at 14.26 A and 7.03 A in the 1–0.2μ fraction to the diffractions at 14.5 A and 7.6 A in the -0.2μ fraction, respectively, may be negligible, if any. The mineral giving such a diffraction pattern as that of the -0.2μ clay has never been reported. The only reference on a similar mineral in two weathered pumices was presented by KANNO et al.[8]. They, however, regarded it as a montmorillonite-like mineral of extremely low crystallinity regardless of a very low molecular ratio of SiO_2 to Al_2O_3. Although detailed chemical and mineralogical properties of the mineral have not been well established, the name "imogolite" was given tentatively by the present authors.

Choyo, Okamoto, and Kawanishi soils (Figure 3) No noticeable diffraction peak was observed in the diffraction patterns of the AK -0.2μ size fraction from the Choyo and Okamoto soils. These fine clays could be regarded also as practically pure allophane. But the AK -0.2μ fraction from the Kawanishi soil exhibited two broad diffraction peaks at 14.26 A and 8.04 A, indicating that an appreciable amount of some layer silicate still remains in this fraction. Further separation of the AK -0.08μ fraction from this soil failed to obtain a perfectly x-amorphous colloid. The diffraction pattern of the AC -0.2μ fractions of the Choyo and Kawanishi soils was very similar to that of the same size fraction of the AC-clay from the Uemura soil. It could be considered that these fractions consist principally of imogolite.

14.26 A and 7.03 A minerals In Figures 2 and 3, clear 14.26 A and 7.03 A diffraction peaks were

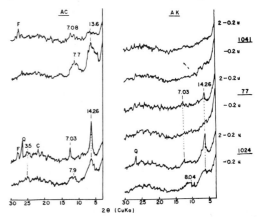

Fig. 3. *X-ray diffraction patterns of Mg-saturated specimens of clay separates from the Choyo, Okamoto, and Kawanishi soils. C: Cristobalite; Q: Quartz; F: Feldspar.*

Table 2. Recovery of Clay Separates.

Sample No.	-2μ* AK+AC (%)	AK**			AC**		
		-2μ (%)	2–0.2μ (%)	-0.2μ (%)	-2μ (%)	2–0.2μ (%)	-0.2μ (%)
905	30	70	23	47	30	10	20
			2–1μ, 13			2–1μ, 8	
			1–0.2μ, 10			1–0.2μ, 2	
1041	28	89	24	65	11	5	6
77	45	100	42	58	—	—	—
1024	31	94	n. d.	n. d.	6	n. d.	n. d.

* On the basis of oven dry fine soil.
** On the basis of oven dry -2μ (AK+AC) clay.

* Imogolite in some Ando Soils. Submitted to Soil Science and Plant Nutrition.

noted in the coarser clay fraction. The AC 1–0.2 μ fraction from the Uemura soil and the AC 2–0.2μ fraction from the Kawanishi soil were investigated on the parallel orientation specimens with Mg-saturation, Mg-saturation and ethylene glycol solvation, K-saturation, and K-saturation and heating, using Geigerflex. The results (not shown) revealed that the 14.26 A spacing arose mainly from 'swelling vermiculite' (AOMINE and KAIDA*) in the Uemura soil, and from 'chlorite-vermiculite intergradient'[6] in the Kawanishi soil, and that the 7.03 A spacing arose as a second order 14 A spacing of chlorite in either case.

Recovery of clay separates The recovery of clay separates was shown in Table 2. As mentioned before, the AK -0.2μ fraction separated from the Uemura, Choyo, and Okamoto soils is nearly pure allophane, and the AC -0.2μ fraction from the Uemura and Choyo soils consists principally of imogolite. The Kawanishi soil also contains an appreciable amount of imogolite in the AC clay fraction as shown in Figure 3. The coarser clay fractions contain considerable amount of primary and secondary crystalline minerals in all samples examined. It is obvious from this table that the simple separation of clay fraction with an acid medium is inadequate for obtaining allophane from these soils.

Properties of Separated Allophane

The -0.2μ size fraction separated in the alkaline medium from the Uemura, Choyo, and Okamoto soils is nearly pure allophane as described before. This fraction hereinafter will be called simply as allophane.

Electron micrograph (Figure 4) The Uemura allophane was mainly composed of fibrous particles of uneven size, with occasional rounded or irregular shape of aggregate masses. The Choyo allophane appeared as a more or less continuous assemblage in which the fibrous particles were seen sporadically. The Okamoto allophane appeared to be composed mainly of minute particles with irregularly shaped particles that looked like aggregate masses.

The shape of allophane varies thus from sample to sample. However, there is a likeness between the allophanes from Uemura and Choyo in that

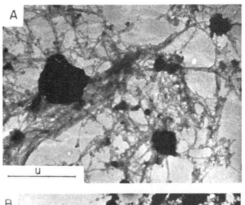

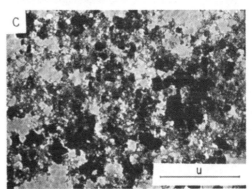

Fig. 4. *Electron micrographs of separated allophane.*
A: Uemura; B: Choyo; C: Okamoto

both of them contain the fibrous particles, and at the same time, there is a distinct difference between these two and the Okamoto allophane in their general appearance. The fibrous particles were also noted in the AK -0.2μ size fraction from the Kawanishi soil (not shown). The fibrous shape of allophane seems to be connected with the

---

* Read at the Meeting of the Society of the Science of Soil and Manure, Japan (April 3-6, 1961, Tokyo).

present of imogolite in the soil. The Uemura, Choyo, and Kawanishi soils contain imogolite and at the same time fibrous particles, while the Okamoto soil contains neither imogolite nor fibrous particles. Imogolite, as will be shown in a subsequent paper, is of thread-like shape with relatively uniform size. It is conceivable that there are intimate interrelationships between fibrous allophane and imogolite in their genesis. On the other hand, the Okamoto soil does not contain imogolite (Figure 3). An appreciable amount of hydrated halloysite has been found by TSUCHIYA and KURABAYASHI[14] in the adjacent lower horizons of this soil. This may suggest that the Okamoto allophane is in a direction of crystallization to halloysite. A comparatively high molecular ratio of silica to alumina (Table 3) may support, to a certain degree, this supposition. Consequently, it may be concluded, so far as the samples of separated allophane are concerned, that the particle shape of allophane is a reflection of the direction of weathering, namely, allophane vs. imogolite or allophane vs. halloysite, to which the sample is subjected in the soil.

X-ray diffraction analysis The x-ray diffraction patterns obtained with random orientation specimens were shown in Figure 5. Two diffraction bands (halos) were noted for all samples at around 3.34 A and 2.25 A. These halos closely resemble those reported by WHITE[16] for Indiana allophane and by MINATO[10] for Bihoro allophane, suggesting that the separated allophane is generally similar to allophane of geological origin.

In addition to the above two halos, the sample exhibited a diffuse band ranging from 4 to 4.6 A. This band is fairly distinct in the Okamoto allophane, but very weak in the other two. Although the position of this band is seemingly identical with that of an hk 0 (02, 11) reflection of layer silicate minerals, the oriented specimens do not show any evidence of these minerals (Figures 2 and 3). Therefore, this band would be regarded as a characteristic of the separated allophane, and its slightly high intensity in the Okamoto allophane might disclose an initial ion-arrangement of layer silicate.

X-ray analysis was carried out also on the samples heated at 1000°C and at about 1200°C for two hours (Figure 6). Upon heating at 1000°C, the Uemura allophane gave two very faint diffranctions at 2.39 A and 1.973 A, which were absent in the other two samples. The spacings are in accord with those of gamma-alumina. Therefore, these faint bands may be attributed to the structural remnants of gamma-alumina formed at about 900°C (Figure 7). On further heating to higher temperature (about 1200°C), alpha-alumina was

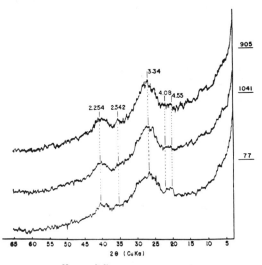

Fig. 5. *X-ray diffraction patterns of separated allophane in random orientation.*

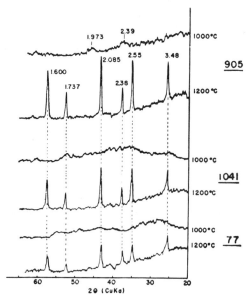

Fig. 6. *X-ray diffraction patterns of separated allophane heated at 1000°C and 1200°C. The product at 1200°C is alpha-alumina.*

produced exclusively for all samples. This may indicate that the conversion of alumina of gamma type to that of alpha type takes place between 1000°C and 1200°C[12]. Mullite and/or gamma-alumina have been regarded as the high temperature products of allophane by a number of investigators[11,15,16]. The reason for alpha-alumina formation in these samples is not clear at present, but, in view of the fact that imogolite produces only mullite at the same temperature (subsequent paper), it appears certain that allophane is different in the inner structure from imogolite.

Thermal analyses The differential thermal curves were shown in Figure 7. The curves resemble closely one another; no marked thermal reaction is observed excepting the strong endothermic peak in the range from 150°C to 170°C and the sharp exothermic peak between 910°C and 915°C. This high temperature exotherm may be attributed to the formation of gamma-alumina (Figure 6). A sluggish exotherm was noted for all samples just after the strong endotherm. This exotherm might be due to organic matter that has remained undecomposed in spite of exhaustive H_2O_2 treatment.

Piperidine treated sample (Uemura) produced

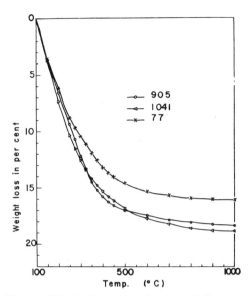

Fig. 8. *Dehydration curves of separated allophane.*

two additional exothermic peaks at 310°C and 470°C. According to ALLAWAY[2], the 470°C exotherm is attributed to the presence of impurities such as hydrous iron oxide. This seems to be true to some extent as shown in the curve of non-deferrated clay fraction (-2μ) of the Uemura soil. However, the sample of separated allophane has undergone deferration treatments twice, and its iron content is very low (Table 3). It is hard to consider that such a small amount of iron causes the relatively intense exothermic reaction when treated with piperidine. Therefore, the exotherm at 470°C, together with that at 310°C, may be considered as a characteristic of allophane as claimed by SUDO[13].

The integral dehydration curves are shown in Figure 8. Very smooth curves were odtained for all samples. The marked water loss varying from 16 to 19 per cent on over dry basis is consistent with the results of chemical analysis (Table 3). Both differential thermal and dehydration curves of the separated allophane are generally similar to those for allophane which have appeared in the literature.

Infrared absorption The infrared absorption spectra for Na- and H-saturated specimens of the Uemura allophane are shown in Figure 9. The marked absorption maxima at about 2.7μ, 2.9μ, 6.2μ, and 10.2μ, and the weak one at about

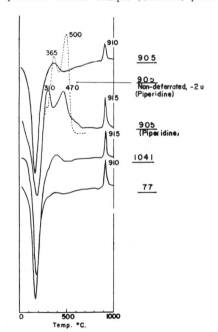

Fig. 7. *Differential thermal curves of separated allophane.*

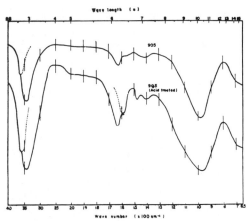

Fig. 9. Infrared absorption curves of separated allophane (Uemura).

7μ are noted for Na-allophane. The positions of these absorption bands are nearly identical to those reported by Adler et al.[1], Fieldes et al.[5], and other investigators for allophane of both geological and soil origin. The H–allophane was prepared by washing with alcohol and acetone after treatment with Na-acetate buffer solution of pH 3.5 in a boiling water bath for 15 minutes. Its absorption maxima due to bonded (2.9μ) and unbonded (2.7μ) OH and absorbed water (6.2μ) increased in intensity as compared with those of the Na-allophane, and two additional bands were produced at about 6.3μ and 6.8μ. The last two maxima may probably come from the CH_3COO^- ions adsorbed on allophane.

Elemental analysis The results of elemental analysis are shown in Table 3. The data show that all samples consist principally of silica, alumina, and water. The high content of Na_2O in all samples may be accounted for mainly as exchangeable cation. Molecular ratios of silica to alumina are about 1.3 for the Uemura and Choyo allophanes, and about 1.7 for the Okamoto allophane. The samples could be classified into two groups with respect to chemical composition: (a) Uemura and Choyo, and (b) Okamoto, in accordance with the grouping with particle-shape by electron microscope. This accordance would not be by chance, but would rather suggest that the difference in the direction of weathering was reflected in the chemical composition. Molecular ratios of water to alumina range from about 5.1 to 5.5 on the basis of air-dryness under the 50 per cent relative humidity. Thus, the chemical composition of the separated allophane can be expressed approximately by the formula that has been presented by OSSAKA[11]:

$$nSiO_2 \cdot Al_2O_3 \cdot 5H_2O \quad (n=1-2).$$

Table 3. Chemical Composition of Allophane.

	905 (%)	1041 (%)	77 (%)
SiO_2	33.28	32.46	37.22
Al_2O_3	42.91	42.01	37.39
Fe_2O_3	0.82	1.02	1.35
CaO	nil.	nil.	nil.
MgO	0.18	0.25	0.39
Na_2O	3.21	4.99	5.98
K_2O	0.68	0.13	0.19
TiO_2	0.21	0.42	0.19
P_2O_5	0.87	0.50	0.50
I.L. (H_2O+)	17.90	18.76	16.31
Total	100.06	100.54	99.52
$H_2O(-)$*	21.08	22.13	19.10
SiO_2/Al_2O_3	1.32	1.31	1.69
H_2O/Al_2O_3**	5.15	5.53	5.35

* Equilibrated at 50 per cent relative humidity.
** Including $H_2O(-)$.

Summary and Conclusions

Two mineral colloids, allophane and a mineral of low crystallinity, were discerned in three Ando soils. Imogolite was proposed as the name of the latter colloid by the present authors. They exhibited a distinct difference with respect to the pH of dispersion: allophane dispersed both in an acid and alkaline media, whereas imogolite only in the acid one. Using this difference in dispersion characteristics, these two colloids were obtained separately.

The nearly pure specimens of allophane separated from three Ando soils with or without imogolite were examined by the techniques of electron microscopy, x-ray diffraction, thermal analysis, infrared absorption, and chemical analysis.

Allophanes which coexist with imogolite bear resemblance to each other both in particle shape and chemical composition. Both Uemura and Choyo allophanes, which coexist with imogolite, contain fibrous particles of uneven size, though their general appearances are not exactly the same, and the molecular ratios of silica to alumina are about 1.3 alike. On the other hand, the Okamoto allophane which does not coexist with imogolite consists mainly of very minute globoids and its molecular ratio of silica to alumina is about 1.7 being different from the Uemura and Choyo allophanes. It exhibits a slight but distinct diffraction at about 4.5 A, suggesting an initial ion-arrangement of a layer silicate, while the other two allophanes do not show any sign of crystallization. These facts have suggested that the characteristics of soil allophane, in general, might vary from sample to sample depending upon the stage or direction of weathering in the soil.

Both x-ray and thermal data exhibited a striking similarity among the samples, and the data, along with the infrared absorption curve, were generally similar to those published by some investigators for the mineral allophane.

Acknowledgment

The authors wish to express their sincere thanks to Mr. TANIGUCHI of Faculty of Agriculture, Kyushu University for his help in the infrared absorption measurement, and to Mr. MIYAUCHI of Faculty of Medicine, Kyushu University and Mr. NANRI of Faculty of Agriculture, Kyushu University for electron micrographs.

References

1) ADLER, H., E. E. BRAY, N. P. STEVENS, J. M. HUNT, W. D. KELLER, E. E. PICKET, and P. F. KERR, A.P.I. Project 49, Clay Min. Standard, Prelim. Rept. No. 8, 146 pp. (1950).
2) ALLAWAY, W. H., Soil Sci. Soc. Am. Proc., **13**, 183 (1948).
3) AOMINE, S., Report of Investigation of a Specific Soil "Imogo", Kumamoto Prefecture, 17 pp. (1954).
4) AOMINE, S. and N. YOSHINAGA, Soil Sci., **79**, 349 (1955).
5) FIELDES, M., I. K. WALKER and P. P. WILLIAMS, New Zealand J. Sci. Tech., **38 B**, 31 (1956).
6) JACKSON, M. L., Proc. 6th Natl. Conf. on Clays and Clay Minerals, 133 pp. (1950).
7) JACKSON, M. L., Soil Chemical Analysis-Advanced Course, Pub. by the author, Dept. of Soils, Univ. of Wis., Madison, Wis. (1956).
8) KANNO, I., Y. KUWANO and Y. HONJO, Advances in Clay Science, Pub. by Clay Research Group of Japan, Vol. 2, 355 pp. (1960).
9) KLUG, H. P. and L. E. ALEXANDER, X-ray Diffraction Procedures for Polycrystalline and Amorphous Materials, John Wiley & Sons, Inc., New York, 300 pp. (1954).
10) MINATO, H., Advances in Clay Science, Pub. by Clay Research Group of Japan, Vol. 2, 350 pp. (1960).
11) OSSAKA, J., Ibid., Vol. 2, 339 pp. (1960).
12) ROOKSBY, H. P., X-ray Identification and Structures of Clay Minerals edited by G. W. BRINDLEY, Mineralog. Soc., London, 245 pp. (1951).
13) SUDO, T., Clay Min. Bull., **2**, 96 (1953).
14) TSUCHIYA, T. and S. KURABAYASHI, J. Geol. Soc. Japan, **64**, 605 (1958).
15) TSUZUKI, Y. and K. NAGASAWA, Advances in Clay Science, Pub. by Clay Research Group of Japan, Vol. 2, 377 pp. (1960).
16) WHITE, W. A., Am. Mineral., **38**, 634 (1953).

IMOGOLITE IN SOME ANDO SOILS*

Naganori YOSHINAGA and Shigenori AOMINE**

Faculty of Agriculture, Kyushu University, Fukuoka,

It was reported in a preceding paper[7] that the Ando soils from Uemura, Choyo, and Kawanishi contained an unknown mineral colloid which was distinctly different in some respects from coexisting allophane. In the Uemura soil, this clay fraction made up more than 20 per cent of the total clay and more than 6 per cent of the soil[7].

Although its chemical and mineralogical properties are not well known at present, it is considered that this mineral has a more ordered structure than allophane. Its occurrence is also considered to be fairly common in most Ando soils and weathered pumices[5]. Therefore, this mineral was tentatively designated as imogolite by the present authors[7].

The objective of this paper is to describe the properties of imogolite as examined by the techniques of electron microscopy, x-ray diffraction, thermal analysis, infrared absorption, and chemical analysis.

Materials

The samples used in this study were the -0.2μ or -0.08μ clay fractions separated in an acid medium from the Uemura, Choyo, and Kawanishi soils[7]. Since allophane has been removed almost exhaustively from the soils prior to the separation of these clay fractions[7], it may be taken that these fractions are free from allophane. Obtained imogolite clay was quite different in appearance from allophane.

Flocculating with NaCl, the imogolite clay formed a voluminous floccule throughout which innumerable small air bubbles were adsorbed; it looked like a drifting cloud (Figure 1). On the contrary, the allophane clay separated from the same soil formed small flakes in NaCl solution and settled readily to the bottom of a container (Figure 1). When collected in a centrifugal tube, imogolite gave a jelly-like appearance, while allophane gave an appearance of white paste.

On drying with a motor fan after washing with water, alcohol, and acetone successively, about 500 mg of imogolite clay required more than 5 hours until apparent air-dryness was attained, while about 15 minutes were sufficient for allophane clay. The air-dried clay cake of imogolite was somewhat greyish, and showed weak elasticity, whereas that of allophane was white, and loose and fragile.

On pulverizing the air-dry clays in an agate mortar, allophane adhered to the mortar and pestle, but imogolite did not.

The imogolite clay, once air-dried, did not swell in water and also in certain neutral or alkaline salt solutions (NaCl, $CaCl_2$, $CaAc_2$, Na_2CO_3). This was the case with the allophane clay. But, when immersed in a salt solution of about pH 3.5 to 4.0 (NaAc, $CaCl_2$), the former easily recovered the property of swelling, while the latter remained just as it was.

Fig. 1. *Imogolite and allophane clays flocculated with normal NaCl solution (-2μ, Uemura).*

Electron Micrograph

The specimens for electron microscope observation were prepared from -0.2μ clay which had been preserved in a normal NaCl solution, according to the procedure described by JACKSON

---

* The results were used by the senior author in partial requirement for the degree of D. Agr., The Kyushu University, 1961.
** Senior author's present address is Faculty of Agriculture, Ehime University, Matsuyama, Japan.

(3, 420pp.). Some specimens were shadowed with chromium at an angle of about 30 degrees.

Electron micrographs were shown in Figure 2. The clays were all composed mainly of thread-like particles of relatively uniform size. Their diameter of cross section is about 100 to 200 Å. The similar shape of particles was found by AOMINE and JACKSON[1] in the Sakae and Choyo clays. The relative uniformity in shape seems to suggest some regularity of atomic arrangement with in the imogolite particles. It is easily supposed that such a shape of particles is responsible for the voluminous floccule of the clay in NaCl solution.

Occasionally the particles of somewhat fluffy appearance were observed in the imogolite clays (not shown). The very similar particles were found by AOMINE and YOSHINAGA[2] in the clay fractions of the Uemura and Kawanishi soils. Although it is not known whether these particles are imogolite or not, the shape is clearly different from that of allophane[7]. More than 12 fields were observed with each specimen; nevertheless, such a shape of particles was not found in the allophane fraction.

X-ray Diffraction Analysis

X-ray diffraction analyses were carried out on parallel and random orientation specimens using a Geiger counter x-ray diffractometer "Geigerflex". The flat specimens of Mg- and K-saturated and Mg-saturated and ethylene glycol solvated clays were prepared according to the procedure of JACKSON (3, 184 pp.). The K-saturated specimen was heated at various temperatures up to 500°C for two hours, and x-rayed immediately and after being allowed to stand in the air for 15 minutes, one hour, and two months. The random orientation specimens of the clays which were air-dried and heated at 1000°C and at about 1200°C for two hours were prepared according to the procedure of MCGREERY[6].

The diffraction patterns obtained with the random orientation specimens of air-dried, Na-saturated -0.2μ clays were shown in Figure 3. The patterns resemble each other closely; all samples exhibited several broad peaks and diffuse bands. Although the peaks at 3.26 Å and 2.26 Å and the band ranging from 4 to 4.5 Å have been observed similarly, but in a less intensity, in the samples of allophane separated from the same soils[7], the general feature of patterns is quite different from that of allophane. It may be considered that imogolite has a more ordered structure than allophane. However, it is doubtful that the band ranging from 4 to 4.5 Å is an evidence of layer silicate minerals, because the samples of allophane also have exhibited the similar diffraction band[7].

The diffraction patterns obtained with the flat

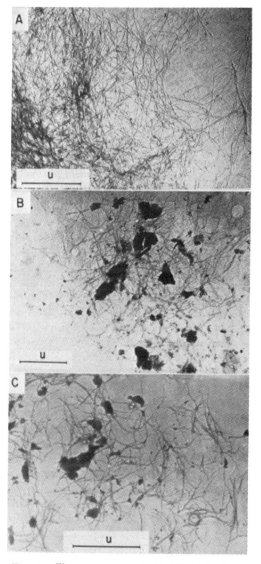

Fig. 2. *Electron micrographs of -0.2μ fractions. A—Uemura; B—Choyo; C—Kawanishi.*

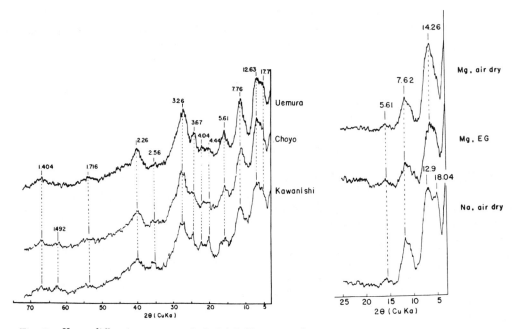

Fig. 3. X-ray diffraction patterns of air-dried, Na-saturated $-0.2\,\mu$ clays in random orientation.

Fig. 4. X-ray diffraction patterns of the Uemura $-0.08\,\mu$ clay.

specimens were shown in Figures 4, 5, and 6. The magnesium-saturated, Uemura $-0.08\,\mu$ clay (Figure 4) gave broad, relatively intense two diffraction peaks at 14.26 Å and 7.62 Å and a very weak diffraction at about 5.6 Å. Ethylene glycol (EG) solvation caused no noticeable change in the pattern. Sometimes the 14 Å broad peak gave two diffraction maxima at about 13 Å and 18 Å as was seen in the pattern of Na-saturated clay (Figure 4), indicating that the 14 Å broad peak is composed of two diffraction components. It seems impossible, however, to attribute the 18 Å maximum to "lattice expansion" caused by the hydration of Na-clay, because the heated specimen produces a very sharp and strong peak at about 17 to 19 Å (Figure 5).

Potassium saturation of the specimen caused a slight decrease of the 14.26 Å spacing (Figure 5). It is doubtful, however, that this shift of spacing is a "lattice contraction" as seen in the usual expandable layer lattice minerals. Subsequent heat treatments caused an interesting behavior of this peak. Upon heating to 100°C, the peak suffered a noticeable change and produced a strong and sharp 18.8 Å peak with an inflection at about 13 Å. This would be another indication that the broad 14 Å peak of Mg-clay is composed of two diffraction components. The 18.8 Å peak increased its intensity progressively with temperature up to 300°C with concurrent decrease of the spacing to 17.63 Å, but was greatly weakened at 350°C and almost disappeared at 400°C leaving a slight inflection at about 17 Å. On the other hand, the broad peaks at 7.8 Å and 5.5 Å hardly underwent any change up to 300°C, but disappeared abruptly at 350°C. These evidences indicate that the structure of imgolite is decomposed almost entirely by heating at 350°C to 400°C.

The clay heated from 100°C to 250°C regained the intensity of reflection at about 13 Å when exposed in the air. This recovery would be attributed to the rehydration of the heated clay, though the mechanism of water retention by imogolite is not known. When heated to 300°C, the recovery became very little. This may suggest that, if the water of original structure is removed completely or nearly so, the ability of rehydration would also be lost almost entirely. A slight recovery of intensity was also noticed for the 17 Å inflection of the specimen heated at 400°C and 450°C, indicating a partial revival

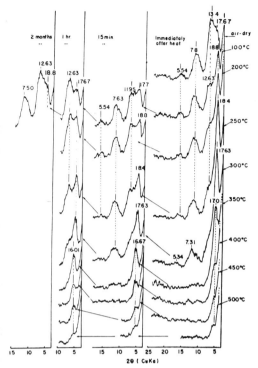

Fig. 5. X-ray diffraction patterns of the Uemura, K-saturated -0.08μ clay.

of imogolite structure.

The diffraction patterns of the clays from Choyo and Kawanishi (Figure 6) were essentially the same as those of the Uemura clay. It may be considered that these clays also consist principally of imogolite. The broad diffraction peak at about 12 Å in the heated Kawanishi clay may probably be attributed to the contamination of chlorite-vermiculite intergradient[7]. Fairly distinct 4.44 Å peak exhibited by the random specimen (Figure 3) may establish the presence of this mineral.

Although no definite conclusion can be reached as yet concerning the structure, the diffraction patterns given by imogolite seem to be made up of two types of diffractions, the broad peak at about 14 Å which consists of 18 Å and 13 Å diffraction components, and the others. A little decrease of 18.8 Å spacing of K-saturated clay caused by heat treatment is somewhat similar to the lattice contraction of layer silicate minerals. In contrast, the 7.8 Å and 5.5 Å peaks are hardly affected by heat treatment; they would be compared to hk0 reflections. Thus, imogolite seems to bear some structural resemblance to the layer silicate mineral. However, the data presented above, together with the absence of prism zone reflections in the patterns of randomly oriented specimen heated at 300°C (not shown in figure), suggest very little possibility of interstratification of common silicate layers. Data of thermal (Figures 8 and 9), infrared absorption (Figure 10), and chemical (Tables 1 and 2) analyses also rule out this possibility. Also the possibility of already-known chain-structure minerals, such as attapulgite, sepiolite, and palygorskite, is very little if any.

KANNO et al.[5] have recently investigated gel-like substances separated from the Kitagami and Akutsu pumice beds. Data of x-ray diffraction, thermal, infrared absorption, and chemical analyses presented by them are nearly the same as those obtained with imogolite clays. It is almost doubtless that imogolite and their gel-like substances are essentially of the same kind. These authors attributed the broad diffractions of the gel-like substances to poorly crystallized montmorillonite. Imogolite, however, is apparently different from montmorillonite.

The diffraction patterns of fired clays were shown in Figure 7. The clays heated at 1000°C gave the patterns which were similar to those of allophane separated from the same soils[7]. Their faint diffractions at 2.38 Å and 1.973 Å

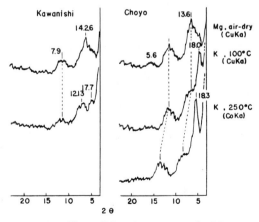

Fig 6. X-ray diffraction patterns of -0.2μ fractions of the Choyo and Kawanishi soils.

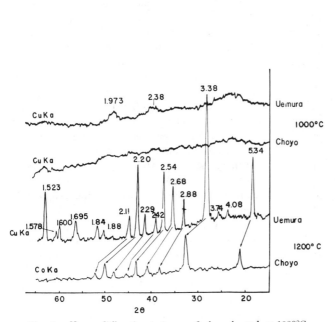
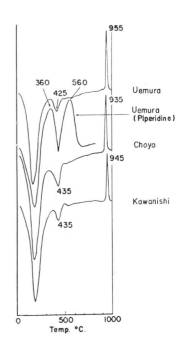

Fig. 7. X-ray diffraction patterns of clays heated at 1000°C and 1200°C. The product at 1200°C is mullite.

Fig. 8. Differential thermal curves of -0.2μ Na-clays.

would be attributed to the structural remnants of gamma-alumina. Upon heating to 1200°C, however, the clays yielded strong diffraction peaks for mullite. This forms a striking contrast to allophane which produces only alpha-alumina at the same temperature[7]. The difference in high temeprature phases may be due to differences in chemical composition and in ionic arrangement between these two materials.

Thermal Analysis

Differential thermal and dehydration analyses were carried out on -0.2μ size fractions by the same method and with the same equipments as used for allophane[7].

Differential thermal curves obtained were shown in Figure 8. A strong endothermic peak is observed in all samples at about 170°C to 190°C. This peak is apparently due to a large amount of hygroscopic water (Table 1). The exothermic peak observed between 935°C and 955°C is probably due to the formation of gamma-alumina. These two peaks were observed similarly in the samples of allophane separated from the same soils[7]. But the exothermic reaction increased its intensity remarkably as compared with allophane, with simultaneous shift of the peak toward high temperature by about 20 to 40°C. This may probably be connected with the difference in the "inner structure" between imogolite and allophane.

Besides the above two peaks, all samples exhibited an endothermic reaction at about 425°C to 435°C. This endotherm has never been observed in the samples of allophane[7]. It is undoubtedly a characteristic of imogolite. The peak temperature coincides fairly well with the temperature where the diffraction lines of imogolite disappear, taking into account the possible delay of peak temperature in differential thermal curves. It seems probable, therefore, that this endothermic peak is indicative of the destruction or dehydroxylation of imogolite.

AOMINE and YOSHINAGA[2] noticed the similar endotherm in nondeferrated clay fractions from the same three soils and their neighboring horizons, and KANNO et al.[5], in the gel-like substances separated from pumice beds. These investigators supposed that the endotherm was attributed to the presence of impurities such as hydrated oxides of iron or aluminum or aluminum

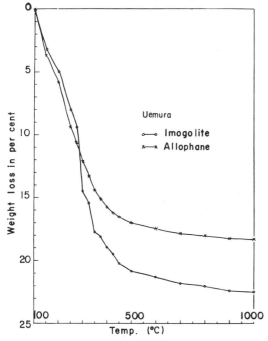

Fig. 9. *Dehydration curves of $-0.2\,\mu$ Na-clays of imogolite and allophane. The curve of allophane was reproduced from the preceding paper[7].*

compounds. However, the imogolite is a purified sample[7], and the peak intensity is nearly equal for all samples. Therefore, it would be reasonable to consider that the peak is a characteristic of imogolite, but not being due to the impurities.

The integral dehydration curve of imogolite, together with that of allophane, was shown in Figure 9. The curve of imogolite shows an inflection indicating a rapid dehydration between 275°C and 350°C. This temperature range also coincides fairly well with that of the disappearance of diffraction peaks in x-ray analysis (Figure 5), taking into account a long time of heating in the dehydration analysis.

The results of thermal analyses indicate that imogolite hold structural water which is released in the neighborhood of 300°C.

Infrared Absorption

Infrared absorption measurement was run with the same method as used for allophane[7].

The curve of imogolite showed four pronounced (2.7μ, 2.9μ, 6.2μ, 10.2μ) and one faint (7μ) absorption maxima (Figure 10). The positions of these maxima are entirely the same as those of allophane. KANNO et al.[5] obtained

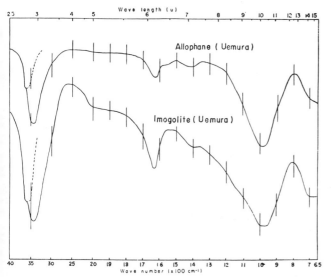

Fig. 10. *Infrared absorption curves of $-0.2\,\mu$ Na-clays of imogolite and allophane. The curve of allophane was reproduced from the preceding paper[7].*

Table 1
Chemical Composition of the Uemura, $-0.2\,\mu$ Na-clay

	%*
SiO_2	29.71
Al_2O_3	47.60
Fe_2O_3	0.57
CaO	0.04
MgO	0.02
Na_2O	0.74
K_2O	0.13
TiO_2	1.04
P_2O_5	0.26
I. L. (H_2O+)	22.05
Total	102.16
$H_2O(-)$**	19.69
SiO_2/Al_2O_3	1.06
H_2O/Al_2O_3***	4.96

* Oven-dry basis.
** Equibrated at 50% relative humidity.
*** Including $H_2O(-)$.

Table 2. Cation-exchange Capacity and CEC-delta value of the Uemura, -0.2μ Fractions

	CEC (pH 7)	CEC		CEC-delta value
		pH 3.5	pH 10.5	
	me/100 g*	me/100 g*	me/100 g*	me/100g*
Imogolite	48	43	113	70
Allophane	151	49	178	129

* Oven-dry basis.

the similar pattern for their gel-like substances, and attributed the band ranging from 7.7 to 12.5μ (maximum; about 10.5 μ) to the presence of allophane. However, it seems that the band comes from both imogolite and allophane in their pattern, since these investigators have not separated allophane from imogolite. Other absorption bands are apparently due to bonded (2.9μ) and unbonded (2.7μ) hydroxyls and absorbed water (6.2μ). It seems that imogolite is indistinguishable from allophane by infrared absorption method, and this suggests that imogolite is a low crystalline mineral in an intermediate stage from amorphous material (allophane) to crystalline clay mineral.

Chemical Composition, Cation-exchange Capacity, and CEC-delta Value

Cation-exchange capacity (CEC) was determined by the Ca saturation-EDTA titration method, and CEC-delta value, by the procedure proposed by AOMINE and JACKSON[1].

The chemical composition of the Uemura, -0.2μ Na-clay was shown in Table 1. The result shows that the clay consists principally of silica, alumina, and water. The molecular ratio of silica to alumina is 1.06, and that of water to alumina is 4.96; thus the ratio of silica : alumina : water is approximately 1 : 1 : 5. The content of titanium dioxide is about one per cent of the clay, but it is not known that this element takes part in the structure of imogolite. Iron is probably in a form of free oxide which has been difficultly soluble against the deferration treatment. The chemical composition is very similar to that presented by KANNO et al[5]. for the gel-like substances of some pumices.

The CEC and CEC-delta values of imogolite were shown in Table 2. The data of allophane separated from the same Uemura soil[7] was added to the table for comparison. The CEC of imogolite clay was 48 me per 100g oven-dry clay. It is doubtful, however, that this is the absolute value for imogolite, because the CEC of allophane is greatly affected by past acid or alkali treatment[1], and this may be true to some extent with a low crystalline material such as imogolite (see CEC-delta value). The CEC-delta value was 70 me per 100g oven-dry clay. This value is about half that of allophane. The difference seems to have some significance in comparing these two minerals for "crystallinity". The low value for imogolite means a lower content of active OH radicals, namely some higher regularity of atomic arrangement within particles. On the other hand, however, this value is remarkably higher than that of common layer silicate minerals[1,4]. This indicates that the "crystallinity" of imogolite is low as compared with these minerals.

Summary

Three Ando soils contained two kinds of mineral colloids, allophane and an unknown mineral of low crystallinity. The latter mineral was designated tentatively as imogolite by the present authors. Fractionated imogolite clays exhibited an extraordinary volume in some aqueous salt solutions; their appearance was almost jelly-like.

In electron micrographs, imogolite appeared as thread-like particles of relatively uniform size.

X-ray diffraction patterns revealed that imogolite has somewhat definite "crystal structure" which bore some resemblance to the layer silicate mineral. This fact was substantiated by differential thermal and dehydration analyses. On the other hand, however, infrared absorption curve, chemical composition, and CEC-delta value bore a close resemblance to allophane.

Although the structure is not exactly known at present, it would be concluded that imogolite is an intermediate weathering product of a metastable state between amorphous material (allophane) and crystalline clay mineral, and that it is formed under a well-drained condition of the Ando soils. It is also suggested that this mineral occurs fairly widely in nature.

Acknowledgement

The authors wish to express their sincere thanks to Mr. TANIGUCHI of Faculty of Agriculture, Kyushu University for his help in the infrared absorption measurement, and to Mr. MIYAUCHI of Faculty of Medicine, Kyushu University and Mr. NANRI of Faculty of Agriculture, Kyushu University for electron micrographs.

References

1) AOMINE, S. and M. L. JACKSON, *Soil Sci. Soc. Am. Proc.*, **23**. 210 (1959).
2) AOMINE, S. and N. YOSHINAGA, *Soil Sci.*, **79**, 349 (1955).
3) JACKSON, M. L., *Soil Chemical Analysis-Advanced Course*, Pub. by the Author, Dept. of Soils, Univ. of Wis., Madison, Wis. (1956).
4) JENNE, E. A., Mineralogical, Chemical, and Fertility Relationships of Five Oregon Coastal Soils. A Thesis Submitted to Oregon State College (1961).
5) KANNO, I., Y. KUWANO, and Y. HONJO, *Advances in Clay Science*, Pub. by Clay Research Group of Japan, **2**, 355 (1960).
6) KLUG, H. P. and L. E. ALEXANDER, X-ray Diffraction Procedures for Polycrystalline and Amorphous Materials, John Wiley & Sons, Inc., New York, 300 (1954).
7) YOSHINAGA, N. and S. AOMINE, *Soil Sci. and Plant Nutrition*, **8**, No. 2, 6 (1962).

21

Copyright © 1968 by Elsevier Scientific Publishing Company
Reprinted from Geoderma 2(2):151-169 (1968)

IMOGOLITE IN VOLCANIC SOILS OF CHILE

EDUARDO BESOAIN

U.N.D.P./S.F. Chilean Soil Survey and Research Project, Santiago (Chile)[1]

(Received August 3, 1967)

SUMMARY

During the course of mineralogical studies of various Chilean andosoils with allophane content it was surprising to find by means of electronic microscopy that a number of clay samples showed a large amount of elongated, thread-like or fibrous particles, together with other spheroidal characteristics of allophane. The fibre-like particles apparently corresponded to the descriptions given by Yoshinaga and Aomine (1962a,b) and Aomine and Miyaushi (1965) for the new mineral called imogolite. It was, therefore, considered interesting to develop a more extensive research in order to verify the existence of imogolite in these soils.

This paper will inform about the existence of imogolite in two Chilean andosoils (trumaos), taken from the southern region of the country. The object is to obtain information which might help in the understanding of problems of genesis and the classification of these volcanic ash soils.

REVIEW OF LITERATURE

Yoshinaga and Aomine (1962b) proposed the name of imogolite for a new mineral found in layers of "imogo", a brownish-yellow soil, derived from volcanic ashes located in the Kanuma Basin, Kumamoto Prefecture, Japan. Based on its behaviour as to the dispersion of pH the authors determined a fractionation procedure for separating imogolite from allophane. Before that Aomine and Yoshinaga (1955) and Aomine and Jackson (1959) had already discovered that some clays of Japanese soils, when examined by electron microscopy showed abundant thread-like particles, and Kanno et al. (1960) obtained from gel materials originating from pumice beds in Japan a number of results which were in accordance with those determined by Yoshinaga and Aomine (1962a). This material also showed characteristic fibrous features. Kanno et al. (1960) were of the opinion that these were particles of allophane in different stages of development.

Today it is believed that imogolite is a common constituent of andosoils, except in extremely young formations (Kawasaki and Aomine, 1966),

---

[1]The analyses were made in the Institut für Bodenkunde der Universität, Bonn (Deutschland).

and that it is a mineral with a low grade of crystallinity, but possessing a certain definite structure which is somewhat similar to the structure of layer silicates. It generally occurs in association with allophane, and from a genetical point of view as well as according to its weathering progress, it is considered to be an intermediate member in the crystallization process from allophane to halloysite. Aomine and Miyaushi (1965) recently determined the existence of two subspecies of imogolite, named imogolite A and B, the latter one corresponding to a more advanced stage of weathering. Jaritz (1967) determined the existence of imogolite in fine-grained pumice tuff of the Taunus Gebirge (Germany), associated to halloysite-metahalloysite, the amount of halloysite increasing with profile depth.

In podsol-braunerde of non-volcanic fine sand from the south of Finland, Alberto and Jaritz (1966) discovered by means of electron microscopy, particles showing a morphology resembling imogolite.

Materials and methods

Soils

Samples were taken from the soil series "Corte Alto" and "Puyehue" derived from andesitic-basaltic volcanic ashes in the southern region of Chile (Fig.1). Soils derived from young ashes are called in Chile "trumaos" which is an indigenous name on well-drained soils with light, dusty properties. The "trumaos" are Holocene andosoils which most conspicuous characteristic is the predominant content of allophane in the clay fraction. A summarizing description of the profiles studied is shown in Table I. As regards the Puyehue soil, imogolite was found only in the lower horizons of the profile (145–210 cm and +210 cm) and because of this the mineralogical study was limited to only those two samples. The upper horizons of this soil show a mineralogy which consists mainly of allophane and some gibbsite. However, data on mechanical composition and electron microscopy are present for the complete profile.

Experimental

Mechanical analysis. Sodium-hexametaphosphate was used as dispersing agent, treating the soils previously with a solution of $N/20$ HCl and washing it afterwards several times with distilled water according to a procedure employed by K.S. Birrell (cited in Wright, 1960). It was found more efficient in some cases to use a solution of N propionic acid as dispersing agent.

Fractioning of imogolite and allophane. The procedure of Yoshinaga and Aomine (1962a) was followed. All samples were deferrated using the methods of Mehra and Jackson (1960).

Chemical analysis. The method of Kolthoff and Sandell (1956) was employed. Cation exchange capacity (C.E.C.) was determined using a conventional procedure given by Jackson (1958): treating the samples with sodium-acetate at pH 8.2; eliminating the sodium excess by washing the samples with a mixture of absolute methanol-acetone, in order to prevent hydrolysis effects and removing sodium by means of ammonium acetate at pH 7 and then, determining sodium by flame photometry.

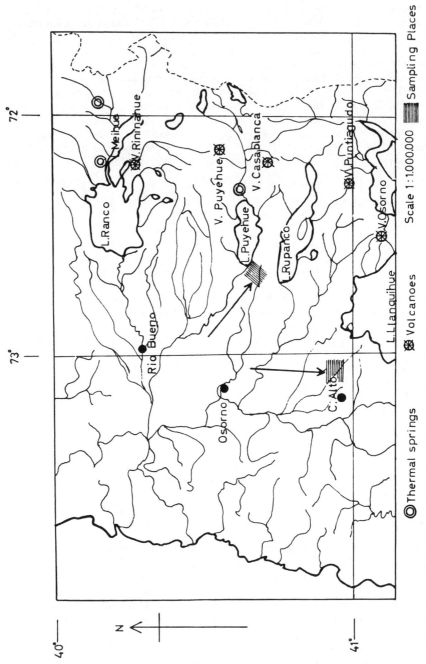

Fig.1. Sampling sites within the lakes region of Chile.

TABLE I

Soil samples

Sample no.	Name	Depth (cm)	pH (KCl)	pH (H₂O)	Organic carbon (%)	Description by color and texture
15	C.Alto	0– 20	4.7	5.7	12.1	dark brown silt-loam, 10YR 3/2
16	C.Alto	20– 70	5.1	5.8	7.3	dark yellowish-brown silt-loam 10YR 3/4
17	C.Alto	70–120	6.0	6.3	1.4	reddish brown silty clay loam 5YR 5/4
18	C.Alto	+120	5.9	6.7	0.5	pale yellowish-brown sandy loam 10YR 6/4
23	Puyehue	0– 22	4.1	5.0	14.8	dark reddish-brown silt-loam 5YR 2/2
24	Puyehue	22– 42	4.4	5.2	6.8	dark reddish-brown loam 5YR 3/2
25	Puyehue	42– 90	5.0	5.7	2.6	very dark brown sandy loam 10YR 3/3
26	Puyehue	90–145	5.2	5.6	1.5	very dark brown sandy loam 10YR 3/3
27	Puyehue	145–210	5.7	5.8	0.6	dark yellowish-brown sandy clay loam 10YR 4/4
28	Puyehue	+210	5.9	6.0	0.2	dark greyish-brown sandy loam 10YR 4/2

Differential thermal analysis (D.T.A.). Thermograms were made in samples the hygroscopic moisture of which was equilibrated at 56% of relative humidity, using a saturated solution of $Mg(NO_3)_2 \cdot 6H_2O$ at 20°C (Mackenzie, 1957). A Linseis oven and a Siemens electronic kompensograph were used. Heating rate: 10°C/min.

Infrared absorption spectrography (I.R). A Perkin Elmer apparatus model 237 was employed. Patterns were obtained with both pellets of KBr and mulls between NaCl plates, using "nujol" (medicinal paraffin) as mulling agent.

X-ray diffraction (X.R.D.). Patterns were obtained using a diffractometer Müller "Micro 111", with K-alpha radiation of Cu.

Electron microscopy. An electron microscope, Zeiss-AEG Em 8, was used. Magnifications are indicated in each micrograph.

(a) *Preparation of clay:* all samples were deferrated using the procedure of Mehra and Jackson (1960).

In the samples of Plate I, A and B of the Puyehue soil it was not intended to separate imogolite from allophane. Samples were dispersed in water, adjusted at pH 4 with HCl, and then a fraction of less than 0.2 μ was separated by centrifugation. In other samples (Plate I, C–I) the imogolite fraction was separated using the procedure of Yoshinaga and Aomine (1962a).

In all samples organic matter was previously eliminated by means of H_2O_2. However, dealing with samples where the organic matter tends to form resistent complexes with amorphous material, additional treatments were performed as regards the clay fraction, in order to eliminate any rest of organic substance. They included: (a) severe treatment in boiling H_2O_2 at 30%; (b) slight treatment with boiling H_2O_2 at 10%; (c) ozonification of the samples, bubbling ozone in a suspension of clay-acetone during 1 hour. Plate I indicates the specific treatment for each micrograph.

(b) *Preparation of samples for electron microscopy:* a small amount of clay was suspended in water until obtaining a lightly opalescent suspension (about 0,002–0,005%). A drop was put on a collodion membrane, supported by a metal screen, and then dried at 30°C.

(c) *Shadow-casting:* only the sample of Plate II, of Corte Alto was shadowed with chromium at an angle of 30°.

Specific surface. The specific surface was determined by means of the glycerol monolayer sorption, according to the procedure of Kinter and Diamond (1956).

RESULTS

Mechanical composition of soils

The results obtained are shown in Table II. The mechanical analysis does not reflect the real composition of these soils, because of the fact that soils containing a high proportion of allophane are difficult to disperse, probably due to a complex interaction between the isoelectric point (I.E.P.) of the allophane and the co-existing sesquioxide systems (Birrell and Fieldes, 1952). Besoain (1967) has shown that a considerable part of the coarse fractions of similar soils consist of particle aggregates smaller than 2 μ, and indirect determinations of clay contents by measuring the specific surface made by the same author show that the real clay content is considerably higher than the one obtained by mechanical analysis.

TABLE II

Mechanical composition of the soil samples (%)*

Particle size	Sample no. 15	16	17	18	23	24	25	26	27	28
2,000–250 μ	15.0	16.4	22.0	27.5	13.5	30.5	29.8	28.2	44.2	40.3
250– 50 μ	13.1	39.4	41.0	29.0	22.3	26.0	31.0	36.4	37.8	38.2
50– 2 μ	61.9	37.0	24.6	19.9	42.0	35.0	31.9	25.6	10.4	12.1
2 μ	10.0	7.2	12.4	23.6	22.2	8.5	7.3	9.8	7.6	9.4

*Oven-dry basis.

156   E. Besoain

TABLE III

Chemical composition of 0.2 μ clay separates (%)*

Sample no. Components	15	16	17	18	27	28
SiO_2	30.72	33.26	35.02	34.72	32.98	31.18
Al_2O_3	44.78	43.16	39.85	40.38	41.80	44.25
Fe_2O_3	1.22	1.41	1.60	1.86	1.39	1.06
TiO_2	1.00	1.57	1.06	1.21	1.07	0.54
CaO	0.59	0.38	0.51	0.65	0.48	0.19
MgO	0.79	0.74	0.64	0.18	0.72	0.26
Na_2O	1.40	0.89	1.75	2.10	1.05	1.65
K_2O	0.31	0.79	0.18	0.14	0.17	0.18
MnO	-	0.22	Tr	-	-	-
P_2O_5	0.46	2.03	0.48	0.56	0.76	0.84
$H_2O(+)$	18.70	16.23	18.85	18.15	19.64	19.82
SiO_2/Al_2O_3	1.16	1.31	1.49	1.46	1.33	1.19
H_2O/Al_2O_3**	4.50	4.40	4.62	4.52	5.36	4.89
C.E.C.	43.40	45.80	54.20	46.90	45.40	48.20
Sp.surf.(m^2/g)	238.00	224.00	208.00	212.00	196.00	192.00

* Oven-dry basis.
** Including H_2O (-).

Chemical composition

The results of chemical analysis of deferrated clays are shown in Table III. The composition of these clays is rather similar as they consist fundamentally of aluminum, silica and water. Silica/alumina ratio varies between 1.16 and 1.49 and that of water/alumina between 4.4 and 5.3. The ratio silica/alumina/water is approximately 1/1/5. It is believed that the iron that remained in the clay is constituted by forms of free, probably amorphous iron, which is resistent to deferration, because no analysis whatsoever showed the existence of a crystalline iron mineral. The chemical composition of both, Corte Alto and Puyehue clays, is, in any case, similar to the composition of the allophane.

Values found for C.E.C. vary between 43.0–54.2 mequiv./100 g, similar to the values found by Yoshinaga and Aomine (1962b). However, as suggested by these authors, it is probable that C.E.C. of imogolite is considerably affected by the extraction treatments as it is the case with allophane.

PLATE I

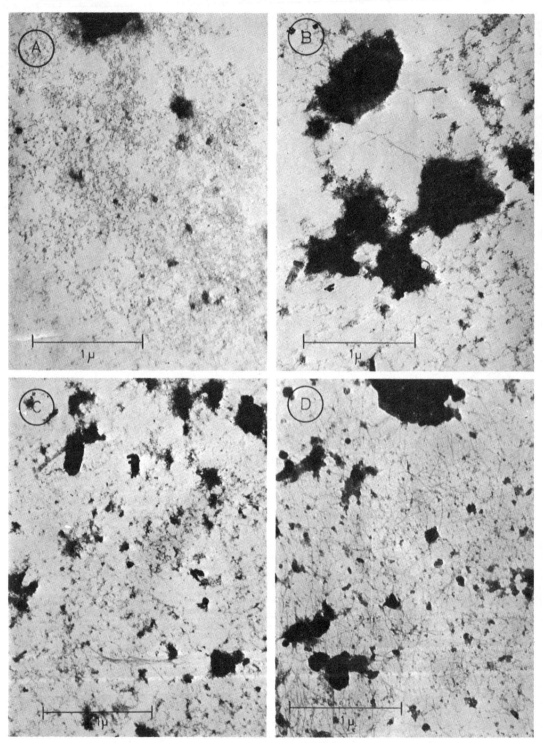

Legend see p.159.

158 E. Besoain

PLATE I (continued)

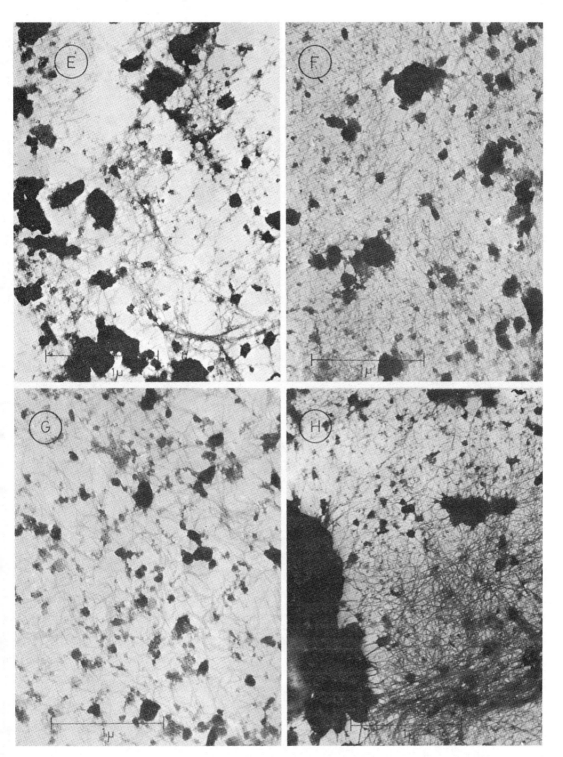

PLATE I (continued)

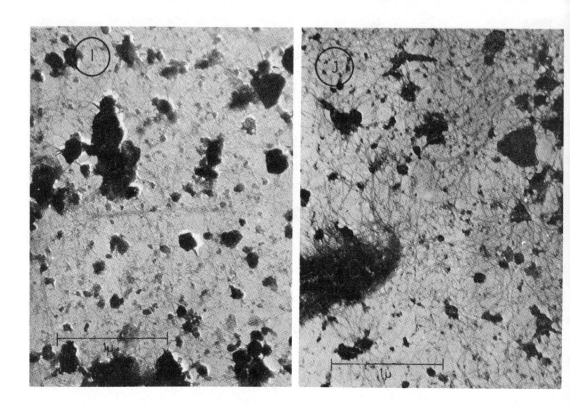

Electron micrographs of $< 0.2\ \mu$ fraction. A–D: Puyehue. E–J: Corte Alto.
A. Sample no.24; 22–42 cm. Treated with diluted H_2O_2. (Without separation of imogolite and allophane.)
B. Sample no.26; 90–145 cm. Treated with diluted H_2O_2. (Without separation of imogolite and allophane.)
C. Sample no.27; 145–210 cm. Treated with diluted H_2O_2.
D. Sample no.28; +210 cm. Treated with diluted H_2O_2.
E. Sample no.17; 70–120 cm. Treated with diluted H_2O_2.
F. Sample no.17; 70–120 cm. Treated with diluted H_2O_2.
G. Sample no.17; 70–120 cm. Treated with ozone.
H. Sample no.18; +120 cm. Treated with concentrated H_2O_2.
I. Sample no.18; +120 cm. Treated with concentrated H_2O_2 and shadowed with chromium.
J. Sample no.18; +120 cm. Treated with ozone.

Electron microscopy

Electron micrographs are shown in Plate I. All observations were made in the fraction smaller than 0.2 μ.

Excepting the upper horizons of Puyehue[1], all samples show characteristic thread-like particles of relatively regular size, the diameter of cross-section being about 60–120Å. Together with fibrous particles, allophane globoids can be observed, specially abundant in the upper horizons. It seems that the imogolite fractioning procedure at pH 3.5 does not prevent a certain amount of allophane being separated with it. On the other hand, micrographs of the allophane fraction (not shown here) practically do not show fibrous particles. In Corte Alto as well as Puyehue separates, the proportion of thread-like particles strongly increases with depth. In Corte Alto, for instance, Plate I, F and G show that the samples are mainly composed of fibrous particles and, as shown in Plate I,E, it can be clearly observed how the allophane spherules dispose themselves over the fibres. Micrographs I and J (Plate I) show scarse allophane globoids. In Puyehue clays the transition from allophane to a dominant fibrous material can be clearly observed[1]. Almost no fibrous particles were detected until a depth of about 90 cm, as shown in Plate I,A. In deeper samples of this soil, with imogolite fractioning, large amounts of thread-like particles, as well as allophane particles, were observed (Plate I, C and D). In some cases it is difficult to decide if allophane spherules had been deposited over a fibre, or if the fibres themselves are really constituted by union of the spherules, streptoid-like, following an unidirectional ordination (Plate I, B). The diameter of the spherules coincides, apparently, with the diameter of the sectional area of the fibrous particles. It could be assumed that the relationships between "thread" and "spherules" are simple and that they may have some implications for the structure and genesis of the imogolite.

It is evident that the thread-like particles of the Corte Alto clays are better developed and defined than those of the Puyehue soil. The treatment given to some samples (H_2O_2; ozone) did not affect in any way the particles' shapes.

Differential thermal analysis

Differential thermal analysis were performed in the fraction smaller than 0.2 μ size. Thermograms (Fig.2) are characterized by one strong endothermic peak at 150–180°C and another exothermic peak around 900–920°C. This behaviour is close to that shown by allophane; however, the exothermic reaction is better developed as compared with allophane, and its intensity increases with soil depth.

---

[1] In samples 23–26 from Puyehue no imogolite fractioning was done.

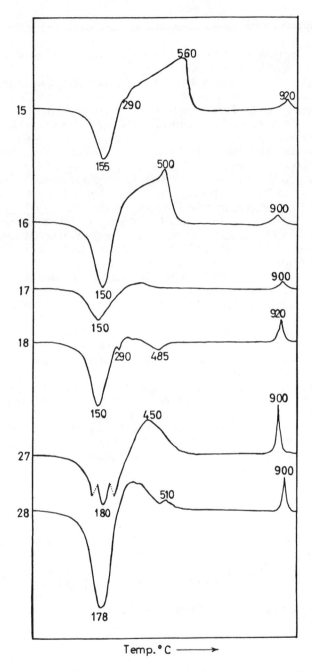

Fig.2. Differential thermal curves of < 0.2 μ clays.

In additon, sample 18 shows a small endothermic reaction at about 290°C which is ascribed to a small amount of gibbsite, and the exothermic peaks at 560, 500 and 450°C of samples 15, 16 and 27 are believed to correspond to stable complexes with O.M. which are able to resist the action of H_2O_2. The small peak at 485°C in sample 18 could be due to small amounts of halloysite, but there is no confirmation of its existence by other methods.

The endothermic reaction about 420°C, which Yoshinaga and Aomine (1962b) had pointed out for imogolite and, more specific, for the subspecies B (Aomine and Miyaushi, 1965) does not appear in our thermograms.

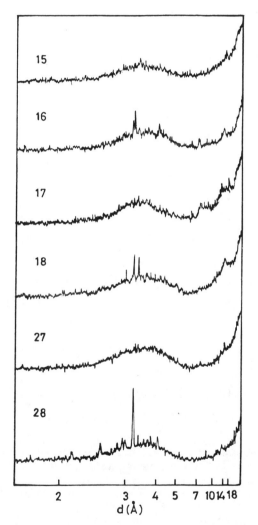

Fig.3. X-ray diffraction patterns of $< 0.2\ \mu$ clays in random orientation.

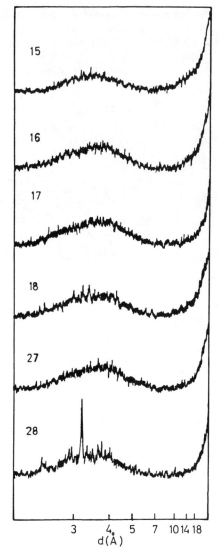

Fig.4. X-ray diffraction patterns of clays heated at 500°C.

X-ray diffraction

X-ray diffraction patterns of sodium-saturated 0.2 μ fraction, were obtained with random orientation of air-dried clays heated at 500, 1,000, and 1,200°C (Fig.3–6). The diffraction patterns are of great importance because, together with electron micrographs, they serve as a criterion for the identification of imogolite.

According to Yoshinaga and Aomine (1962b), imogolite is characterized for showing peaks around 14.2 and 7.6 Å, and another weak one at 5.6 Å.

The peak at 14 Å would be likely to decompose into two diffraction maxima, one around 13 Å and the other around 18 Å. Recently Aomine and Miyaushi (1965), on the basis of the stability of the lattice spacing against heating, determined two subspecies of imogolite: form A, with a lattice spacing about 13 Å, with a partial collapse at 100°C and a complete one at 300°C, and form B, with lattice spacing at about 18 Å, stable until 350°C and an almost complete collapse at 500°C. In addition to these characteristics imogolite B shows an endothermic peak around 420°C in the D.T.A. diagrams, as well as in I.R.-spectra with a diffuse absorption band at 1,140–1,220 cm^{-1}.

All diagrams of Corte Alto are very similar showing weak diffraction peaks at 17.7–14.2 Å with exception of sample 17 which shows peaks at 18.39 and 13.02 Å. Weak peaks at 7.2, 7.4 or about 5.5 Å could be connected

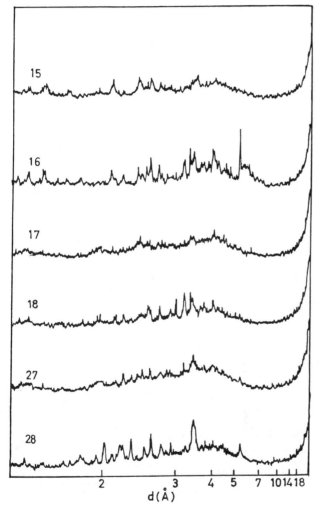

Fig.5. X-ray diffraction patterns of clays heated at 1,000°C.

Imogolite in Volcanic Soils of Chile 165

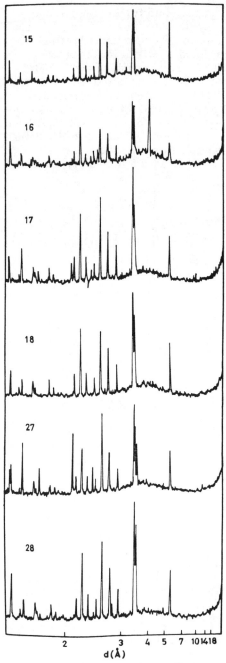

Fig.6. X-ray diffraction patterns of clays heated at 1,200°C.

with the former peaks. As for the Puyehue clays, sample 27 shows weak peaks at 14.71, 7.46, and 5.12 Å and sample 28 at 13.28, 8.53 and 5.33 Å.

In clays from both Corte Alto and Puyehue, feldspar lines (4.07, 3.36, 3.229, 2.844 Å) can be observed, specially intense in samples 16, 18 and 28.

In clays from Corte Alto, as well as in those from Puyehue, with the exception of feldspar lines, the other peaks are small, with an intense background, resembling the diffraction patterns of allophane. It is not believed that spacing around 7.3 Å corresponds to halloysite as neither peaks at 4.43 Å or other are present, nor is there any confirmation by other methods, with the exception of a doubtful one given by D.T.A. Peaks about 13 Å or 14 Å do not seem to correspond to vermiculite chlorite minerals according to its behaviour on heating. We suggest, therefore, that these peaks are probably due to imogolite and that the variations observed when comparing them with the parameters given by Yoshinaga and Aomine (1962b) can be attributed to a peculiar variation of the mineral itself or to variations influenced by allophane contamination.

When heated at 500°C (Fig.4), peaks at 18, 14, 13, 7 or 5 Å disappear, but some feldspar lines, however, remain. This indicates a complete structural collapse below 500°C.

Heating the samples at 1,000°C (Fig.5) we obtained patterns which are little representative and are probably products of transition to mullite. Samples 17, 18 and 28 present spacings about 1.97 Å or 1.39 Å which might be attributed to residues of gamma-alumina.

When heated at 1,200°C (Fig.6) strongs peaks are obtained (5.41, 4.502, 3.403, 2.900, 2.704 Å, etc.), which might be attributed to mullite, considering the small angular differences and the range of intensities with this species. Additional peaks at 4.11 Å or 3.34 Å were not identified. According to Yoshinaga and Aomine (1962b), mullite formation is characteristic of imogolite, forming a sharp contrast to allophane, which, when heated at 1,200°C, produces only alpha-alumina.

Infra-red absorption spectra

I.R. patterns are shown in Fig.7. All patterns show maxima absorption at 3,650, 3,400, 1,625, 1,000 and a small band at 1,410/cm. All these bands appear also in the allophane patterns. Aomine and Miyaushi (1965), when examining samples with high contents of imogolite, found the presence of a diffused band between 1,140–1,220/cm, probably due to Si–O linkage, and they determined it to be a characteristic feature of imogolite. In our patterns we did not detect such a diffused band, even not in samples with high amounts of fibres or threads, clearly visible by electron microscopic examination.

Specific surface

Surface determinations, evaluated as "internal surface" and obtained by the method of glycerol monolayers sorption (Kinter and Diamond, 1956), show lower values than the allophane separates obtained from the same samples (Table II). Values obtained for allophane were 320 m^2/g on the

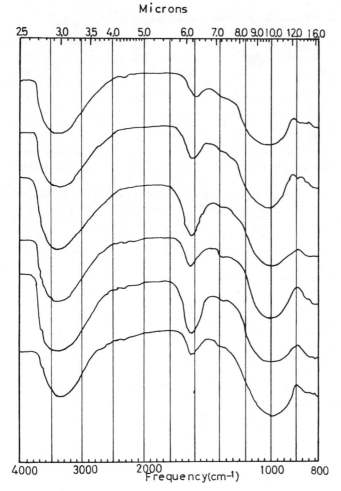

Fig.7. Infrared absorption curves of 0.2 μ clays of separated imogolite (Corte Alto and Puyehue).

average, while corresponding imogolite separates showed an average of 208 m²/g. This behaviour could be related to the development of a more ordered arrangement, with reduction of "internal access", and, consequently, with reduction of the specific surface of imogolite separates. However, we consider that a value of 208 m²/g is too high and is certainly influenced by the allophane contamination.

DISCUSSION OF RESULTS

The results of the electron microscopy and X-ray diffraction seem to indicate the existence of imogolite in clays of Corte Alto and Puyehue by comparing them to the parameters given by Yoshinaga and Aomine (1962b). There is, however, no definite relationship established between the amount of fibres observed by electron microscopy and the intensity and constancy of the X-ray peaks. This behaviour is probably attributable to proper variations of this mineral or to a certain degree of allophane contamination. Assuming that the weathering process of volcanic glass, passing from allophane to halloysite, is a continuous process, it could be conceived as a wide intermediate metastable term between the two extremes. Consequently it would, therefore, not be surprising to find mineralogical variations in imogolites formed in different environmental conditions.

The absence of diffraction maxima at 18 Å in our samples (except in sample 17), of an endothermic peak at 420°C and of a diffused band between 1,140–1,220/cm make us consider imogolite to be a species not well developed and more related to the subspecies A, described by Aomine and Miyaushi (1965). In addition, if diffraction peaks are little pronounced, we are not quite sure if the forms of threads or fibres are sufficient to characterize imogolite or if allophane with a similar morphology could exist.

The specific-surface measurements indicate that a considerable retraction of about 35% occurs in imogolite separates as regards to allophane, which is probably directly connected to the increase of structural order of imogolite separates. However, we assume that allophane contamination in clay fraction of Corte Alto and Puyehue, has influenced this result.

The better development and the main proportion of imogolite fibres of Corte Alto, compared with Puyehue, is probably connected to the soil age. Although we do not dispose of radio-carbon datation, geological and geomorphological facts and the opinion of the above cited Japanese authors (Aomine and Miyaushi, 1965) lead us to believe that Corte Alto is not older than 15,000 years, the Puyehue soil being younger than Corte Alto.

The fact that imogolite occurs fairly widely in nature as a member of the weathering sequence of volcanic ashes seems to be confirmed. It is probable that a number of andepts or andosols contain imogolite as a dominant species of the clay fraction instead of allophane, which would obviously exceed the limits imposed by the definition of andepts given by the Soil Survey Staff U.S.D.A. (1960). It would, in any case, be necessary to consider the imogolite existence as a criterion in volcanic ash soils classification and even more so, when having more knowledge about its frequency, structure and properties.

The probable existence of imogolite in soils of nonvolcanic origin leads to the question wether imogolite is necessarily associated with the weathering of volcanic ashes, or if it could be formed from other parent materials under certain environmental conditions, like those which lead to the formation of podsols, as in the case of Finland.

Conclusion

The evidence given by the different analyses performed shows the existence of imogolite in andosoils of Chile. The apparent proportion of this mineral increases strongly with depth, until it became predominant in the lower layers of the soils. We find an increased amount and better development of fibres in the Corte Alto soil, supposed to be older than the

Puyehue soil. No definite relationships were found between the intensity of
X-ray peaks and the amounts of fibres. D.T.A. patterns, I.R. spectra and
results of chemical analysis were very close to those of allophane. The
specific-surface values, however, of imogolite separates, were about 35%
lower than those typical for allophane, and are probably related to the
development of a more ordered structure, as compared with allophane.

REFERENCES

Alberto, G.F. und Jaritz, G., 1966. Amorphe und kristalline Bestandteile einiger typischen Bodenbildungen Skandinaviens. Z. Pflanzenernähr., Düng., Bodenk., 114(1): 27–46.
Aomine, S. and Jackson, M.L., 1959. Allophane determination in andosoils by cation exchange capacity delta value. Soil Sci. Soc. Am. Proc., 23(3): 210–214.
Aomine, S. and Miyaushi, N., 1965. Imogolite of imogo-layers in Kyushu. Soil Sci. Plant Nutrition, 11(5): 28–35.
Aomine, S. and Yoshinaga, N., 1955. Clay minerals of some welldrained volcanic-ash soils of Japan. Soil Sci., 79(5): 349–358.
Besoain, E., 1967. Mineralogía de las fracciones limo de dos andosoles de Chile. Proyecto Estudios y Reconocimiento de Suelos, Santiago, Chile.
Birrell, K.S. and Fieldes, M., 1952. Allophane in volcanic ash soils. Soil Sci., 2: 156–166.
Jackson, M.L., 1958. Soil Chemical Analysis. Prentice Hall, Englewood Cliffs, N.Y., 498 pp.
Jaritz, G., 1967. Ein Vorkommen von Imogolit in Bimsböden, Westdeutschland. Im Druck.
Kanno, I., Kuwano, I. and Honjo, Y., 1960. Clay minerals of gellike substances in Pumice Beds. Adv. Clay Sci., 2: 335–365.
Kawasaki, H. and Aomine, S., 1966. So-called 14A clay minerals in some andosoils. Soil Sci. Plant Nutrition, 12(4): 18–24.
Kinter, B. and Diamond, S., 1956. Gravimetric determination of monolayers glycerol complexes of clay minerals. Proc. Natl. Conf. Clays Clay Minerals, 5th–Natl. Acad. Sci. Natl. Res. Council, Publ., 318–333.
Kolthof, I.M. y Sandell, E.B., 1956. Tratado de Química Analítica Cuantitativa. Librería y Editorial Nijar, S.R.L. Buenos Aires, 917 pp.
Mackenzie, R. (Editor), 1957. The Differential Thermal Investigations of Clays. Mineralog. Soc., London, 456 pp.
Mehra, O.P. and Jackson, M.L., 1960. Clays and clay minerals. Proc. Natl. Conf. Clays Clay Minerals, 7th– Natl. Acad. Sci. Natl. Res. Council, Publ., 317–327.
Soil Survey Staff U.S.D.A., 1960. Soil Classification. A Comprehensive System, 7th Approximation. U.S. Government Printing Office, Washington, D.C., 265 pp.
Wright, Ch.S., 1960. The Volcanic Ash Soils of Chile. F.A.D. Tech. Assist. Mission Chile, 1337 pp.
Yoshinaga, N. and Aomine, S., 1962a. Allophane in some andosoils. Soil. Sci. Plant Nutrition, 8(2): 6–13.
Yoshinaga, N. and Aomine, S., 1962b. Imogolite in some andosoils. Soil Sci. Plant Nutrition, 8(3): 22–29.

22

Copyright © 1972 by the Mineralogical Society
Reprinted from Clay Minerals **9**:281–285 (1972)

MORPHOLOGY OF ALLOPHANE, IMOGOLITE AND HALLOYSITE

H. Eswaran

Geological Institute, Rozier 44, Ghent, Belgium

(*Received* 28 *September* 1971)

ABSTRACT: Volcanic ash soils were studied with the scanning electron microscope to observe the morphology and occurrence of allophane, imogolite and halloysite. Allophane and imogolite in general were randomly distributed indicating a process of dissolution and recrystallization. Halloysite is typically associated with feldspars showing that it is formed by direct alteration.

The imogolite fibres studied with the SEM show them to be much thicker than currently reported in the literature and it is thought that pretreatment of the clays for TEM measurements results in a partial destruction. Halloysite consists typically of short rigid tubes and thus differs from imogolite which tends to curl.

INTRODUCTION

Allophane, imogolite and halloysite are the three secondary clay minerals frequently associated with the recent alteration of volcanic ash materials. Imogolite was identified by Yoshinaga & Aomine (1962) who considered it to be soil allophane of low crystallinity and a fibrous morphology. Aomine & Miyauchi (1965), Miyauchi & Aomine (1966), Wada (1966) and Jaritz (1967) and others, have reported on the occurrence of this fibrous mineral in volcanic ash soils. Detailed studies on imogolite have also been made by Wada (1967), Yoshinaga, Yotsumoto & Ibe (1968) and Russell, McHardy & Fraser (1969). With the great number of studies on imogolite, it can be considered as a crystallochemically distinct mineral.

Most or all of the studies of these clay minerals in soils have been made after separation of the clay fraction which necessitated pretreatment with consequent disturbance. The scanning electron microscope (SEM), however, enables the study of these soil minerals with a minimum of pretreatment. Eswaran & De Coninck (1971) have employed SEM in studying the alteration of feldspars to halloysite, kaolinite and montmorillonite. Borst & Keller (1969) and Bohor & Hughes (1971) have provided some excellent photographs of reference clay minerals.

In this study, the morphology of allophane, imogolite and halloysite was determined with the SEM.

EXPERIMENTAL

The samples employed and their locations are given below:

Sample no. 1. A horizon (topsoil), 0–34 cm.
   Locality: Ue-mura, Kuma-gun, Miyazaki Prefecture, Kyushu.

Sample no. 2. IIB horizon ('Imogo' or 'Akahoya' layer), 34–60 cm.
   Locality: as no. 1.

Sample no. 3. IIB2 horizon ('Imogo' or 'Akahoya' layer), 51–93 cm.
   Locality: Kobayashi-shi, Miyazaki Prefecture, Kyushu.

Sample no. 4. IIB horizon ('Imogo' or 'Akahoya' layer), 35–60 cm.
   Locality: Kawaminami-cho, Miyazaki Prefecture, Kyushu.

Sample no. 5. Kodonbaru (sample 905 in publications of Aomine). Imogolite, C horizon.

Sample no. 6. Choyo (sample 1041 in publications of Aomine). Allophane, B horizon.

Sample no. 7. Puy de Dome, France. Halloysite and allophane, B horizon: (Sample collected by the author).

The air-dried samples were first studied with a reflecting microscope, using fracture surfaces of soil peds or rock fragments. Thin sections of similar materials were simultaneously studied with a polarizing microscope. Prior to these examinations, the clays were separated, and studied by X-ray diffraction, thermal analysis and transmission electron microscopy.

The microscopic studies determined the nature of the surfaces that were to be examined by SEM. A sketch was made of the fracture surfaces showing the location of glass fragments, feldspars and any other minerals that could be identified. Voids and other special features were also indicated.

The minerals and features were first scanned, followed by the matrix of the fracture surface.

The samples were mounted onto aluminium stubs and coated first with carbon and then with gold. This double coating proved to be superior to coating either with carbon or gold alone. Each specimen was studied at magnifications ranging from 500–20,000.

RESULTS

Allophane

Allophane was observed in practically all the samples. The typical form is shown in Plate 1 (a, b). The allophane occurs as fluffy aggregates of varying diameters, which do not show any particular arrangement patterns in the soil materials.

Under low magnifications (about 1000), the surface of the aggregates was smooth with a gel-like appearance. Under the reflecting microscope (magnification ×40), the matrix of the soil material was brownish. At magnifications of about 10,000

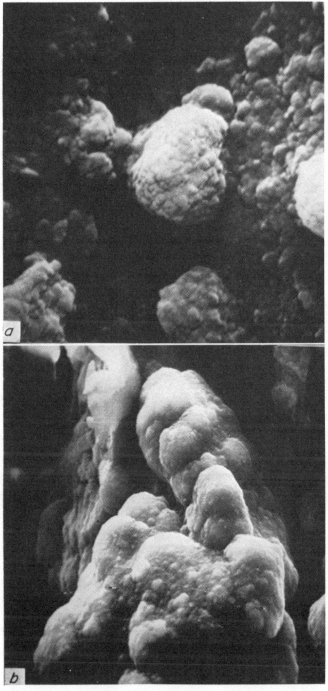

(a) SEM photograph of allophane, from sample No. 1 (Ue-mura), in soil material. × 10,000 (b) SEM photograph of allophane, from sample No. 6 (Choyo), in soil material. × 10,050.

Eswaran: Plate II

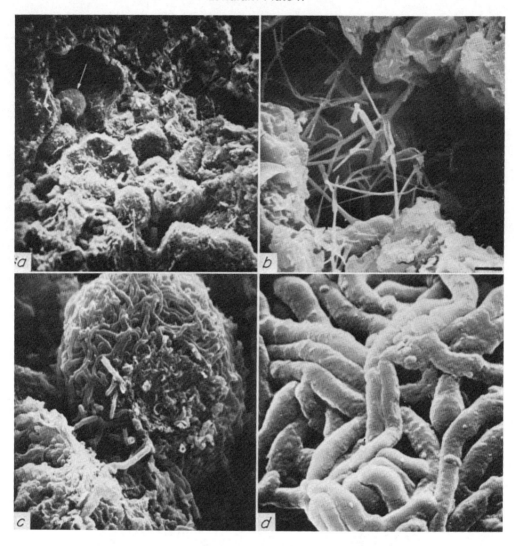

SEM photograph of imogolite from Kodonbaru sample. (a) × 187·5. (b) × 862·5. (c) × 862·5. (d) × 1875·0.

(Plate 1) an irregularity of the surface was observed. Under higher magnifications (50,000), each of the larger aggregates was seen to be composed of smaller, rounded aggregates.

Some of the samples showed volcanic glass which could be relatively easily observed under the reflecting microscope. Under the scanning microscope, these were present as sharp-edged structures, morphologically resembling feldspar crystals.

Imogolite

Preliminary studies of the soil fracture surfaces with the reflecting microscope showed gelatinous globular material. The globules are scattered randomly in the matrix of the fracture surface. Under the reflecting microscope they frequently glisten white. According to Miyauchi & Aomine (1966), the filmy, gel-like substances in pumice beds are identifiable as imogolite. Fracture surfaces with a dominance of the globules were examined by SEM. The Kodonbaru sample (no. 5) showed the greatest amount of the globular material. Plate 2(a) is a low-magnification SEM photograph of this sample, in which arrows point to the globular material. In the void and parts of the fracture surface a network of imogolite can be seen. In Plate 2(b) the void is magnified and the very erratic and random branching of the imogolite fibres are visible. Such a distribution pattern of imogolite has not been encountered in the other samples, the globules being more common. One of the globules is enlarged in Plate 2(c) and is a very compact and interlaced mass, resembling a ball of thread. Plate 2(d) is a higher magnification of the same globule showing the individual strands with a rounded appearance. A cross-section was not available for study.

Plates 3(a) and (b) show similar features in the Kobayashi-shi sample (no. 3), illustrating the uniformity of morphology. The individual strands (cf. Plate 3(b) with 2(d)) of the imogolite in sample 3 however, are less rounded.

In the sample (no. 5) from Kawaminami-cho, the imogolite is different, the characteristic white globules being absent. The imogolite is considerably sparser, and occurs as worm-like strands, growing from the edges of the volcanic glass particles (Plate 3c). This seems to indicate that the imogolite here is a direct alteration product of volcanic glass. The planar surface of the glass has a spongy texture due to the presence of allophane. In the other samples (Plates 2 and 3a, b), the imogolite has been formed by the reprecipitation of silica and alumina released by the weathering of primary minerals. In the Choyo soil, where allophane is the dominant clay, high magnification of some of the allophane aggregates shows knobbly surfaces. Some strand-like features are also present, but poor resolution at high magnifications prevented any definite conclusions.

From these SEM studies, it appears that the imogolite fibres are much thicker than those reported in the literature (20–30 Å). Measurements from SEM photos suggest a range of 0·02–0·8 μm, with the larger sizes being more frequent. This raises the problem of the nature of these fibres. Two possibilities arise: (i) that they are strands composed of very fine fibres, the latter not being evident at the magnifications employed. The effect of the pretreatments employed for preparation of the clay for TEM is to break up the strands into their individual components. (ii) The fibres are of the dimensions as observed in SEM. Pretreatment breaks up the fibres along planes of weakness. The latter is more probable as studies of individual fibres at a magnification of 50,000 show that the surface is smooth and not striated as would be expected if they were composed of thinner fibres. These observations need further verification.

Halloysite

Neoformation of halloysite is observed in the sample (no. 7) from France. Plate 4(a) shows the alteration of feldspars to halloysite. Characteristically they consist of short tubes about 1–2 μm long, with a diameter of 0·02–0·16 μm, and a circular cross-section. In contrast to imogolite the halloysite tubes are always rigid and show no tendency to curl up. Tropical alteration of feldspars to halloysite (Eswaran & De Coninck, 1971) show similar morphological features.

In the top left hand corner of Plate 4(a), an initial stage in the alteration of the plagioclase feldspar can be seen. A general reorganization of the crystal lattice of the feldspar seems to be taking place during its alteration to halloysite. Hence the process involved is not one of dissolution and recrystallization, as observed in some of the allophane and imogolite specimens.

Halloysite is present very locally in this sample, the dominant clay mineral being allophane. Imogolite is absent. The halloysite is of the $2H_2O$ variety as shown by X-ray diffraction studies.

The morphology of the endellite, or the $4H_2O$ variety, is distinctly different (Plate 4(b)). This shows a triclinic symmetry with distinct edges. The sample is from Ward's Scientific Agency (halloysite no. 3, Eureka, Utah).

DISCUSSION AND CONCLUSION

From these studies it is evident that morphologically, allophane, imogolite and halloysite are distinct mineral species. The International Mineralogical Association (Fleischer, 1963) suggested that imogolite need not be considered as a distinct mineral type. From previous and present work on imogolite, it is evident that it is morphologically and crystallochemically a distinct mineral species.

It has been shown that allophane and imogolite frequently occur as aggregates or globules respectively. Such a habit will be important in their separation from soil material for granulometric analysis or more detailed studies on these clays. Kobo & Oba (1961) and Kobo, Oba & Oishi (1962) report that with normal dispersing agents, the aggregates are not completely separated and as a result, a part of the clay fraction is included in sand and silt fractions. Consequently an ultrasonic technique is indispensable in order to loosen the aggregates for clay dispersion. Most of these aggregates are effectively broken up by ultrasonic treatment employing a 10 KC frequency and 500 W output for 10–15 min.

Halloysite, on the other hand, does not present such problems. It appears to be a brittle mineral, easily breaking up into small pieces.

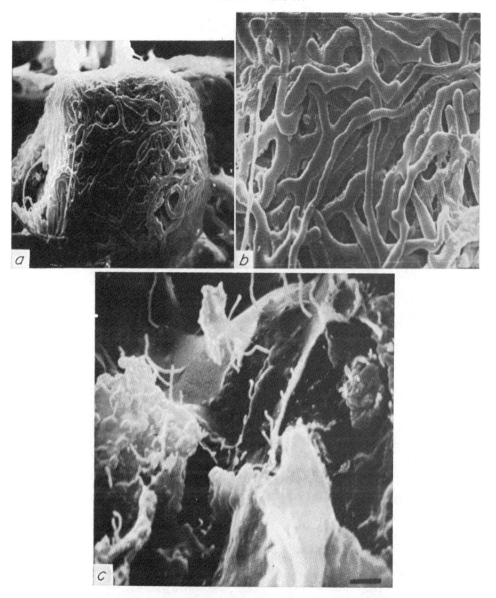

Imogolite from Kobayashi-shi sample. (a) × 787·5. (b) × 1575·0. (c) Imogolite from Kawaminami-cho sample. × 5100.

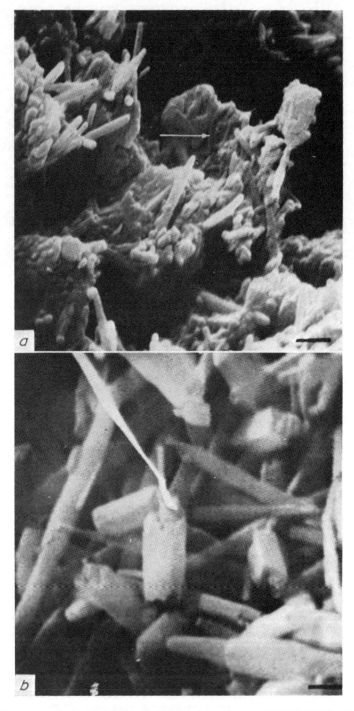

(a) Alteration of feldspars to halloysite. Sample from France. × 10,000. (b) Morphology of endellite from Utah. × 38.000.

The genesis of these minerals needs further study. From their morphology and occurrence in the matrix of the soil and rock fragments, the neoformation of allophane and imogolite suggests a dissolution and precipitation mechanism. Direct alteration from primary minerals is also possible. This was only observed in the case of imogolite.

Halloysite, as observed in this study and by Eswaran & De Coninck (1971), seems to originate by direct alteration. Parham (1964) has shown this experimentally with his studies on the alteration of albite. Secondary halloysite is always associated with feldspars as shown in these SEM studies. Volcanic glass does not seem to alter to halloysite, unlike imogolite which was observed as a direct alteration product.

Little or no evidence has been observed to suggest the frequently reported sequence of alteration: allophane → imogolite → halloysite. The globules of imogolite could arise as a result of in-situ alteration of the allophane aggregates, though this was not observed.

All three minerals coexist side by side in the matrix of the soil. The factors governing the formation of these minerals are worthy of further study. The role of the microenvironment also needs greater consideration.

ACKNOWLEDGMENT

The author wishes to thank Professor Aomine and Dr Kanno for the samples of the Japanese volcanic ash soils. The author is grateful to Professor Lagasse for the facilities of the Electronmicroscopic Centre. Thanks are also due to Professor Tavernier, and my colleagues for valuable discussions.

REFERENCES

AOMINE S. & MIYAUCHI N. (1965) *Soil Sci. Pl. Nutr., Tokyo*, **11**, 212.
BOHOR B.F. & HUGHES R.F. (1971) *Clays Clay Miner.* **19**, 49.
BORST R.L. & KELLER W.D. (1969) *Proc. Int. Clay Conf., Tokyo*, 871.
ESWARAN H. & DE CONINCK F. (1971) *Pédologie, Gand*, **21**, 181.
FLEISCHER M. (1963) *Am. Miner.* **48**, 434.
JARITZ G. (1967) *Z. Pfl-Ernähr. Düng Bodenk.* **117**, 65.
KOBO K. & OBA Y. (1961) *Trans. Congr. Sci. Soil and Manure, Japan*, **7**, 35.
KOBO K., OBA Y. & OISHI K. (1962) *Trans. Congr. Sci. Soil and Manure, Japan*, B, 15.
MIYAUCHI N. & AOMINE S. (1966) *Soil Sci. Pl. Nutr. Tokyo*, **12**, 187.
PARHAM W.E. (1964) *Clays Clay Miner.* **17**, 13.
RUSSELL J.D., MCHARDY W.J. & FRASER A.R. (1969) *Clay Miner.* **8**, 87.
WADA K. (1966) *Soil Pl. Fd., Tokyo*, **12**, 176.
WADA K. (1967) *Am. Miner.* **52**, 690.
YOSHINAGA N. & AOMINE S. (1962) *Soil Sci. Pl. Nutr., Tokyo*, **8**, 114.
YOSHINAGA N., YOTSUMOTO H. & IBE K. (1968) *Am. Miner.* **53**, 319.

Part VII

BIOLOGICAL CHARACTERISTICS

Editor's Comments
on Papers 23 and 24

23   MURAYAMA
   The Monosaccharide Composition of Polysaccharides in Ando Soils

24   MARTINEZ and RAMIREZ
   Study of the Microfungal Community of an Andosol

Except for organic matter content and nature of humic and fulvic acids, little is known about the biological properties of Andosols. The magnitude of humus accumulation in Andosols has baffled many soil scientists for a long time, and a lot of effort has been spent to find an explanation for this problem. Currently a number of reasons have been presented, but all of the theories pointed to the presence of allophane as the major reason for the large accumulation of organic matter in Andosols. The opinion is that most of the humus is preserved in these soils because of its close interaction with allophane and/or aluminum (Tokudome and Kanno, 1965a, 1965b; Goh, 1980). The mineral allophane is often considered to be a major source of the aluminum that can be chelated by the organic fraction.

The charge characteristics of Andosols are known to be variable or pH dependent and are usually considered caused by the presence of allophane. In view of the exceptionally high humus content, it is not surprising that a major proportion of this variable charge comes from the dissociation of functional groups, such as carboxyl and phenolic OH groups, in the soil humus.

The interaction between humus and allophane also brings changes in many other soil properties. Evidence has been reported several times that the formation of clay-organic complexes contributes to development of stable soil aggregates (Theng, 1974; Baver, 1968). Not only will the latter result in an improvement of the physical condition of the soil—such as increased water-holding capacity—it also increases the soil's resistance to erosion.

The nature of the humus in Andosols is different from that of other soils. The humus fraction of Andosols has been reported to contain more humic than fulvic acids (Tokudome and Kanno, 1965a, 1965b; Tan, 1959). Todudome and Kanno (1968) indicated that the humic-fulvic acid ratio was on the average 1.2 and that this ratio increased with an increase in allophane content. The authors believed that the accumulation of humic acids reached a maximum in humid subtropical climates, where formation of allophane was favorable. A number of Japanese scientists postulated that the transformation of fulvic acid into humic acid was accelerated by the catalytic action of allophane (Kyuma and Kawaguchi, 1964). Attempts have been made to express the degree of humification in terms of $\Delta \log K = \log K_{400}\,m\mu - \log K_{600}\,m\mu$, where K = absorbance (Kumada, 1955). In general $\Delta \log K$ values for the humic fractions in Andosols are on the average 0.356. This value is lower than the ΔK values of humus fractions in other soils.

Another distinguishing feature of the humic fraction of Andosols often reported in the Japanese literature is its aromaticity (Tokudome and Kanno, 1968). Humic acids in Andosols possess a higher degree of aromaticity than humic acids in other soils. The degree of aromaticity is considered the reason for the refractory nature of humic acids (Goh, 1980). Humic acid with a highly aromatic nucleus is believed to be very active in interacting with a number of metals and organic compounds, such as polysaccharides, proteins, amino acids, and phenolic compounds (Cranwell and Haworth, 1975; Haworth, 1971).

Because of the differences in the nature and content of humic acids and in clay mineralogy, the possibility arises that Andosols may be different in composition from other soils with respect to other organic compounds and microorganism population. It is currently known that the clay fraction of soils can affect microorganisms by affecting soil pH, nutrient supply, and concentration of toxic substances, by acceleration or inhibiting interactions between cells, by adsorbing cells, ensuring in this way their survival under adverse conditions (Stout and Lee, 1980), and by altering their growth rate (Haider, Filip, and Martin, 1970). Different kinds of clays are expected to have different effects on these factors. The specific environment created in Andosols by their allophane and peculiar humic acid content tends to lend support for the presence of different organic compounds and development of microorganisms specific to Andosols. Paper 23 outlines the kinds of sugars found in Andosols. It reports that the monosaccharide composition in Andosols is characterized by higher contents of mannose, fructose, and ribose compared to other soils. According to the author, such a composition is the result of the breakdown of microbially produced polysaccharides and is not the result of the effect of climate and vegetation.

Because soils can be characterized by different physical and chemical characteristics, Ishizawa (1964) believed that each soil also can be characterized by a specific group of microrganisms. He indicated that the microflora in Andosols differed from that of nonvolcanic ash soils. Since he provided no details, Paper 24 has been included to provide a more detailed insight on the microfungal population in Andosols.

Not only is it possible that a specific physical and chemical soil condition may favor the development of specific microorganisms, but the latter may also be able to interact with the mineral soil constituents in a similar fashion as humic acid interacting with clay. Such an interaction between microorganisms and clay may also result in changing the soil properties, especially the charge characteristics of soils. Most soil scientists are used to the idea that the charge characteristics of soils are confined mainly to the clay and humic fractions; however, currently it is known that cells of living organisms possess a significant amount of variable charges. This cell charge will be affected by the charges of the clays in such a way that an interaction can take place, producing cell-clay complexes (Stout and Lee, 1980). Such an interaction is similar to the interaction between humic acid and clays. The difference with humic acid is that bacteria, for example, are composed of cells that are too large to classify as colloids but nevertheless behave as do colloids. The colloidal behavior is attributed to repulsion among the negatively charged bacterial cells, causing them to disperse in aqueous solutions. Such suspension is a property known to be exhibited only by colloids. Notwithstanding the fact that bacteria are living cells, which possess the capability to grow, exhibit metabolism processes, and can move independently, Marshall (1976) postulated that the interaction between bacteria and clay, forming complexes, can be described in physicochemical terms as follows:

$$xB + yCl \rightarrow B_xCl_y$$

where B = bacterial cell, Cl = clay mineral, and x and y are the numbers of B and Cl, respectively. If this is valid, then in my opinion, the equilibrium constant, K, can be calculated. Using the mass action law:

$$K = \frac{(B_xCl_y)}{(B)^x (Cl)^y}$$

In log form this equation becomes:

$$\log K = \log (B_xCl_y) - x\log (B) - y\log (Cl).$$

Log K is considered the stability constant of the complexes formed. The larger the value for K, the larger the value of the stability constant and hence the greater the affinity of B for interaction with Cl. No investigations have been done in this respect.

The interaction between bacterial cells and soil clays is affected by the charge densities of the cells and the clay mineral. Some bacterial species (such as Rhizobia spp.) have two types of cells, each characterized by different charge densities. One group of cells has charges attributed to carboxyl groups only, and the other group has charges created by complex combinations of amino and carboxyl groups. The latter yields smaller charge densities than the single carboxyl groups (Marshall, 1968) and is expected to have a weaker reaction with clay.

Most of the investigations on the interaction between bacteria and clay have been conducted with crystalline clays—such as montmorillonite (smectite) and kaolinite (Filip, 1979; Stotzky, 1972)—and little has been done yet with allophane. Of course, not all organisms will react in such a way on the electrical charges of the clay. The larger soil flora and fauna are generally too large and well insulated to be affected by the charge characteristics of the mineral fraction in soils. In analogy to the preservation of organic matter in soils as a result of complex formation with clay, this interaction between bacterial cells and clays also ensures the survival and accumulation of a specific group of bacteria, their enzymes and metabolic products. What such an interaction will do to the activity of the microorganisms and to many other biochemical reactions in soils is open to further investigation.

REFERENCES

Baver, L. D., 1968, The Effect of Organic Matter on Soil Structure, Semaine d'Etude sur la Matiere Organique et Fertilité du Sol, *Pontificiae Academiae Scientiervm Scripta Varia* **32:**384-413.

Cranwell, P. A., and R. D. Haworth, 1975, The Chemical Nature of Humic Acids, in *Humic Substances, Their Structure and Function in the Biosphere,* B. Povoledo and H. L. Golterman, eds., Centre Agric. Public. Docum., Wageningen, the Netherlands, pp. 13-18.

Filip, Z., 1979, Wechselwirkung von Mikroorganismen und Tonmineraleneine Ubersicht, *Z. Pfanzenernahr. Düng. Bodenk.* **142:**375-386.

Goh, K. M., 1980, Dynamics and Stability of Organic Matter, in *Soils with Variable Charge,* B. K. G. Theng, ed., New Zealand Soc. Soil Sci., Lower Hutt, New Zealand, pp. 373-393.

Haider, K., Z. Filip, and J. P. Martin, 1970, Einfluss von Montmorillonite auf die Bildung von Biomasse und Stoffwechselzwischenprodukten durch einige Mikroorganismen, *Archiv. Mikrobiol.* **73:**201-215.

Haworth, R. D., 1971, The Chemical Nature of Humic Acid, *Soil Sci.* **111:**71-79.

Ishizawa, I., 1964, Microbiology of Volcanic Ash Soils in Japan, Meeting on the Classification and Correlation of Soils from Volcanic Ash, Tokyo, Japan, June 11-27, 1964, *Food and Agric. Org., United Nations, World Soil Resources Rept.* **14**:87-88.

Kumada, K., 1955, Studies on the Colour of Humic Acids, Part 1, On the Concepts of Humic Substances and Humification, *Soil Sci. Plant Nutr. (Japan)* **11**:151-156.

Kyuma, K., and K. Kawaguchi, 1964, Oxidative Changes of Polyphenols as Influenced by Allophane, *Soil Sci. Soc. Am. Proc.* **28**:371-374.

Marshall, K. C., 1968, Interaction between Colloidal Montmorillonite and Cells of Rhizobium Species with Different Ionogenic Surfaces, *Biochim. Biophys. Acta.* **156**:179-186.

Marshall, K. C., 1976, *Interfaces in Microbial Ecology,* Harvard University Press, Cambridge, 156p.

Stotzky, G., 1972, Activity, Ecology, and Population Dynamics of Microorganisms in Soil, *CRC Critical Rev. Microbiology*: 59-137.

Stout, J. D., and K. E. Lee, 1980, Ecology of Soil Micro- and Macro-Organisms, in *Soils with Variable Charge,* B. K. G. Theng, ed., New Zealand Soc. Soil Sci., Lower Hutt, New Zealand, pp. 353-372.

Tan, K. H., and J. Van Schuylenborgh, 1959, On the Classification and Genesis of Soils Derived from Andesitic Volcanic Material under a Monsoon Climate, *Netherlands Jour. Agric. Sci.* **7**:1-22.

Theng, B. K. G., 1974, *The Chemistry of Clay-Organic Reactions,* John Wiley & Sons, New York, 343p.

Tokudome, S., and I. Kanno, 1968, Nature of the Humus of Some Japanese Soils, *Intern. Congress Soil Sci. Trans.* (Adelaide, Australia) **3**:163-173.

Tokudome, S., and I. Kanno, 1965a, Nature of the Humus of Humic Allophane Soils in Japan, Part 1, Humic Acids (Ch)/Fulvic acids (Cf) Ratios, *Soil Sci. Plant Nutr. (Japan)* **11**:185-192.

Tokudome, S., and I. Kanno, 1965b, Nature of the Humus of Humic Allophane Soils in Japan, Part 2, Some Physico-Chemical Properties of Humic and Fulvic Acids, *Soil Sci. Plant Nutr. (Japan)* **11**:193-199.

23

Copyright © 1980 by Blackwell Scientific Publications
Reprinted from *Jour. Soil Science* 31:481-490 (1980)

THE MONOSACCHARIDE COMPOSITION OF POLYSACCHARIDES IN ANDO SOILS

SHIGETOSHI MURAYAMA

(*National Institute of Agricultural Sciences, Yatabe-machi, Tsukuba-gun, Ibaraki-ken, 305, Japan*)

Summary

The monosaccharide composition of Ando soils, which originate from volcanic ash and have high organic matter content (8–21% carbon), was quite different from that of non-volcanic ash soils (1.2–1.9% carbon), being richer in mannose, fucose and ribose, whereas there was less glucose in cellulose-like form, arabinose, xylose and rhamnose. The Ando soils were also characterized by a lower percentage of organic carbon in the form of saccharide (4.4–7.4%) in comparison with non-volcanic ash soils (10.5%), though the former soils contain a greater amount of saccharides.

The monosaccharide composition of Ando soils was unrelated to the vegetation, land usage, or climatic conditions, and is presumed to be a soil characteristic resulting from the preferential accumulation of microbial polysaccharides.

Introduction

IT HAS been suggested that the composition of polysaccharides might be similar in different soils and that it is only the amount of polysaccharides which differs (Oades 1972). Japanese non-volcanic paddy soils (Murayama, 1977*b*) have been shown to have a similar monosaccharide composition to that of some grass and arable cropped soils of Great Britain (Whitehead *et al.*, 1975), with glucose the most and xylose the second most abundant component. However, mannose and galactose were more abundant than xylose in thirteen different American profiles of prairie and forest soils (Folsom *et al.*, 1974), in various profiles in Canadian soils (Gupta *et al.*, 1963), in an Ando soil from Australia (Oades, 1972), and in a volcanic ash soil from Japan (Murayama, 1977*a*).

The present study reports some distinctive features of the monosaccharide composition of several Ando soils compared with those of non-volcanic ash soils. Plant debris collected from Ando soils, and rice and barley straws were also analyzed for their monosaccharide composition for comparison with soil.

Materials and Methods

Samples used are listed in Table 1. Choyo, Kodonbaru, Utsunomiya and Kuriyagawa soils are Ando soils developed in deposits of volcanic ash. The sampling sites are distributed from the southern to the northern part of Japan, which volcanic ash soils widely cover, and are different from each other in climate and topography. The surface layer of Choyo and Kodonbaru virgin soils was excluded as sod structure was well developed in these profiles. The volcanic ash parent material of Kuriyagawa is basaltic

TABLE 1
Sample descriptions

Sample name	Locality	Land use, Vegetation	Horizon, sampling depth (cm)	Size of sample analyzed (mm)	pH	Total C (%) oven-dry	Total N (%) material	C/N	References
(A) Ando soils developed from volcanic ashes
Choyo	Choyomura, Kumamoto	Virgin (mountainside), Bamboo grass (Sasa)	A (15–30)	0.5	5.5	8.3	0.51	16	Aomine and Yoshinaga (1955)
Kodonbaru (a)	Uemura, Kumamoto	Virgin (basin flat), Bamboo grass (Sasa) Miscanthus, Imperata cylindrica	A (5–20)	2	4.5	20.8	0.66	31	Aomine and Yoshinaga (1955)
Kodonbaru (b)	Uemura, Kumamoto	Paddy field, Rice	Ap (0–10)	2	5.7	13.1	0.52	25	
Kuriyagawa (a)	Morioka, Iwate	Virgin (mountainside), Deciduous trees, Miscanthus	A (2–17)	0.5	5.6	14.4	0.81	18	Yamada (1978)
Kuriyagawa (b)	Morioka, Iwate	Upland field, Upland crops	Ap (0–15)	0.5	5.7	11.8	0.72	16	
Utsunomiya	Utsunomiya, Tochigi	Paddy field, Rice	Ap (0–11)	2	6.2	9.0	0.49	18	Yamada and Shoji (1975)
(B) Non-volcanic ash soils
Hiratsuka (a)	Hiratsuka, Kanagawa	Paddy field, Rice	Ap (0–14)	2	6.5	1.2	0.11	11	
Hiratsuka (b)	Hiratsuka, Kanagawa	Upland field, Upland crops	Ap (0–15)	2	5.0	1.4	0.12	12	
Nagano	Nagano, Nagano	Paddy field, Rice	Ap (0–15)	2	5.6	1.9	0.19	10	
(C) Plant debris collected from the soil samples
Choyo	Roots of Bamboo grass					49.6	0.49	101	
Kodonbaru	Roots of Bamboo grass, Miscanthus and Imperata cylindrica					41.7	0.44	95	
Kuriyagawa	Roots of Deciduous trees and Miscanthus					37.9	0.64	59	
(D) Straw
Rice tops (matured)						38.7	0.56	69	
Barley tops (matured)						47.2	0.36	131	

(Yamada, 1978), and that of the other three soils is andesitic (Wada and Tokashiki, 1972; Shoji *et al.*, 1974). The predominant clay mineral of these Ando soils is allophane, and all the soils have a black A horizon, both value and chroma being less than 2 in Munsell notation. Organic matter content and carbon-nitrogen ratio (C/N) are much higher than those of non-volcanic ash soils.

Hiratsuka and Nagano soils are non-volcanic ash soils developed in alluvial deposits. Hiratsuka is a grey lowland soil, and Nagano, a greyish brown soil (Matsuzaka, 1969). The predominant clay minerals of Hiratsuka and Nagano soils are kaolin minerals and montmorillonite, respectively (Watanabe, 1979).

Plant debris was collected from Choyo, Kodonbaru and Kuriyagawa virgin soils when the soil samples were screened through a 2 mm sieve. It was washed on a 100 mesh sieve, air-dried, and finely ground. The debris was mainly plant or tree roots with a small proportion of partially decomposed plant fragments. The straw samples (including leaves) were whole tops of mature rice and barley, and were chopped and finely ground for analysis.

Chemical analysis. Total carbon and nitrogen were determined using a carbon-nitrogen analyzer of Yanagimoto Co. Ltd. (MT 500). The pH of the air-dried soil was measured in a 1:2.5 suspension in distilled water.

Hydrolysis. For the release of monosaccharides all samples were hydrolyzed with sulphuric acid, according to the procedure of Oades *et al.* (1970). The air-dried samples were refluxed with 2.5 M H_2SO_4 for 20 min. The primary hydrolysate was filtered through a glass fibre filter supported on a sintered glass filter. The dried residue was soaked for 16 h at room temperature in 13 M H_2SO_4 and subsequently refluxed for 5 h after diluting to 0.5 M H_2SO_4 with water (the secondary hydrolysis). The hydrolysates were neutralized with finely ground $Ba(OH)_2$ powder and $Ba(OH)_2$ solution. Neutralization of some hydrolysates was made with sodium hydroxide solution.

Determination of monosaccharides by gas–liquid chromatography. Amounts of individual monosaccharides in the hydrolysates were determined by gas–liquid chromatography of the derived alditol acetates, using *myo*-inositol as an internal standard, according to the procedure of Cheshire *et al.* (1973). Gas–liquid chromatographic separation was made with a Hitachi 163 gas chromatograph with a flame ionizing detector, using a glass column 3 m long by 3 mm internal diameter, packed with Gas Chrome Q (100–120 mesh) coated with 3 per cent (w/s) ECNSS–M.

Results

Amounts of monosaccharides in acid hydrolysates of soils and plant materials

Galactose, glucose, mannose, arabinose, ribose, xylose, fucose and rhamnose were identified and determined in the primary hydrolysates of both Ando soils and non-volcanic ash soils (Table 2). The secondary hydrolysate contained predominantly glucose and small amounts of galactose, mannose, arabinose and xylose. Ribose, fucose and rhamnose were present in only trace amounts in the secondary hydrolysates. Greater proportions (76–84%) of the glucose of Ando soils were released by the primary hydrolysis, than with non-volcanic ash soils (56–67%). Most of the

TABLE 2
Amounts of monosaccharides released from soils

Sample	Hydro-lysis*	Galac-tose	Glucose	Mannose	Arabi-nose	Ribose	Xylose	Fucose	Rham-nose	Total	Sugar-C / Total-C (%)	H/P,** DH/P & Xyl/Man
				mg/100 g oven-dry soil and molar per cent								
Choyo (Virgin)	(a)	338	797	564	141	17	184	182	114	2337		5.0
	(b)	5	181	21	18	tr	8	tr	tr	233		0.7
	Sum	343	978	585	159	17	192	182	114	2570	12.5	0.4
	%	13	37	22	7.1	0.8	8.6	7.5	4.7			
Kodonbaru (Virgin)	(a)	333	1264	817	196	17	309	128	113	3177		4.2
	(b)	7	407	25	29	tr	11	tr	tr	479		0.4
	Sum	340	1671	842	225	17	320	128	113	3656	7.1	0.5
	%	9.0	44	22	7.1	0.5	10	3.7	3.3			
Kodonbaru (Paddy field)	(a)	252	810	543	139	16	189	95	84	2128		4.1
	(b)	4	190	12	17	tr	4	tr	tr	227		0.5
	Sum	256	1000	555	156	16	193	95	84	2355	7.2	0.4
	%	10	41	23	7.7	0.8	9.5	4.3	3.9			
Kuriyagawa (Virgin)	(a)	190	511	317	120	16	113	57	82	1406		3.6
	(b)	4	128	17	14	tr	7	tr	tr	170		0.5
	Sum	194	639	334	134	16	120	57	82	1576	4.4	0.4
	%	12	39	20	10	1.2	8.8	3.8	5.5			

Site		1	2	3	4	5	6	7	8	Sum		
Kuriyagawa (Upland field)	(a)	227	562	406	137	18	151	95	97	1693		3.3
	(b)	3	109	15	17	tr	9	tr	tr	153		0.5
	Sum	230	671	421	154	18	160	95	97	1846	6.3	0.5
	%	12	35	22	10	1.1	10	5.4	5.5			
Utsunomiya (Paddy field)	(a)	168	396	293	105	11	112	61	69	1215		3.2
	(b)	2	76	10	12	tr	3	tr	tr	103		0.5
	Sum	170	472	303	117	11	115	61	69	1318	5.9	0.5
	%	12	34	22	10	0.9	10	4.9	5.5			
Hiratsuka (Paddy field)	(a)	42	67	43	39	1	49	10	21	272		1.5
	(b)	tr	35	3	13	tr	1	tr	tr	52		0.3
	Sum	42	102	46	52	1	50	10	21	324	10.6	1.3
	%	12	29	13	18	0.4	17	3.2	6.7			
Hiratsuka (Upland field)	(a)	46	73	48	45	1	45	8	20	286		1.8
	(b)	tr	55	3	11	tr	1	tr	tr	70		0.2
	Sum	46	128	51	56	1	46	8	20	356	10.5	1.1
	%	12	34	13	18	0.4	15	2.4	5.8			
Nagano (Paddy field)	(a)	55	133	64	47	2	72	14	32	419		2.1
	(b)	tr	67	5	7	tr	3	tr	tr	82		0.3
	Sum	55	200	69	54	2	75	14	32	501	10.6	1.3
	%	10	38	13	12	0.4	17	2.9	6.6			

*(a), Saccharides released by 20 min hydrolysis with 2.5 M H_2SO_4; (b), Saccharides released by 13 M/0.5 M H_2SO_4 treatment on the residue of (a).

**H/P, DH/P and Xyl/Man are the molar ratios of hexoses to pentoses, deoxyhexoses to pentoses and xylose to mannose, respectively.

TABLE 3
Amounts of monosaccharides released from plant debris and straw

Sample	Hydrolysis	Galactose	Glucose	Mannose	Arabinose	Ribose	Xylose	Fucose	Rhamnose	Total	Sugar-C / Total-C (%)	H/P, DH/P & Xyl/Man
		mg/10 g oven-dry material and molar per cent										
Plant debris												
Choyo	(a)	78	222	17	135	tr	798	9	4	1263		2.4
	(b)	0	2521	0	22	0	20	0	0	2563		0.01
	Sum	78	2743	17	157	—	818	9	4	3826	30.8	59
	%	1.9	68	0.4	4.7		24	0.2	0.1			
Kodonbaru	(a)	68	356	52	149	tr	1333	6	10	1974		1.2
	(b)	0	1705	7	15	0	11	0	0	1738		0.01
	Sum	68	2061	59	164	—	1344	6	10	3712	35.6	27
	%	1.7	51	1.5	4.9		40	0.1	0.2			
Kuriyagawa	(a)	143	297	78	201	tr	630	14	19	1382		1.5
	(b)	0	948	11	12	0	6	0	0	977		0.04
	Sum	143	1245	89	213	—	636	14	19	2359	24.9	8.6
	%	5.7	49	3.5	10		30	0.6	0.8			
Straw												
Rice tops	(a)	87	958	24	283	2	1333	3	9	2699		1.8
	(b)	0	2542	6	23	0	49	0	0	2620		0.01
	Sum	87	3500	30	306	2	1382	3	9	5319	55.0	55
	%	1.5	62	0.5	6.5	0.04	29	0.1	0.2			
Barley tops	(a)	77	441	13	339	1	1782	3	12	2668		1.5
	(b)	0	3428	8	22	0	77	0	0	3535		0.01
	Sum	77	3869	21	361	1	1859	3	12	6203	52.6	103
	%	1.2	58	0.3	6.5	0.02	34	0.05	0.2			

galactose, mannose, arabinose and xylose was released in the first hydrolysate with both types of soil.

Ando soils contained 1320–3660 mg saccharides per 100 g, accounting for 4.4–7.2 per cent of the organic matter. Saccharides of Choyo Ando soil accounted for 12.5 per cent, a much higher value than other Ando soils. Non-volcanic ash soils contained much lower quantities of saccharides than Ando soils, 320–500 mg per 100 g, accounting for 10.5 per cent of soil organic matter.

Seven saccharide components excluding ribose were determined in the acid hydrolysates of plant debris collected from Ando soils (Table 3). In rice and barley straw ribose was present, but in very small amount. Most (73–92%) of the glucose was released by the secondary hydrolysis of plant debris and straw.

Plant debris contained 236–383 mg saccharides per 1 g, accounting for 25–36 per cent of the organic matter. For the straw these values were 532–620 mg and 53–55 per cent respectively.

Monosaccharide composition of soils, plant debris and straws

Monosaccharide composition was calculated on a molar basis (Table 2 and 3). Glucose was the predominant component in both types of soil, accounting for between 34 to 44 per cent of the total saccharides of Ando soils, and between 29 to 38 per cent of those of non-volcanic ash soils. Mannose was the second most common sugar in Ando soils. The third was galactose (except for Kodonbaru virgin soil in which xylose was the third, present in slightly greater amount than galactose), closely followed by xylose and arabinose. Except for Kuriyagawa virgin soil, the amount of fucose was almost equal to or greater than rhamnose in all Ando soils. Of non-volcanic ash soils, the second most common component was not mannose, but arabinose and xylose, followed by mannose and galactose. The amount of rhamnose was greater than that of fucose.

The monosaccharide composition of plant materials was quite different from those of soils. In plant debris collected from Ando soils glucose accounted for more than half the total saccharides. Xylose was the next most common at 24–40 per cent of the total. Third was arabinose at less than 10 per cent of the total. Mannose and galactose were very minor components, fucose and rhamnose were negligible, at less than one per cent.

In the rice and barley straw hydrolysates, glucose accounted for 58–62 per cent of the total saccharides, and xylose for 29–34 per cent. Together with arabinose, these three components accounted for more than 97 per cent of the total.

Molar ratios of hexoses to pentoses were 3.2–5.0 for Ando soils, which were greater than those for non-volcanic ash soils, 1.5–2.1. For plant materials this ratio was 1.2–2.4. Molar ratios of deoxyhexoses to pentoses for Ando soils were 0.4–0.7, which were greater than those for non-volcanic ash soils, 0.2–0.3. The ratios for plant materials were very small, ranging from 0.01 to 0.04.

Molar ratios of xylose to mannose for Ando soils were 0.4–0.5, which were much smaller than those for non-volcanic ash soils, 1.1–1.3. Plant debris collected from soils and straws had much greater ratios than soils;

Choyo 59, Kodonbaru 27, Kuriyagawa 8.6, rice straw 55, and barley straw 103.

Discussion

Ando soils originating from volcanic ash are very common in Japan and are also known to be distributed around the world (Thorp and Smith, 1949; Aomine and Yoshinaga, 1955; FAO/UNESCO, 1964). The soil is quite different from other types of soil in clay minerals, organic matter, and in other chemical and physical properties (Wada and Harward, 1974; Shoji and Ono, 1978).

This study has shown that the monosaccharide composition of Ando soils is quite different from that of non-volcanic ash soils, being richer in mannose, fucose and ribose, whereas they contained less arabinose, xylose and rhamnose. Much less glucose in cellulose-like form, which was released after soaking with a strong acid, was another feature of the monosaccharide composition of this soil. The monosaccharide composition of Ando soils appeared to be independent of differences in climate, land usage, or rock type of volcanic ash.

The organic matter in Ando soils in Japan is thought to be derived from the grass vegetation, *Miscanthus*, cogongrass (*Imperata cylindrica*) or bamboo grass (Takehara, 1964), and this is the present vegetation of the virgin sites used here. The monosaccharide composition of plant debris collected from the virgin soil samples was quite different from those of the Ando soils. Consequently the feature of the monosaccharide composition of Ando soils could be related neither to the saccharide constituents of the present vegetation, the previous one, nor to those of any other.

Based on the conclusions of Cheshire *et al.* (1969, 1971, 1973, 1976, 1977, 1978), that the hexoses mannose and galactose and the deoxyhexoses fucose and rhamnose in soil are mainly of microbial origin, and the pentoses arabinose and xylose are mainly of plant origin, the saccharides of Ando soils appear to have more of a microbial origin than those of non-volcanic ash soils.

A comparative study of Broadbent *et al.* (1964) indicated that the rates of decomposition of added clover and Sudan grass were similar in volcanic and non-volcanic ash soils. It is therefore speculated that the difference in saccharides between Ando soils and non-volcanic ash soils is not due to different rates of decomposition of plant materials added to soil, but to the accumulation of microbial polysaccharides.

Possibly, soil polysaccharides are protected to some extent in their native environment. Among the reasons offered for this by Cheshire (1977) are: (i) adsorption by clays, (ii) a preservative effect from tanning by humic substances, (iii) the formation of complexes with metals. Amorphous clay materials which are dominant in the clay fractions of most Ando soils have a greater specific surface than crystalline clay minerals in non-volcanic ash soils (Kubota, 1976). As a whole, volcanic ash soils *per se* also have a greater specific surface than non-volcanic ash soils (Tada *et al.*, 1963; Kubota, 1976). This greater specific surface would be very favourable for the accumulation of polysaccharides in Ando soils, as adsorption of polysaccharides seems to be dependent primary on physical adsorption forces (Greenland, 1965).

Ando soils contain humic substances in much greater amount than non-volcanic ash soils. Higher contents of carboxylic and phenolic hydroxyl groups in humic acid of Ando soils than those of non-volcanic ash soils (Kosaka, 1963; Tsutsuki and Kuwatsuka, 1978) may favour the tanning reaction in the former soils, resulting in a greater stability of saccharides.

The possibility that polysaccharides containing mannose are stabilized by the ftrmation of complexes with metals was noted by Martin (1971). The protective effect of metal complexing on the stability of polysaccharides would be expected to be greater in Ando soils than in non-volcanic ash soils, as the dominant clay materials allophane and allophane-like of the former soils would be a source of aluminium, iron, and sesquioxidic constituents (Wada and Harward, 1974). In fact, a much greater part of the saccharides was released by the primary hydrolysis implying that the polysaccharides in Ando soils exist as organo-mineral complexes.

The meaning of the ratio of hexoses to pentoses has not been well established, but a soil or a soil fraction poor in plant debris, such as the subsoil or the 'heavy' fraction, has a relatively high value (Folsom *et al.*, 1974; Whitehead *et al.*, 1975; Murayama, 1977*b*; Molloy *et al.*, 1977). The ratios of Ando soils were consistently higher than those of non-volcanic ash soils.

The molar ratio of xylose to mannose is another representative value of the monosaccharide composition of soil polysaccharides (Murayama, 1977*b*). The ratios for the Ando soils were much smaller than those for non-volcanic ash soils, reflecting less undecomposed plant saccharides in the former.

Acknowledgements

The author wishes to thank Drs. A. Inoko and Y. Harada, The National Institute of Agricultural Sciences, and Professor Dr. S. Shoji, Tohoku University, for their helpful comments and criticism of the manuscript; and Dr. M. V. Cheshire, The Macaulay Institute for Soil Research, for his kind and critical reading of the manuscript.

REFERENCES

AOMINE, S. and YOSHINAGA, N. 1955. Clay minerals of some well-drained volcanic ash soils in Japan. Soil Science **79**, 349–358.

BROADBENT, F. E., JACKMAN, R. H., and McNICOLL, J. 1964. Mineralization of carbon and nitrogen in some New Zealand allophanic soils. Soil Science **98**, 118–128.

CHESHIRE, M. V., 1977. Origins and stability of soil polysaccharide. Journal of Soil Science **28**, 1–10.

CHESHIRE, M. V., GREAVES, M. P., and MUNDIE, C. M. 1976. The effect of temperature on the microbial transformation of ^{14}C glucose during incubation in soil. Journal of Soil Science **27**, 75–88.

CHESHIRE, M. V., MUNDIE, C. M., and SHEPHERD, H. 1969. Transformation of ^{14}C glucose and starch in soil. Soil Biology and Biochemistry **1**, 117–130.

CHESHIRE, M. V., MUNDIE, C. M., and SHEPHERD, H. 1971. The origin of the pentose fraction of soil polysaccharide. Journal of Soil Science **22**, 222–236.

CHESHIRE, M. V., MUNDIE, C. M., and SHEPHERD, H. 1973. The origin of soil polysaccharide: Transformation of sugars during the decomposition in soil of plant material labelled with ^{14}C. Journal of Soil Science **24**, 54–68.

CHESHIRE, M. V., SPARLING, G. P., MUNDIE, C. M., SHEPHERED, H., and MURAYAMA, S. 1978. Effect of temperature and soil drying on the transformation of (^{14}C) glucose in soil. Journal of Soil Science **29**, 360–366.

FAO/UNESCO 1964. *Meeting on the classification and correlation of soils from volcanic ash.* World Soil Resources Report 14, ed. by FAO/UNESCO.

FOLSOM, B. L., WAGNER, G. H., and SCRIVNER, C. L. 1974. Comparison of soil carbohydrate in several prairie and forest soils by gas–liquid chromatography. Soil Science Society of America Proceedings **38**, 305–309.

GREENLAND, D. J. 1965. Interaction between clays and organic compound in soils. Part II. Adsorption of soil organic compounds and its effect on soil properties. Soils and Fertilizers **28**, 521–532.

GUPTA, U. C., SOWDEN, F. J., and STOBBE, P. C. 1963. The characterization of carbohydrate constituents from different soil profiles. Soil Science Society of America Proceedings **27**, 380–382.

KOSAKA, J. 1963. Study on the process of humification in upland soils. Bulletin of the National Institute of Agricultural Sciences. Series B. **13**, 253–352.

KUBOTA, T. 1976. Surface chemical properties of volcanic ash soil – especially on phenomenon and mechanism of irreversible aggregation of the soil by drying. Bulletin of the National Institute of Agricultural Sciences, Series B. **28**, 25–26.

MARTIN, J. P. 1971. Decomposition and binding action of polysaccharides in soil. Soil Biology and Biochemistry **3**, 33–41.

MATSUZAKA, Y. 1969. Study on the classification of paddy soils in Japan. Bulletin of the National Institute of Agricultural Sciences, Series B. **20**, 155–349.

MOLLOY, L. F., BARBARA A. BRIDGER, and ANNETTE CAIRNS 1977. Studies on a climosequence of soils in tussock grasslands. 13. Structural carbohydrates in tussock leaves, roots, and litter and in the soil light and heavy fractions. New Zealand Journal of Science **20**, 443–451.

MURAYAMA, S. 1977a. An automated anion-exchange chromatography for the estimation of saccharides in acid hydrolysates of soil. Soil Science and Plant Nutrition **23**, 247–252.

MURAYAMA, S. 1977b. Saccharides in some Japanese paddy soils. Soil Science and Plant Nutrition **23**, 479–489.

OADES, J. M., 1972. Studies on soil polysaccharides. III. Composition of polysaccharides in some Australian soils. Australian Journal of Soil Research **10**, 113–126.

OADES, J. M., KIRKMAN, M. A., and WAGNER, G. H. 1970. The use of gas–liquid chromatography for determination of sugars extracted from soils by sulphuric acid. Soil Science Society of America Proceeding **34**, 230–235.

SHOJI, S., and ONO, T. 1978. Physical and chemical properties and clay mineralogy of Andosols from Kitakami, Japan, Soil Science **126**, 297–312.

SHOJI, S., YAMADA, I., and MASUI, J. 1974. Soils formed from the andesitic and basaltic volcanic ashes. I. The nature of the parent ashes and soil formation. Tohoku Journal of Agricultural Research **25**, 104–112.

TADA, A., TAKENAKA, H., SOMA, K., KUROBE, T., and HAYAMA, Y. 1963. On the characteristics of the particles of Kanto loam volcanic ash soil. Transactions of the Agricultural Engineering Society, Japan **7**, 14–21.

TAKEHARA, H. 1964. In *Volcanic ash soils in Japan.* Chapter VIII. Classification. 135–154. Ed. by Ministry of Agriculture and Forestry, Japanese Government.

THORP, J., and SMITH, G. D. 1949. Higher categories of soil classification: Order, suborder, and great soil groups. Soil Science **67**, 117–126.

TSUTSUKI, K., AND KUWATSUKA, S. 1978. Chemical studies on soil humic acids. II. Composition of oxygen-containing functional groups of humic acids. Soil Science and Plant Nutrition **24**, 547–560.

WADA, K., and HARWARD, M. E. 1974. Amorphous clay constituents of soils. Advances in Agronomy **26**, 211–260.

WADA, K., and TOKASHIKI, Y. 1972. Selective dissolution and difference infrared spectroscopy in quantitative mineralogical analysis of volcanic ash soil clays. Geoderma **7**, 199–213.

WATANABE, Y., 1979. Unpublished.

WHITEHEAD, D. C., HAZEL BUCHAN, and HARTELY, R. D. 1975. Components of soil organic matter under grass and arable cropping. Soil Biology and Biochemistry **7**, 65–71.

YAMADA, I. 1978. Studies on properties and fertilities of Ando soils of Tohoku district. PhD thesis, Faculty of Agriculture, Tohoku University.

YAMADA, I., and SHOJI, S. 1975. Soils formed from the andesitic and basaltic volcanic ashes. II. Soil properties and soil fertility problems. Tohoku Journal of Agricultural Research **26**, 102–116.

24

Copyright © 1979 by Blackwell Scientific Publications
Reprinted from *Jour. Ecology* **67**:305-319 (1979)

STUDY OF THE MICROFUNGAL COMMUNITY OF AN ANDOSOL

A. T. MARTÍNEZ AND C. RAMÍREZ

Instituto 'Jaime Ferrán' de Microbiología, Consejo Superior de Investigaciones Científicas, Joaquín Costa 32, Madrid-6, Spain

SUMMARY

(1) Variations in the composition of the microfungal community of an andosol under a beech forest in northern Spain, both during the course of a year and between the different soil horizons, were studied. Microfungi were recovered by soil dilution, by soil washing and from the surface of leaves.

(2) The species-diversity decreased with increasing depth, and was associated with increasing dominance of *Aureobasidium pullulans* (de Bary) Arnold in the lower soil horizons. *Mortierella minutissima* van Tieghem was characteristically present in spring isolates, while *Trichoderma pseudokoningii* Rifai was characteristic of summer isolates.

(3) The similarities between microfungal recoveries from the various soil horizons and from different seasons of the year were compared by means of three-dimensional ordinations and dendrograms.

(4) Similarity between the different horizons at one particular season was greater than within the same horizon at different seasons, and the floras of subhorizons A_{11} and A_{12} were always similar. Winter recoveries differed widely from those of the rest of the year.

(5) The microfungal community colonizing the surfaces of fallen leaves in autumn and winter was quite different from the soil community.

INTRODUCTION

This paper reports the results of a survey of the microfungi of an andosol under a beech (*Fagus sylvatica* L.) forest, in terms of the component species and their behaviour throughout the course of the year and within the different horizons of the soil profile. Previous studies carried out on the same soil which complement the present one include a general microbiological analysis (Moriyón, Martinez & Rodriguez-Burgos 1978) and an investigation of the variations of fungal biomass and numbers of spores and propagules (Martínez & Ramírez 1978a). A number of studies of the microfungi of beech forest soils, including the superficial leaf litter, either in the Northern Hemisphere under *Fagus* forests or in the Southern Hemisphere under *Nothofagus* forests, have also been published. Krzemieniewska & Badura (1954a, b), Peyronel (1961), Jensen (1962), Badura (1963a, b), Caldwell (1963), Carré (1964), Hogg (1966), Hogg & Hudson (1966), Malan, Ambrosoli & Alessandria (1969) and Lindgreen & Jensen (1973) investigated the microfungal flora under *Fagus sylvatica* forests in Europe, while surveys of the microfungal communities under *Fagus crenata* Bl. were carried out in Japan by Saito (1956, 1958, 1960, 1965 and 1966), and under *Nothofagus truncata* (Col.) Ckn. in New Zealand by Dutch & Stout (1968) and Ruscoe (1971a, b). Among the few authors who have investigated andosols microbiologically, Peña-Cabriales & Valdés (1974) are the only ones to have reported the isolation of a few species of microfungi.

MATERIAL AND METHODS

The study area

The main characteristics of the climate, vegetation and soil have been described in a previous work (Martínez & Ramírez 1978a). The study area is located near the Oroquieta Pass, province of Navarra (northern Spain), at 800 m above sea level (latitude 43°02'N, longitude 1°45'W). The vegetation comprises a beech forest dominated by *Fagus sylvatica*. The climate is mild temperate with high rainfall, resulting in few periods of soil-water deficit during the year. The soil under study may be classified as a typical Andaquepts (Soil Conservation Service 1975), with two horizons, A_{00} and A_1, differentiated within it; at least two subhorizons may be recognized in the A_1. Some of the main physicochemical features of these A_{11} and A_{12} subhorizons, as given by Iñiguez & Barragán (1974), are as follows: pH_{H_2O} 3·85 and 4·50 respectively, organic matter (as carbon) 8·26 and 5·61% dry wt, C/N 20·7 and 23·2, and cation-exchange-capacity 33·7 and 30·9 m-equiv. per 100 g dry wt.

Methods of study

Preparation of samples

Sampling was carried out aseptically over an area of 100 × 100 m with apparently homogeneous soil and vegetation characteristics. Four sites (2 × 2 m) were chosen at random, and 50–100-g samples were collected in each season of the year (21 January, 24 April, 17 August and 2 November 1974) from horizons A_{00}, A_{11} and A_{12}, and processed for analysis within 24 h. Several different isolation techniques were used in order to obtain as much information as possible on the composition of the microfungal community:—

(1) By plating soil dilutions. Decimal dilutions of soil samples (10 g) in a sterile, aqueous, 0·01% Tween 80% solution were inoculated onto an agar medium of the following composition: 2% glucose, 0·5% yeast extract (Difco), 1·5% agar (pH 6·6); immediately before use an aqueous solution of oxytetracycline hydrochloride (Pfizer), was added to the molten agar medium maintained at 45 °C, to give a concentration of 0·01%. In each season of the year four replicates (one from each site) from each soil horizon were analysed separately.

(2) By plating washed soil particles. The apparatus described by Bisset & Widden (1972) was employed for this method. In each season 2·5 g from each site were mixed and washed jointly. The washing procedure was repeated sixty times, and the soil particles were dried at 37 °C before plating upon the previously-described agar medium.

(3) By plating spores from the leaf surfaces, using the Last (1955) technique. In each season ten beech leaves from the A_{00} horizon at each site were suspended from the inside of the lid of Petri-dishes over the agar medium described. After 48 h they were removed, and the dishes were incubated.

All plates were incubated at 27 °C.

Analysis of results

Brillouin's index (1962) was used to assess diversity:

$$\text{diversity} = (1/N)\log_2 \frac{N!}{n_a!n_b!\ldots n_s!}$$

where $n_a, n_b, \ldots n_s$ are the numbers of individuals of each species, and N is the total number of individuals.

Czechanovski's index (1913) was used to measure the similarities between the community composition for the different soil horizons and for different seasons of the year:

$$\text{similarity between A and B} = 100\frac{2c}{a+b}$$

where a is the number of species in A, b the number of species in B, and c the number of species common to A and B.

Three-dimensional models of similarity values were then prepared following a modification of the Bray & Curtis (1957) ordination method described by Christensen (1969) in her study of microfungi in Wisconsin soils. To check the validity of the models, the linear correlation coefficients between the dissimilarity values ($= 100 - $ similarity) and the distances in the three-dimensional ordinations were calculated. A model was considered valid when a significant correlation was shown. Dendrograms comparing similarities were also prepared by the centroid grouping method of Williams, Lambert & Lance (1966).

RESULTS

Table 1 shows the relative abundance (percentage of total colonies) of species isolated from soil dilutions of samples taken from each horizon throughout the year. Species encountered on leaf surfaces are also listed, but without abundance values. The total numbers of propagules (per 1 mg of soil dry weight) as determined by the different dilution isolates, and the total number of colonies counted on each occasion for the estimation of the percentage (the total number of colonies examined is very high) are also given. Table 2 shows the relative abundance of species isolated from washed soil particles (percentage of particles colonized for each species), the total number of colonies counted, and the percentage of sterile particles. Species whose relative abundance is not greater than 5% in any one sample are listed in the Appendix.

Table 3 shows the diversity-index values, calculated from the abundances of the different species.

Three-dimensional ordination diagrams and dendrograms were prepared to show the similarities between the different horizons throughout the year. Figure 1 shows a three-dimensional ordination for the soil-dilution and leaf-surface isolates. The linear correlation coefficient between dissimilarities and distances within the tridimensional diagram (120 pairs of values) is 0·71, which is significant at the 0·1% level. Figure 2 shows the three-dimensional diagram for soil-washing isolates. In this case the linear correlation coefficient value for the 66 pairs of values is 0·52, which is also significant at the 0·1% level. Figures 3 and 4 show the corresponding dendrograms for soil-dilution and leaf-surface isolates and for soil-washing isolates. Similarity coefficients between each pair of fusing groups, and the characteristic species of each fusion (i.e. species absent except in the fusing groups) in order of their importance, are indicated.

TABLE 1. Relative abundance (percentage of total colonies) of species isolated by the soil-dilution method at different seasons of the year, from three soil horizons and from leaf surfaces (L); abundance data are not available for the leaf-surface isolates

Species	Winter				Spring				Summer				Autumn			
	L	A_{00}	A_{11}	A_{12}	L	A_{00}	A_{11}	A_{12}	L	A_{00}	A_{11}	A_{12}	L	A_{00}	A_{11}	A_{12}
Hyaline sterile mycelia	+	17	17	17	+	4	26	12	+	2	9	6	.	8	25	19
Penicillium nigricans (Bainier) Thom	.	2	1	0*	.	1	1	1	+	1	8	6	.	1	9	2
Aureobasidium Viala et Boyer sp.	+	1	.	15	.	0	4	39	.	1	9	31	.	0	2	13
Trichoderma hamatum (Bonorden) Bainier	+	.	.	.	.	.	1	0	+	35	12	2	+	.	2	7
Acremonium Link ex Fresenius sp.	++	2	3	.	.	26	.	.	.	5	2	.	+	5	18	10
Trichoderma harzianum Rifai	++	3	6	1	.	1	1	.	.	.	.	0	+	1	1	.
T. polysporum (Link ex Persoon) Rifai	+	.	.	0	.	.	2	3	.	1	2	1	.	8	1	4
Dark sterile mycelia	+	5	.	.	.	5	.	1	+	.	.	.	.	11	2	2
Cladosporium Link ex Fresenius sp.	.	8	20	.	.	.	.	.	.	2	.	12	.	1	.	.
Phoma Saccardo sp.	+	1	.	.	.	9	.	.	++	8	.	1	.	7	0	.
Mortierella isabellina Oudemans et Koning	+	3	.	.	.	1	.	.	+	9	1	.	.	.	.	.
AMF 254[1]	.	0	.	.	.	2	11	3	+	0	.	.	.	.	.	.
Cylindrocarpon Wollenweber sp.	.	.	.	.	.	1	.	.	.	.	.	.	.	.	.	.
Aspergillus niger van Tieghem	.	.	3	6	.	.	1	.	++	.	2	1	+	1	1	0
Acremonium Link ex Fresenius[2]	.	.	.	5	.	1	1	.	.	11	2	.	.	2	2	0
Oidiodendron Robak sp.	.	.	.	.	.	.	2	.	+	9	2	12	.	.	.	.
Trichoderma koningii Oudemans	+	12	3	2	.	0	.	.	.	.	.	.	.	12	.	1
Rhizopus arrhizus Fischer	.	0	10	.	+	.	.	.	.	.	.	.	.	.	.	.
Aspergillus versicolor Tiraboschi ex Vuillemin	.	.	.	1	.	5	.	6	.	2	6	12	.	.	.	.
Penicillium notatum Westling	.	3	1	0	.	.	.	.	.	.	.	.	.	.	.	.

	1	2	3	4	5	6	7	8	9	10	11	12
Aspergillus flavus Link	+	.	.	.	.	.	.	.	.	.	.	.
AMF 72[3]	.	.	.	.	.	.	.	.	.	.	1	4
Sporothrix Hektoen et Perkins sp.	1	5	5	.	.	.	.	.	.	.	.	.
AMF 349[3]	.	4	6	.	.	.	.	.	.	.	.	.
Aspergillus fumigatus Fresenius	.	.	1	.	15	.	.	.	1	+	.	.
Penicillium spinulosum Thom	.	.	.	+	0	34	.	.	.	.	.	.
Phialophora Medlar[2]	.	.	.	+	3	.	.	.	.	+	.	.
Fusarium Link ex Fresenius sp.	.	.	.	.	8	.	−	.	.	.	.	.
Penicillium purpurescens (Sopp) Raper et Thom	+	6	.	.	.	.	.	.	.	.	.	26
	.	3	25	.	.	23	.	.	.	.	.	.
P. decumbens Thom	5	12	1	.	.	.	.	.	.	.	.	.
P. brevicompactum Dierckx	.	1	6	.	.	.	.	.	.	.	.	.
P. casei Staub	11	1	.	.	.	.	.	.	.	.	.	.
Codinea Maire sp.	6	.	.	.	.	.	.	.	.	.	.	.
Penicillium notatum Westling, deficient growth	.	.	.	.	8	6	+	.	.	.	.	.
Trichoderma pseudokoningii Rifai	.	.	.	.	.	.	+	6	1	.	7	4
Penicillium chrysogenum Thom	.	.	.	.	.	.	.	1	24	.	17	.
P. corylophilum Dierckx	.	.	.	.	.	.	.	1	.	20	.	.
Aspergillus parasiticus Speare	.	.	.	.	.	.	.	.	.	+	.	.
Mucor petrinsularis Naumov	.	.	.	.	.	.	.	.	.	10	.	.
Colonies counted	1116	654	1240	758	189	278	1058	268	560	647	367	246
Propagules per mg	111·6	65·4	12·4	758·0	189·0	27·8	105·8	26·8	5·6	64·7	36·7	2·5

[1] Unidentified Oomycetale; [2] Atypical form which could be an imperfect Ascomycete; [3] Unidentified Moniliale; * Percentages of c. 0·5 are indicated as 0.

TABLE 2. Relative abundance (percentage of total soil particles colonized) of species isolated by the soil-washing method at different seasons of the year from three soil horizons

	Winter			Spring			Summer			Autumn		
	A_{00}	A_{11}	A_{12}	A_{00}	A_{11}	A_{12}	A_{00}	A_{11}	A_{12}	A_{00}	A_{11}	A_{12}
Hyaline sterile mycelia	33	72	66	1	1	22	12	13	4	3	41	9
Aureobasidium Viala et Boyer sp.	.	1	13	.	.	2	8	23	56	.	20	63
Penicillium nigricans (Bainier) Thom	.	3	1	.	14	.	1	6	.	2	6	.
Trichoderma harzianum Rifai	7	.	1	4	.	.	.	1	1	2	.	.
Penicillium thomii Maire	8	2	.	4	.	.	7	1	.	19	.	.
Trichoderma koningii Oudemans	7	.	.	.	.	1	1	.	.	.	.	.
Mucor hiemalis Wehmer	18	19	16	6	.	1	.	.	.	13	2	13
Trichoderma hamatum (Bonorden) Bainier	.	.	.	1	10	14	22	1	1	3	8	3
Scytalidium Pesante sp.	.	.	.	.	.	12	1	3	.	5	2	.
Trichoderma polysporum (Link ex Persoon) Rifai	.	.	.	1	1	.	7	.	.	9	.	.
T. aureoviride Rifai	14	.	.	5	.	1	14	.	.	2	.	.
Dark sterile mycelia	.	.	.	.	.	.	.	.	.	.	2	6
Penicillium daleae Zaleski	1	.	3	.	.	.	.	.	.	5	2	.
Aspergillus niger van Tieghem	.	.	.	8	5	3	5	.	.	5	.	.
Gliocladium Corda sp.	7	.	.	.	.	.	4	3	.	2	.	.
Discosia Maire sp.	.	.	.	.	.	35	.	.	.	.	.	.
Mortierella minutissima van Tieghem	.	.	.	44	68	.	.	.	.	.	.	.
Mucor ambiguus Vuillemin	.	.	.	11	.	.	.	.	.	.	.	.
Mortierella nana Linnemann	.	.	.	.	.	.	5	17	12	.	.	.
Trichoderma pseudokoningii Rifai	.	.	.	.	.	.	5	17	1	.	.	.
Sterile Trichoderma Persoon	.	.	.	.	.	.	4	9	18	5	3	.
Aureobasidium Viala et Boyer 2	.	.	.	.	.	.	.	.	.	.	.	.
Total number of colonies	108	180	174	140	147	116	80	122	138	67	79	52
Sterile particles (percent of total)	0	0	0	0	0	3	0	0	0	0	0	54

TABLE 3. Values of Brillouin's diversity index for species isolated by soil-dilution and soil-washing techniques; (a) for the various soil horizons at different seasons, (b) for the horizons and seasons separately

(a)

	Winter			Spring			Summer			Autumn		
	A_{00}	A_{11}	A_{12}	A_{00}	A_{11}	A_{12}	A_{00}	A_{11}	A_{12}	A_{00}	A_{11}	A_{12}
Soil dilution	4·0	3·7	3·4	3·6	2·7	2·6	3·4	3·6	3·3	3·6	3·2	3·1
Soil washing	2·6	1·2	1·4	2·7	1·4	2·4	3·3	2·9	1·9	3·6	2·6	1·6

(b)

	A_{00}	A_{11}	A_{12}	Winter	Spring	Summer	Autumn
Soil dilution	4·7	4·2	3·8	4·3	4·0	3·9	4·0
Soil washing	4·4	3·3	2·9	2·3	2·8	3·2	3·9

DISCUSSION

For any given soil horizon the diversity values for the soil-dilution isolates are higher than those for the soil-washing isolates. The difference is probably due to the fact that with the second method spores were eliminated and only active forms remained. Consequently, it could be that the diversity values calculated for the soil-washing data are more representative of the real structure of the community. Diversity generally decreased with increasing soil depth. The more abundant species in the various horizons and at different seasons of the year (those with a relative abundance greater than 10%) are listed in Table 4. The reduction in diversity with increasing depth can be seen to be associated with increasing predominance of the most abundant species. This fact is

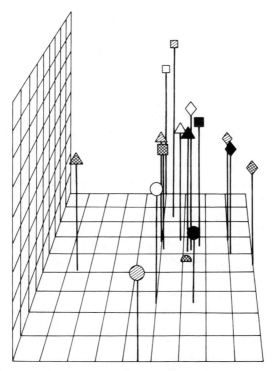

FIG. 1. Three-dimensional ordination of soil-dilution and leaf-surface isolates. ○ winter, □ spring, ◇ summer, △ autumn; ▩ leaf surfaces, ■ A_{00}, ▨ A_{11}, □ A_{12}

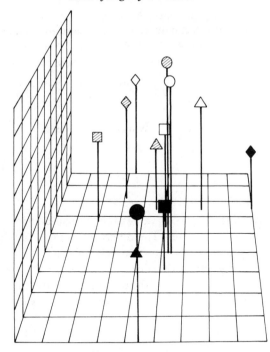

FIG. 2. Three-dimensional ordination of soil-washing isolates. For symbols see Fig. 1.

probably related to the existence of more pronounced limiting factors with increasing depth.

As regards the behaviour of individual species, twelve species were found to show a well-marked distribution with depth down the profile by both techniques. The relative abundance of *Aureobasidium* sp. and *Scytalidium* sp. increased with depth, while that of *Mucor hiemalis, Mortierella isabellina, Penicillium thomii, Penicillium frequentans, Trichoderma aureoviride, Trichoderma longibrachiatum, Oedocephalum* sp., *Aureobasidium* 2, *Discosia* sp. and dark sterile mycelia diminished with depth. *Aureobasidium* sp. (mostly typical *Aureobasidium pullulans*) was clearly the most abundant species at depth with both techniques. On the other hand, the most abundant species at the soil surface differed according to the isolation technique employed. With soil dilution, *Acremonium* sp. and *Phoma* sp. were the most abundant species, while *Mucor hiemalis* and *Penicillium thomii* were the most abundant when soil washing was used. *Aureobasidium pullulans* belongs to the primary saprophytic group that colonizes leaves when still attached to the tree, diminishing rapidly in importance when leaves fall to the ground (Kendrick & Burges 1962; Hayes 1965, 1967; Saito 1966; Hogg & Hudson 1966; Frankland 1966; Parkinson & Balasooriya 1969; Remacle 1970; Eicker 1973). However, some authors emphasize its importance not only in the litter but also in the deeper horizons (Jensen 1962; Reddy & Knowles 1965; Wicklow, Bollen & Denison 1974). In two instances the relative importance of *Aureobasidium* increased with increasing depth, but in none of these cases was the increase so well-marked as in the present study.

In our study four species were found showing similar behaviour all the year round, whatever the isolation technique employed: *Trichoderma harzianum*, with diminishing abundance from winter to autumn, *Trichoderma polysporum*, with opposite behaviour,

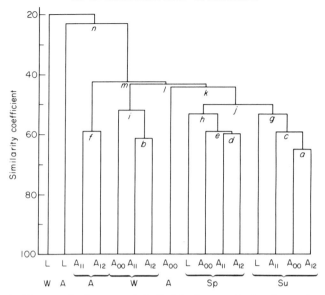

FIG. 3. Centroid-group dendrogram of similarities between leaf-surface and soil-dilution isolates. W = winter, Sp = spring, Su = summer, A = Autumn, L = leaf surfaces. Characteristic species at each fusion (those found within the fusing groups but not in the remaining ones) are as follows: b = *Penicillium brevicompactum*, *P. cyclopium*; c = *Aspergillus terricola*; e = *Penicillium notatum*, deficient growth; g = *P. granulatum*, *Trichoderma pseudokoningii*, *Penicillium chrysogenum*; h = *Mortierella minutissima*, *Gliocladium* sp., *Epicoccum* sp.; i = *Penicillium decumbens*, *Paecilomyces* sp., *Penicillium casei*; j = *Trichoderma viride*, AMF 349, *Discosia* sp.; k = *Mortierella ramanniana*, cf. *Phialophora*, *Mammaria* sp.; l = *Cladosporium* sp., AMF 254, *Geotrichum* sp.; m = *Penicillium nigricans*, *Aureobasidium* sp., cf. *Acremonium*; n = *Trichiderma polysporum*, *T. aureoviride*, *Aspergillus niger*.

and *Trichoderma hamatum* and *Aureobasidium* sp., with increasing abundance from winter to summer followed by a decrease from summer to autumn.

No species was found in any instance which was confined to a particular soil horizon, though certain species characteristic of a particular season (species isolated only in that season and along the whole profile) were encountered. *Mortierella minutissima* appeared as a characteristic spring form, while *Trichoderma pseudokoningii* was characteristic of the summer. During the winter, characteristic species appeared only when soil-dilution isolates were involved. On the other hand, the autumn samples were devoid of characteristic forms.

Similarity coefficients calculated from soil-dilution data were always higher than those obtained by soil-washing, probably because a greater number of colonies originated by spores in the former method; spores are more easily transported from one horizon to another than vegetative forms and may remain viable for longer periods of time.

On the basis of the three-dimensional ordinations and the dendrograms, certain conclusions may be reached on the influence of season and soil horizon on the microfungal community. In the three-dimensional model which represents similarities between soil-dilution isolates and leaf-surface isolates (Fig. 1) seasonal grouping can be seen. Leaf-surface isolates, with the exception of the spring sample, are well separated from the other isolates; the most remote of all is the one taken in autumn. Winter isolates are segregated from isolates from other seasons of the year. Summer isolates and those from

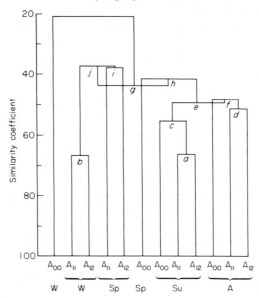

FIG. 4. Centroid-group dendrogram of similarities between soil-washing isolates. W = winter, Sp = spring, Su = summer, A = autumn. Characteristics species at each fusion are as follows: a = *Melanospora* sp.; c = *Mortierella nana*, *Trichoderma pseudokoningii*, sterile *Trichoderma*; e = *Botrytis* sp., *Graphium* sp., *Eladia* sp.; f = *Aureobasidium* 2, *Penicillium melinii*, h = *Trichoderma longibrachiatum*; i = AMF 254; j = *Cylindrocarpon* sp.

the two deeper horizons in spring are also grouped. Finally, the remaining isolates are gathered at the centre of the diagram. In the three-dimensional ordination for the soil-washing isolates (Fig. 2), clustering is poorly-developed and evident only for isolates from subhorizons A_{11} and A_{12} grouped in winter and summer. However, isolates from the four A_{00} horizons are well separated from the remainder.

TABLE 4. Species recorded with a relative abundance of more than 10% by soil-washing and soil-dilution techniques, (a) from the three different soil horizons, (b) from isolates at different seasons

	Soil washing		Soil dilution	
(a)				
A_{00}	Hyaline sterile mycelia *Mortierella minutissima*	23%	*Acremonium* sp. AMF 349	31%
A_{11}	Hyaline sterile mycelia *Mortierella minutissima*	59%	Hyaline sterile mycelia *Aspergillus fumigatus*	43%
A_{12}	*Aureobasidium* sp. Hyaline sterile mycelia *Mortierella minutissima*	68%	*Aureobasidium* sp. *Fusarium* sp. Hyaline sterile mycelia	58%
(b)				
Winter	Hyaline sterile mycelia *Trichoderma koningii*	70%	Hyaline sterile mycelia *Cladosporium* sp.	28%
Spring	*Mortierella minutissima*	49%	*Acremonium* sp. AMF 349	32%
Summer	*Aureobasidium* sp. *Mortierella nana* Sterile *Trichoderma*	53%	*Trichoderma hamatum*	29%
Autumn	*Aureobasidium* sp. Hyaline sterile mycelia	38%	*Penicillium corylophilum* Hyaline sterile mycelia	29%

Three-dimensional ordinations are useful for the information they convey on the totality of relationships between isolates. On the other hand, dendrograms are more distinct and easier to interpret, but by considering only the most notable similarities other less important ones which can be appreciated in the three-dimensional diagrams may be disregarded.

The dendrograms in Figs 3 and 4 show that the closest similarities between isolates always correspond to the various horizons at a particular season of the year, with similarities between A_{11} and A_{12} being the most marked. The various horizons of each season are grouped together most clearly when soil-dilution data are considered (Fig. 3). When the microfungal community of leaves is included in the dendrogram, it shows homogeneity with the rest of the soil community in spring and summer (i.e. the leaf-surface isolates are grouped with the corresponding soil isolates), while in autumn and winter it becomes quite different (Fig. 3). The reason for this discrepancy is that autumn leaves carry with them when they fall a characteristic microfungal flora which is retained for a while. The dendrogram of the data collected from soil-washing isolates (Fig. 4) again groups together the three horizons in spring and winter, but in summer and autumn horizon A_{00} was quite different from the other two.

It may be suggested that these seasonal groupings were partly the result of the great activity of the soil fauna bringing about a homogenization of the species-composition along the whole profile. Spores would be more affected than mycelia, thus influencing the results to a greater degree when the soil-dilution data were processed.

Considerable similarities existed between both the summer and the autumn microfungal communities, while the winter community differed greatly from the rest.

It remains finally to consider the species of microfungi isolated in the present study. We isolated from litter all those species considered as primary saprophytic species common to leaves (Hudson 1968), and a great number of those known as secondary saprophytic species of beech litter. The majority of species reported by other authors as being the most abundant in beech forest soils were also isolated in the present survey. Among the species that have not hitherto been reported in studies of beech leaves and beech forest soils, *Absidia californica*, *Alternaria consortiale*, *Codinea* sp., *Geniculosporium* sp., *Helminthosporium* sp., *Mucor petrinsularis*, *Nodulisporium* sp., *Penicillium multicolor*, *Penicillium estinogenum*, *Penicillium stoloniferum*, *Penicillium velutinum*, *Penicillium kapuscinskii* and *Trichoderma longibrachiatum*, were isolated by us from beech litter. We also found twenty-one species in the mineral horizons that are newly recorded for beech soils, although we think they have no significance for the great diversity of soils that may exist under such vegetation.

REFERENCES

Badura, L. (1963a). Ricerche sulla microflora del suolo sotto i faggi dell'orto botanico dell'Università di Torino. *Alliona*, **9**, 65–74.

Badura, L. (1963b). Fungilli nuovi, rari o critici isolati del suolo sotto i faggi dell'orto botanico di Torino. *Alliona*, **9**, 175–186.

Bisset, J. & Widden, P. (1972). An automatic multichamber soil-washing apparatus for removing fungal spores from soil. *Canadian Journal of Microbiology*, **18**, 1399–1404.

Bray, J. R. & Curtis, J. T. (1957). An ordination of the upland forest communities of southern Wisconsin. *Ecological Monographs*, **27**, 325–349.

Brillouin, L. (1962). *Science and Information Theory*. Academic Press, New York.

Caldwell, R. (1963). Observations of the fungal flora of decomposing beech litter in soil. *Transactions of the British Mycological Society*, **46**, 249–261.

Carre, C. J. (1964). Fungus decomposition of beech cupules. *Transactions of the British Mycological Society*, **47**, 437–444.

Christensen, M. (1969). Soil microfungi of dry to mesic conifer–hardwood forests in Northern Wisconsin. *Ecology*, **50**, 9–27.

Czechanovski, J. (1913). *Zays Metod Statystycznych*. Warsaw.

Dutch, M. E. & Stout, J. D. (1968). The carbon cycle in a beech forest ecosystem in relation to microbial and animal populations. *Transactions of the 9th International Congress of the Soil Science Society (Adelaide)*, Vol. 2, pp. 37–48. Angus & Robertson, London.

Eicker, A. (1973). The mycoflora of *Eucalyptus maculata* leaf litter. *Soil Biology and Biochemistry*, **5**, 441–448.

Frankland, J. (1966). Succession of fungi on the decaying petioles of *Pteridium aquilinum*. *Journal of Ecology*, **54**, 41–63.

Hayes, A. J. (1965). Some microfungi from Scots pine litter. *Transactions of the British Mycological Society*, **48**, 179–185.

Hayes, A. J. (1967). Biology of forest soils: mycology of Scots pine litter. *Report of Forest Research*, **19**, 147–148.

Hogg, B. M. (1966). Microfungi on leaves of *Fagus silvatica*. II. Duration of survival of spore viability and cellulolytic activity. *Transactions of the British Mycological Society*, **49**, 193–204.

Hogg, B. M. & Hudson, H. J. (1966). Microfungi on leaves of *Fagus silvatica*. I. The microfungal succession. *Transactions of the British Mycological Society*, **49**, 185–192.

Hudson, H. J. (1968). The ecology of fungi on plant remains above the soil. *New Phytologist*, **67**, 837–874.

Iñiguez, J. & Barragan, E. (1974). Andosuelos desarrollados sobre filitas en Ulzama (Navarra). *Anales de Edafología y Agrobiología*, **33**, 1055–1069.

Jensen, V. (1962). Studies on the microflora of Danish beech forest soils. V. The microfungi. *Zentralblatt für Bakteriologie Parasitenkunde, II Abt.*, **117**, 167–179.

Kendrick, W. B. & Burges, A. (1962). Biological aspects of the decay of *Pinus sylvestris* leaf litter. *Nova Hedwigia*, **4**, 313–342.

Krzemieniewska, H. & Badura, L. (1954a). Z badań nad mikoflora lasu bukowego. *Acta Societatis Botanicorum Poloniae*, **23**, 545–587.

Krzemieniewska, H. & Badura, L. (1954b). Przyczynek do znajomósci mikroorganizmów ściółki i gleby lasu bukowego. *Acta Societatis Botanicorum Poloniae*, **23**, 727–781.

Last, F. T. (1955). Seasonal incidence of *Sporobolomyces* on cereal leaves. *Transactions of the British Mycological Society*, **38**, 221–239.

Lindgreen, H. B. & Jensen, V. (1973). Microbiological examination of a forest soil profile. *Kongelige. Veterinaer- og Landbohøgskole Aarsskrift*, **57**, 147–159.

Malan, C. E., Ambrosoli, R. & Alessandria, G. (1969). Intervento di comuni Ifomiceti saprofite nella humificazione della copertura morta della faggeta alpina. *Alliona*, **15**, 133–154.

Martínez, A. T. & Ramírez, C. (1978a). Microfungal biomass and numbers of propagules in an andosol during the year and in the whole profile. *Soil Biology and Biochemistry*, **63**, 57–59.

Martínez, A. T. & Ramírez, C. (1978b). *Penicillium fagi* sp. nov., isolated from beech leaves. *Mycopathologia*, **63**, 57–59.

Moriyón, I., Martinez, A. & Rodriguez-Burgos, A. (1978). Microbiological study of an andosol. *Anales de Edafología y Agrobiología*, **37**, 478–483.

Parkinson, D. & Balasooriya, I. (1969). Studies on fungi in a pinewood soil. IV. Seasonal and spatial variations in the fungal populations. *Revue d'Écologie et de Biologie du Sol*, **6**, 147–153.

Peña-Cabriales, J. J. & Valdes, M. (1974). Rhizosphèrc du sapin (*Abies religiosa*). I. Microbiologie et activité microbienne. *Revista Latinoamericana de Microbiología*, **17**, 25–31.

Peyronel, B. (1961). Funghi del suolo di un bosco di faggio dell'Aspromonte. *Alliona*, **7**, 27–38.

Reddy, T. K. R. & Knowles, R. (1965). The fungal flora of a boreal forest raw humus. *Canadian Journal of Microbiology*, **11**, 837–843.

Remacle, J. (1970). La chênaie à *Galeobdolon* et *Oxalis* de Mesnil-Église (Ferage). 16: La microflore des litières. *Bulletin de la Societé de Botanique de Belgique*, **103**, 83–96.

Ruscoe, Q. W. (1971a). The soil mycoflora of a hard beech forest. *New Zealand Journal of Science*, **14**, 554–567.

Ruscoe, Q. W. (1971b). Mycoflora of living and dead leaves of *Nothofagus truncata*. *Transactions of the British Mycological Society*, **56**, 63–74.

Saito, T. (1956). Microbiological decomposition of beech litter. *Ecological Review*, **14**, 141–147.

Saito, T. (1958). The characteristic features of fungi taking part in the decomposition of beech litter. *Science Report of the Tôhoku University, Series 4*, **24**, 73–79.

Saito, T. (1960). An approach to the mechanism of microbial decomposition of beech litter. *Science Reports of the Tôhoku University, Series 4*, **26**, 125–231.

Saito, T. (1965). Coactions between litter-decomposing hymenomycetes and their associate microorganisms during decomposition of beech litter. *Science Reports of the Tôhoku University, Series 4*, **31**, 255–273.

Saito, T. (1966). Sequential pattern of decomposition of beech litter with special reference to microbial succession. *Ecological Review*, **16**, 245–254.

Soil Conservation Service (1975). *Soil Taxonomy*. Agriculture Handbook No. 436. Soil Conservation Service, Washington D.C.

Wicklow, M. C., Bollen, W. B. & Denison, W. C. (1974). Comparison of soil microfungi in 40-year-old stands of pure alder, pure conifer and alder-conifer mixtures. *Soil Biology and Biochemistry*, **6**, 73–78.

Williams, W. T., Lambert, J. M. & Lance, G. N. (1966). Multivariate methods in plant ecology. V. Similarity analyses and information-analysis. *Journal of Ecology*, **54**, 427–445.

(*Received* 23 *May* 1978)

APPENDIX

Alphabetical list of species with a relative abundance not exceeding 5% in any one sample, with an indication of their presence in the various materials (W = winter, Sp = spring, Su = summer, A = autumn, L = leaf surface, O = A_{00}, 1 = A_{11}, 2 = A_{12})

Soil dilution isolates

Absidia cylindrospora Hagem	W2, A1
A. glauca Hagem	AL
Alternaria consortiale Hughes ex Thün	SuO
A. tenuis Nees	W1, SpL, Su2, AL
Aspergillus repens Saccardo ex Corda	A1
A. terricola Marchal	SuO, Su1, Su2
A. sydowi Church	W1
Aureobasidium Viala et Boyer, red form	AL
Aureobasidium Viala et Boyer 2[1]	Su1, AO, A2
Chloridium Link ex Saccardo sp.	SpO, SuO, Su2, A1
Corynespora Güssov sp.	Su1
Chrysosporium Corda sp.	WO
Discosia Libert sp.	SpL, SpO, SuO
Epicoccum Link ex Fresenius sp.	SpL, Sp1
Geniculosporium Chesters sp.	SuO
Geotrichum Link ex Persoon sp.	W2, Sp1, Sp2, SuO, Su1, Su2, AO
Gliocladium Corda sp.	SpL, SpO, Sp1, Sp2
Humicola Traaen sp.	SuO
Mammaria Cesati sp.	Sp2, AO
Monodyctis Hughes sp.	A2
Mortierella Coemans sp.	W1
M. minutissima van Tieghem	SpL, SpO, Sp1, Sp2
M. nana Linnemann	W1, W2, SuO, Su1, Su2
M. pusilla Oudemans	SpO
M. ramanniana Linnemann ex Möller	SpO, SuL, SuO
M. ramanniana var. *angulispora* Linnemann ex Naumov	SpO, Sp1, A1
M. subtillissima Oudemans et Konig	WO, W1, SpO

318   *Microfungi of an andosol*

Mucor hiemalis Wehmer	WL, WO, SpL
Nodulisporium Preuss sp.	SuO
Oedocephalum Preuss sp.	WO
Paecilomyces Bainier sp.	WO, W2
Penicillium brevicompactum Dierckx, thermosensitive	WO
P. cyclopium Westling	W1, W2
P. daleae Zaleski	WO, SpO, Sp1
P. digitatum Saccardo	W2
P. diversum Raper et Fennell	AO, A1
P. estinogenum Komatsu et Abe	AL
P. fagi sp. nov. (Martínez & Ramírez 1978b)	AL
P. frequentans Westling	SuO
P. granulatum Bainier	SuL, SuO, Su1, Su2
P. jenseni Zaleski	Su1
P. kapuscinskii Zaleski	SpO
P. melinii Thom	SuO, Su2, A2
P. multicolor Grigorieva-Manoilova et Poradielova	SuL, SuO
P. paxilli Bainier	SuL, SuO, AL, AO
P. purpurogenum Stoll	W1
P. p. var. *rubri-esclerotium* Thom	SuL
P. simplicissimum (Oudemans) Thom	SuL, Su1
P. steckii Zaleski	AO
P. stoloniferum Thom	SuO, AO
P. thomii Maire	WL, WO, SpL, SpO, SuL, SuO, Su1, AO, A1, A2
P. urticae Bainier	WO, W1, W2, AL
P. variabile Sopp	Su1
P. velutinum van Beyma	WO
P. viridicatum Westling	W1, SpO
P. waksmani Zaleski AMF 299[2]	WL, Sp2
Scytalidium Pesante sp.	Sp1, Sp2, Su1, Su2, AO, A1, A2
Sordaria Cesati et de Notaris sp.	SuL
Talaromyces flavus var. *flavus* (Klöcker) Stolk et Samson	Su2
Trichoderma aureoviride Rifai	W1, SpL, SpO, Sp1, SuL, SuO, Su2, AL, AO, A2
T. longibrachiatum Rifai	AO
T. viride Persoon ex Gray	SpL, SpO, Sp1, SuO, Su2
Trichosporiella Kamyschko ex Gams et Domsch sp.	Su1

Soil washing isolates

Absidia californica Ellis et Hesseltine	SpO
A. cylindrospora Hagem	W1, SpO, AO, A1
Acremonium Link ex Fresenius sp.	Sp2, SuO, A1
Acremonium Link ex Fresenius-like[3]	Sp2, AO
Alternaria tenuis Nees	AO
AMF 72[4]	WO
AMF 254[5]	Sp1, Sp2

396

Aspergillus fumigatus Fresenius	SuO, AO, A1
A. terricola Marchal	SuO
Aureobasidium Viala et Boyer, red form	SpO
Botrytis Persoon ex Fresenius sp.	SuO, AO
Chaetomium Kunze ex Fresenius sp.	AO
Cladosporium Link ex Fresenius sp.	AO
Cylindrocarpon Wollenweber sp.	W1, Sp2
Eladia Smith sp.	Su2, A1, A2
Graphium Corda sp.	SuO, AO
Helminthosporium Link ex Fresenius sp.	SuO
Mammaria Cesati sp.	Sp2
Melanospora Corda sp.	Su1, Su2
Mortierella isabellina Oudemans et Koning	SpO
M. ramanniana Linnemann ex Möller	SpO, Su1
M. r. var. *angulispora* Linnemann ex Naumov	Su2
Oedocephalum Preuss sp.	WO
Penicillium casei Staub	W2
P. diversum Raper et Fennell	AO
P. frequentans Westling	AO
P. granulatum Bainier	Su1
P. melinii Thom	AO, A1, A2
P. purpurogenum Stoll	AO
P. simplicissimum (Oudemans) Thom	SpO, AO, A1
P. spinulosum Thom	SpO
P. waksmani Zaleski AMF 299	W1
Phialophora Medlar[1]	A1
Rhizopus arrhizus Fischer	WO
Sordaria Cesati et de Notaris sp.	SpO
Sporothrix Hektoen et Perkins sp.	AO
Trichoderma longibrachiatum Rifai	SpO, SuO, AO
T. viride Persoon ex Gray	SpO, SuO, Su1
Verticillium Nees ex Link sp.	WO

[1] Different from all described species; [2] Strain with characteristics intermediate with *Penicillium corylophilum* Dierckx; [3] Atypical form, could be an imperfect Ascomycete; [4] Unidentified Monilial; [5] Unidentified Oomycetale.

BIBLIOGRAPHY

Adachi, T., 1971, The Area and Humus Content of Ando Soils in Japan, *Jour. Sci. Soil and Manure* (Japan) **42**:309-313.

Adachi, T., 1974, Characterization of the Humus of Ando Soils in Japan, *Japan Agric. Res. Quart.* **8**:13-18.

Alvarado, A., 1975, Fertility of Some Andepts under Pasture in Costa Rica, *Turrialba* **25**:265-270.

Alvarado, A., and S. W. Buol, 1975, Toposequence Relationships of Dystrandepts in Costa Rica, *Soil Sci. Soc. Am. Proc.* **39**:932-937.

Alvarado, A., C. W. Berish, and F. Peralta, 1981, Leaf-cutter Ant (Atta cephalotes) Influence on the Morphology of Andepts in Costa Rica, *Soil Sci. Soc. Amer. Jour.* **45**:790-794.

Amano, Y., 1982, Some Characteristics of Andosols in Japan, Proceedings of the International Symposium on the Distribution, Characterization and Utilization of Problem Soils, *Trop. Agric. Res. Series* **15**:265-273.

Andriesse, J. P., H. A. Van Rosmalen, and A. Muller, 1976, On the Variability of Amorphous Materials in Andosols and Their Relationship to Irreversible Drying and P-Retention, *Geoderma* **16**:125-138.

Aomine, S., and M. L. Jackson, 1959, Allophane Determination in Ando Soils by Cation-Exchange Capacity Delta Value, *Soil Sci. Soc. Am. Proc.* **23**:210-214.

Aomine, S., and N. Miyauchi, 1965, Imogolite in Imogolayers in Kyushu, *Soil Sci. Plant Nutr.* (Japan) **11**:28-35.

Aomine, S., and N. Yoshinaga, 1955, Clay Minerals of Some Well-Drained Volcanic Ash Soils in Japan, *Soil Sci.* **77**:349-358.

Asghar, M., V. Balasubramanian, and Y. Kanihero, 1979, Transformation of Applied Nitrogen in some Hydrandepts of Hawaii, *Agrochimica* **23**:427-435.

Balasubramanian, V., and Y. Kanihero, 1978, Surface Chemistry of the Hydrandepts and Its Relation to Nitrate Adsorption as Affected by Profile Depth and Dehydration, *Jour. Soil Sci.* **29**:47-57.

Barragan, E., and J. Iniguez, 1976, Weathering of Clay Minerals in a Navarre Andosol (Spain), *Clay Minerals* **11**:269-272.

Birrell, K. S., 1961, The Adsorption of Cations from Solution by Allophane in Relation to Their Effective Size, *Jour. Soil Sci.* **12**:307-316.

Bornemisza, E., M. Constenia, A. Alvarado, E. J. Ortega, and A. J. Vasquez, 1979, Organic Carbon Determination by the Walkley-Black and Dry Combustion Methods in Surface Soils and Andept Profiles from Costa Rica, *Soil Sci. Soc. Amer. Jour.* **43**:78-83.

Briones, A. A., 1982, Characteristics and Fertilization of Andepts in the Philippines (Volcanic Ash Soils), *Trop. Agric. Res. Series* **15**:251-264.

Bibliography

Cabezas Viano, O., J. Hernandez Moreno, M. L. Tejedor Salguero, and E. Fernandez Caldas, 1977, Mineralogical Characteristics of the Sand Fraction in a Chronological Sequence of Andepts in the Canary Island, *Anales de Edafologia y Agrobiologia* **36:**787-803.

Calhoun, F. G., 1974, Taxonomy of Oxisols, Ultisols and Andepts, *Soil and Crop Sci. Soc. Florida Proc.* **33:**108-111.

Calhoun, F. G., and V. W. Carlisle, 1977, Microfabric Characteristics and Pedogenesis of a Colombian Andosol Climosequence, Informes de Conferencias Cursos y Reuniones (IICA), *Instituto Interamericano de Ciencas Agricolas Turrialba* **82:**209-224.

Campbell, A. S., A. W. Young, L. G. Livingstone, M. A. Wilson, and T. W. Walker, 1977, Characterization of Poorly-Ordered Alumino Silicates in a Vitric Andosol from New Zealand, *Soil Sci.* **123:**362-368.

Coutinet, S., 1967, Contribution to the Study of Allophane Soils, *Agron. Trop.* (Paris) **22:**1157-1175.

Cradwick, P. D. G., V. C. Farmer, J. D. Russell, C. R. Masson, K. Wada, and N. Yoshinaga, 1973, Imogolite, A Hydrated Aluminum Silicate Mineral of Tubular Structure, *Nature Phys. Sci.* **240:**187-189.

Egawa, T., 1964, A Study on Coordination Number of Aluminium in Allophane, *Clay Sci.* **2:**1-7.

El Swaify, S. A., and A. H. Sayegh, 1975, Charge Characteristics of an Oxisol and an Inceptisol in Hawaii, *Soil Sci.* **120:**49-56.

Espinoza, G. W., and F. E. Riquelme, 1976, Chemical Characteristics of Two Andepts (Trumaos) in the Nuble Province (Chile): Arrayan and Santa Barbara, *Agri. Technica* **36:**49-58.

Espinoza, W., R. H. Rust, and R. S. Adams, Jr., 1975, Characterization of Mineral Forms in Andepts from Chile, *Soil Sci. Soc. Am. Proc.* **39:**556-561.

Every, J. P., 1981, The Effect of Magnesium Fertilizers Applied to a Pumice Soil on Magnesium Concentration in Grass Herbage, *New Zealand Jour. Expt. Agric.* **9:**251-254.

Fieldes, M., 1962, The Nature of the Active Fraction of Soils, *Trans. Joint Meeting Comm. IV and V, Intern. Soil Sci. Soc.,* Massey College, Palmerston North, New Zealand, pp. 62-78.

Fieldes, M., 1966, The Nature of Allophane in Soils. I. Significance of Structural Randomness in Pedogenesis, *New Zealand Jour. Sci.* **9:**599-607.

Fieldes, M., I. K. Walker, and P. P. Williams, 1956, Clay Mineralogy of New Zealand Soils. 3. Infrared Absorption Spectra on Soil Clays, *New Zealand Sci. Tech.* **38:**31-43.

Flach, K. W., 1977, The Differentiation of the Cambic Horizon of Andepts from Spodic Horizon, *Informes de Conferencias Cursos y Reuniones (IICA), Instituto Interamericano de Ciencas Agricolas Turrialba (Costa Rica)* **82:**127-138.

Flach, K. W., and R. Tavernier, 1978, Soils of Humid Mesothermal Climates, *Intern. Congress Soil Sci. Trans.,* 11th, Edmonton, Canada, **2:**148-165.

Furuno, S., M. Nakano, T. Yamamoto, and T. Takahashi, 1977, Effect of Heavy Mixing Application of Phosphate in Plow Layer of Ando Soils in Japan, *Proc. Intern. Seminar Soil Environment and Fertility Management in Intensive Agriculture (SEFMIA),* Japan. Soc. Sci. Soil and Manure, pp. 103-111.

Galindo, G. G., and F. T. Bingham, 1977, Homovalent and Heterovalent Cation Exchange Equilibria in Soils with Variable Charge, *Soil Sci. Soc. Am. Jour.* **41:**883-886.

Gautheyrou, J., M. Gautheyrou, and F. Colmet Daage, 1976, *Chronobibliographie Signa letique et Analytique des Sols a' Allophane,* Office de la Recherche Scientifique et Technique Outre-Mer, Centre des Antilles, Pointe-a-Pitre Cedex, Guadaloupe, pt. 1, 158p., pt. 2, 331p.

Gunjigake, N., and K. Wada, 1981, Effects of Phosphorus Concentration and pH on Phosphate Retention by Active Aluminum and Iron of Ando Soils, *Soil Sci.* **132:**347-352.

Harada, Y., and K. Wada, 1974, Effects of Previous Drying on the Measured Cation- and Anion-Exchange Capacities of Ando Soils, *Intern. Congress Soil Sci. 10th Trans.,* **11:**248-256.

Higashi, T., and K. Wada, 1977, Size Fractionation, Dissolution Analysis and Infrared Spectroscopy of Humus Complexes in Ando Soils, *Jour. Soil Sci.* **28:**653-663.

Iniguez, J., and R. M. Val, 1982, Variable Charge Characteristics of Andosols from Navarre, Spain, *Soil Sci.* **133:**390-396.

Inoue, K., and N. Yoshida, 1981, Physical, Chemical, and Clay Mineralogical Properties of the A1 and Buried A Horizons of Ando Soils from the Upper Kitakami River Basin, Japan, *Soil Sci. Plant Nutr.* (Japan) **27:**523-534.

Jackman, R. H., 1955, Organic Phosphorus in New Zealand Soils under Pasture, II, Relation between Organic Phosphorus Content and Some Soil Characteristics, *Soil Sci.* **79:**292-299.

Jenkins, D. A., 1977, Preliminary Observations on the Distribution of Trace Elements in Four Andosols from the Cauca Region of Columbia, *Informes de Conferencias Cursos y Reuniones (IICA), Instituto Interamericano de Ciencas Agricolas Turrialba* (Costa Rica) **82:**93-96.

Kanehiro, Y., and L. D. Whittig, 1961, Amorphous Mineral Colloids of Soils of the Pacific Region and Adjacent Areas, *Pacific Sci.* **15:**477-482.

Kanno, I., 1955, Glassy Volcanic-ash Soils in Japan, *Soil and Plant Food* **1:**1-2.

Kato, Y., and T. Matsui, 1979, Some Application of Paleopedology in Japan, *Geoderma* **22:**45-60.

Kawai, K., 1969a, Changes in Cation Exchange Capacity of Some Ando Soils with Dithionite-citrate Treatment, *Soil Sci. Plant Nutr.* (Japan) **15:**97-103.

Kawai, K., 1969b, Micromorphological Studies of Andosols in Japan, *Nat. Inst. Agric. Sci. Bull.* (Tokyo) **B20:**77-154.

Kawai, K., 1977a, Estimation of the Amount of Amorphous Materials for Characterizing Andosols, *Soil Sci. Soc. Am. Jour.* **41:**1171-1175.

Kawai, K., 1977b, Method of Determining Amorphous Material for Characterizing Andosols, *Informes de Conferencias Cursos y Reuniones (IICA), Instituto Interamericano de Ciencas Agricolas Turrialba* **82:**83-92.

Kawai, K., 1980a, Amorphous Materials of Andosols (Kuroboku) in Japan, *Japan Agric. Res. Quart.* **12:**132-137.

Kawai, K., 1980b, The Relationship of Phosphorus Adsorption to Amorphous Aluminum for Characterizing Andosols, *Soil Sci.* **129:**186-190.

Kitagawa, Y., 1976, Specific Gravity of Allophane and Volcanic Ash Soils Determined with a Pycnometer, *Soil Sci. Plant Nutr.* (Japan) **22:**199-202.

Kitagawa, Y., 1977, Determination of Allophane and Amorphous Inorganic

Matter in Clay Fraction of Soils, II, Soil Clay Fractions, *Soil Sci. Plant Nutr.* (Japan) **23**:21-31.

Kumada, K., and K. Aizawa, 1958, The Infrared Spectra of Humic Acids, *Soil Plant Food* (Japan) **3**:152-159.

Langohr, R., 1976, The Volcanic Ash Soils of the Central Valley of Central Chile, II, The Parent Materials of the Trumao and Nadi Soils of the Lake District in Relation with the Geomorphology and Quaternary Geology, *Pédologie* **24**:238-255.

Loganathan, P., and L. D. Swindale, 1969, Properties and Genesis of 4 Middle-Altitude Dystrandept Volcanic Ash Soils from Mauna Kea, Hawaii, *Pacific Sci.* **23**:161-171.

Martinez, A. T., and C. Ramirez, 1978, Microfungal Biomass and Number of Propagules in an Andosol, *Soil Biol. Biochem.* **10**:529-531.

Martini, J. A., 1970, Caracterizacion del Estado Nutricional de los Principales Andosoles de Costa Rica, mediante la Tecnica del Elemento Faltante el Invernadero, *Turrialba* **20**:72-84.

Martini, J. A., and J. A. Palencia, 1975, Soils Derived from Volcanic Ash in Central America: I. Andepts, *Soil Sci.* **120**:278-287.

Martini, J. A., and A. Suarez, 1975, Potassium Status of Some Costa Rican Latosols and Andosols and Their Response to Potassium Fertilization under Greenhouse Conditions, *Soil Sci. Soc. Am. Proc.* **39**:74-80.

Martini, J. A., and A. Suarez, 1977, Potassium Supplying and Fixing Capacity of Some Costa Rican Latosols and Andosols Determined by Successive Cropping, Extractions and Incubations, *Soil Sci.* **123**:37-47.

Matsui, T., 1965, An Application of Soil Stratigraphy to the Quaternary Geology and Land Development in Kyushu, *Japan. Proc. 7th INQUA Congress, "Quaternary" Soils,* **9**:206-219.

Matsumoto, Y., 1980, Effect of a Heavy Application of Hog Feaces on the Surface Dryness of an Ando Soil, *Jour. Sci. Soil and Manure* (Japan) **51**:175-178.

Miyazawa, K., 1966, Clay Mineral Composition of Andosols in Japan with Reference to Their Classification, *Natl. Inst. Agric. Sci. Bull.* (Tokyo) **B17**:1-100.

Mizota, C., 1976, Relationships between the Primary Mineral and the Clay Mineral Compositions of Some Recent Ando Soils, *Soil Sci. Plant Nutr.* (Japan) **22**:257-268.

Mizota, C., 1977, Phosphate Fixation by Ando Soils Different in Their Clay Mineral Composition, *Soil Sci. Plant Nutr.* (Japan) **23**:311-318.

Mizota, C., 1978, Clay Mineralogy of the A Horizons of Seven Ando Soils, Central Kyushu, *Soil Sci. Plant Nutr.* (Japan) **24**:63-73.

Mizota, C., and K. Wada, 1980, Implications of Clay Mineralogy to the Weathering and Chemistry of Ap Horizons of Ando Soils in Japan, *Geoderma* **23**:49-63.

Mizota, C., M. A. Carrasco, and K. Wada, 1982, Clay Mineralogy of Ando Soils Used for Paddy Rice in Japan, *Geoderma* **27**:225-237.

Mohr, E. C. J., and F. A. Van Baren, 1960, *Tropical Soils: A Critical Study of Their Genesis as Related to Climate, Rock and Vegetation,* Les Editions A. Manteau S.A., Bruxelles, 498p.

Munevar, F., and A. G. Wollum, 1977, Effects of the Addition of Phosphorus

and Inorganic Nitrogen on Carbon and Nitrogen Mineralization in Andepts from Colombia, *Soil Sci. Soc. Am. Jour.* **41**:540-545.
Neall, V. E., 1977, Genesis and Weathering of Andosols in Taranaki, New Zealand, *Soil Sci.* **123**:400-408.
New Zealand Commonwealth Bureau of Soils, 1978, *Soils from Volcanic Ash and Pumice, 1972-1977*, 48p.
Ohba, K., and Y. Fujita, 1978, The Effect on the Nitrification of "Kuroboku" (Ando) Soil with the Fumigation of Di-Trapex, *Jour. Sci. Soil and Manure* (Japan) **49**:426-428.
Otsuka, H., and K. Kumada, 1978, Studies on a Volcanic Ash Soil at Ohnobaru, Tarumizu City, Kagoshima Prefecture. 1. Status Accumulated in the Profile, *Soil Sci. Plant Nutr.* (Japan) **24**:265-276.
P. Quantin and G. G. C. Claridge, trans., 1974, Andosols: Bibliographic Review of Present Day Knowledge, *New Zealand Record* 38, 48p.
Read, N. E. (ed.), 1974, *Soil Groups of New Zealand, Part I, Yellow Brown Pumice Soils*, New Zealand Soc. Soil Sci., Lower Hutt, New Zealand, 251p.
Saigusa, M., S. Shoji, and T. Kato, 1978, Origin and Nature of Halloysite in Ando Soils from Towada Tephra, Japan, *Geoderma* **20**:115-129.
Saigusa, M., S. Shoji, and T. Takahashi, 1980, Plant Root Growth in Acid Ando Soils from Northeastern Japan. 2. Exchange Acidity Y_1 as a Realistic Measure of Aluminum Toxicity Potential, *Soil Sci.* **130**:242-250.
Saunders, W. M. H., 1959, Effect of Phosphate Topdressing on a Soil from Andesitic Volcanic Ash. II. Effect on Distribution of Phosphorus and on Related Chemical Properties, *New Zealand Agric. Res.* **2**:445-462.
Shoji, S., and J. Masui, 1971, Opaline Silica of Recent Volcanic Ash Soils in Japan, *Jour. Soil Sci.* **22**:101-108.
Shoji, S., and M. Saigusa, 1977, Amorphous Clay Materials of Towada Ando Soils, *Soil Sci. Plant Nutr.* (Japan) **23**:437-455.
Shoji, S., and M. Saigusa, 1978, Occurrence of Laminar Opaline Silica in Some Oregon Andosols, *Soil Sci. Plant Nutr.* (Japan) **24**:157-160.
Shoji, S., and I. Yamada, 1977, Soil Mineralogy and Fertility of Ando Soils in Japan, *Proc. Intern. Seminar Soil Environment and Fertility Management in Intensive Agriculture (SEFMIA), Soc. Sci. Soil and Manure (Japan)*, pp. 96-102.
Shoji, S., M. Saigusa, and T. Takahashi, 1980, Plant Root Growth in Acid Andosols from Northeastern Japan. I. Soil Properties and Root Growth of Burdock, Barley and Orchardgrass, *Soil Sci.* **130**:124-131.
Shoji, S., I. Yamada, and K. Kurashima, 1981, Mobilities and Related Factors of Chemical Elements in the Top Soils of Ando Soil in Tohoku, Japan. 2. Chemical and Mineralogical Compositions of Size Fractions and Factors Influencing the Mobilities of major Chemical Elements, *Soil Sci.* **132**:330-346.
Shoji, S., Y. Fuyiwara, I. Yamada, and M. Saigusa, 1982, Chemistry and Clay Mineralogy of Ando Soils, Brown Forest Soils and Podzolic Soils Formed from recent Towada Ashes, *Northeastern Japan. Soil Sci.* **133**:69-86.
Sudo, T., 1954, Clay Mineralogical Aspects of the Alteration of Volcanic Glass in Japan, *Clay Minerals* **2**:96-106.

Bibliography

Tan, K. H., 1959, On the Classification of Black Colored Soils in Humid Regions of Indonesia (in Indonesian, with English summary), *Tehnik Pertanian* **8:**217-223.

Tan, K. H., 1963, Differential Thermal Analysis of Andosols in Indonesia, *Ministry of Higher Educ. and Sci. Indonesia Res. Jour.* **1B:**11-20.

Tan, K. H., 1966, On the Pedogenetic Role of Organic Matter in Volcanic Ash Soils under Tropical Conditions, *Soil Sci. Plant Nutr.* (Japan) **12:**34-38.

Tan, K. H., 1968, The Genesis and Characteristics of Paddy Soils in Indonesia, *Soil Sci. Plant Nutr.* (Japan) **14:**117-121.

Tan, K. H., 1969, Chemical and Thermal Characteristics of Allophane in Andosols of the Tropics, *Soil Sci. Soc. Am. Proc.* **33:**469-472.

Tan, K. H., and H. F. Massey, 1964, Effect of Site on the Pulpwood Productive Capacity of *Pinus merkusii*. I. The Relation of Site Quality to Soil Factors. *Ministry Higher Educ. and Science, Indonesia, Res. Jour.* **1B:**88-102.

Tan, K. H., and P. S. Troth, 1982, Silica-sesquioxide Ratios as Aids in Characterization of Some Temperate Region and Tropical Soil Clays, *Soil Sci. Soc. Am. Jour.* **46:**1109-1114.

Tan, K. H., H. F. Perkins, and R. A. McCreery, 1970, The Nature and Composition of Amorphous Material and Free Oxides in Some Temperate Region and Tropical Soils, *Soil Sci. Plant Anal.* **1:**227-238.

Tan, K. H., H. F. Perkins, and R. A. McCreery, 1975, Amorphous and Crystalline Clays in Volcanic Ash Soils of Indonesia and Costa Rica, *Soil Sci.* **119:**431-440.

Takahashi, T., 1982, Problems and Improvement of Volcanic Ash Soils in Japan with Reference to the Weathering Sequence, Proc. Intern. Symp. Distribution, Characterization and Utilization of Problem Soils, *Trop. Agric. Res. Series* **15:**275-286.

Takahashi, Y., and K. Wada, 1975, Weathering Implications of the Mineralogy of Clay Fractions of Two Ando Soils, Kyushu, *Geoderma* **14:**47-62.

Tokudome, S., and I. Kanno, 1963, Characteristics of Humus in Humic-Allophane Soils of Japan, Preliminary Report, *Pedologist* **7:**82-95.

Wada, K., 1959, Reaction of Phosphate with Allophane and Halloysite, *Soil Sci.* **87:**325-330.

Wada, K., 1977, Characterization and Determination of "Amorphous" Clay Constituents in Volcanic Ash Soils, *Informes de Conferencias Cursos y Reuniones (IICA), Instituto Interamericano de Ciencas Agricolas Turrialba* **82:**295-307.

Wada, K., and H. Ataka, 1958, The Ion Uptake Mechanism of Allophane, *Soil Plant Food* (Japan) **4:**12-18.

Wada, K., and N. Gunjigake, 1979, Active Aluminum and Iron and Phosphate Adsorption in Ando Soils, *Soil Sci.* **128:**331-336.

Wada, K., and T. Higashi, 1976, The Categories of Aluminum- and Iron-Humus Complexes in Ando Soils determined by Selective Dissolution, *Jour. Soil Sci.* **27:**357-368.

Wada, K., and T. Kawano, 1978, Use of Jeffries Acid Oxalate Treatment in Particle Size Analysis of Ando Soils, *Geoderma* **20:**215-224.

Wada, K., and Y. Okamura, 1980, Electric Charge Characteristics of Ando A1 and Buried A1 Horizon Soils, *Jour. Soil Sci.* **31:**307-314.

Wada, K., and Y. Tokashiki, 1972, Selective Dissolution and Difference

Infrared Spectroscopy in Quantitative Mineralogical Analysis of Volcanic Ash Soil Clays, *Geoderma* **7:**199-213.

Wada, K., and S. Wada, 1976, Clay Mineralogy of Two Andepts, an Oxisol and an Unidentified Soil in Hawaii, *Proc. Intern. Clay Conf.*, Mexico City, July 16-23, 1975, pp. 657-658.

Wada, S. I., and A. Eto, 1979, Synthetic Allophane and Imogolite (Clay Constituents in Ando Soils or Andepts), *Jour. Soil Sci.* **30:**347-355.

Wada, S. I., and K. Wada, 1980, Formation, Composition and Structure of Hydroxy-aluminosilicate ions (in Ando Soils), *Jour. Soil Sci.* **31:**457-467.

Watanabe, Y., 1963, Study of Soil Clays by Means of the Electron Microscope, 2, Allophane and Synthetic Gels, *Soil Sci. Plant Nutr.* (Japan) **9:**11-14.

Yamada, I., and S. Shoji, 1982, Retention of Potassium by Volcanic Glasses of the Topsoils of Andosols in Tohoku, Japan, *Soil Sci.* **133:**208-212.

Yamada, I., M. Saigusa, and S. Shoji, 1978, Clay Mineralogy of Hijiori and Numazawa Ando Soils, *Soil Sci. Plant Nutr.* (Japan) **24:**75-89.

Yamane, I., 1973, The Significance of *Miscanthus sinensis* in the Formation of Kuroboku Soils, *Pedologist* **17:**84-94.

Yoshinaga, N., 1966, Chemical Composition and Some Thermal Data of Eighteen Allophanes from Ando Soils and Weathered Pumices, *Soil Sci. Plant Nutr.* (Japan) **12:**47-54.

Yoshinaga, N., J. M. Tait, and R. Soong, 1973, Occurrence of Imogolite in Some Volcanic Ash Soils of New Zealand, *Clay Minerals* **10:**127-130.

Yost, R. S., and R. L. Fox, 1981, Partitioning Variation in Soil Chemical Properties of some Andepts using Soil Taxonomy, *Soil Sci. Soc. Am. Jour.* **45:**373-377.

Yuan, T. L., 1974, Chemistry and Mineralogy of Andepts, *Soil Crop Sci. Soc. Florida Proc.* **33:**101-108.

AUTHOR CITATION INDEX

Abe, K., 302
Adachi, T., 179, 298, 399
Adams, F., 75
Adams, R. S., Jr., 179, 209, 400
Adamson, A. W., 209
Adler, H., 329
Agar, G. E., 209
Aguilera, N., 45, 119
Ahmad, N., 179
Aida, S., 182, 198
Aizawa, K., 402
Akatsuka, K., 298, 300
Alberto, G. F., 307, 356
Alessandria, G., 394
Alexander, L. E., 329, 337
Alexander, L. T., 59, 98, 223
Allaway, W. H., 329
Alvarado, A., 45, 204, 399
Amano, Y., 247, 399
Ambo, F., 180, 300
Ambrosoli, R., 394
Anderson, M. W., 181
Ando, H., 197
Andreux, F., 98
Andriesse, J. P., 399
Aomine, S., 7, 48, 59, 63, 83, 121, 179, 197, 298, 299, 301, 302, 308, 309, 329, 337, 356, 365, 381, 399
Arimura, G., 299
Arimura, S., 120
Arizumi, A., 179
Asghar, M., 399
Association of Official Agricultural Chemists, 59
Ataka, H., 198, 232, 302, 404
Atkinson, R. J., 232, 236, 240
Austin, M. E., 17, 59
Aylmore, L. A. G., 236

Baak, J. A., 307
Baba, H., 179
Baba, N., 180
Babcock, K. L., 209
Badura, L., 393, 394
Baker, B. H., 75
Baker, R. T., 247

Balasooriya, I., 394
Balasubramanian, V., 399
Baldwin, M., 17
Bar-Yosef, B., 240
Barkoff, E., 300
Barragan, E., 394, 399
Barrow, N. J., 236
Bascomb, C. L., 98
Baver, L. D., 197, 371
Behr, B., 236
Bellis, E., 76
Berish, C. W., 399
Berkelhamer, L. H., 319
Berlage, H. P., Jr., 63
Besoain, E., 45, 119, 356
Bingham, F. T., 401
Birch, W. R., 76
Birrell, K. S., 6, 13, 45, 47, 59, 63, 119, 179, 180, 204, 223, 240, 298, 308, 356, 399
Bisque, R. E., 98
Bisset, J., 393
Black, C. A., 82, 197, 209, 247
Blakemore, L. C., 247
Blyholder, G., 209
Boersma, L., 179
Bohor, B. F., 365
Bollen, W. B., 395
Bonfils, P., 179
Borchardt, C. A., 120, 179
Bornemisza, E., 204, 236, 399
Borst, R. L., 365
Bowden, J. W., 204
Bower, C. A., 223
Boyd, C. C., 236
Bray, E. E., 329
Bray, J. R., 393
Bremner, J. M., 209
Brewer, R., 98, 145, 179, 298
Bridger, B. A., 382
Brillouin, L., 393
Brindley, G. W., 329
Briones, A. A., 399
Broadbent, F. E., 381
Brown, G., 308
Brunauer, S., 223
Brydon, J. E., 98, 236, 308

Buchan, H., 382
Bullock, P., 98
Buol, S. W., 45, 399
Buondonno, C., 46
Burges, A., 394
Butler, J. N., 236, 240

Cabezas Viano, O., 400
Cairns, A., 382
Caldwell, R., 394
Calhoun, F. G., 197, 400
Campbell, A. S., 400
Carlisle, V. W., 197, 400
Carrasco, M. A., 402
Carre, C. J., 394
Chang, M. L., 236, 240
Chang, S. C., 298
Chao, T. T., 236
Chenery, E. M., 59
Cheshire, M. V., 381
Chichester, F. W., 179
Christensen, M., 394
Claridge, G. G. C., 179, 403
Cloos, P., 231
Cochran, P. H., 179
Cole, C. V., 240
Coleman, J. D., 179
Coleman, N. T., 236, 240
Colmet-Daage, F., 46, 179, 180, 401
Conea, A., 32, 46
Constenia, M., 399
Cope, J. T., 75
Corching, W. E., 46
Coutinet, S., 400
Cox, J. E., 7, 14, 23, 59
Cradwick, P. D. G., 400
Cranwell, P. A., 371
Croney, D., 179
Cucalon, F., 46, 179
Curtis, J. T., 393
Czechanovski, J., 394

D' Hoore, J. L., 59
Daikuhara, G., 298
Dames, T. W. G., 22, 82
Daniels, R. B., 98

Author Citation Index

Davies, B., 320
Davies, E. B., 179, 319
Davila, A., 32
Davis, J. F., 197
Day, J., 59
Day, P. R., 197
de Bruyn, P. L., 209
De Coninck, F., 6, 98, 365
de Kimpe, C., 46, 179
DeMent, J. A., 59
Dean, L. A., 59, 98, 240
Delaune, M., 46, 179
Denison, W. C., 395
Diamond, S., 300, 356
Dion, H. G., 319
Dixon, J. B., 204
Driuf, J. H., 32
Druif, J. H., 6, 63
Duchaufour, P., 17, 32, 46
Dudal, R., 13, 32, 46, 59, 82
Dutch, M. E., 394
Dyrness, C. T., 182

Egashira, K., 179
Egawa, T., 46, 120, 231, 247, 298, 299, 301, 400
Eicker, A., 394
Eigen, M., 236
El-Swaify, S. A., 179, 400
Emmett, P. M., 223
Eschena, T., 32, 46
Espinal, S., 120
Espinoza, W., 179, 209, 400
Eswaran, H., 365
Eto, A., 405
Every, J. P., 400
Ewing, W. W., 223

Fang, S. C., 236
FAO/UNESCO, 46, 75, 381
Farmer, V. C., 400
Farstad, L., 59
Fassbender, H. W., 240
Fernandez Caldas, E., 32, 33, 46, 400
Fieldes, M., 22, 46, 59, 63, 98, 120, 179, 222, 223, 231, 298, 299, 308, 329, 356, 400
Filip, Z., 371
Fitzpatrick, E. A., 6, 82, 145
Flach, K. W., 46, 98, 120, 179, 400
Fleischer, M., 365
Folsom, B. L., 382
Food and Agriculture Organization, 17
Forsythe, W. M., 46, 179
Foster, P. K., 180
Fox, R. L., 236, 405
Frankland, J., 394
Franzmeier, D. P., 98
Fraser, A. R., 365
Frei, E., 46, 59, 75

Freud, R., 75
Fripiat, J. J., 231
Frost, R. J., 179
Fuerstenau, D. W., 209
Fujioka, Y., 180
Fujisawa, T., 300
Fujita, Y., 403
Fujiwara, H., 180
Funabiki, S., 299
Furkert, R. J., 182, 231, 298, 299, 308
Furrer, G., 75
Furuno, S., 400
Furusaka, C., 298, 299, 302
Fusil, G., 46, 179
Fuyiwara, Y., 403

Galindo, G. G., 401
Galindo-Griffith, G. G., 180
Gallardo, J. F., 46
Galvan, J., 32
Gamble, E. E., 98
Garcia Sanchez, A., 46
Gardner, W. H., 197
Gardner, W. R., 197
Gautheyrou, J., 46, 179, 180, 401
Gautheyrou, M., 46, 179, 180, 401
Gavande, S. A., 46, 179
Gebhardt, H., 236, 240
Gelaude, F., 98
Gerasimov, I. P., 22
Gessa, C., 32, 46
Gibbs, H. S., 46
Goh, K. M., 371
Golterman, H. L., 371
Gonzalez, C., 181
González, M. A., 46, 179
Gould, R. F., 209
Gradwell, M. W., 145, 180, 298
Grange, L. I., 320
Greaves, M. P., 381
Greenland, D. J., 198, 209, 308, 382
Griggs, R. F., 59
Grim, R. E., 180
Guerra Delgado, A., 33
Guerrero, R., 32
Gunjigake, N., 47, 401, 404
Gupta, U. C., 382
Gutierrez Rodriguez, E., 32

Haantjens, H. A., 46
Hahasa, G., 120
Haider, K., 371
Hajek, B. F., 75, 98
Hakoishi, T., 302
Hanada, S., 247
Hanai, H., 299
Hanson, S. M., 236
Harada, M., 59
Harada, T., 299

Harada, Y., 198, 302, 401
Hardy, F., 320
Hartely, R. D., 382
Hartman, M. A., 120
Harward, M. E., 99, 120, 179, 182, 198, 236, 382
Hashimoto, I., 98
Hattori, T., 299
Haworth, R. D., 371
Hayama, Y., 382
Hayashi, T., 299
Hayes, A. J., 394
Hayes, M. H. B., 98, 308
Henmi, T., 182, 197, 198, 204, 308
Herbillon, A., 231
Hernandez Moreno, J., 400
Hétier, J. M., 46
Higashi, A., 180
Higashi, T., 47, 76, 98, 99, 197, 299, 401, 404
Highway Research Board, 180
Hingston, F. J., 232, 236, 240
Hogg, B. M., 394
Holmgren, G. G. S., 98
Holzhey, C. S., 98
Honda, C., 23, 120, 299, 300
Hong, C. H., 247
Honjo, I., 120
Honjo, Y., 299, 329, 337, 356
Hori, S., 301
Hosoda, K., 22, 59, 299
Hudson, H. J., 394
Hughes, I. R., 180
Hughes, R. F., 365
Hunt, J. M., 329
Hurni, H., 75
Hutagalung, O., 63

Ibe, K., 182, 198, 302, 365
ICOMAND, 46
Igarashi, T., 301
Iimura, K., 197, 298, 299
Ikawa, H., 120, 181
Ikeda, H., 197
Ikegami, M., 180
Imp. Bur. Soil Sci., 320
Imura, J., 232
Iniguez, J., 399, 401
Iñiguez, J., 394
Iniquez, J., 204
Inostroza, O., 32
Inoue, K., 401
Inoue, T., 299, 302
Inter-American Institute of Agricultural Sciences, 46
Iseaki, A., 23
Iseki, A., 300
Ishii, J., 299
Ishikawa, T., 180
Ishizawa, I., 6, 372
Ishizawa, S., 299, 301
Ishizuka, Y., 82

Author Citation Index

Ito, K., 180, 300
Ito, M., 180
Ivanova, Ye. N., 23
Iwata, S., 180, 299
Izeki, A., 120

Jackman, R. H., 381, 401
Jackson, M. L., 6, 63, 98, 197, 209, 236, 240, 298, 308, 329, 337, 356, 399
Jahja, H., 32
Japanese Ministry of Agriculture and Forestry, 59
Jaritz, G., 307, 356, 365
Jeffries, C. D., 320
Jenkins, D. A., 401
Jenne, E. A., 337
Jenny, H., 75
Jensen, V., 394
Johannesson, B., 59
Joyner, L. G., 223

Kafkafi, U., 240
Kamoshita, Y., 23, 59, 299
Kanaya, H., 198
Kanehiro, Y., 209, 232, 399, 401
Kanno, I., 7, 23, 59, 63, 82, 120, 180, 299, 301, 302, 329, 337, 356, 372, 401, 404
Karayeva, Z. S., 59
Karin, M., 236
Kasubuchi, T., 180
Kato, T., 304
Kato, Y., 59, 197, 299, 300, 401
Kawaguchi, K., 180, 299, 300, 372
Kawai, K., 46, 98, 120, 299, 401
Kawajiri, M., 301
Kawamura, K., 23, 299
Kawano, T., 404
Kawasaki, H., 299, 308, 356
Keller, W. D., 329, 365
Kellogg, C. E., 17, 59
Kendrick, W. B., 394
Kenya Soil Survey, 75
Kerr, P. F., 320, 329
Khan, S. U., 98
Kidachi, M., 299
Kilmer, V. J., 59, 98
Kingo, T., 240
Kinjo, T., 209
Kinter, E. B., 300, 356
Kira, Y., 180, 300
Kirkman, M. A., 382
Kishita, A., 181, 300, 301
Kita, D., 180
Kitagawa, Y., 180, 308, 401
Kitagishi, K., 300
Kittrick, J. A., 240
Klötzli, F., 75
Klug, H. P., 329, 337
Knowles, R., 394
Kobayashi, S., 197

Kobayashi, Y., 298
Kobo, K., 180, 181, 197, 300, 365
Koda, T., 299
Kodama, H., 7, 308
Kodama, I., 298
Kodani, Y., 180
Kolthof, I. M., 356
Komamura, M., 180
Kon, T., 180
Kondo, Y., 299
Konno, T., 181
Kono, H., 180
Kononova, M. M., 23, 98
Kosaka, J., 23, 46, 120, 300, 382
Koyanagawa, T., 301
Krzemieniewska, H., 394
Kubiena, W. L., 145, 300
Kubota, T., 180, 382
Kumada, K., 300, 372, 402, 403
Kumano, Y., 120
Kuno, G., 180
Kurabayashi, S., 300, 302, 329
Kurashima, K., 403
Kurobe, T., 59, 300, 382
Kuroda, M., 180
Kutsuna, K., 299, 300
Kuwano, I., 356
Kuwano, Y., 299, 329, 337
Kuwatsuka, S., 382
Kyuma, K., 299, 300, 372

Lahav, N., 240
Lai, C. Y., 32, 46
Lai, S. -H., 232
Laird, D. G., 59
Lambert, J. M., 395
Lance, G. N., 395
Land Survey Section, Economic Planning Agency, Japan, 197
Langohr, R., 402
Last, F. T., 394
Lavkulich, L. M., 209
Lee, K. E., 372
Legg, J. O., 247
Leonard, A. J., 231
Leung, K. W., 32, 46
Lietzke, D. A., 197
Lind, E. M., 75
Lindgreen, H. B., 394
Liu, F. W. J., 223
Liverovskii, Yu A., 46
Livingstone, L. G., 400
Llanos, R., 236
Loganathan, P., 402
Lucus, R., 197
Lulli, L., 46
Luna, C. Z., 46, 120, 197

McConaghy, S., 47, 120
McCreery, R. A., 82, 404
McHardy, W. J., 365
McKeague, J. A., 98

Mackenzie, R., 356
McNicoll, J., 381
Maeda, T., 145, 146, 180, 181, 182, 197
Makower, B., 223
Malan, C. E., 394
Mariano, J. A., 47
Marshall, C. E., 209
Marshall, K. C., 372
Martin, J. P., 371, 382
Martinez, A. T., 394, 402
Martini, J. A., 32, 47, 120, 204, 402
Massey, H. F., 63, 404
Masson, C. R., 400
Masui, J., 197, 198, 300, 301, 302, 308, 309, 382, 403
Masujima, H., 180
Masujima, T., 300
Matsuaka, Y., 120
Matsubara, I., 198, 302
Matsui, T., 59, 120, 300, 401, 402
Matsumoto, Y., 402
Matsuno, K., 198
Matsuo, Y., 300
Matsuzaka, Y., 181, 197, 247, 382
Mattson, S., 300, 320
Maucorps, J., 98
Mehlich, A., 63
Mehra, O. P., 197, 356
Messerli, B., 75, 76
Metson, A. J., 223
Michell, B. D., 198
Miles, N. M., 98
Minato, H., 329
Ministry of Agriculture, Forestry and Fisheries, 247
Ministry of Agriculture and Forestry, Japan, 26, 47, 120, 197
Mishustin, E. M., 300
Misono, S., 181, 197, 300, 301
Mitchell, B. D., 308
Miyasato, S., 302
Miyauchi, N., 248, 301, 308, 356, 365, 399
Miyazawa, K., 181, 247, 301, 402
Miyoshi, Y., 301
Mizota, C., 47, 98, 197, 248, 308, 402
Mohr, E. C. J., 32, 63, 146, 308, 402
Moinereau, J., 179
Molloy, L. F., 382
Montenegro, E., 120
Montgomery, C. W., 223
Moreau, B., 46, 179
Moreau, J. P., 231
Mori, H., 180
Mori, N., 300
Mori, T., 180
Moriyón, I., 394

Author Citation Index

Morrison, M. E. S., 75
Mortland, M. M., 197
Mouthaan, W. L. P. J., 46
Muljadi, D., 240
Muller, A., 399
Mundie, C. M., 381
Munevar, F., 402
Murayama, S., 382

N.Z. Soil Bureau, 47
Nachod, F. C., 223
Nagahori, K., 180
Nagai, T., 299, 301
Nagarajah, S., 240
Nagasawa, K., 329
Nagasawa, T., 182
Nagata, N., 181
Nakamura, T., 120
Nakano, M., 181, 400
National Institute of Agricultural Sciences, The Third Division of Soils, 248
National Land Agency, 248
Neall, V. E., 403
New Zealand Commonwealth Bureau of Soils, 403
Newman, A. C. D., 308
Nishimune, A., 299
Nishimura, T., 298
Nishio, D., 302
Nitta, K., 301
Nomoto, K., 300
Northrey, R. D., 181
Nygard, I. J., 59

Oades, J. M., 382
Oba, Y., 47, 181, 197, 365
Ogasawara, K., 302
Ohba, K., 403
Ohba, Y., 300, 301
Ohmasa, M., 7, 26, 82, 120
Oishi, K., 365
Okamura, Y., 404
Okita, T., 300
Okuda, A., 301
Olsen, S. R., 120, 240
Ong, H. L., 98
Onikura, Y., 248, 299, 300, 302, 309
Ono, T., 197, 198, 382
ORSTOM, 45
Ortega, E. J., 399
Ossaka, J., 329
Otowa, M., 198
Otsuka, H., 403

Packard, R. Q., 181
Palencia, J. A., 402
Panel on Volcanic Ash Soils in Latin America, 232
Parfitt, R. L., 181, 204, 308
Parham, W. E., 365

Parkinson, D., 394
Parks, G. A., 209
Pask, J. A., 320
Patterson, S. H., 198
Pauling, L., 209
Peech, M., 59, 209
Peech, M. A., 98
Peña-Cabriales, J. J., 394
Peralta, F., 204, 399
Perkins, H. F., 82, 404
Perrott, K. W., 46, 98, 198, 298, 308
Pettijohn, F. J., 120
Peyronel, B., 394
Picket, E. E., 329
Pohlen, I. J., 27
Ponomareva, V. V., 23
Pope, R. J., 181
Posner, A. M., 204, 232, 236, 240
Povoledo, B., 371
Prashad, S., 179
Pratt, P. F., 209, 240
Pullar, W. A., 47

Quantin, P., 47, 403
Quirk, J. P., 204, 232, 236, 240

Rach, K., 46
Rajan, S. S. S., 198, 204
Ramírez, C., 394, 402
Rayner, J. H., 308
Read, N. E., 403
Reddy, T. K. R., 394
Reed, J. F., 59, 98
Reeve, N. G., 232
Reiserauer, H. M., 236
Remacle, J., 394
Reynders, J. J., 46
Rich, C. I., 209
Richardson, E. A., 209
Richardson, J. P., 223
Rico, M., 47
Rieger, S., 17, 47, 59
Righi, D., 98
Rijkse, W. C., 46
Riquelme, F. E., 400
Ritchie, T. E., 17
Robertson, R. H. S., 32
Robin, A. M., 98
Rodrigues, G., 320
Rodriguez-Burgos, A., 394
Rodriguez Pascual, C., 32
Rooksby, H. P., 329
Ross, C. S., 320
Ross, G. J., 308
Rousseaux, J. M., 47, 146, 181
Rozov, N. N., 23
Rubins, E. J., 240
Ruscoe, Q. W., 394
Russell, J. D., 365, 400
Russell, M. B., 301

Rust, R. H., 179, 209, 400
Rutherford, G. K., 17, 59

Saavedra, J., 46
Saigusa, M., 197, 198, 403, 405
Saito, K., 197
Saito, T., 394, 395
Sakai, H., 301, 302
Sanchez Dias, J., 33
Sandell, E. B., 356
Sänger, A. M. H., 121
Sasaki, S., 180, 181, 182
Sasaki, T., 47, 180, 181, 182, 299
Saski, K., 302
Sato, A., 298
Sato, K., 180
Sato, O., 299
Sato, T., 182
Saunders, W. M. H., 198, 403
Sawada, Y., 301
Sayegh, A. H., 400
Schafer, G. J., 181
Schalscha, E. B., 181
Schatz, A., 181
Schnitzer, M., 7, 98, 99
Schofield, R. K., 231
Schollenberger, C. J., 223, 301
Scotter, D. R., 181
Scrivner, C. L., 382
Seki, T., 23, 59, 248, 301
Sekiya, K., 247, 298, 301
Sen, B. C., 7
Seo, H., 248
Shaw, T. M., 223
Shaw, W. M., 223
Shepherd, H., 381
Sherman, G. D., 47, 120, 181, 308
Shiga, H., 300
Shiina, K., 181
Shimada, A., 302
Shimoda, S., 308
Shin, Y. H., 47
Shinagawa, A., 299, 301
Shioiri, M., 301
Shoji, S., 197, 198, 300, 301, 302, 308, 309, 382, 403, 405
Sieffermann, G., 46, 47, 179
Simon, R. H., 301
Simon, R. N., 223
Simonson, C. H., 98
Simonson, R. W., 17, 47, 59
Singh, B. R., 209, 232
Singh, S. S., 236
Smith, B. F. L., 198
Smith, G. D., 7, 17, 23, 59, 121, 301, 382
Sóma, K., 180
Soepraptohardjo, M., 59
Soil Conservation Service, 47, 395
Soil Survey Laboratory Staff, 98

Author Citation Index

Soil Survey Staff, 17, 23, 47, 59, 76, 98, 198
Soil Survey Staff, Japan, 301
Soil Survey Staff, U.S.D.A., 356
Sokolov, I. A., 59
Soma, K., 145, 180, 181, 300, 382
Soong, R., 405
Souchier, B., 17, 32, 46
Sowden, F. J., 382
Sowers, G. F., 181
Sparling, G. P., 381
Stähli, P., 75
Steele, J. G., 59
Stevens, N. P., 329
Stobbe, P. C., 382
Stotzky, G., 372
Stout, J. D., 372, 394
Suarez, A., 402
Sudo, S., 181, 301
Sudo, T., 120, 301, 329, 403
Sumner, M. E., 232
Supraptohardjo, M., 32
Survey of Kenya, 76
Suzuki, A., 181
Suzuki, T., 181, 299, 301
Swift, R. S., 98
Swindale, L. D., 7, 13, 22, 47, 63, 179, 181, 223, 232, 402
Syers, J. K., 248

Tabuchi, K., 181
Tabuchi, T., 181
Tachiiri, M., 180
Tada, A., 181, 382
Tait, J. M., 308, 405
Takada, H., 299
Takahashi, T., 301, 400, 403, 404
Takahashi, Y., 404
Takata, H., 22
Takehara, H., 382
Takenaka, H., 179, 180, 181, 182, 197, 300, 301, 382
Tamura, T., 63
Tan, K. H., 7, 27, 33, 47, 59, 63, 82, 120, 204, 308, 372, 404
Tanabe, I., 301
Tanoue, M., 300
Tate, K. R., 7
Tatsukawa, R., 300
Tavernier, R., 400
Taylor, N. H., 7, 14, 23, 27, 59, 121, 320
Tejedor Salguero, M. L., 32, 46, 400
Teller, E., 223
Terasawa, S., 181, 301
Tercinier, G., 47
Theisen, A. A., 179
Theng, B. K. G., 7, 198, 204, 308, 309, 371, 372
Thomas, G. W., 236, 240

Thorp, J., 7, 17, 23, 59, 76, 121, 301, 382
Tiurin, I. V., 23
Tokashiki, Y., 47, 99, 198, 382, 404
Tokudome, S., 82, 301, 302, 372, 404
Tokunaga, K., 180, 181, 182
Torres, C., 32
Toyoda, K., 299
Trapnell, C. G., 76
Troth, P. S., 404
Truog, E., 223, 302
Tsuchiya, T., 300, 302, 329
Tsujinaka, N., 182
Tsukidate, K., 180
Tsuru, N., 302
Tsutsuki, K., 382
Tsuzuki, Y., 329
Twyford, I. T., 47
Tyurin, I. V., 302

Uchida, K., 180
Uchiyama, N., 300, 302, 309
Uehara, G., 181
Ulrich, H. P., 59
Umeda, Y., 182
U.S.D.A., Soil Conservation Service, 27
U.S. Soil Survey Staff, 7, 82, 146

Val, R. M., 204, 401
Valdés, A., 47, 121
Valdes, M., 394
Van Baren, F. A., 63, 308, 402
Van Bemmelen, J. M., 121
Van Olphen, H., 7, 209
Van Raij, B., 209
Van Rosmalen, H. A., 399
Van Rummelen, F. F. F. E., 82
Van Schuylenborgh, J., 7, 33, 59, 63, 82, 120, 121, 182, 308, 372
Vasquez, A. J., 399
Vergara, I., 181
Violante, A., 47
Violante, P., 47
Vucetich, C. G., 47

Wada, H., 248
Wada, K., 7, 47, 76, 83, 99, 121, 146, 182, 197, 198, 204, 209, 232, 240, 298, 299, 302, 308, 309, 365, 382, 400, 401, 402, 404, 405
Wada, S. I., 182, 198, 405
Wagner, G. H., 382
Walker, I. K., 329, 400
Walker, T. W., 248, 400
Walkley, A., 99
Warkentin, B. P., 47, 145, 146, 180, 181, 182, 197
Watanabe, S. R., 240

Watanabe, Y., 298, 302, 405
Weed, S. B., 204
Weinberger, E. B., 223
Weir, A. H., 308
Wells, N., 47, 182
Wendt, H., 236
Wesley, L. D., 182
White, L. P., 47
White, W. A., 329
Whitehead, D. C., 382
Whiteside, E. P., 197
Whittig, L. D., 401
Wicklow, M. C., 395
Widden, P., 393
Wiens, J. H., 209
Williams, C. H., 320
Williams, G., 300
Williams, J. D. H., 248
Williams, P. P., 329, 400
Williams, W. T., 395
Wilson, M. A., 400
Wollum, A. G., 402
World Soil Resources Report, 14, 48
Wright, A. C. S., 7, 14, 27, 33, 47, 48, 59, 83, 121, 302, 309
Wright, Ch. S., 356
Wright, J. R., 99
Wunderlich, R. E., 17, 59

Yabe, M., 180
Yakuwa, R., 182
Yamada, I., 197, 198, 382, 403, 405
Yamada, S., 48, 83
Yamada, Y., 48
Yamagata, U., 302
Yamamoto, T., 301, 302, 400
Yamanaka, K., 48, 182, 198
Yamane, I., 247, 405
Yamanouchi, T., 182
Yamazaki, F., 181, 182
Yasuhara, K., 182
Yasutomi, R., 181, 182
Yawata, T., 181
Yazawa, M., 182
Yoshida, M., 198, 248, 302
Yoshida, N., 401
Yoshida, T., 302
Yoshinaga, N., 48, 59, 63, 121, 182, 198, 298, 302, 308, 309, 329, 337, 356, 365, 381, 399, 400, 405
Yost, R. S., 405
Yotsumoto, H., 182, 198, 302, 365
Young, A. W., 400
Youngberg, C. T., 179, 182
Yuan, T. L., 405

Zachariae, G., 99
Zavaleta, A., 48, 121
Zurbuchen, M., 75

SUBJECT INDEX

Acric, 134
Actinomycetes, 292
Activity values, 154
Adhesion, 178
Adsorption isotherms, 214, 218, 230
Aeration, 279
Agglomeratic fabric, 145
Agglomeroplasmic fabric, 145
Aggregation
    polyvinyl alcohol in, 162
    sodium alginate in, 162
Alfisols, 6, 11, 31, 79
Allic, 135
Allophane, 4, 25, 78, 89, 304, 310, 357
    A, 115, 255, 306
    adsorption with alizarin red-S, 316
    B, 115, 255, 306
    base exchange of, 315
    composition of, 115
    definition of, 4
    differential thermal analysis (DTA) of, 61, 62, 115, 327
    elemental analysis of, 328
    field test of, 41–42
    -humic acid complexes, 3
    infrared spectra of, 335, 354
    morphology of, 304, 357–365
    physical properties of, 149
    properties of, 258
    scanning electron microscopy of, 357–359
    separation of, 322–323
    silica-sesquioxide ratio of, 10, 16
    soils, 142
        physical properties of, 147
    structure of, 4, 115
    surface acidity of, 4
    water relation of, 163
    X-ray diffraction of, 323–326
Al-organic ratio, 3, 4
Altic, 135
Alvic soil, 24–25
Alvisols, 24, 100
Amo-fulvic, 24–25
Amorphous clay, 3
Amorphous material, 210
    diagnostic properties of, 85
Anaerobic decomposition, 80

Andepts, 13, 17, 24–25, 34, 36, 52, 64, 100
    chemical and physical characteristics of, 93
    classification of, 78, 84, 92
    climate of, 89
    composition of, 56, 73
    diagnostic features of, 84
    environment of, 53
    morphology of, 53, 71–72, 78
    parent materials of, 66
    soil formation factors of, 64–65
    soil genesis of, 64–65, 78–79, 84, 89
    vegetation of, 67, 84, 89
Andept-spodosol relation, 58
Andesite, 1, 18–19, 305, 310
Andesito-dacitic tuff, 60
Andic epipedon, 82
Andisols, 34, 81
    central concept of, 127
    color of, 39
    consistence of, 41
    definition of, 36, 126–127
    density of, 42
    distribution of, 37
    key to great groups of, 130
    key to suborders of, 129
    morphology of, 43
    organic carbon of, 41
    structural stability of, 42
    texture of, 41
    water retention of, 42
Ando podzolic, 16
Ando soils, 2, 10, 17, 34, 52
Andosols, 100
    acreages of, 31
    active fraction of, 61
    AEC of, 62
    aquic, 184
    base saturation of, 10, 20
    biological properties of, 290–294, 367–397
    CEC of, 62
    CECp of, 62
    CECv of, 62
    central concept of, 26, 118
    chemical characteristics of, 199, 280
    classification of, 117
    climatic conditions for, 1, 2, 60, 79, 101

Subject Index

correlation of, 24
electric charges in, 61
fertility of, 243
genesis of, 79, 243
geographic distribution of, 2, 241
highland, 25, 31, 61
high mountain, 31, 61, 64, 100
history of, 52
humic, 73
humus characteristics of, 20
land use of, 62-63, 247
microfungal community in, 368, 383-397
   list of species, 388, 395-397
   relative abundance, 386
micromorphology of, 119, 143-144, 274-275
mineralogy of, 2, 4, 107, 109, 111, 113, 251-253, 304-365
moisture regimes of, 2
morphology of, 185
organic carbon content of, 3
organic matter content of, 2
origin of name, 10-28
parent materials of, 60, 65
pH of, 10, 16
phosphate fixation, 61
physical characteristics of, 142, 222, 276
physical and chemical characteristics of, 104, 106, 109, 111, 113
soil profile of, 19, 102-103, 185
topography of, 65
vegetation of, 101
Anion adsorption, 200, 228-237
Anion exchange, 203
Anion uptake, 221
Ankantsu-shoku-do, 100
Anmoor-like fabric, 145
Anshoku, 16
Anshokudo, 16, 34
Antagonism, 203
Aqaunds, 44, 129, 130, 133
Aquic, 44, 135
Aquods, 94
   physical and chemical properties of, 95
Argillic, 36
Argillization, 25
Aridic moisture regime, 35
Ash, 124
  acid, 304
  primary, 19
  volcanic, 1, 305
    age of, 1
    basic, 40
Ashy, 125
Ashy-pumiceous, 125
Ashy-skeletal, 125
Aspergillus, 295
Atterberg limits, 152
Azotobacter, 292

Bacteria, 293
Basaltic, 18-19, 60
Base saturation, 35, 52, 122, 188, 280
Biological properties, 6
Black dust soil, 5, 24, 100
Black forest soil, 15
Black soils, 15
Black volcanic ash soils, 24
Borands, 44, 129-130, 133
Braunerde, 52
Brillouin's index, 385, 389
Brønsted acid, 201
Brown forest soil, 15, 18, 20, 24, 52
Brunauer, Emmett, and Teller (BET), 212, 215, 217, 219
Brunizem, 52
Bulk density, 4-5, 37, 40, 92, 142-143, 149, 157, 159

Cambic, 36, 80
Capillary potential, 279
Carbon limit, 37
Carbon-nitrogen (C/N) ratio, 20, 40-41, 80, 86, 268
Carboxyl groups, 86
Cation exchange capacity (CEC), 35, 122, 188, 224-225, 259, 285
CEC delta value, 336
Cell charge, 370
Cell-clay complexes, 370
Charge density of cells, 371
Chelation complexes, 86
Chernozems, 20
Chloride adsorption, 228
Cinders, 80, 124, 128
Cindery, 125
Clastic, 124
Coadsorption, 237
Cohesion, 177-178
Cohesive volcanic ash soil, 171
Compaction, 172-174
Complexes
  humus-allophane, 89
  humus-aluminum 89
Compressibility, coefficient, 178
Consolidation, 177-178
  coefficient, 177
Coordination complexes, 86
Coordination number, 264
Co-precipitation, 5
Coulombic attraction, 263
Cryandept, 26, 52, 59, 73, 117
Cryic, 136
Cryoborand, 130
Cryohumod, 87, 90
Cutans, 73

Dacite, 1, 40, 304
Darcy flow, 169
Dendogram, 392-393
Desorption, 220, 230, 239
Dessication, 86
Differential thermal analysis (DTA), 266, 312-313, 327. *See also* Allophane; Imogolite

Subject Index

Dispersion, 142, 150
Duriaquands, 130
Duric great group, 45
Duripan, 36, 128
Durixerands, 131
Durustands, 131
Dystrandept, 73, 117

Earthy braunlehm, 145
Elalvic soil, 26
Electric double layer, 205
Electron micrographs, 192, 321, 325, 330-331. *See also* Imogolite
Electrophoresis, 266
Eluvial horizon, 86
Entic, 136
Entisol, 79
Eutrandept, 37, 117, 128
Eutrochrept, 15
Evaporation, 170
Exchangeable Al, 90
Exchangeable bases, 52, 188, 222
Exchangeable hydrogen, 222, 226
Exchange acidity, 283

Fabric, 40, 145
Feldspar, 5, 305
Felsic glass, 256
Ferrods, 94
Ferromagnesians, 5
Field capacity, 169, 278
Fourteen (14) Å mineral, 202, 256
Fragipan, 25
Fulvic acid, 21, 40, 60
Fulviform, 118

Gehlenite, 178
Gibbsite, 20, 61, 89
Gouy-Chapman theory, 205
Grain size distribution, 150

Halloysite, 4, 20, 89, 202, 357
  scanning electron microscopy of, 357, 364
  tubular, 307, 364
Haplaquands, 130
Haplaquods, 94
Haploborands, 130, 133
Haplotropands, 132
Hapludands, 132
Hapludoll, 15
Haplustands, 131
Heat
  of adsorption, 155, 220
  capacity, 158
  of wetting, 155
Histic, 36
Histosol, 65
Hornblende, 5, 251, 305
Humic allophane soil, 2, 11, 18, 21-25, 52, 60, 117
  base saturation of, 20
  humus characteristics of, 20
  soil profile of, 19

Humic-fulvic acid ratio, 20, 25-26, 40, 60, 115, 118-119, 272, 369
Humification, 41, 268-270, 272, 369
  degree of, 369
  delta K of, 369
Humod, 87, 94
Humus
  Dauer, 268
  Nähr, 268
Hydrandepts, 25, 92, 123-124
Hydraulic conductivity, 168-169
Hydric, 136
Hydric great soil group, 45
Hydrocatena, 245
Hydrol-alvic soil, 25
Hydrol humic latosol, 24, 44
Hydrotropands, 45, 131
Hydrous class, 97, 125
Hydrous elalvic, 26
Hydrous-skeletal, 125
Hydrudands, 45, 132

ICOMAND, 34, 37
Illuvial material, 86
Illuviation, 25
Imogo, 322, 338
Imogolite, 4, 41, 115, 148, 202, 254, 304, 357
  chemical composition of, 343
  differential thermal analysis (DTA) of, 334, 347-348
  electron micrographs of, 334-346
  infrared spectra of, 192
  fractionation of, 339
  morphology of, 304, 322, 330-337, 338-356, 357-365
  scanning electron microscopy of, 357, 360, 363
  structure, 160, 304-306
X-ray diffraction of, 332, 334, 349-352
Inceptisols, 25, 34, 52
  classification of, 96
Infiltration, 170
Infrared spectra, 192. *See also* allophane; Imogolite
Inorganic P, 287
Inside negative charge, 202, 284
Intermolecular coagulation, 86
Ion exchange, 200, 210
  anion, 201
Island very fine sandy loam, 56
Isoelectric point, 318
Isohyperthermic, 40
Isomorphous substitution, 202
Isopachs, 241
Isothermic, 40

Jeffries method, 314
Jeltozem-forest soil, 22

Kachemal silt loam, 55
Kaolinite, 4, 21, 202
Kodiak silt loam, 53
Kuroboku, 2, 24, 26, 39, 44, 89, 118
Kurotsuchi, 24

415

Subject Index

Lahar
  andesito-dacitic, 60
  basalto-andesitic, 60
Langmuir equation, 230
Lapilli, 78, 124, 128
Latosols, 31, 100
Leaching, 19
Ligand exchange, 203
Liquid limit, 149, 153, 154
Lithic contact, 36, 69, 128, 136
Lowland andosols, 25, 31, 61

Macrostructure, 275
Mafic glass, 256
Magma
  acid, 1, 5
  basic, 1, 5
Magnetite, 5
Margalite soil, 18, 60
Mean residue time (MRT), 87
Medial, 125
Medial-pumiceous, 125
Medial-skeletal, 125
Melanic great soil group, 44
Melanaquands, 44
Melanization, 25
Melanoborand, 130, 133
Melanudands, 132
Metal-humic acid complexes, 3
Microaggregates, 160
Microbial activities, 295
Micromorphology, 40, 274
Micropeds, 40
Micropores, 143
Microprobe, 88
Microstructure, 143, 145, 274-275
Mineral(s)
  density of, 156
  primary, 5, 251
Mineralization of soil N, 295, 297
Mineralogy
  of clay fraction, 190, 252, 304
  of sand fraction, 5, 304-305
Miscanthus sinensis, 20, 40, 79
Moisture, retention of, 35, 123
Mollandepts, 25-26
Mollic, 36, 82
Mollisols, 6, 11, 81
Monosaccharide
  in andosols, 368, 373
  in plant debris, 379
  in soils, 379
  in straw, 379
Montmorillonite, 22, 202, 256, 371
Mountain granulation, 5

Natric, 36
Negative charge, 201
Nitrate retention, 200, 205-206
Nitrogen relation with organic matter, 62-63
Noncrystalline clay, 3
Nonspecific adsorption, 4, 203
Normal andosol, 184

Nothofagus, 383
N-value, 175

Ochrandepts, 25
Ochric, 82
Olivine, 251, 256
Onji soils, 15, 52
Opaline silica, 306
Optimum water content, 172
Organans, 86
Organic matter, 267
  relation with N content, 62-63
Organic P, 287
Organo-mineral complexes, 202
Orthoeluvium, 19
Outside negative charge, 202, 284
Oxic, 36, 137
Oxisols, 1, 6, 31, 79-80

Pachic, 137
Palagonite, 210
Paralithic contact, 36, 128
Pardo-forestal soil, 100, 117
Particle size distribution, 150
Pedotubules, 87
Pergelic, 137
Permanent charges, 6
Permeability, 173-174
Perudic moisture regime, 123
pF, 42, 151, 166, 278-279
Phase relations, 272
pH, 188
  sodium fluoride (NaF), 90, 189, 196
pH dependent CEC, 122, 284
pH dependent charge, 201-202
Phenolic hydroxyls, 86
Phosphate fixation, 4, 6, 244
Phosphate sorption, 160, 188, 190, 197, 237-240, 285-290
Phosphorus, 52
  in andosols, 244
  forms of, 287-289
  organic, 246
  status, 200, 241-248
Placic, 36, 92
Placic great soil group, 45
Placoborands, 130, 133
Placotropands, 131
Placudands, 132
Plagioclase, 21
Plasma, 40
Plasticity, 5, 150, 152
  activity value of, 154
  cassagrande, 153
  index, 149, 154
  limit, 153
Plating methods, 384
Pleistocene sediment, 21
Podiform, 118
Podzol, 15, 20, 52
Polysaccharide in Andosols, 373
Pore size distribution, 276
Porosity, 4-5, 159

Subject Index

Potentiometric titration, 206
Preferential adsorption, 284
Psammic, 138
Pseudosand, 143
Pumice, 78, 124, 128
Pumice-like, 124, 128
Pumiceous, 124
Pyroclastic material, 1, 34, 78, 124
Pyroxenes, 5, 257, 305
Pythium, 295

Quaternary period, 1, 78

Red Yellow soils, 20
Rheotropy, 161
Rhyolite, 1, 18–19, 126, 304, 310
Ruptic-placic, 137

Scoria, 78
Sesquioxides, 61
Shear strength, 175–176
Shrinkage curves, 164–165
Shrinkage limit, 164
Silica-alumina ratio, 20, 306
Silica-sesquioxide ratio, 16, 20, 92, 306
Siloxane bonds, 306
Skeletiform, 118
Smectite, 371
Sodium fluoride (NaF) pH, 43, 90, 128, 306
Soil engineering, 171
Soil respiration, 296
Specific heat, 157
Specific surface, 353
Spodic, 36, 86–87
Spodosols, 1, 6, 11, 31, 52, 79
    classification of, 92
    chemical and physical characteristics of, 91
    genesis of, 85–86
Spruce, 58
Stability constant, 371
Stabilization, 178
Stickiness, 5
Strength, 174
Stress, 176
Structure, 159, 274
    macro, 143–145, 275
    micro, 143–145, 274
Sulphate adsorption, 233–236
Surface acidity, 4
Surface area, 155, 160
Synergism, 203

Talpete soils, 24, 117
Tapeta soils, 100
Taupo soils, 43–44
Tectonic zones, 37
Tension-moisture curve, 69
Tephra, 241
Tephramorphology, 43
Tephrastratigraphy, 43
Tetrahedrally coordinated Al, 4
Thapto-histic, 139
Thermal conductivity, 156, 158

Thermal diffusivity, 157–158
Thermic region, 40
Thin sections, 144
Thixotropy, 35, 97, 122, 125–126, 161
Titration curves, 261
Titration of allophane clays, 221
Tropands, 129, 131, 134
Tropic, 138
Trumao soils, 24, 39, 338
Tuffs, 1
    andesito-dacitic, 60
Tundra, 52

Udands, 44, 129, 132, 134
Udic moisture regime, 43
Ultisols, 1, 6, 11, 31, 79–80
Ultrasonic dispersion, 151–152
Umbrandepts, 20, 25
Umbric, 36–37, 69, 73, 82
Ustand, 45, 129, 131, 134
Ustic, 79, 128, 138
Ustollic, 138

Variable charges, 6, 25, 118, 201
    in cells, 370
    in living organisms, 370
Variscite, 237
Vitrandepts, 36
Vitraquands, 130, 133
Vitriborands, 130, 133
Vitric great soil groups, 138
Vitric volcanic ash, 36, 41, 80, 94, 124, 128
Vitritropands, 131
Vitrixerands, 131
Vitrudands, 44, 132
Vitrustands, 131
Void ratio, 122
Volcanic glass, 5, 21
Volcaniclastic, 122, 124
Volume changes, 164

Water
    availability of, 170–171
    content, 15-bar, 123
    relations, 276
    retention, 37, 123, 126, 128, 150, 165
        at different suctions, 167
        influence of organic matter on, 168
    transmission, 168
Waterholding capacity, 4, 5, 40, 142–143
Waxy pans, 310
Weathering, 19, 89, 253, 256, 258

Xerands, 45, 129, 131, 134
Xeric, 139
X-ray diffraction, 195, 265, 312, 326, 331–334. See also Allophane; Imogolite

Yellow brown loams, 11, 13, 18, 24, 100

Zero point of charge (ZPC), 5, 205
Zero point of titration (ZPT), 205

About the Editor

KIM HOWARD TAN is professor of agronomy at the University of Georgia. He received the M.S. in tropical agriculture and the Ph.D. in soils from the University of Indonesia. Dr. Tan has received the Outstanding Teacher Award (1972), the Distinguished Faculty Award for Teaching (1972), and the D. W. Brooks Award for Excellence in Teaching (1982) while at the University of Georgia. He is well-known for his research in soil organic matter and clay mineralogy and is considered an authority on humic acids. He has authored or coauthored over 115 articles, including chapters of books. In addition, he has written and published two books. He is a member of the American Society of Agronomy, Soil Science Society of America, International Society of Soil Science, Association internationale pour l'etude des argilles, and Gamma Sigma Delta.